Undergraduate Texts in Mathematics

Readings in Mathematics

Graduate Texts in Mathematics
Readings in Mathematics

Ebbinghaus/Hermes/Hirzebruch/Koecher/Mainzer/Neukirch/Prestel/Remmert: *Numbers*
Remmert: *Theory of Complex Functions*

Undergraduate Texts in Mathematics
Readings in Mathematics

Hämmerlin/Hoffmann: *Numerical Mathematics*
Samuel: *Projective Geometry*

Günther Hämmerlin
Karl-Heinz Hoffmann

Numerical Mathematics

Translated by Larry Schumaker

With 76 Illustrations

Springer-Verlag
New York Berlin Heidelberg London
Paris Tokyo Hong Kong Barcelona

Günther Hämmerlin
Mathematisches Institut
Ludwig-Maximilians-Universität
W-8000 München 2
FRG

Karl-Heinz Hoffmann
Mathematisches Institut
Universität Augsburg
W-8900 Augsburg
FRG

Larry L. Schumaker (Translator)
Department of Mathematics
Vanderbilt University
Nashville, TN 37235
USA

This book is a translation of *Numerische Mathematik*, Grundwissen Mathematik 7, Springer-Verlag, 1989.

Library of Congress Cataloging-in-Publication Data
Hämmerlin, G. (Günther), 1928-
[Numerische Mathematik. English]
Numerical mathematics / Günther Hämmerlin, Karl-Heinz Hoffmann ; translated by Larry Schumaker.
p. cm. – (Undergraduate texts in mathematics. Readings in mathematics)
Translation of: Numerische Mathematik.
Includes bibliographical references and index.
1. Numerical analysis. I. Hoffmann, K.-H. (Karl-Heinz) II. Title. III. Series.
QA297.H2513 1991
519.4–dc20 90-22552

Printed on acid-free paper.

Camera-ready copy provided by the translator.
Printed and bound by R.R. Donnelley & Sons, Harrisonburg, VA.
Printed in the United States of America.

9 8 7 6 5 4 3 2 1

ISBN 0-387-97494-6 Springer-Verlag New York Berlin Heidelberg
ISBN 3-540-97494-6 Springer-Verlag Berlin Heidelberg New York

Preface

> "In truth, it is not knowledge, but learning, not possessing, but production, not being there, but travelling there, which provides the greatest pleasure. When I have completely understood something, then I turn away and move on into the dark; indeed, so curious is the insatiable man, that when he has completed one house, rather than living in it peacefully, he starts to build another."
>
> Letter from C. F. Gauss to W. Bolyai on Sept. 2, 1808

This textbook adds a book devoted to applied mathematics to the series "Grundwissen Mathematik." Our goals, like those of the other books in the series, are to explain connections and common viewpoints between various mathematical areas, to emphasize the motivation for studying certain problem areas, and to present the historical development of our subject.

Our aim in this book is to discuss some of the central problems which arise in applications of mathematics, to develop constructive methods for the numerical solution of these problems, and to study the associated questions of accuracy. In doing so, we also present some theoretical results needed for our development, especially when they involve material which is beyond the scope of the usual beginning courses in calculus and linear algebra. This book is based on lectures given over many years at the Universities of Freiburg, Munich, Berlin and Augsburg. Our intention is not simply to give a set of recipes for solving problems, but rather to present the underlying mathematical structure. In this sense, we agree with R. W. Hamming [1962] that the purpose of numerical analysis is "insight, not numbers."

In choosing material to include here, our main criterion was that it should show how one typically approaches problems in numerical analysis. In addition, we have tried to make the book sufficiently complete so as to provide a solid basis for studying more specialized areas of numerical analysis, such as the solution of differential or integral equations, nonlinear optimization, or integral transforms. Thus, cross-connections and open questions have also been discussed. In summary, we have tried to select material and to organize it in such a way as to meet our mathematical goals, while at the same time giving the reader some of the feeling of joy that Gauss expressed in his letter quoted at the beginning of this preface.

The amount of material in this book exceeds what is usually covered in a two semester course. Thus, the instructor has a variety of possibilities for selecting material. If you are a student who is using this book as a supplement to other course materials, we hope that our presentation covers all of the material contained in your course, and that it will help deepen your understanding and provide new insights. Chapter 1 of the book deals with the basic question of arithmetic, and in particular how it is done by machines. We start the book with this subject since all of mathematics grows out of numbers, and since numerical analysis must deal with them. However, it is not absolutely necessary to study Chapter 1 in detail before proceeding to the following chapters. The remaining chapters can be divided into two major parts: Chapters 4 – 7 along with Sections 1 and 2 of Chapter 8 deal with classical problems of numerical analysis. Chapters 2, 3 and 9, and the remainder of Chapter 8 are devoted to numerical linear algebra.

A number of our colleagues were involved in the development and production of this book. We thank all of them heartily. In particular, we would like to mention L. Bamberger, A. Burgstaller, P. Knabner, M. Hilpert, E. Schäfer, U. Schmid, D. Schuster, W. Spann and M. Thoma for suggestions, for reading parts of the manuscript and galley proofs, and for putting together the index. We would like to thank I. Eichenseher for mastering the mysteries of TEX; C. Niederauer and K. Bernt for preparing the figures and tables; and H. Hornung and I. Mignani for typing parts of the manuscript. Our special thanks are due to M.-E. Eberle for her skillful preparation of the camera-ready copy of the book, and her patient willingness to go through many revision with the authors.

Munich and Augsburg — G. Hämmerlin

December, 1988 — K.-H. Hoffmann

Note to the reader: This book contains a total of 270 exercises of various degrees of difficulty. These can be found at the end of each section. Cross references to material in other sections or subsections of a given chapter will be made by referring only to the section and subsection number. Otherwise, the chapter number is placed in front of them. We use square brackets [·] to refer to the papers and books listed at the end.

Translator's Note: This book is a direct translation of the first German edition, with only very minor changes. Several misprints have been corrected, and some English language references have been added or substituted for the original German ones. I would like to thank my wife, Gerda, for her help in preparing the translation and the camera-ready manuscript.

Munich, July, 1990 — L. L. Schumaker

Contents

Chapter 1. Computing

Chapter 2. Linear Systems of Equations

Chapter 3. Eigenvalues

Chapter 4. Approximation

Chapter 5. Interpolation

Chapter 6. Splines

Chapter 7. Integration

1
Computing

As we have already mentioned in the preface to this book, we consider numerical analysis to be the mathematics of constructive methods which can be realized numerically. Thus, one of the problems of numerical analysis is to design computer algorithms for either exactly or approximately solving problems in mathematics itself, or in its applications in the natural sciences, technology, economics, and so forth. Our aim is to design *algorithms* which can be programmed and run on a calculator or digital computer. The key to this approach is to have an appropriate way of *representing numbers* which is compatible with the physical properties of the memory of the computer. In a practical computer, each number can only be stored to a finite number of digits, and thus some way of *rounding off* numbers is needed. This in turn implies that for complicated algorithms, there may be an accumulation of errors, and hence it is essential to have a way of performing an *error analysis* of our methods. In this connection there are several different kinds of error types, which in addition to the *roundoff error* mentioned above, include *data error* and *method error*.

It is the goal of this chapter to present the basics of machine calculation with numbers. Armed with this knowledge, we will be in a position to realistically evaluate the possibilities and the limits of machine computation.

1. Numbers and Their Representation

In numerical computations, numbers are the carriers of information. Thus, the questions of representing numbers in various number systems and dealing with them in a computer are of fundamental importance. A detailed discussion of the development of our present-day concept of numbers can be found in the book *Numbers* by H.-D. Ebbinghaus et al. [1990], and thus we restrict our historical remarks in this chapter to an outline of the main developments as they pertain to computers.

1.1 Representing Numbers in Arbitrary Bases. We are all used to working with real-valued numbers in decimal form. A study of the historical development of number systems clearly shows, however, that the decimal system is not the only reasonable one, and in fact, for the standpoint of practical applications in computers, is not necessarily the most useful one. In this section we discuss representing numbers using an arbitrary base $B \geq 2$.

Example. Consider representing the periodic decimal number $x = 123.\overline{456}$ in the binary system, that is with respect to the basis $B = 2$. Clearly, we can decompose x into the two parts $x_0 = 123$ and $x_1 = 0.\overline{456}$, where $x_0 \in \mathbb{Z}_+$ and $x_1 \in \mathbb{R}_+$ with $x_1 < 1$.

We do not go into any detail on how to find the representation of x_0 in the binary system; the result is $x_0 = 1111011$. Now the decimal fraction x_1 can be converted to a binary fraction by repeatedly doubling it as follows:

$$\begin{aligned}
x_1 \cdot 2 &= x_2 + x_{-1}, & x_2 &:= 0.\overline{912}, & x_{-1} &:= 0\\
x_2 \cdot 2 &= x_3 + x_{-2}, & x_3 &:= 0.\overline{825}, & x_{-2} &:= 1\\
x_3 \cdot 2 &= x_4 + x_{-3}, & x_4 &:= 0.\overline{651}, & x_{-3} &:= 1\\
x_4 \cdot 2 &= x_5 + x_{-4}, & x_5 &:= 0.\overline{303}, & x_{-4} &:= 1\\
x_5 \cdot 2 &= x_6 + x_{-5}, & x_6 &:= 0.\overline{606}, & x_{-5} &:= 0\\
x_6 \cdot 2 &= x_7 + x_{-6}, & x_7 &:= 0.\overline{213}, & x_{-6} &:= 1\\
&\vdots & &\vdots & &\vdots
\end{aligned}$$

This immediately leads to the binary representation $x_1 \doteq 0.011101$, and it follows that $x \doteq 1111011.011101$. This can also be written in the *normalized form*

$$x \doteq 2^7 \cdot 0.1111011011101.$$

Remark. *We use the symbol $\doteq$ in connection with numbers to mean that all of the digits up to the last one are exact, while the last digit is rounded. In tables, we do not distinguish between exact and rounded numbers.*

The general situation is described by the following

Theorem. *Let B be an integer with $B \geq 2$, and let x be a real-valued number, $x \neq 0$. Then x can always be represented in the form*

$$(*) \qquad x = \sigma\, B^N \sum_{\nu=1}^{\infty} x_{-\nu} B^{-\nu}$$

with $\sigma \in \{-1, +1\}$, $N \in \mathbb{N}$, and $x_{-\nu} \in \{0, 1, \ldots, B-1\}$. Moreover, there is a unique representation of this type with $x_{-1} \neq 0$ and with the property that for every $n \in \mathbb{N}$, there exists a $\nu \geq n$ such that

$$(**) \qquad x_{-\nu} \neq B - 1.$$

Proof. Let $x \in \mathbb{R}$, $x \neq 0$. Then the numbers $\sigma \in \{-1, +1\}$ and $N \in \mathbb{N}$ are uniquely determined by $\sigma := \operatorname{sign} x$ and $N := \min\{\kappa \in \mathbb{N} \mid |x| < B^\kappa\}$. Now we set

$$x_1 := B^{-N}|x|,$$

and apply the method of the example, using the base B instead of 2.

The definition of N implies $B^{N-1} \le |x| < B^N$, and so $0 < x_1 < 1$. Generalizing the method used in the example, we now consider the rule

$$x_\nu \cdot B = x_{\nu+1} + x_{-\nu}, \quad \nu \in \mathbb{Z}_+,$$

where $x_{-\nu}$ is the largest integer which does not exceed $x_\nu \cdot B$. This gives rise to sequences of numbers $\{x_\nu\}_{\nu\in\mathbb{N}}$ and $\{x_{-\nu}\}_{\nu\in\mathbb{N}}$ with the properties

$$0 \le x_\nu < 1,$$
$$x_{-\nu} \in \{0, 1, \ldots, B-1\}, \quad \nu \in \mathbb{Z}_+.$$

This is easy to check for $\nu = 1$, since in this case we already have shown that $0 < x_1 < 1$, and the asserted property of x_{-1} is implied by $0 < x_1 B < B$. The proof for arbitrary $\nu \in \mathbb{Z}_+$ follows by induction.

Our inductive argument shows that for arbitrary $n \in \mathbb{Z}_+$, x_1 can be expanded in the form

$$(***) \qquad x_1 = \sum_{\nu=1}^{n} x_{-\nu} B^{-\nu} + B^{-n} x_{n+1}$$

with $x_{-\nu} \in \{0, 1, \ldots, B-1\}$ and $0 \le x_{n+1} < 1$. This implies that for every $n \in \mathbb{Z}_+$,

$$0 \le x_1 - \sum_{\nu=1}^{n} x_{-\nu} B^{-\nu} < B^{-n}.$$

Passing to the limit as $n \to \infty$ leads to the representation

$$x_1 = \sum_{\nu=1}^{\infty} x_{-\nu} B^{-\nu}.$$

The number N was chosen precisely so that $x_{-1} \ne 0$.

It remains to check the property (**). We assume that it does not hold. Then there exists an $n \in \mathbb{Z}_+$ such that $x_{-\nu} = B - 1$ for all $\nu \ge n+1$, and it follows that

$$x_1 = \sum_{\nu=1}^{n} x_{-\nu} B^{-\nu} + (B-1) \sum_{\nu=n+1}^{\infty} B^{-\nu} = \sum_{\nu=1}^{n} x_{-\nu} B^{-\nu} + B^{-n}.$$

Comparing this equality with the formula (***), we see that $x_{n+1} = 1$. But this contradicts the fact that $0 \le x_{n+1} < 1$.

To complete the proof of this theorem, it remains to establish the uniqueness of the representation (*). Suppose

$$x_1 = \sum_{\nu=1}^{\infty} x_{-\nu} B^{-\nu} \quad \text{and} \quad y_1 = \sum_{\nu=1}^{\infty} y_{-\nu} B^{-\nu}$$

are two expansions. Set $z_{-\nu} := y_{-\nu} - x_{-\nu}$. Then $0 = \sum_{\nu=1}^{\infty} z_{-\nu} B^{-\nu}$. Now suppose that x_1 and y_1 are not the same; i.e., there is some first index $n-1$ with $z_{-n+1} \neq 0$. Clearly, we can assume that $z_{-n+1} \geq 1$. Now it follows from

$$z_{-n+1} B^{-n+1} = \sum_{\nu=n}^{\infty} (-z_{-\nu}) B^{-\nu} \leq \sum_{\nu=n}^{\infty} |z_{-\nu}| B^{-\nu} \leq \sum_{\nu=n}^{\infty} (B-1) B^{-\nu} =$$

$$= \lim_{m\to\infty} \sum_{\nu=n}^{m} (B^{-\nu+1} - B^{-\nu}) = B^{-n+1} - \lim_{m\to\infty} B^{-m} = B^{-n+1}$$

that the reverse estimate $z_{-n+1} \leq 1$ holds, and consequently $z_{-n+1} = 1$. This means that in the last inequality, we must have equality everywhere, which implies, in particular, that

$$z_{-\nu} = -B + 1$$

for all $\nu \geq n$, and so $y_{-\nu} = 0$ and $x_{-\nu} = B-1$ for all $\nu \geq n$. This contradicts the assumption that x_1 satisfies (**), and so y_1 must be the same as x_1; see also Problem 1. □

Given a number x which is expanded as in (*) with respect to the basis B, we can now write it in coded form as

$$x = \sigma B^N \cdot 0.x_{-1} x_{-2} x_{-3} \ldots,$$

where the $x_{-\nu}$ are the integers arising in the formula (*). Each of these is in the set $\{0, 1, \ldots, B-1\}$, and is called a *digit*. The digits characterize the number.

The most used bases are 2, 8, 10, 16. The following table shows the usual symbols used to represent the digits in these systems:

System	Basis B	Digits
binary	2	0, 1
octal	8	0, 1, 2, 3, 4, 5, 6, 7
decimal	10	0, 1, 2, 3, 4, 5, 6, 7, 8, 9
hexadecimal	16	0, 1, 2, 3, 4, 5, 6, 7, 8, 9, A, B, C, D, E, F

The enormous simplification which arises when using the binary system for computing was already recognized by Leibniz. A disadvantage of the system is the fact that many numbers have very long representations, and are hard to recognize. But the binary system has become of great practical importance with the advent of electronic computers, since in such machines, any representation scheme has to be based on the ability to distinguish

Digits	Octal	Decimal			Hexadecimal
		Direct code	3-excess Stibitz-Code	Aiken-Code	
0	000	0000	0011	0000	0000
1	001	0001	0100	0001	0001
2	010	0010	0101	0010	0010
3	011	0011	0110	0011	0011
4	100	0100	0111	0100	0100
5	101	0101	1000	1011	0101
6	110	0110	1001	1100	0110
7	111	0111	1010	1101	0111
8		1000	1011	1110	1000
9		1001	1100	1111	1001
A					1010
B					1011
C					1100
D					1101
E					1110
F					1111

between two states; i.e., it has to involve a *binary* coding. If we identify these two states with the digits 0 and 1, we immediately get a one-to-one mapping between numbers represented in binary form and the states of the computer. If we want to use some other base, then each of the corresponding digits has to be coded in binary form. When the basis B is a power of 2, then this is especially simple. For example, in the octal system we need *triads* (= blocks of 3 digits), and in the hexadecimal system, we need *tetrads* (= blocks of 4 digits) in order to code each digit. Tetrads are also needed for the binary coding of the digits in the decimal system, although in this case six of the possible tetrads are not used. This implies some degrees of freedom remain, and we say that the code is *redundant*. The above table shows three of the known codes for the decimal system.

We note that in the 3-excess and Aiken codes, the nine compliment of a digit can be obtained by exchanging the zeros and ones.

1.2 Analog and Digital Computing Machines. Computing machines can be divided into two rather different classes depending on the way in which they store and work with numbers: analog and digital machines. The following table lists examples of both.

Digital Computers	Analog Computers
Tables	Slide Rule
Mechanical calculator	Nomogram
Pocket calculator	Mechanical Analog Computer
Electronic Digital Computer	Electronic Analog Computer

Analog computers use continuous physical quantities such as the length of a rod, voltage, etc. to represent numbers. Mathematical problems can be solved by using a corresponding device which simulates the problem. Then the solution of the problem can be interpreted as the result of a physical experiment. For analog devices, the accuracy of the numbers being stored and manipulated depends very heavily on the exactness of the measurements being taken. We will not discuss analog computers further. Their present-day use is limited to some special applications.

Digital computers represent numbers by a finite sequence of (discrete) physical states; in fact it suffices to use two states such as yes/no. Here the accuracy of a number is not constrained by how exact a physical measurement can be made, but rather by the length of the sequences being used.

All computing machines have their roots in the various forms of the *abacus* which were invented by different civilizations. It is known from several sources that the abacus was already being used at the time of the Greek empire. Versions of the abacus also apparently developed independently in Russia and Asia, and have been used very heavily from ancient times right up to the present day. The origin of the asiatic abacus was most likely in China, where its present-day form, the *Suanpan*, uses two beads for carrying tens. It was exported to Japan in the 16-th Century, where it is known as the *Soroban*. This device is very similar to the Roman abacus, and uses only one bead for carrying tens. The abacus used in Russia is called the *Stchoty*, and with its ten beads on each rod is very similar to devices used as late as the middle of this century in elementary schools in Europe to learn arithmetic. It is interesting to note that, despite the wide availability of electronic pocket calculators, in asiatic countries such as Japan and China, various forms of the abacus are still heavily used, especially by tradespeople.

Books on arithmetic appeared in the middle ages, and served to explain the passage from using a mechanical device like an abacus to written computation. For those who could read, these books taught arithmetic rules in the form of algorithms. Following these developments, and inspired by the book on logarithms of the Scotsman LORD NAPIER OF MERCHISTON (1550–1617), the Englishman EDMUND GUNTER (1581–1626) developed the first slide rule in 1624. This analog device was still being used in the 1960's, especially by engineers, and was only replaced with the advent of the cheap electronic pocket calculator. Lord Napier also developed a simple multiplication device capable of carrying out single digit multiplication, and where the carryover of a ten had to be especially noted. WILHELM SCHICKARD (1592 – 1635), a professor of biblical languages at Tübingen in Germany, and one of the great scholars of his time, is regarded today as the father of the mechanical calculator. He was later also a professor of mathematics and astronomy, and was active in geodesy and as an artist and copperplate engraver. He was a friend of KEPLER, and it is clear from their correspondence that Schickard had built a functioning machine capable of all four arithmetic operations: addition, subtraction, multiplication and division. Unfortunately, his machine has not survived. The chaos of the thirty year's war no doubt helped prevent Schickard's idea from becoming widely known. He died in 1635 from the plague.

The idea of a mechanical calculating machine was popularized by an invention of the famous French mathematician BLAISE PASCAL (1623 – 1662). As a twenty-year old, Pascal developed an eight place machine capable of addition and subtraction, which was particularly useful for his father, who was the tax collector in Normandy. Because of his entree to the higher social circles, and the widespread discussion of his ideas, he was greatly admired. He built approximately seven copies of his machine, which he either sold or gave away.

Another essential step in the mechanization of calculation was provided by an invention of GOTTFRIED WILHELM LEIBNIZ (1646 – 1716), who was a philosopher, mathematician, and perhaps the last universal genius. Without knowing about Schickard's earlier work, he also constructed a machine capable of all four arithmetic operations. In a letter to Duke Johann Friedrich of Hannover in 1671, he wrote *"In Mathematicis und Mechanicis habe ich vermittels Artis Combinatoriae einige Dinge gefunden, die in Praxi Vitae von nicht geringer Importanz zu achten, und ernstlich in Arithmeticis eine Maschine, so ich eine lebendige Rechenbank nenne, dieweil dadurch zu wege gebracht wird, daß alle Zahlen sich selbst rechnen, addieren, subtrahieren, multipliciren, dividiren ..."* (from L. v. Mackensen: Von Pascal zu Hahn. Die Entwicklung der Rechenmaschine im 17. und 18. Jahrhundert, p. 21 – 33. In: M. Graef (ed.): 350 Jahre Rechenmaschinen. Vorträge eines Festkolloquiums veranstaltet vom Zentrum für Datenverarbeitung der Universität Tübingen. Hanser Verlag, München 1973). The mechanical principles used in Leibniz' machine were used for many more years in the further development of computing machines. Carry-over of digits was accomplished using staggered cylinders, and the carry-over of tens was done in parallel. Moreover, the machine ran in both directions; that is, addition and subtraction differed only in the direction in which the cylinders were turned. Multiplication and division were accomplished, for the first time, by successive addition and subtraction with the correct decimal point. Leibniz also had plans to develop a machine using binary numbers, but it was never built.

Of those who constructed mechanical calculators in the 17-th and 18-th centuries, we mention here only PHILIP MATTHÄUS HAHN (1739 – 1790), a pastor who built about a dozen machines based on the use of staggered cylinders. It should be noted, that in those times, mechanical calculators were constructed more as curiosities than as practical machines for business use. The possibility of their construction was also used from time to time as proof of the validity of various philosophical hypotheses. Indeed, Hahn was inspired by theology. He wrote in his diary on August 10, 1773: *" What computer, what astronomical clock, what garbage! But to help spread the gospel, I am ready to bear the yoke a little longer."* (From L. v. Mackensen, loc.cit.).

The production of mechanical calculators in large numbers began in the nineteenth century. CHARLES XAVIER THOMAS (1785 – 1870) from Colmar used the drum principle of Leibniz to build an *Arithmometer*, which for the first time, solved the carry-over problem exactly. Approximately 1500 copies of this machine were produced. In 1884, the American WILLIAM SEWARD BURROUGHS developed the first printing adding machine with keys. Using a patent of the Swede WILLGODT THEOPHIL ODHNER, the Brunsviga company in Braunschweig, Germany, began producing a machine using gears in 1892. All together, they sold more than 200,000 machines. Machines from this company were still being used in the 1960's.

The methods discussed in this book are designed for digital computers, since they are the only machines capable of solving large scale numerical problems.

1.3 Binary Arithmetic. In the binary system we have just the two digits 0 and 1. Hence, the addition and multiplication tables are especially simple:

+	0	1
0	0	1
1	1	10

×	0	1
0	0	0
1	0	1

These operations behave in the same way as certain operations in the *Boolean algebra* used in Logic.

Definition. A *binary Boolean Algebra A* is a set of two elements, which we denote by 0 and 1, and three binary operations defined on the set called *Negation* = *not* (written $\neg$), *conjunction* = *and* (written $\wedge$), and *disjunction* = *or* (written $\vee$), defined by the following tables:

¬	
0	1
1	0

∧	0	1
0	0	0
1	0	1

∨	0	1
0	0	1
1	1	1

Disjunction and conjunction are commutative, associative and distributive operations, and the elements of A are idempotent with respect to both.

Now let x and y be two binary digits (abbreviated *Bit*) which are to be added. Then the result consists of a *sum bit* s and a *carry bit* u, where

$$s := (\neg x \wedge y) \vee (x \wedge \neg y),$$

$$u := x \wedge y.$$

The operation leading to the sum bit s is called *disvalence*.

The following symbols are traditionally used to represent circuits capable of performing these binary operations:

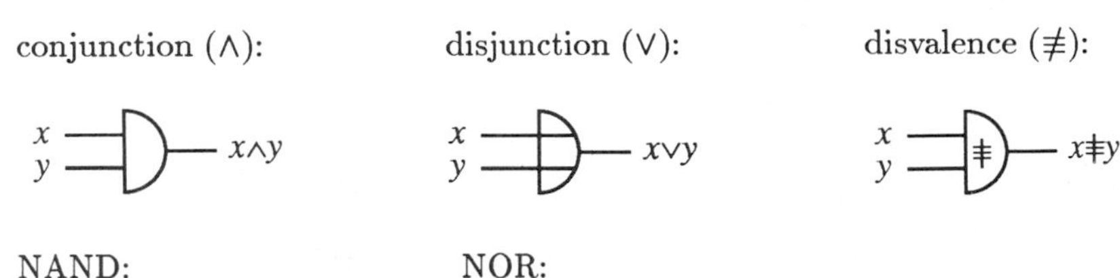

NAND: NOR:

x, y — $\neg(x \wedge y)$ x, y — $\neg(x \vee y)$

The combination

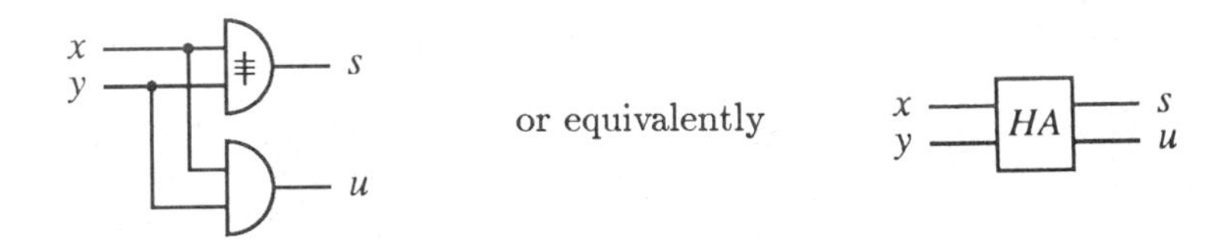

is called a *half-adder*.

Addition of two binary numbers can be accomplished by successive use of half-adders. Suppose

$$x = \sum_{\nu=1}^{n} x_{-\nu} 2^{-\nu}, \qquad y = \sum_{\nu=1}^{n} y_{-\nu} 2^{-\nu}$$

are two n-digit binary numbers and let

$$z = x + y = \sum_{\nu=0}^{n} z_{-\nu} 2^{-\nu}$$

be their sum. Then the logical circuit shown on the following page leads to the digits $z_{-\nu}$ of the binary number z. We shall not present the circuit for performing multiplication. Note that the intermediate information, in this case the binary numbers $.x_{-1}\, x_{-2} \cdots x_{-n}$ and $.y_{-1}\, y_{-2} \cdots y_{-n}$, which we have on hand as bit chains, have to be *stored* somewhere in the computer. This is done in *registers* which have a given capacity, called the *word length*. This is the length of the bit chains which can be simultaneously manipulated by the machine. For example, the word length in an IBM 360/370 machine is 32 bits, consisting of 4 *bytes*, each with 8 bits. The word length constrains the length of the binary numbers which can be directly manipulated by the computer without additional organization. This means that all arithmetic operations have to be done using a restricted set of numbers called the *set of machine numbers*. Thus, the expansion given in Theorem 1.1 for a real-valued number x has to be truncated to

$$(*) \qquad x = \sigma B^{N} \sum_{\nu=1}^{t} x_{-\nu} B^{-\nu},$$

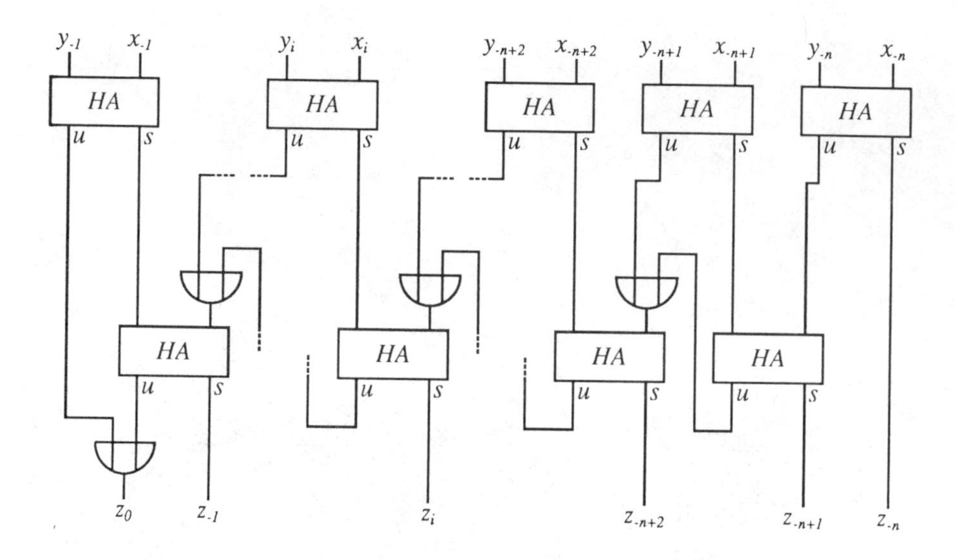

where $t \in \mathbb{N}$ is fixed. The number $m := \sum_{\nu=1}^{t} x_{-\nu} B^{-\nu}$ is called the *mantissa* of x, and t is called the *length of the mantissa*. We call σ the *sign* and N the *exponent* of the number x.

1.4 Fixed-Point Arithmetic. In fixed-point arithmetic, we work with the set of numbers which can be written in the form $(*)$ with a fixed N, and where we allow $x_{-1} = 0$. Since N is fixed, it does not require any place in memory.

Example. For $N = 0$, the formula $(*)$ represents numbers x with $0 \leq |x| < 1$. For $N = t$, it represents all integers x with $|x| \leq B^t - 1$. In the latter case we can write

$$x = \sigma \sum_{\nu=0}^{t-1} \overline{x}_\nu B^\nu,$$

where the coefficients here are related to those in the formula $(*)$ by $x_{-\nu+t} := \overline{x}_\nu$.

The fixed-point representation of numbers is used in some calculators – for example in the business world – and in the internal management of computers, for example to handle INTEGER-variables. The fixed-point representation is not suitable for scientific calculation, since physical constants take on values over several orders of magnitude. For example,

Mass of an electron	$m_0 \doteq 9.11 \cdot 10^{-28}\text{g}$,
Speed of light	$c \doteq 2.998 \cdot 10^{10}\text{cm/sec}$.

1.5 Floating-Point Arithmetic. In floating-point arithmetic, we work with numbers of the form (*) in 1.3 with fixed mantissa length $t > 0$, and with integer bounds $N_- < N_+$ for the exponent N. We require that

$$\begin{aligned} &x_{-\nu} \in \{0, 1, \ldots, B-1\}, \quad 1 \le \nu \le t; \\ &x_{-1} \ne 0, \quad \text{if } x \ne 0; \\ &\sigma = \pm 1 \text{ and } N_- \le N \le N_+. \end{aligned}$$

The set of numbers $x \ne 0$ which can be represented in this form lie in the range

$$B^{N_- - 1} \le |x| < B^{N_+}.$$

If $|x| < B^{N_- - 1}$, then it is replaced by zero. Numbers whose value is greater than B^{N_+} cannot be handled. We describe both cases as an *exponential overflow.* Thus in implementing numerical methods, we have to be careful that no overflow occurs. This can, in general, be achieved.

As we have already seen in 1.1, in view of the fact that the binary digits 0 and 1 can be interpreted as physical states, it is appropriate to choose the basis $B = 2$.

Integers are usually stored directly in binary form. For floating-point numbers, however, the binary system has the disadvantage that very large numbers N_- and N_+ have to be chosen for the exponents in order to be able to represent a sufficiently large set of numbers. Hence, most computers use a base B for floating-point numbers which is a power of two; e.g., $B = 8$, (octal system) or $B = 16$, (hexadecimal system). Then each of the digits $x_{-\nu}$ can be written as a binary number. For example, if $B = 2^m$, then we need m bits for the representation of $x_{-\nu}$ (cf. Section 1.1).

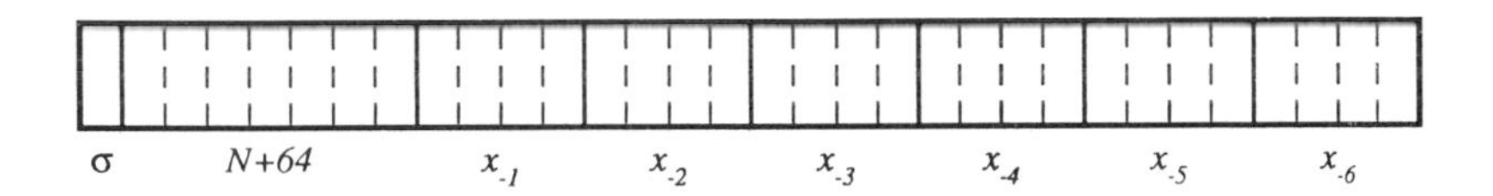

Example. As an example, we discuss the IBM 360. For this machine, we have $B = 16 = 2^4$. For floating-point numbers in single precision, we have 32 bits = 4 bytes available. We use one byte for sign (1 bit) and exponent (7 bits). This corresponds to chosing $N_- = -64$, $N_+ = 63$, and we use the 7 bits to store the number $N + 64$ with $0 \le N + 64 \le 127 = 2^7 - 1$. The remaining 3 bytes can be used for $t = 6$ hexadecimal digits. The sign bit is interpreted as "+" if it is 0, and as "−" if it is 1.

As an example, the number

$$\begin{aligned} x = 123.75 &= 7 \cdot 16^1 + 11 \cdot 16^0 + 12 \cdot 16^{-1} \\ &= 16^2(7 \cdot 16^{-1} + 11 \cdot 16^{-2} + 12 \cdot 16^{-3}) \end{aligned}$$

is stored in memory as

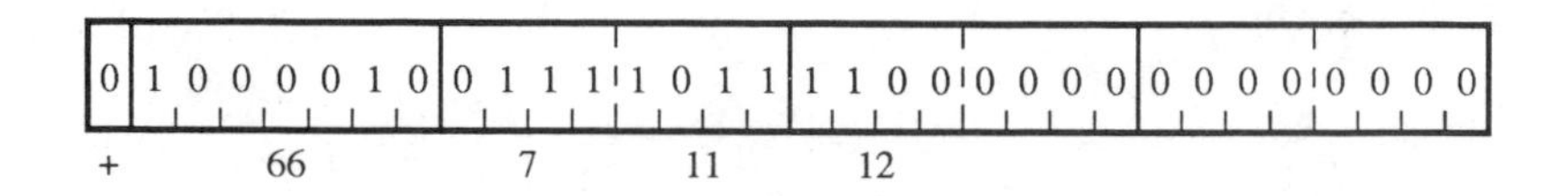

Double precision floating-point numbers use 8 bytes. As before, the first byte is used for the sign and exponent, and now 7 bytes remain for the mantissa ($t = 14$).

1.6 Problems. 1) Find a number whose representation (*) in 1.3 is not unique if the condition "$x_{-m} \neq B - 1$ for some $m \geq n$ and every $n \in \mathbb{N}$" is dropped. Note that even in this case, there cannot be more than two representations of this kind.

2) Find out how numbers are stored in your computer, and what is their accuracy. What are the smallest and largest positive machine numbers?

3) Rewrite the decimal numbers $x = 11.625$ and $y = 2.41\overline{6}$ in binary, octal, and hexadecimal form.

4) Let t_2 and t_{10}, respectively, be the mantissa lengths of the binary and decimal representations of an integer n. Show:

$$[t_{10}/log_{10}2] - 3 \leq t_2 \leq [t_{10}/log_{10}2] + 1.$$

Here $[a]$ denotes the largest integer less than or equal to a.

5) Negative numbers can be coded in complementary form. The coding of a number x in base B of the form $x = \sigma \cdot 0.x_{-1}x_{-2}\cdots x_{-n}$ is then replaced by

$$x \to (B^n + x)mod(B^n) \qquad (B\text{-complementary map})$$

or by

$$x \to (B^n - 1 + x + u)mod(B^n) \qquad ((B-1)\text{-complementary map}),$$

where

$$u = \begin{cases} 1 & \text{if } x \geq 0 \\ 0 & \text{otherwise.} \end{cases}$$

Show:

a) The B-complementary map does not change positive numbers, while negative numbers are replaced by their complement with respect to B^n.

b) Given two numbers with the same absolute value, how can one distinguish the positive one from the negative one?

c) What happens to positive and negative numbers under the $(B-1)$-complementary map? How is zero represented?

d) How must the adding circuit be modified for the B-complementary map and for the $(B-1)$-complementary map in order to always produce the correct results?

2. Floating-Point Arithmetic

The set of all numbers which can be represented with a finite mantissa length t is also finite, of course. Thus, in general, a real number x has to be replaced by an approximation $\tilde{x}$ which has such a representation. This process is called *rounding*, and involves introducing errors.

Notation. Let $x, \tilde{x} \in \mathbb{R}$, where $\tilde{x}$ is an approximation to x. Then

(i) $x - \tilde{x}$ is called the *absolute error*,

(ii) if $x \neq 0$, then $\frac{x-\tilde{x}}{x}$ is called the *relative error.*

Throughout this section we restrict our attention to floating-point numbers, and assume that in all calculations, the condition $N_- \leq N \leq N_+$ remains satisfied; i.e., there is no overflow.

2.1 The Roundoff Rule. Let $B \geq 2$ be an even integer, $t \in \mathbb{Z}_+$, and $x \in \mathbb{R} \setminus \{0\}$ with $x = \sigma\, B^N \sum_{\nu=1}^{\infty} x_{-\nu} B^{-\nu}$, $(\sigma = \pm 1)$. Then we define:

$$Rd_t(x) := \begin{cases} \sigma B^N \sum_{\nu=1}^{t} x_{-\nu} B^{-\nu} & \text{if } x_{-t-1} < \frac{B}{2}, \\ \sigma B^N (\sum_{\nu=1}^{t} x_{-\nu} B^{-\nu} + B^{-t}) & \text{if } x_{-t-1} \geq \frac{B}{2}. \end{cases}$$

$Rd_t(x)$ is called the *value of x rounded to t digits.*

It is easy to see that applied to the decimal system, this rule reduces to the usual "rounding" process of arithmetic.

Theorem. *Let $B \in \mathbb{N}$, $B \geq 2$, be even, and let $t \in \mathbb{Z}_+$. Suppose $x \neq 0$ has the representation*

$$x = \sigma\, B^N \sum_{\nu-1}^{\infty} x_{-\nu} B^{-\nu}.$$

Then:

(i) $Rd_t(x)$ has an expansion of the form $Rd_t(x) = \sigma\, B^{N'} \sum_{\nu=1}^{t} x'_{-\nu} B^{-\nu}$.

(ii) The absolute error satisfies $|Rd_t(x) - x| \leq 0.5\, B^{N-t}$.

(iii) The relative error satisfies $|\frac{Rd_t(x)-x}{x}| \leq 0.5\, B^{-t+1}$.

(iv) The relative error with respect to $Rd_t(x)$ satisfies $|\frac{Rd_t(x)-x}{Rd_t(x)}| \leq 0.5\, B^{-t+1}$.

Proof. (i) There is something to prove only for the case $x_{-t-1} \geq 0.5\, B$. We distinguish two possibilities: Either there exists a $\nu \in \{1, 2, \ldots, t\}$ with $x_{-\nu} < B - 1$, or for all such ν, $x_{-\nu} = B - 1$.

In the first case we set $N' := N$, $x'_{-\nu} := x_{-\nu}$ for $1 \leq \nu \leq l - 1$, $x'_{-l} := x_{-l} + 1$, and $x'_{-\nu} = 0$ for $l + 1 \leq \nu \leq t$. Here the index l is defined by $l := \max\{\nu \in \{1, 2, \ldots, t\} \mid x_{-\nu} < B - 1\}$.

In the second case we take $N' := N + 1$, $x'_{-1} := 1$, and $x'_{-\nu} = 0$ for $2 \leq \nu \leq t$.

(ii) For $x_{-t-1} < 0.5\,B$, we have

$$-\sigma(Rd_t(x) - x) = B^N \sum_{\nu=t+1}^{\infty} x_{-\nu}B^{-\nu} = B^{N-t-1}x_{-t-1} + B^N \sum_{\nu=t+2}^{\infty} x_{-\nu}B^{-\nu}$$
$$\leq B^{N-t-1}(0.5\,B - 1) + B^{N-t-1} = 0.5\,B^{N-t}.$$

On the other hand, if $x_{-t-1} \geq 0.5\,B$, then

$$\sigma(Rd_t(x) - x) = B^{N-t} - B^N x_{-t-1}B^{-t-1} - B^N \sum_{\nu=t+2}^{\infty} x_{-\nu}B^{-\nu} =$$
$$= B^{N-t-1}(B - x_{-t-1}) - B^N \sum_{\nu=t+2}^{\infty} x_{-\nu}B^{-\nu} \leq$$
$$\leq 0.5B^{N-t}.$$

But $B^{N-t-1} \leq B^{N-t-1}(B - x_{-t-1})$ and $B^N \sum_{\nu=t+2}^{\infty} x_{-\nu}B^{-\nu} < B^{N-t-1}$ imply $\sigma(Rd_t(x) - x) > 0$.

(iii) We always have $x_{-1} \geq 1$. From this it follows that $|x| \geq B^{N-1}$, and using (ii), that

$$\left|\frac{Rd_t(x) - x}{x}\right| \leq \frac{1}{2}B^{N-t}B^{-N+1} = 0.5\,B^{-t+1}.$$

(iv) The roundoff rule implies $|Rd_t(x)| \geq x_{-1}B^{N-1} \geq B^{N-1}$, and applying (ii) leads to

$$\left|\frac{Rd_t(x) - x}{Rd_t(x)}\right| \leq \frac{1}{2}B^{N-t} \cdot B^{-N+1} = 0.5\,B^{-t+1}. \qquad \square$$

If we set $\varepsilon := \frac{Rd_t(x)-x}{x}$ and $\eta := \frac{Rd_t(x)-x}{Rd_t(x)}$, then we immediately get the

Corollary. If the assumptions of the theorem hold, then

$$\max\{|\varepsilon|, |\eta|\} \leq 0.5 \cdot B^{-t+1} \quad \text{and} \quad Rd_t(x) = x(1+\varepsilon) = \frac{x}{1-\eta}.$$

The number $\tau := 0.5\,B^{-t+1}$ is called the *relative accuracy* of t-digit floating-point arithmetic.

Example. On the IBM 360, all real-valued numbers x are stored with a relative error less than or equal to $\tau = 0.5 \cdot 16^{-5} < 0.5 \cdot 10^{-6}$. Thus, it makes little sense to input or output numbers with more than seven digits in the mantissa. For double precision, $\tau = 0.5 \cdot 16^{-13} < 0.5 \cdot 10^{-15}$.

In the decimal system, one can also measure the accuracy of an approximation $\tilde{x}$ to a real-valued number x by the number of digits which coincide.

Notation. Let $x = \sigma \cdot 10^N \cdot m$ with $0.1 \le m < 1$ and $\tilde{x} = \sigma \cdot 10^N \cdot \tilde{m}$ with $\tilde{m} \in \mathbb{R}$ be given. Then the number of digits of the approximation $\tilde{x}$ which coincide with those of x is given by

$$s = \max\{t \in \mathbb{Z} \mid |m - \tilde{m}| \le 0.5 \cdot 10^{-t+1}\}.$$

We say that $\tilde{x}$ has $s - 1$ ***significant digits.***

Example. Let $\tilde{x} = 10^2 \cdot 0.12415$ be an approximation to $x \doteq 10^2 \cdot 0.12345$. Then the associated mantissas m and $\tilde{m}$ satisfy

$$0.5 \cdot 10^{-3} < |m - \tilde{m}| < 0.5 \cdot 10^{-2},$$

and so $\tilde{x}$ has $s - 1 = 2$ significant digits.

2.2 Combining Floating-Point Numbers. In this section we use the symbol $\Box$ to denote one of the arithmetic operations $+, -, \cdot, \div$. If x and y are two floating-point numbers with t-digit mantissas, then, in general, $x \Box y$ is not necessarily representable using a t-digit mantissa. For example, suppose $t = 3$, and consider $x = 0.123 \cdot 10^4$ and $y = 0.456 \cdot 10^{-3}$. Then $x + y = 0.123\,000\,0456 \cdot 10^4$.

Thus, in general, after performing an elementary arithmetic operation $\Box$, it will be necessary to round off the result. We think of this as happening in two steps:

(a) Compute $x \Box y$ to as much accuracy as possible;

(b) round off the result to t digits.

We denote the result of these two steps by

$$Fl_t(x \Box y).$$

We shall assume that the arithmetic performed by our computer is such that for any two t-digit floating-point numbers x and y,

$$(*) \qquad Fl_t(x \Box y) = Rd_t(x \Box y).$$

Then by Corollary 2.1,

$$Fl_t(x \Box y) = (x \Box y)(1 + \varepsilon) = \frac{x \Box y}{1 - \eta}, \qquad |\varepsilon|, |\eta| \le \tau.$$

We now show how addition in the decimal system can be organized so that $Fl_t(x+y) = Rd_t(x+y)$. Let $x = \sigma_1 \cdot 10^{N_1} \cdot m_1$ and $y = \sigma_2 \cdot 10^{N_2} \cdot m_2$ with $0 \le m_1, m_2 < 1$ and $N_2 \le N_1$ be two decimal numbers in floating-point form. The usual approach to adding x and y is to write both numbers as $2t$-digit floating-point numbers with the same exponents (i.e. as double precision numbers), and then to add them. The result is then *normalized* so that the mantissa m of the sum satisfies $0 \leq m < 1$, and then the result is rounded off to t digits. When $N_1 - N_2 > t$, this rule always gives $Fl_t(x+y) = x$.

Example 1. Let $B = 10, t = 3$ and suppose $x = 0.123 \cdot 10^6$ and $y = 0.456 \cdot 10^2$. Then $Fl_3(x+y) = Rd_3(x+y) = 0.123 \cdot 10^6$.

We now illustrate the case $0 \leq N_1 - N_2 \leq t$ with several examples.

Example 2. Let $B = 10, t = 3$.

(i) $x = 0.433 \cdot 10^2$, $y = 0.745 \cdot 10^0$.

$$\begin{array}{l} 0.433\,000 \cdot 10^2 \\ \underline{+0.007\,450 \cdot 10^2} \\ 0.440\,450 \cdot 10^2 \end{array} \quad \Rightarrow Fl_3(x+y) = 0.440 \cdot 10^2$$

(ii) $x = 0.215 \cdot 10^{-4}$, $y = 0.998 \cdot 10^{-4}$

$$\begin{array}{l} 0.215\,000 \cdot 10^{-4} \\ \underline{+0.998\,000 \cdot 10^{-4}} \\ 1.213\,000 \cdot 10^{-4} \end{array} \quad \Rightarrow Fl_3(x+y) = 0.121 \cdot 10^{-3}$$

(iii) $x = 0.1000 \cdot 10^1$, $y = -0.998 \cdot 10^0$.

$$\begin{array}{l} 0.100\,000 \cdot 10^1 \\ \underline{-0.099\,800 \cdot 10^1} \\ 0.000\,200 \cdot 10^1 \end{array} \quad \Rightarrow Fl_3(x+y) = 0.200 \cdot 10^{-2}.$$

We now examine case (iii) in more detail. Suppose the numbers $Fl_3(x) = 0.100 \cdot 10^1$ and $Fl_3(y) = -0.998 \cdot 10^0$ are the floating-point representations of the numbers $x = 0.9995 \cdot 10^0$ and $y = -0.9984 \cdot 10^0$. Then

$$Fl_3(Fl_3(x) + Fl_3(y)) = (Fl_3(x) + Fl_3(y))(1+\varepsilon) =$$

$$= (x(1+\varepsilon_x) + y(1+\varepsilon_y))(1+\varepsilon) = (x+y) + E$$

with an absolute error E of

$$E = x(\varepsilon + \varepsilon_x(1+\varepsilon)) + y(\varepsilon + \varepsilon_y(1+\varepsilon)).$$

$$E = 0.9995 \cdot 0.5003 \cdot 10^{-3} + 0.9984 \cdot 0.4006 \cdot 10^{-3} =$$

$$= 0.9000 \cdot 10^{-3}.$$

Now the relative error satisifies

$$Fl_3(Fl_3(x) + Fl_3(y)) = (x+y)(1+\rho), \quad \text{and so} \quad \rho = \frac{E}{(x+y)}.$$

Substituting the above values, we get $\rho = 0.82$. Thus, the relative error of this calculation is 82 %, although the floating-point addition of $Fl_3(x)$ and $Fl_3(y)$ was exact, and $Fl_3(x)$ and $Fl_3(y)$ differ from x and y by only about 0.05 % and 0.04 %, respectively. Clearly, the reason for this is the fact that two numbers of approximately the same size but with opposite signs have been added, resulting in a *cancellation of digits.*

In general, if we suppose that the mantissa length satisfies $t \geq 2$, so that $\tau = 0.5 \cdot 10^{-t+1} \leq 0.05$, then Corollary 2.1 implies

$$|E| \leq |x|(\tau + 1.05|\varepsilon_x|) + |y|(\tau + 1.05|\varepsilon_y|).$$

It follows that

$$|\rho| \leq \frac{|x|}{|x+y|}(\tau + 1.05|\varepsilon_x|) + \frac{|y|}{|x+y|}(\tau + 1.05|\varepsilon_y|).$$

We now distinguish three cases:

(a) $|x+y| < \max(|x|, |y|)$; i.e., in particular, $\operatorname{sgn}(x) = -\operatorname{sgn}(y)$. Then $|\rho|$ is, in general, larger than $|\varepsilon_x|$ and $|\varepsilon_y|$ (cf. the example above).

In this case the calculation is *numerically instable.*

(b) $\operatorname{sgn}(x) = \operatorname{sgn}(y)$. Then $|x+y| = |x| + |y|$, and thus it follows that $|\rho| \leq \tau + 1.05 \max(|\varepsilon_x|, |\varepsilon_y|)$.

In this case, the error is of the same order as the larger of $|\varepsilon_x|$ and $|\varepsilon_y|$.

(c) $|y| \ll |x|$.

In this case, the order of the error ρ is primarily determined by the error of x. We call this *error damping.*

2.3 Numerically Stable vs. Unstable Evaluation of Formulae. The numerical evaluation of complicated mathematical formulae reduces to performing a sequence of elementary operations. To assure the stability of the overall process, we must make sure that each individual step is stable.

Example. Suppose we want to solve the quadratic equation

$$ax^2 + bx + c = 0,$$

where $|4ac| < b^2$. It is well-known that there are two solutions given by

$$x_1 = \frac{1}{2a}(-b - \operatorname{sgn}(b)\sqrt{b^2 - 4ac}), \quad x_2 = \frac{1}{2a}(-b + \operatorname{sgn}(b)\sqrt{b^2 - 4ac}).$$

Now if $|4ac| \ll b^2$, then there will be some instability in computing x_2 of the type discussed in (2.2, case (a)), while in the computation of x_1, the error will be of the same order as that of the original numbers (2.2, case (b)). In this case it is recommended that x_2 be computed by using the identity $x_1 \cdot x_2 = \frac{c}{a}$; i.e.,

$$x_2 = \frac{2c}{-b - \operatorname{sgn}(b)\sqrt{b^2 - 4ac}}.$$

The following example also illustrates how an inappropriate choice of the order in which the individual steps are carried out can lead to a completely incorrect result.

Example. Suppose we want to compute the integrals

$$I_n = \int_0^1 \frac{x^n}{x+5} dx$$

for $n = 0, 1, 2, \cdots, 20$. It is easy to see that the numbers I_n satisfy the recursion

$$I_n + 5I_{n-1} = \int_0^1 \frac{x^n + 5x^{n-1}}{x+5} dx = \int_0^1 x^{n-1} dx = \frac{1}{n}.$$

Starting with the value $I_0 = ln\ \frac{6}{5}$, this recursion can theoretically be used to find all of the numbers $I_n = \frac{1}{n} - 5I_{n-1}$. But if we carry out this process, it turns out that after only a few steps we already have wrong results, and that after a few more steps we even get negative numbers. It is clear from the recursion that whatever roundoff error is made in computing I_0 will be multiplied by a factor (-5) at each step. After $n = 20$ steps, this gives the rather poor estimate $|\varepsilon_n| \leq 5^n \cdot 0.5 \cdot 10^{-t+1}$ for the accumulated error. On the other hand, if we rewrite the recurrence formula in the form $I_{n-1} = \frac{1}{5n} - \frac{1}{5} I_n$, then in calculating I_{n-1} from I_n, the error is reduced by a factor $(-\frac{1}{5})$. Starting with the approximate value $I_{30} = \frac{1}{280}$, it turns out that the computation of the numbers $I_{20}, I_{19}, \cdots, I_1, I_0$ is extremely stable, giving results which are exact to 10 digits.

n	$I_n = -5I_{n-1} + \frac{1}{n}$ $I_0 = ln\frac{6}{5}$	$I_{n-1} = \frac{1}{5}(-I_n + \frac{1}{n})$ $I_{30} = \frac{1}{280}$
1	0.088 392 216	0.088 392 216
2	0.058 038 919	0.058 038 919
3	0.043 138 734	0.043 138 734
4	0.034 306 327	0.034 306 329
5	0.028 468 364	0.028 468 352
6	0.024 324 844	0.024 324 905
7	0.021 232 922	0.021 232 615
8	0.018 835 389	0.018 836 924
9	0.016 934 162	0.016 926 489
10	0.015 329 188	0.015 367 550
11	0.014 263 149	0.014 071 338
12	0.012 017 583	0.012 976 639
13	0.016 835 157	0.012 039 876
14	-0.012 747 213	0.011 229 186
15	0.130 402 734	0.010 520 733
16	-0.589 513 672	$9.896\ 332\ 328 \cdot 10^{-3}$
17	3.006 391 892	$9.341\ 867\ 770 \cdot 10^{-3}$
18	$-1.497\ 640\ 391 \cdot 10^{1}$	$8.846\ 216\ 703 \cdot 10^{-3}$
19	$7.493\ 465\ 113 \cdot 10^{1}$	$8.400\ 495\ 432 \cdot 10^{-3}$
20	$-3.746\ 232\ 556 \cdot 10^{1}$	$7.997\ 522\ 840 \cdot 10^{-3}$

We consider the question of the numerical stability of numerical methods in more detail in the next section.

2.4 Problems. 1) In calculating $\sum_{\nu=1}^{n} a_\nu$ in fixed-point arithmetic, we can get an arbitrarily large relative error. If, however, all a_ν are of the same sign, then it is bounded. Neglecting terms of higher order, derive an upper bound in this case.

2) Rearrange the following expressions so that their evaluation is stable:
a) $\frac{1}{1+2x} - \frac{1-x}{1+x}$ for $|x| \ll 1$; b) $\frac{1-\cos x}{x}$ for $x \neq 0$ and $|x| \ll 1$.

3) Suppose the sequence (a_n) is defined by the following recurrence relation:

$$a_1 := 4, \; a_{n+1} := \frac{\sqrt{1 + a_n^2/2^{2(n+1)}} - 1}{a_n} \cdot 2^{2(n+1)+1}.$$

a) Rewrite the recurrence in an equivalent but stable form.
b) Write a computer program to compute a_{30} using both formulae, and compare the results.

4) Prove that the sequence of numbers $y_n = e^{-1} \int_0^1 e^x x^n dx$ can be computed using the recurrence

$$(*) \qquad y_{n+1} + (n+1)y_n = 1 \text{ for } n = 0, 1, 2, \ldots \qquad \text{and} \qquad y_0 = \frac{1}{e}(e-1).$$

a) Using (*), compute the numbers $y_0, \ldots, y_{30}$ and interpret the results.
b) Prove the sequence of numbers in (*) converges to 0 as $n \to \infty$. Thus y_0 can be computed by working backwards, starting with the approximation $y_n = 0$ for a given n. Carry out this process for $n = 5, 10, 15, 20, 30$, and explain why it gives such a good result for y_0.

3. Error Analysis

As we saw in 2.3, in general, there may be several different ways of arranging the computation leading to a solution of a given problem. Competing algorithms can be compared in terms of their complexity (the number of arithmetic operations needed), the amount of storage space required for the input and all intermediate results, and the error bounds which can be established for the final result. In this section we discuss three different types of errors:

Data Error. Before we can start computing, we have to input data, generally in the form of numbers. Often these data come from physical measurements or empirical studies, and are therefore subject to measurement errors or simple mistakes.

Method Error. The formulation and solution of many mathematical problems involve taking limits, which, of course, is something which cannot be done on a computer. Thus, for example, derivatives are replaced by difference quotients, and iterations have to be stopped after a finite number of steps. The resulting errors are called method errors.

Roundoff Error. Since we are working with a finite set of machine numbers, each step in a computation where roundoff occurs produces an error. The accumulation of such roundoff errors can lead to a completely incorrect final result.

We begin by discussing the effect of data errors on the solution of a problem.

3.1 The Condition of a Problem. A mathematical problem is called *well-conditioned* provided that small changes in the data leads only to small changes in the (exact) solution. If this is not the case, we call the problem *ill-conditioned.*

In 2.2 we saw that the subtraction of two floating-point numbers can lead to a result with a relative error which is much larger than the relative errors of the input data (numerical instability!). This raises the question of how to decide if a given problem is well-conditioned or not.

To discuss this question, suppose D is an open subset of $\mathbb{R}^n$, and that

$$\varphi : D \to \mathbb{R}$$

is a two-times continuously differentiable mapping. The problem is to compute

$$(*) \qquad y = \varphi(x), \quad x \in D.$$

The vector $x = (x_1, x_2, \cdots, x_n) \in D$ represents the data vector, and φ represents the set of (generally rational) operations which have to be performed on the data to get the result y. We now study how errors in x effect the result y.

Dropping terms of higher order, the Taylor expansion leads to

$$\delta y := \sum_{\nu=1}^{n} \frac{\partial \varphi}{\partial x_\nu}(x)(\tilde{x}_\nu - x_\nu),$$

which is a first order approximation to the absolute error $\Delta y := \varphi(\tilde{x}) - \varphi(x)$. Then a first order approximation to the relative error is given by

$$\frac{\delta y}{y} = \sum_{\nu=1}^{n} \frac{x_\nu}{\varphi(x)} \frac{\partial \varphi}{\partial x_\nu}(x) \frac{(\tilde{x}_\nu - x_\nu)}{x_\nu}.$$

Definition. The numbers $\frac{x_\nu}{\varphi(x)} \frac{\partial \varphi}{\partial x_\nu}(x)$, $1 \le \nu \le n$, are called *condition numbers* of the problem $(*)$.

Remark. If the absolute values of the condition numbers are less than or equal to 1, then our problem is a well-conditioned problem; otherwise it is poorly-conditioned.

In terms of this definition we have the

Proposition. The arithmetic operations □ satisfy:

(i) $\cdot, \div$ are well-conditioned operations.

(ii) $+$ and $-$ are well-conditioned as long as the summands have the same or opposite signs, respectively.

(iii) $+$ and $-$ are poorly-conditioned whenever the two summands are of approximately the same absolute value, but have opposite or the same signs, respectively.

Proof. We examine the condition numbers. In case (i) the condition numbers have the value 1, and errors are not magnified. In cases (ii) and (iii), the sum or difference of the two numbers, respectively, appears in the denominator of the formulae for the condition numbers. □

We return now to the considerations in 2.3, and analyse the first example there using the concept of the condition of a problem. The problem is to determine the largest root

$$\varphi(p,q) := -p + \sqrt{p^2+q}$$

of a quadratic equation

$$x^2 + 2px - q = 0.$$

Here we assume that $p, q > 0$ and $p \gg q$, where the symbol $\gg$ means that p is large compared with q. The computation now proceeds as follows: Set $s := p^2$, and successively find $t := s + q$ and $u := \sqrt{t}$. Then as in 2.3, we distinguish

Method 1: $y := \varphi_1(u) := -p + u$,

and

Method 2: $v := -p - u$ and $y := \varphi_2(v) = -\frac{q}{v}$.

We begin by showing that the problem of finding the number $\varphi(p,q)$ is well-conditioned. Consider the relative error

$$\frac{\delta y}{y} = \frac{p}{\varphi(p,q)}\frac{\partial\varphi}{\partial p}\varepsilon_p + \frac{q}{\varphi(p,q)}\frac{\partial\varphi}{\partial q}\varepsilon_q =$$

$$= \frac{p}{-p+(p^2+q)^{\frac{1}{2}}}\left(-1+\frac{p}{(p^2+q)^{\frac{1}{2}}}\right)\varepsilon_p + \frac{q}{-p+(p^2+q)^{\frac{1}{2}}}\cdot\frac{1}{2(p^2+q)^{\frac{1}{2}}}\varepsilon_q$$

$$= -\frac{p}{(p^2+q)^{\frac{1}{2}}}\varepsilon_p + \frac{p+(p^2+q)^{\frac{1}{2}}}{2(p^2+q)^{\frac{1}{2}}}\varepsilon_q.$$

The coefficients in front of the relative data errors ε_p and ε_q in the data p and q are smaller than one in absolute value, and so the problem is well-conditioned. Moreover, if we replace the absolute error Δy by the expression δy which approximates it to first order, then we see that the value of the relative error of the result $\frac{\Delta y}{y}$ is no larger than the sum of the absolute values of the data errors.

We now consider the two methods for computing a solution to the problem, and carry out a similar analysis of the corresponding functions φ_1 (Method 1) and φ_2 (Method 2).

Method 1:

$$\frac{\delta y}{y} = \frac{u}{-p+u}\varepsilon_u = \frac{(p^2+q)^{\frac{1}{2}}}{-p+(p^2+q)^{\frac{1}{2}}}\varepsilon_u = \frac{1}{q}(p(p^2+q)^{\frac{1}{2}} + p^2 + q)\varepsilon_u.$$

Since $p, q > 0$ and $p \gg q$ (i.e. p is large relative to q) then the coefficient of ε_u satisfies the inequality $|\frac{1}{q}(p(p^2+q)^{\frac{1}{2}} + p^2 + q)| > \frac{2p^2}{q} \gg 1$, and so an error in the data u is magnified in the relative error $\frac{\Delta y}{y}$ of the result. Thus, Method 1 turns out to be *numerically instable.*

Method 2: A similar calculation leads to

$$\frac{\delta y}{y} = -\frac{(p^2+q)^{\frac{1}{2}}}{p+(p^2+q)^{\frac{1}{2}}}\varepsilon_u.$$

Since the coefficient of ε_u has absolute value less than one, we see that Method 2 is *numerically stable.*

In summary, we have seen that in solving a well-conditioned problem, then depending on how it is designed, a numerical method can be either stable or unstable; i.e., it may or may not magnify the data errors. On the other hand, if the the problem itself is poorly-conditioned, then no method can dampen out data errors. (cf. Problem 1).

We have seen that computing the size of the condition number of a problem gives us a tool for predicting what effect data errors will have: a multiplying effect or a dampening effect.

3.2 Forward Error Analysis. In performing a forward error analysis, we go through each step of an algorithm, estimating the roundoff error at each step. In general, this method will only lead to a qualitative assertion about which of the factors in a formula have the largest influence on the accuracy of the final result. Quantitatively, forward error analyses usually significantly overestimate the error.

Example. Suppose we want to compute the value of the determinant of the matrix

$$A = \begin{pmatrix} a & b \\ c & d \end{pmatrix} = \begin{pmatrix} 5.7432 & 7.3315 \\ 6.5187 & 8.3215 \end{pmatrix}$$

using floating-point arithmetic with a mantissa length of $t = 6$. To do this, we need to compute each of the arithmetic expressions $a \cdot d$, $b \cdot c$ and $ad - bc$. The following table gives the exact results, along with upper and lower bounds on each of these quantities, assuming they are calculated using the rules in 2.2 for floating-point arithmetic.

	exact value	rounded value	error interval
$a \cdot d$	47.7920 3880	47.7920	[47.7920, 47.7921]
$b \cdot c$	47.7918 4905	47.7918	[47.7918, 47.7919]
$ad - bc$	$0.189750 \cdot 10^{-3}$	$0.20000 \cdot 10^{-3}$	$[1 \cdot 10^{-4}, 3 \cdot 10^{-4}]$

The actual relative error is on the order of 5%, while the lower and upper bounds are in error by 47 % and 58 %, respectively.

In addition to the fact that it usually significantly overestimates the error, the forward error analysis method is also very tedious for complicated expressions. We illustrate this by estimating the roundoff error in calculating the value of a *finite continued fraction.*

Definition. Let $n \in \mathbb{Z}_+$, and suppose b_0, a_ν, b_ν, $1 \le \nu \le n$, are given real or complex numbers. Given $x \in \mathbb{C}$, we call the rational expression

$$(*) \qquad k(x) = b_0 + \cfrac{a_1 x}{b_1 + \cfrac{a_2 x}{b_2 + \cfrac{a_3 x}{b_3 + \cfrac{a_4 x}{\ddots \, \cfrac{a_n x}{b_n}}}}}$$

a *finite continued fraction of order* n, assuming it is well-defined. This is the case whenever all of the numbers

$$b_n, b_{n-1} + \frac{a_n x}{b_n}, \quad b_{n-2} + \frac{a_{n-1} x}{b_{n-1} + \frac{a_n x}{b_n}}, \quad \ldots$$

appearing in the denominators are nonzero.

It is common to write continued fractions as in $(*)$ in the abbreviated form

$$k(x) = b_0 + \frac{a_1 x|}{|\,b_1} + \frac{a_2 x|}{|\,b_2} + \cdots + \frac{a_n x|}{|\,b_n}.$$

In general, continued fractions are more difficult to deal with than polynomials or power series. Nevertheless, they play an important role in approximating elementary functions; in particular, in pocket calculators where a high accuracy is required. They are also of use in the evaluation of infinite series, since infinite continued fraction expansions often converge much faster than the corresponding series. Finally, finite continued fractions can also be used to construct interpolating rational functions. It is beyond the scope of this book, however, to give a full treatment of the theory. For more details, see the monograph of G. A. Baker, Jr. and P. Graves-Morris ([1981], Chap. 4).

To evaluate a continued fraction $(*)$ for fixed $x \in \mathbb{R}$, we may successively evaluate each of the following rational expressions:

$(**)$

$$k^{(n)} := b_n, \quad k^{(n-1)} := b_{n-1} + \frac{a_n x}{k^{(n)}}, \quad k^{(n-2)} := b_{n-2} + \frac{a_{n-1} x}{k^{(n-1)}}, \quad \ldots,$$
$$k(x) = k^{(0)} := b_0 + \frac{a_1 x}{k^{(1)}},$$

making sure in each step that none of the intermediate values $k^{(\mu)}$ vanishes. This procedure is similar to the evaluation of a polynomial using the Horner scheme (cf. 5.5.1).

Another possible way to evaluate the continued fraction $(*)$ for fixed x in $\mathbb{R}$ is based on a recurrence formula which goes back to L. Euler and J. Wallis. Define approximate numerators $P_\mu(x)$ and approximate denominators $Q_\mu(x)$ by

$$\frac{P_\mu(x)}{Q_\mu(x)} = r_\mu(x) := b_0 + \frac{a_1 x|}{| b_1} + \frac{a_2 x|}{| b_2} + \cdots + \frac{a_\mu x|}{| b_\mu}, \quad 0 \le \mu \le n.$$

Then we have the

Recurrence Formulae of Euler and Wallis. The approximate numerators $P_\mu(x)$ and denominators $Q_\mu(x)$ can be computed recursively from the formulae

$$P_\mu(x) := P_{\mu-1}(x) \cdot b_\mu + P_{\mu-2}(x) a_\mu x, \; P_0 := b_0, \; P_1(x) := P_0 \cdot b_1 + a_1 x;$$
$$Q_\mu(x) := Q_{\mu-1}(x) \cdot b_\mu + Q_{\mu-2}(x) a_\mu x, \; Q_0 := 1, \; Q_1 := b_1$$

for $2 \le \mu \le n$.

Proof. We prove the recurrence formulae by complete induction. First note that the expressions P_0, P_1 and Q_0, Q_1 are all obviously correct. Now to pass from $r_{\mu-1}(x)$ to $r_\mu(x)$, we replace $b_{\mu-1}$ by $b_{\mu-1} + \frac{a_\mu x}{b_\mu}$. This leads to the formula

$$\begin{aligned} r_\mu(x) &= \frac{(b_{\mu-1} + \frac{a_\mu x}{b_\mu}) P_{\mu-2}(x) + a_{\mu-1} x P_{\mu-3}(x)}{(b_{\mu-1} + \frac{a_\mu x}{b_\mu}) Q_{\mu-2}(x) + a_{\mu-1} x Q_{\mu-3}(x)} \\ &= \frac{b_\mu P_{\mu-1}(x) + a_\mu x P_{\mu-2}(x)}{b_\mu Q_{\mu-1}(x) + a_\mu x Q_{\mu-2}(x)} = \frac{P_\mu(x)}{Q_\mu(x)}. \end{aligned}$$ □

We note that the recurrence formulae also hold for $\mu = 1$, provided we set $P_{-1} := 1$ and $Q_{-1} := 0$.

The recurrence formulae of Euler and Wallis can immediately be translated into a calculation scheme for evaluating a finite continued fraction $k(x)$.

Clearly, one must be careful to check for overflows, since the numerators and denominators can become very large even though the quotient $P_\mu(x)/Q_\mu(x)$ remains of a reasonable size.

Example. Let $b_0 = 0$, $b_\mu = 2\mu - 1$, $1 \le \mu \le 10$, and let $a_1 = 4$, $a_\mu = (\mu-1)^2$, $2 \le \mu \le 10$. Compute $k(1)$. The following table shows the results of computing $k(1)$ using both the method (**), and the recurrence formulae of Euler and Wallis. All computations were carried out in floating-point arithmetic with a mantissa length of $t = 7$.

μ	$k^{(\mu)}$		μ	$P_\mu(1)$	$Q_\mu(1)$	$P_\mu(1)/Q_\mu(1)$
10	19.000000		0	0	1	0.000000
9	21.263159		1	4	1	4.000000
8	18.009901		2	12	4	3.000000
7	15.720726		3	76	24	3.166667
6	13.289970		4	640	204	3.137255
5	10.881118		5	6976	2220	3.142342
4	8.470437		6	92736	29520	3.141464
3	6.062519		7	1456704	463680	3.141615
2	3.659792		8	26394624	8401680	3.141589
1	1.273240		9	541937664	172504080	3.141593
0	3.141593		10	12434780160	3958113600	3.141593

It seems clear from the results that the continued fraction provides an approximation to the number π. Indeed, $\arctan(z^2)/z = k(z^2)/4$ which implies that $k(1) = 4\arctan(1) = \pi$ (see e.g. Baker and Graves Morris [1981], p. 139). Despite the large sizes of the intermediate values $P_\mu(1)$ and $Q_\mu(1)$, using the recurrence formulae has the advantage that successive values of the quotients $P_\mu(1)/Q_\mu(1)$ and $P_{\mu+1}(1)/Q_{\mu+1}(1)$ always bracket the number π. In comparison, the method (**) only gives an acceptable result in the last step.

In both cases, a complete forward analysis of the roundoff error is extremely complicated. The analysis is possible in certain special cases; we now treat the case arising in the example above where $x = 1$.

In carrying out the method (**) in floating-point arithmetic (with mantissa length t and base B), at each step we have to compute expressions of the form

$$\begin{aligned}\tilde{k}^{(\mu-1)} &= Fl_t(b_{\mu-1} + Fl_t(\frac{a_\mu}{\tilde{k}^{(\mu)}})) = \\ &= (b_{\mu-1} + \frac{a_\mu}{\tilde{k}^{(\mu)}}(1+\varepsilon_\mu))(1+\delta_\mu),\end{aligned}$$

where ε_μ and δ_μ satisfy $|\varepsilon_\mu|, |\delta_\mu| \le 0.5 \cdot B^{-t+1}$. If $|a_\mu/\tilde{k}^{(\mu)}| \ll |b_{\mu-1}|$, then only the addition error has an effect on the result. Thus, the method (**) will perform well whenever

$$|k^{(\mu)}| \gg \frac{|a_\mu|}{|b_{\mu-1}|}.$$

This condition is generally satisfied as soon as we have $|a_\mu| \ll |b_{\mu-1}|$ and $|b_0| \ll |b_1| \ll \cdots \ll |b_n|$. In this case the denominators $k^{(\mu)}$ can never vanish.

In many cases, the method based on the recurrence formulae of Euler and Wallis can be carried out without any roundoff error. This is the case, for example, when the coefficients a_μ, b_μ are integers, and the approximate numerators P_μ and denominators Q_μ are not too large. In general, however, this method involves larger roundoff errors than the method (**). In this method, an approximation to the numerator P_μ is computed as

$$\begin{aligned}\tilde{P}_\mu &= Fl_t(Fl_t(\tilde{P}_{\mu-1} \cdot b_\mu) + Fl_t(\tilde{P}_{\mu-2} a_\mu)) \\ &= [(\tilde{P}_{\mu-1} \cdot b_\mu)(1+\beta_\mu) + (\tilde{P}_{\mu-2} a_\mu)(1+\alpha_\mu)](1+\delta_\mu)\end{aligned}$$

with $|\alpha_\mu|$, $|\beta_\mu|$, $|\delta_\mu| \leq 0.5 \cdot B^{1-t}$. Now if the values $|a_\mu|$ are small compared to the values $|b_\mu|$, and the numbers $|P_\mu|$ grow rapidly, then we can make similar assertions about the error propogation as for the method (**). It is obvious, however, that an exact forward error analysis is very complicated.

In general, it is only possible to give an exact forward error analysis in those cases where the formulae have a linear structure (cf. e.g. the Horner scheme in 5.5.1).

3.3 Backward Error Analysis. In a backward error analysis, we start with the result of a calculation $Fl_t\, \varphi(x_1, \ldots, x_n)$ and with the input data $x_1, x_2, \ldots, x_n$, and determine a set of perturbed data $x_1 + \varepsilon_1$, $x_2 + \varepsilon_2$, $\ldots, x_n + \varepsilon_n$ which would lead to the given result, assuming the computation is done exactly:

$$\varphi(x_1 + \varepsilon_1, \ldots, x_n + \varepsilon_n) = Fl_t\, \varphi(x_1, \ldots, x_n).$$

This method has applications when the input data comes from physical measurements. For example, if the input data has a relative exactness of 1%, while a backward analysis shows that the numerical result can be regarded as coming from an exact calculation using input data which vary by at most 0.5 %, then the method can be deemed acceptable. We now illustrate this process for the problem of computing the sum of a set of numbers using floating-point arithmetic.

Example. Let $\varphi_n(x_1, x_2, \ldots, x_n) := \sum_{\nu=1}^{n} x_\nu$. The evaluation of the function φ_n can be done in a variety of ways. The result obtained depends on the order in which the individual numbers are added. We proceed as follows:

$$\varphi_1(x_1, \ldots, x_n) := x_1$$

and

$$\varphi_k(x_1, \ldots, x_n) := \varphi_{k-1}(x_1, \ldots, x_n) + x_k, \quad (2 \leq k \leq n).$$

Let $x_1, x_2, \ldots, x_n$ be floating-point numbers. Then $Fl_t\,\varphi_1(x_1, \ldots, x_n) = x_1$, and

$$Fl_t\,\varphi_k(x_1, \ldots, x_n) = Fl_t(Fl_t\,\varphi_{k-1}(x_1, \ldots, x_n) + x_k) =$$

$$= (Fl_t\,\varphi_{k-1}(x_1, \ldots, x_n) + x_k)(1 + \varepsilon_k)$$

with $|\varepsilon_k| \leq \tau := 0.5 \cdot B^{-t+1}$. Using complete induction, we can easily show that

$$Fl_t\,\varphi_n(x_1, \ldots, x_n) = x_1 \cdot \prod_{\mu=2}^{n}(1 + \varepsilon_\mu) + \sum_{\nu=2}^{n} x_\nu \prod_{\mu=\nu}^{n}(1 + \varepsilon_\mu).$$

Abbreviating the products in this formula as

$$1 + \eta_\nu := \prod_{\mu=\nu}^{n}(1 + \varepsilon_\mu), \quad 2 \leq \nu \leq n,$$

we get the estimates

$$(1 - \tau)^{n+1-\nu} \leq 1 + \eta_\nu \leq (1 + \tau)^{n+1-\nu}, \quad 2 \leq \nu \leq n.$$

The bounds for the factors $1+\eta_\nu$ obviously depend on the order in which the addition is carried out. Clearly, the products multiplying the terms x_ν with smallest index involve a larger number of factors, and thus are more likely to have a larger absolute value. Thus, it makes sense to order the sequence so that the largest summand is multiplied by the smallest product; i.e., we can optimize the backward error analysis if the addition is carried out according to the size of the absolute values of the numbers, starting with the smallest. Since our considerations are based only on error bounds, there is no guarantee that in practice this will always lead to the smallest possible error. We leave it to the reader to find an example to illustrate this point.

In general, a backward analysis is much easier to carry out then a forward analysis. However, it leads only to a qualitative estimate of the accuracy of a numerical result.

3.4 Interval Arithmetic. The search for a way to systematize the forward error analysis and automatically compute associated bounds led to the development of *interval arithmetic.* This arithmetic works with the set of closed intervals of real numbers.

Let $\mathrm{I\!I\!R} := \{\mathrm{I} \subset \mathbb{R} \mid \mathrm{I} := [a, b],\ a \leq b\}$ be the set of closed intervals in $\mathbb{R}$. We define

$$\mathrm{I}\rfloor := \max_{x \in I} x, \qquad \lfloor \mathrm{I} := \min_{x \in I} x.$$

We now define an arithmetic on the elements of $\mathrm{I\!I\!R}$. Given $\mathrm{A}, \mathrm{B} \in \mathrm{I\!I\!R}$, we define:

Addition: $\mathrm{X} = \mathrm{A} + \mathrm{B}$, $\mathrm{X} := \{x \in \mathbb{R} \mid x = a + b \text{ with } a \in \mathrm{A} \text{ and } b \in \mathrm{B}\}$, i.e. $\mathrm{X} = [\lfloor \mathrm{A} + \lfloor \mathrm{B}, \mathrm{A}\rfloor + \mathrm{B}\rfloor]$;

Subtraction: $X = A - B$, $X := \{x \in \mathbb{R} \mid x = a - b \text{ with } a \in A \text{ and } b \in B\}$, i.e. $X = [\lfloor A - B\rfloor, A\rfloor - \lfloor B]$;

Multiplication: $X = A \cdot B$, $X := \{x \in \mathbb{R} \mid x = a \cdot b \text{ with } a \in A \text{ and } b \in B\}$;

Division: $X = A/B$, if $0 \notin B$,
$X := \{x \in \mathbb{R} \mid x = a/b \text{ with } a \in A \text{ and } b \in B\}$.

If we replace $\mathbb{R}$ by the set of machine numbers, we get a corresponding set of machine intervals. In this case we can define arithmetic operations as above, except that we have to take account of roundoff, which means that the result intervals must be enlarged.

Interval arithmetic has been heavily studied, and many of the methods in numerical analysis have been built into the theory. The main problem is to develop methods such that the intervals in the calculation do not grow too much. This requires a clever combination of conventional techniques with those from interval analysis. For details, see the extensive literature and the book of R. E. Moore [1966].

In present-day computers, which are capable of carrying out millions of arithmetic operations per second, it is essential to have an effective control over roundoff errors. Interval arithmetic provides the possibility of doing this, especially since there now exist computers with special hardware for carrying out the arithmetic. Moreover, there are compilers which lead to programs which can be executed in sufficiently high precision arithmetic to provide rigorous error bounds along with the result.

3.5 Problems. 1) Determine the maximal (absolute and relative) error in $y = x_1 x_2^2 \sqrt{x_3}$ for $x_1 = 2.0 \pm 0.1$, $x_2 = 3.0 \pm 0.2$, $x_3 = 1.0 \pm 0.1$ using an error analysis with differentials (cf. 3.1). Compute the condition numbers. Which variable contributes the most to the error?

2) Consider the linear system of equations

$$a_{11}x_1 + a_{12}x_2 = b_1,$$
$$a_{21}x_1 + a_{22}x_2 = b_2,$$

with $a_{\mu\nu}, b_\mu \in \mathbb{R}$.

a) Suppose the coefficients $a_{11} = a_{22} = 1.9$, $a_{12} = a_{21} = -1.7$ and the right-hand sides $b_1 = 1.2$, $b_2 = 1.5$ are subject to errors whose size is not larger than $5 \cdot 10^{-2}$. Find the sharpest possible bounds for the solution.

b) Consider the solution $x = (x_1, x_2)$ as a function of the coefficients and the right-hand side:

$$\begin{pmatrix} x_1 \\ x_2 \end{pmatrix} = \varphi(a_{11}, a_{12}, a_{21}, a_{22}, b_1, b_2).$$

Compute the condition numbers of this problem, and give sufficient conditions for it to be well-conditioned and poorly-conditioned, respectively.

c) What are the condition numbers corresponding to the values given in part a) of this problem?

3) The condition number of a problem of the form $y = \varphi(x)$ defined by $\varphi : \mathrm{D} \subset \mathbb{R}^n \to \mathbb{R}^m$ can also be determined experimentally (ignoring roundoff errors) by approximating the differential quotient

$$\frac{\Delta_{\mu\nu}(\varepsilon)}{\varepsilon} := \frac{\varphi_\mu(x_1, \ldots, x_{\nu-1}, x_\nu + \varepsilon, x_{\nu+1}, \ldots, x_n) - \varphi_\mu(x)}{\varepsilon}.$$

For example, to do this, we should choose ε so that $|\Delta_{\mu\nu}(\varepsilon) - \Delta_{\mu\nu}(-\varepsilon)| \ll 1$. Apply this method to the linear system of equations in 2a), and compare with the results from 2c).

4) Suppose we want to compute the product $P_n := \prod_{\mu=1}^n a_\mu$ of real numbers a_μ using the following recurrence:

$$P_1 := a_1,$$
$$P_\nu := P_{\nu-1} \cdot a_\nu, \quad 2 \le \nu \le n.$$

Carry out an exact forward analysis, assuming the calculation is done with floating-point arithmetic with base B and mantissa length t. Is there possibly a better way to compute the product?

5) Let x and y be vectors in $\mathbb{R}^n$. Carry out a forward error analysis for the problem of computing the scalar product

$$\langle x, y \rangle := \sum_{\nu=1}^n x_\nu \cdot y_\nu.$$

What does the result say for $n = 3$?

6) Carry out a backward analysis for the computation of the product $P_n := \prod_{\mu=1}^n a_\mu$ using the method in Problem 4).

7) Let $\mathrm{A}, \mathrm{B}, \mathrm{C} \in \mathbb{IR}$ be closed intervals in $\mathbb{R}$. Show:

a) The subdistributivity law

$$\mathrm{A} \cdot (\mathrm{B} + \mathrm{C}) \subset \mathrm{A} \cdot \mathrm{B} + \mathrm{A} \cdot \mathrm{C}$$

holds.

b) If $\mathrm{B} \cdot \mathrm{C} > 0$ (i.e., all elements of $\mathrm{B} \cdot \mathrm{C}$ are positive), then the distributivity law

$$\mathrm{A} \cdot (\mathrm{B} + \mathrm{C}) = \mathrm{A} \cdot \mathrm{B} + \mathrm{A} \cdot \mathrm{C}$$

holds.

8) Use interval arithmetic to find bounding intervals for the values of the following functions:

a) $f(x) = x(1-x), \quad 0 \le x \le 1$; b) $f(x) = x/(1-x), \quad 0 \le x \le 1$;
c) $f(x) = x^7 + x^3 - 6x^2 + 0.11x - 0.006, \quad 0 \le x \le 0.2$.

4. Algorithms

In the previous sections we have already presented several computational algorithms, albeit in a rather informal way. To describe an algorithm in a form which can be executed by a computer, we will have to be more precise. The explosive development of programmable computers was, in fact, preceeded in the 1930's by a period of intensive mathematical research on how to precisely formalize the concept of an algorithm. Today the theory of algorithms is an important part of mathematics and computer science.

The word algorithm is derived from the name of the Persian mathematician Abu Jafar Mohammed ibn Musa al-Khowarizmi, who worked in Baghdad around 840 and wrote a collection of problems concerning the laws of inheritance. The city of Khowarizmi, after which he was named, is now called Khiva and lies in the Soviet Union. The meaning of the word algorithm has changed over time. As typical dictionary entries, we have found "This concept unites the four types of arithmetic calculations, namely addition, multiplication, subtraction and division" and "(Arabic + Greek), shortened name of Alchwarism = a set of computational instruction which can be automatically carried out."

Clearly, algorithms are closely connected with the development of computers capable of following a set of instructions. In the following section we discuss a typical prototype of an algorithm, the Euclidean Algorithm, and also give a short historical discussion of programmable computers.

4.1 The Euclidean Algorithm. The Euclidean algorithm for determining the greatest common divisor of two positive integers was described already around 325 B.C., and can be found in Euclid's "Elements", Book 7, Propositions 1 and 2.

Suppose we are given two positive integers m and n with $m \geq n$. The problem is to find the largest positive integer which divides both m and n without remainder. We write GCD(m, n) for this number. Euclid's algorithm for finding GCD(m,n) can be expressed as follows:

Input: $m, n \in \mathbb{N}$
Output: $\mathrm{GCD}(m,n) \in \mathbb{N}$
Steps of the algorithm: $m' := m$, $n' := n$

(i) Determine the remainder:
Divide m' by n':
Let the integer r, $0 \leq r < n'$ be the remainder.

(ii) Test:
Is $r = 0$?
If $r = 0$, set $\mathrm{GCD}(m,n) := n'$.
Stop the calculation.

(iii) Reset the starting value:
Set $m' := n'$ and $n' := r$.
Return to step (i).

We now show that this algorithm actually finds the greatest common divisor GCD(m, n). After carrying out step (i), we get nonnegative integers q and r, $0 \leq r < n'$, with $m' = qn' + r$. In step (ii) we check the remainder r. If it is zero, then m' is a multiple of n', and it is obvious that GCD$(m, n) = n'$. If $r \neq 0$, we claim that m' and n' have the same common divisors as n' and r. Indeed, if s is a divisor of m' and n', then since $r = m' - qn'$, it follows that s also divides r. Conversely, if s is a divisor of both n' and r, then $m' = qn' + r$ implies that it is also a divisor of m'. Now step (iii) reduces the problem to an equivalent one, but with smaller numbers. Since n is finite, we get the result after at most n steps.

Example of the Euclidean Algorithm. Suppose we want to find the greatest common divisor of the numbers $m = 753$ and $n = 325$. Then the above algorithm gives:

(i)$_1$	$q = 2,\ r = 103$	(i)$_4$	$q = 2,\ r = 2$
(ii)$_1$	$r \neq 0$	(ii)$_4$	$r \neq 0$
(iii)$_1$	$m' := 325,\ n' := 103$	(iii)$_4$	$m' := 7,\ n' := 2$
(i)$_2$	$q = 3,\ r = 16$	(i)$_5$	$q = 3,\ r = 1$
(ii)$_2$	$r \neq 0$	(ii)$_5$	$r \neq 0$
(iii)$_2$	$m' := 103,\ n' := 16$	(iii)$_5$	$m' := 2,\ n' := 1$
(i)$_3$	$q = 6,\ r = 7$	(i)$_6$	$q = 2,\ r = 0$
(ii)$_3$	$r \neq 0$	(ii)$_6$	$r = 0 \Rightarrow GCD(753, 325) = 1.$
(iii)$_3$	$m' := 16,\ n' := 7$		

We have found that the numbers m and n have no common divisor.

This algorithm illustrates several typical properties: the steps of the algorithm are clearly defined, a certain block of steps of the algorithm has to be repeated, and only a finite number of such blocks need to be executed. Once the data have been input to the algorithm, it runs automatically until an answer is produced and sent to the output. Clearly, such an algorithm could be executed by a programmable machine. The development of such machines must have been one of the dreams of early mathematics, which, however, was only to be realized at the beginning of the 20-th century.

A computing machine which could be controlled by a set of instructions along the lines of our modern computers was designed by the Englishman CHARLES BABBAGE (1791–1871). As a youth, Babbage was already fascinated with mathematical ideas and the problem of realizing them on a mechanical machine. While dealing with function tables in the early years of his study of mathematics at Cambridge, he came up with the idea of building a machine which could interpolate and extrapolate in such tables. Since his machine worked with finite differences, he called it a ***difference engine.*** Indeed, the basis of the machine was the fact that the difference of n-th order of a polynomial of n-th degree is a constant (cf. 5.3.4, Problem 2). After finishing his studies, Babbage published some mathematical papers, and became sufficiently well-known that in 1828 he was offered the Lucas Chair for Applied Mathematics at the University of Cambridge. This is particularly remarkable, since this chair had earlier been occupied by Isaac Newton. Babbage remained a faculty member at Cambridge until 1839, although he never lived there, and never gave a course. Starting in 1833, Babbage began the development of his ***Analytical Engine***, a machine which included all the elements of a modern computer:

- A memory unit,
- an arithmetic unit,
- a control unit based on punched cards which directed the computational process,
- input and output units.

The analytical engine was conceived so that it could carry out any computation, no matter how long or complex. Because of the technical limitations of the time, the machine was unfortunately only partially realized. In 1840, Babbage gave a course in Turin. One of his students was ADA AUGUSTA, COUNTESS OF LOVELACE (1815–1852). She was the daughter of the poet Lord Byron, and became a trusted coworker of Babbage. Her understanding of the principle of the analytical engine and its mathematical basis were remarkable. She is responsible for a detailed writeup of Babbage's lectures in Turin, as well for the first computer program, a program which computed the Bernoulli numbers. The Countess can in fact be considered to be the first programmer of all time. Babbage died at the age of 79, disappointed and unappreciated. His exceptional ideas were about a century too early.

Another milestone in the development of externally controlled computers was the invention of HERMANN HOLLERITH (1860–1929). He developed a method of storing information on punched cards which was used in place of the usual questionnaire for the eleventh US census in 1890. The information of interest was entered by punching holes at well-defined places on the card. The data was then later read with the help of a counting machine (the Hollerith machine). The idea was quickly adopted elsewhere around the world. Many large companies used them for storing and sorting data. Hollerith himself formed his own company, the Tabulating Machine Company, in 1896. After merging with two other companies in 1914, it became the Computing-Tabulating-Recording Company, and later changed its name again to the International Business Machines Corporation (IBM). Hollerith remained a consulting engineer until his death.

The modern development of computers began in 1934/1935 in Berlin, and is connected with the name KONRAD ZUSE. Zuse studied mechanical engineering at the Technical University of Berlin, and, in addition, was heavily involved in the development of computers. After graduation, he was able to put together a programmable computer in the apartment of his parents, despite the very limited facilities. His machine was called the Z1, and used the binary system. All four arithmetic operations were realized by the logical operations of *and*, *or*, and *negation*, although the fundamental papers of C. E. SHANNON were not to appear until 1938. The external control of the machine was accomplished with the help of perforated film strips. Zuse was not aware of the ideas of Babbage at this point in time.

Zuse planned to build an improved version of the Z1, to be called the Z2, and based on electromechanical relays, but the beginning of the war in 1939 interfered. In 1941 Zuse completed work on the Z3, a fully functional relay-based machine. The machine featured an instruction unit utilizing 8-channel paper strips, one address commands, a memory with 2,000 relays holding 64 numbers with 22 binary digits, and an arithmetic unit with 600 relays. The machine did 15–20 additions and subtractions per second, and a complete multiplication in about four to five seconds. The Z3 was damaged in a bombing attack on Berlin, but by 1945 Zuse had already finished his model Z4. This machine, after being moved to Göttingen and later to the Allgäu region, survived until the end of the war. It was later expanded, and put into service at the Swiss Federal Polytechnic Institute (ETH) in Zürich. During 1951–1956, it was the only functioning computer in Europe. At the beginning of his scientific career, one of the authors of this book actually carried out some calculations on this machine. We should also mention that Zuse also was a pioneer in the field of software. Already in 1945 he developed the concept of a higher level programming language which he called *Plankalkül*. He regarded it as an extension of Hilbert's predicate calculus. His work, however, did not play a role in the later development of the languages Fortran, Algol and Cobol.

Independently of the developments in Germany, and about three years later, work began in the USA on the construction of a modern computer. HOWARD HATHAWAY AIKEN (1900–1973), who in 1941 became a Professor of Applied Mathematics at Harvard University, finished his first model *MARK I* in 1944. This machine used punched cards, relays, and electrical connections. It had 70,000 parts, 3,000 ball bearings, and 80 km of wire. It was some 15 meters long and 2.5 meters high, and weighed over five tons. It took approximately 0.3 second for an addition, six seconds for a multiplication, and 11 seconds for a division.

The first completely electronic computer was built in 1946 at the University of Pennsylvania by JOHN PRESPER ECKERT and JOHN W. MAUCHLY. This machine was designed to solve the special problem of solving differential equations by iterative methods. The builders christened it the *ENIAC* (Electronic Numerical Integrator and Computer). The machine employed more than 18,000 electronic tubes and 1,500 relays, and required 150 KW to run. Since electronic tubes often failed, the machine was frequently "down." This remained a problem until machines based on transistor technology were built.

In addition to the work in Germany and the USA, development of a computer was also underway in England already during the Second World War. A functioning model with the name *COLOSSUS* was in use as early as 1943. This computer had 1500 electronic tubes, and utilized binary arithmetic. Its development was partially based on ideas of ALAN M. TURING (1912–1954), who had studied the theoretical problems of computability.

As one of the fathers of the modern computer, we must also mention JOHN VON NEUMANN (1903–1957). He was one of the most important mathematicians of this century. Von Neumann made essential contributions to many areas, including quantum mechanics, operator theory, ergodic theory, and game theory. His conception of a computational automata remains the basic blueprint for our modern computers. The essential new idea in his work was that of an internally controlled computer. The control program, which earlier was contained on paper tape or on punched cards, was now stored internally in the computer, and could therefore be modified like any other data. The associated flexibility led directly to universal programming in various languages. Research to expand on von Neumann's ideas is still in full swing.

4.2 Evaluation of Algorithms. Our study of the Euclidean Algorithm in 4.1 has given us some idea of what an algorithm is, what essential properties it has, and what criteria can reasonably be applied to judge its performance.

Notation. An *algorithm* is a rule consisting of a set of unique instructions. These specify a finite sequence of operations, which when carried out, lead to the solution of a problem lying in a special class of problems.

The Euclidean Algorithm in 4.1 shows the characteristic

Form of an Algorithm. The *Input* consists of starting values which must be prescribed before the individual steps of the algorithm can be carried out. It consists of certain subsets of prescribed sets.

For the Euclidean Algorithm, the input consists of two integers m and n, taken from the set $\mathbb{N}$.

The *Output* consists of one or more quantities which have a special relationship to the input values. This relationship is uniquely defined by the steps of the algorithm.

The Euclidean Algorithm outputs the greatest common divisor of m and n. This is the number n' determined in step (ii) when the algorithm comes to a stop.

The *Steps of the Algorithm* or *Procedure* prescribes the sequence of arithmetic operations to be performed, where the following properties are required:

> *Definiteness*: Every step of the procedure must be precisely and unambiguously defined. Every possible situation must be accounted for.
>
> *Finiteness*: The process must stop after a finite number of steps. In the case of the Euclidean Algorithm, this requirement is satisfied, since for every loop through the steps of the algorithm, the number n is reduced.
>
> *General Applicability*: The algorithm should work on an entire class of problems, where the solutions of specific problems in the class arise solely by changing the input.

We regard the basic arithmetic operations to be the elementary arithmetic operations $+$, $-$, $\cdot$, $\div$, along with the comparison operations $<$, $\leq$, and the replacement operation $:=$. Frequently, to simplify the description of algorithms, one also uses the operations $\sqrt{\cdot}$, $|\cdot|$, sin, cos, exp, or even entire subalgorithms, such as a linear system solver. The requirement of definiteness of an algorithm has led to the development of precise languages for describing algorithms. Our description of the Euclidean Algorithm 4.1 is not in one of these languages, but rather in an *informal form* which we shall use throughout the remainder of this book.

Algorithms can be precisely described using *flow charts*. There is a standard set of symbols which we now describe:

Symbol	Meaning	Symbol	Meaning
□	General computation	——	Computational path

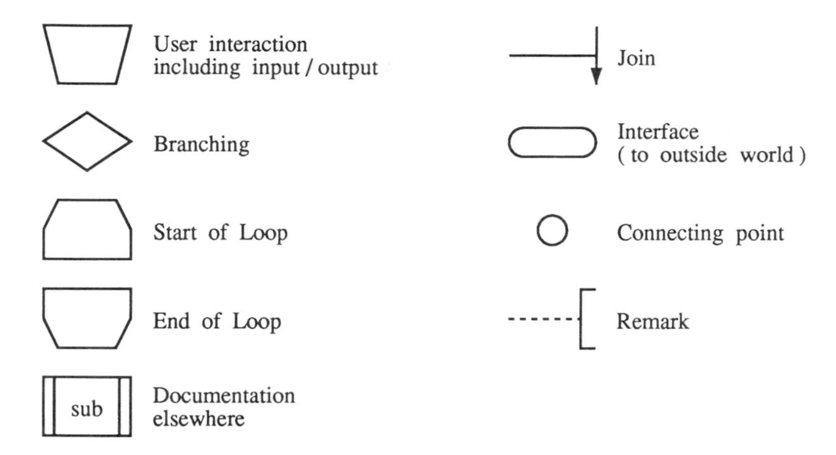

The flow chart of the Euclidean Algorithm is as follows:

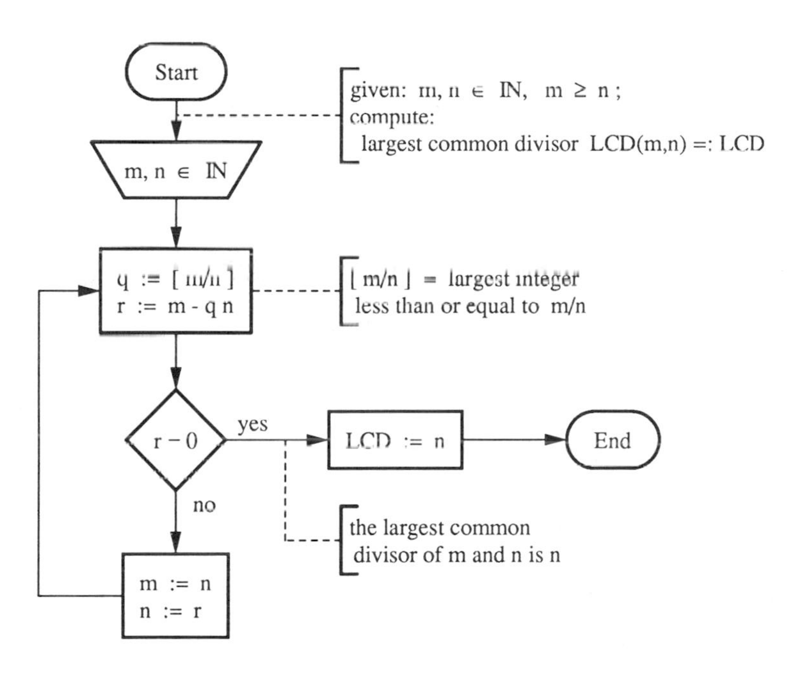

In order to overcome the difficulties associated with formulating an algorithm precisely, formal programming languages have been developed. The formulation of a set of computational steps in a computer language is called a *program*. A program is the most precise form of an algorithm. Using a prescribed syntax, the computational steps are described in such a way that the computer itself can generate a *machine program.* A program which performs this translation is called a *compiler.* There are a large number of programming languages and associated compilers available, both for general scientific purposes and for special applications. The Euclidean Algorithm can be written in PASCAL as follows:

```
FUNCTION ggt(m, n: Integer): Integer;
VAR q, r: Integer;
BEGIN
  REPEAT
    q := m DIV n;
    r := m - q * n;
    IF r <> 0 THEN
    BEGIN
      m := n;
      n := r
    END
  UNTIL r = 0;
  ggt := n
END;
```

In general, there may be several algorithms which accomplish the same task. For example, we could also solve the problem of finding the greatest common divisor of integers m and n ($m \geq n$), by simply dividing both m and n by all integers from 2 to n, and checking for the largest divisor for which there is no remainder for either m or n. Obviously, this process also satisfies all of the requirements of an algorithm. It requires, however, many more operations than the Euclidean Algorithm, and so, in this sense, is inefficient. In order to judge the efficiency of an algorithm, we need some way of measuring performance. In response to this need, the subject of complexity has developed as a part of computer science. In the next section we give a brief introduction to this rather extensive subject.

4.3 Complexity of Algorithms. Since there are often many different algorithms which can be used to solve the same problem, we need some criteria for comparing them. These criteria should be independent of particular implementations on particular computers, and hence should permit objective assertions to be made. We need a mathematical theory giving answers to the following questions:

- How can the qualilty and performance of an algorithm be quantitatively analysed?
- What criteria can be constructed to compare algorithms?
- How can existing algorithms be improved?
- In what sense can one prove that an algorithm is best possible?

– Are "best" algorithms of any practical use?

These questions are dealt with in complexity theory, which we now briefly discuss.

There are two forms of complexity; static and dynamic. A *static complexity measure* involves, for example, the length of a program, the number of instructions, or other similar measures for the efficiency of a program. Since these quantities are independent of the characteristics of the input, static complexity measures have little importance from a mathematical point of view. A *dynamic complexity measure* involves the running time and memory requirements of a program. These do depend on the amount of input data, and hence are of much more practical significance. Moreover, dynamic complexity measures also have the advantage that they can be treated mathematically, and so we now restrict our attention to them.

The following examples illustrate the point that the amount of input data is a measure of the size of a problem.

Example (Max-min Search). Given a set of n integers, find the largest and the smallest. The size of this problem depends directly on the size of n.

Example (Matrix Multiplication). Let A and B be matrices of real numbers of size $m \times n$ and $n \times r$, respectively. The problem is to compute the matrix product $C := A \cdot B$. The integer $r = \max(m, n)$ is a measure of the size of the problem.

Time complexity and *memory complexity* are especially interesting in the limit; i.e., as the size of the problem grows without bound. The amount of time required to solve a problem will in general be proportional to the number of elementary arithmetric operations which the algorithm has to carry out. This leads to the following

Definition of Complexity. Let A be an algorithm for solving a problem P with $n \in \mathbb{N}$ input data. The mapping $T_A : \mathbb{N} \to \mathbb{N}$ from the number of input data to the number of basic operations carried out by the algorithm is called the *complexity of* A.

This definition of complexity does not take into account everything that it should. For example, runtime depends not only on the number of input data, but also in an essential way on the way these numbers are coded. In this sense, the runtime of an algorithm also depends on the type of machine used. For the time being, we shall assume that we are working on some universal machine, say a Turing machine. Later we shall introduce a sharper form of complexity which will take account of the coding of the algorithm.

To describe the behavior of complexity functions for large n, it is useful to introduce the

Landau Symbols. Consider two functions $f,\ g : \mathrm{D} \to \mathbb{R}$, $\mathrm{D} \subset \mathbb{R}$, where $g(x) \neq 0$ for $x \in \mathrm{D}$.

1. We say that f is *of the order "large O" with respect to g as x goes to* x_0, provided that there is some constant $C > 0$ and a $\delta > 0$, such that

$$\left|\frac{f(x)}{g(x)}\right| \leq C$$

for all $x \in \mathrm{D}$ with $x \neq x_0$ and $|x - x_0| < \delta$. We write $f(x) = O(g(x))$ as $x \longrightarrow x_0$.

2. We say that f is *of the order "small o" with respect to g as x goes to* x_0, provided that for every constant $C > 0$, there exists $\delta > 0$ such that

$$\left|\frac{f(x)}{g(x)}\right| \leq C$$

for all $x \in \mathrm{D}$ with $x \neq x_0$ and $|x - x_0| < \delta$. We write $f(x) = o(g(x))$ as $x \longrightarrow x_0$.

It is easy to check that the Landau symbols O and o possess the following properties as $x \longrightarrow x_0$:

(i) $f(x) = O(f(x))$;
(ii) $f(x) = o(g(x)) \Rightarrow f(x) = O(g(x))$;
(iii) $f(x) = K \cdot O(g(x))$ for some $K \in \mathbb{R} \Rightarrow f(x) = O(g(x))$;
(iv) $f(x) = O(g_1(x))$ and $g_1(x) = O(g_2(x)) \Rightarrow f(x) = O(g_2(x))$;
(v) $f_1(x) = O(g_1(x))$ and $f_2(x) = O(g_2(x)) \Rightarrow$
$\Rightarrow f_1(x) \cdot f_2(x) = O(g_1(x) \cdot g_2(x))$;
(vi) $f(x) = O(g_1(x) g_2(x)) \Rightarrow f(x) = g_1(x) \cdot O(g_2(x))$.

The analogs of properties (iii) - (vi) also hold for the symbol "o."

Examples of the Landau symbols.

(1) Let $f : [0,1] \longrightarrow \mathbb{R}$ be a function with $f(0) = 0$. If f is continuous or continuously differentiable on the interval $[0,1]$, then $f(x) = o(1)$ and $f(x) = O(x)$ as $x \longrightarrow 0$, respectively.
(2) Let (a_μ) be a sequence of real numbers, and suppose a constant K exists such that $|a_{\mu+1} - a_\mu| \leq K$ for all $\mu \in \mathbb{N}$. Then $a_\mu = O(\mu)$ as $\mu \longrightarrow \infty$.
(3) Suppose a function $f : [0,\infty) \longrightarrow \mathbb{R}$ is such that

$$|f(x)| \leq K + \int_0^x k\,|f(t)|dt$$

for all $x \in [0,\infty)$, where K and k are nonnegative constants. Then by the *Gronwall inequality*,

$$|f(x)| \leq K\,e^{kx}$$

for all $x \in [0,\infty)$. This can be written as $f(x) = O(g(x))$ as $x \longrightarrow \infty$ with $g(x) := e^{kx}$. (T. H. Gronwall: Note on the derivatives with respect to a parameter of the solutions of a system of differential equations. Ann. Math. 20 (1918), 292–296).

There is a useful *discrete form of the Gronwall inequality*, which we now give for completeness. Let (f_μ) be a sequence of real numbers such that

$$|f_\mu| \le K + k \sum_{\nu=0}^{\mu-1} |f_\nu|$$

for all $\mu \in \mathbb{N}$, where K and k are nonnegative constants. Then

$$|f_\mu| \le K\, e^{\mu k}$$

for all $\mu \in \mathbb{N}$.

We now come back to the definition of the complexity of an algorithm. In view of our requirement in 4.2 that algorithms should be applicable to entire classes of problems, i.e., in particular to an entire set $\Omega := \{w_1, w_2, w_3, \dots\}$ of input data, it is natural to try to determine the complexity with respect to all input data sets Ω with a fixed size $g(\omega_i) = n$. We can now define the time complexity $T_A^S(n)$ in the "worst case" as

$$T_A^S(n) := \sup\{T_A(w) \mid w \in \Omega, g(w) = n\},$$

or in the "average case" as

$$T_A^M(n) := E\{T_A(w) \mid w \in \Omega, g(w) = n\},$$

where E is the expected-value operator over the conditional distribution $W(w|g(w) = n)$. This immediately leads to the question of what probability density W we should use. This question has only been adequately answered for a few algorithms. We return to this point in our treatment of the simplex method in Chapter 9.

We illustrate these ideas with the following

Example for the Euclidean Algorithm.
Fix n, and let m run over the positive integers. How often must step (i) of the algorithm be carried out for the worst case, and for the average case? First we note that the number n actually determines the size of the problem for all values of $m \in \mathbb{N}$, since after the first time that step (i) is executed where m is divided by n, then only the remainder r is relevant, and this number lies between 0 and n.

In the worst case, step (i) will obviously be executed n times; i.e., $T_A^S(n) = n$. The problem of finding the average complexity is not so simple, and will only be illustrated here with an example. Let $n = 7$. As already mentioned, in this case we only need to count the calls of step (i) for $m = 1, 2, \dots, 7$.

m	number of calls of step (i)
1	2
2	3
3	3
4	4
5	4
6	3
7	1

This gives $T_A^M(7) = \frac{20}{7} < 2.86$.

It has been shown that for large n,

$$T_A^M(n) = \frac{12 ln\, 2}{\pi^2} ln\, n = O(ln\, n).$$

In the remainder of this chapter we restrict our attention to complexity in the worst-case sense.

4.4 The Complexity of Some Algorithms. In this sections we compute the time complexitity $T_A^S(n)$ of several of the algorithms presented above.

Example of the naive evaluation of a polynomial.

The naive algorithm for the evaluation of a polynomial of degree n at a point $\alpha \in \mathbb{R}$ is as follows:

Input: $n \in \mathbb{N}$, $(a_0, a_1, \ldots, a_n) \in \mathbb{R}^{n+1}$, $\alpha \in \mathbb{R}$
Output: $p := \sum_{i=0}^{n} a_i \alpha^i$
Computational steps:
(i) $p := a_0$,
(ii) For $i = 1, 2, \ldots, n$:
$b := a_i$;
for $j = 1, 2, \ldots, i$:
$b := b\alpha$;
$p := p + b$.

Counting the additions and multiplications, this algorithm has complexity

$$T_A^S(n) = \frac{1}{2} n(n+3) = O(n^2).$$

Example of the evaluation of a polynomial by Horner's scheme (cf. 5.5.1).

Input: $n \in \mathbb{N}$, $(a_0, a_1, \ldots, a_n) \in \mathbb{R}^{n+1}$, $\alpha \in \mathbb{R}$
Output: $p := \sum_{i=0}^{n} a_i \alpha^i$
Computational steps:
(i) $p := a_n$,
(ii) For $i = (n-1), (n-2), \ldots, 0$:
$p := a_i + \alpha p$.

Counting the additions and multiplications, this algorithm has complexity

$$T_A^S(n) = 2n = O(n).$$

Example. Find the maximum of a set of numbers

Input: $n \in \mathbb{N}$, $(f_1, f_2, \ldots, f_n) \in \mathbb{R}^n$
Output: $\max := \max_{1 \le j \le n} f_j$
Computational steps:
(i) $\max := f_1$,
(ii) For $i = 2, 3, \ldots, n$:
$\max := f_i$, in case $f_i > \max$.

Counting the number of comparison operations, the complexity of this algorithm is

$$T_A^S(n) = n - 1 = O(n).$$

Example. Find both the maximum and minimum of a set of numbers

<u>Input:</u> $n \in \mathbb{N}$, $(f_1, f_2, \ldots, f_n) \in \mathbb{R}^n$
<u>Output:</u> $\min := \min_{1 \le i \le n} f_i$, $\max := \max_{1 \le i \le n} f_i$
<u>Computational steps:</u>
(i) If $f_1 < f_2$: $\min := f_1$, $\max := f_2$;
otherwise: $\min := f_2$, $\max := f_1$,
(ii) For $i = 3, 4, \ldots, n$:
If $f_i > \max$: $\max := f_i$;
if $f_i < \min$: $\min := f_i$.

Counting the number of comparison operations, this algorithm has complexity

$$T_A^S(n) = 2n - 3.$$

Example. Matrix multiplication.
It is straightforward to check that computing the product C of two $n \times n$ matrices A and B is of complexity

$$T_A^S(n) = O(n^3).$$

It is now natural to ask if the algorithms presented in the above examples are optimal in the sense of time complexitity; i.e., are there other algorithms which accomplish the same tasks but with fewer operations? For matrix multiplication, it is easy to give a lower bound for the complexity. Since an $n \times n$ matrix has n^2 elements, certainly there cannot exist an algorithm with complexity better than $O(n^2)$ as $n \to \infty$. It is an open problem whether there exists an algorithm achieving this order.

In the following section we study a general approach which can be used in many cases to improve an algorithm.

4.5 Divide and Conquer. The basic idea which we want to discuss in this section for improving the complexity of an algorithm is very simple: We divide the original problem into a sequence of smaller problems, each of which can be quickly solved. We refer to this idea as the "*divide and conquer method*". We now illustrate it for the algorithm for finding a maximum and minimum presented in 4.4:

<u>Input:</u> $k \in \mathbb{N}$, $F := (f_1, f_2, \ldots, f_{2^k}) \in \mathbb{R}^{2^k}$;
<u>Output:</u> $\max := \max_{1 \le i \le 2^k} f_i$, $\min := \min_{1 \le i \le 2^k} f_i$
<u>Computational steps:</u>
(i) If $k = 1$:
If $f_1 < f_2$, set $\min := f_1$, $\max := f_2$;
otherwise $\min := f_2$, $\max := f_1$,
(ii) Divide F into two vectors F_1 and F_2:

$F_1 := (f_1, f_2, \ldots, f_{2^{k-1}}), F_2 := (f_{2^{k-1}+1}, f_{2^{k-1}+2}, \ldots, f_{2^k})$ and find the maximum and minimum in F_1 and F_2 by applying the algorithm of 4.4. Suppose the results are $\max_1$, $\min_1$ with respect to F_1, and $\max_2$, $\min_2$ with respect to F_2.

(iii) Set $F := (\max_1, \max_2)$, resp. $F := (\min_1, \min_2)$, and repeat (i) with F.

To compute the complexity $T_A^S(n)$ of this algorithm, we take $n = 2^k$ and use the recurrence formula:

$$T_A^S(n) = \begin{cases} 1 & \text{if } n = 2 \quad \text{(Step (i))}, \\ 2T_A^S(\frac{n}{2}) + 2 & \text{if } n > 2 \quad \text{(Steps (ii) and (iii))}. \end{cases}$$

It is easy to check that this recurrence is satisfied by the function

$$T_A^S(n) = \frac{3}{2}n - 2.$$

A comparison with the complexity of the original algorithm in 4.4 for finding both the maximum and mininum shows that the factor multiplying n has been reduced from 2 to 3/2. This basic idea can be generalized as follows:

Principle of "Divide and Conquer." *Let α, b_ν, $0 \le \nu \le r$, be nonnegative constants. Suppose the function T_A^S satisfies the recurrence*

$$T_A^S(2n) \le \alpha \cdot T_A^S(n) + \sum_{\nu=0}^{r} b_\nu n^\nu, \quad n \in \mathbb{Z}_+,$$

with $T_A^S(1) > 0$, $b_r > 0$. Then for all numbers n of the form $n = 2^k$, $k \in \mathbb{N}$,

$$T_A^S(n) = \begin{cases} O(n^r) & \text{if } \alpha < 2^r, \\ O(n^r \log_2 n) & \text{if } \alpha = 2^r, \\ O(n^{\log_2 \alpha}) & \text{if } \alpha > 2^r, \end{cases}$$

as $n \to \infty$.

To prove this theorem, we first establish the

Lemma. *Suppose the sequence (s_k) of real numbers is such that*

$$s_0 \le a,$$

$$s_{k+1} \le q \cdot s_k + \sum_{\nu=0}^{r} b_\nu q_\nu^k \quad \text{for } k \ge 0,$$

where q, a and b_ν, q_ν for $0 \le \nu \le r$ are arbitrary nonnegative numbers. Then

$q \neq q_\nu$ *for all* $0 \leq \nu \leq r$ *implies*

$$(*) \qquad s_k \leq a \cdot q^k + \sum_{\nu=0}^{r} \frac{b_\nu}{q_\nu - q}(q_\nu^k - q^k),$$

and if $q = q_\mu$ *for exactly one* μ, $0 \leq \mu \leq r$, *we have*

$$(**) \qquad s_k \leq a \cdot q^k + \sum_{\substack{\nu=0 \\ \nu \neq \mu}}^{r} \frac{b_\nu}{q_\nu - q}(q_\nu^k - q^k) + b_\mu \cdot k \cdot q^{k-1}$$

for $k > 0$.

The inequalities $(*)$ and $(**)$ can easily be established using induction on k. We leave the details to the reader.

Proof of the theorem. By the hypothesis, we have the inequality

$$T_A^S(2^{k+1}) \leq \alpha \cdot T_A^S(2^k) + \sum_{\nu=0}^{r} b_\nu (2^\nu)^k$$

for $k \geq 0$. In order to apply the Lemma, we set $s_k := T_A^S(2^k)$, $q := \alpha$, $q_\nu := 2^\nu$ and $a := T_A^S(1)$. Now suppose $q > 2^r$. Then in particular $q > q_\nu$ for $\nu = 0, 1, \ldots, r$, and $(*)$ can be applied:

$$\begin{aligned} T_A^S(2^k) &\leq T_A^S(1) \cdot \alpha^k + \sum_{\nu=0}^{r} \frac{b_\nu}{2^\nu - \alpha}((2^\nu)^k - \alpha^k) \leq \\ &\leq T_A^S(1) \cdot \alpha^k + \alpha^k \sum_{\nu=0}^{r} \frac{b_\nu}{\alpha - 2^\nu} \leq C\,\alpha^{\log_2 n} \leq C\,n^{\log_2 \alpha}, \end{aligned}$$

where C is a positive constant. This implies

$$T_A^S(n) = O(n^{\log_2 \alpha}).$$

Next we suppose $q = q_r$. In this case, we have to use the inequality $(**)$ with $\mu = r$:

$$\begin{aligned} T_A^S(2^k) &\leq T_A^S(1)\alpha^k + \sum_{\nu=0}^{r-1} \frac{b_\nu}{2^\nu - \alpha}((2^\nu)^k - \alpha^k) + b_r \cdot k \cdot q^{k-1} \leq \\ &\leq C(n^{\log_2 \alpha} + \alpha^{k-1} \log_2 n) = C(n^{\log_2 \alpha} + \frac{1}{\alpha} n^r \log_2 n) = O(n^r \log_2 n). \end{aligned}$$

Finally, suppose $q < q_r$. Then two subcases can occur. Either $q \neq q_\nu$ for all $0 \leq \nu \leq r$, or $q = q_\mu$ for some $0 \leq \mu < r$. In the first case we can again apply the inequality $(*)$ of the Lemma:

$$\begin{aligned} T_A^S(2^k) &\leq T_A^S(1)\alpha^k + \sum_{\substack{\nu=0 \\ q_\nu < q}}^{r} \frac{b_\nu}{2^\nu - \alpha}((2^\nu)^k - \alpha^k) + \\ &+ \sum_{\substack{\nu=0 \\ q_\nu > q}}^{r} \frac{b_\nu}{2^\nu - \alpha}((2^\nu)^k - \alpha^k) \leq C_1 \alpha^k + C_2 (2^r)^k \leq C(n^{\log_2 \alpha} + n^r) = O(n^r). \end{aligned}$$

In the second case we use the inequality (**):

$$\begin{aligned} T_A^S(2^k) \leq & \; T_A^S(1)\alpha^k + \sum_{\substack{\nu=0 \\ q_\nu < q}}^{r} \tfrac{b_\nu}{2^\nu - \alpha}((2^\nu)^k - \alpha^k) + \\ & + \sum_{\substack{\nu=0 \\ q_\nu > q}}^{r} \tfrac{b_\nu}{2^\nu - \alpha}((2^\nu)^k - \alpha^k) + b_\mu k q^{k-1} \leq \\ \leq & \; C(n^{\log_2 \alpha} + n^r + \log_2 n \alpha^{k-1}) \leq \\ \leq & \; C(n^{\log_2 \alpha} + n^r + \tfrac{1}{2}\log_2 n \cdot n^{r-1}) = O(n^r). \end{aligned}$$

This completes the proof. □

4.6 Fast Matrix Multiplication. In 4.4 we have seen that the multiplication of two $n \times n$ matrices has complexity $O(n^3)$. We now show how to use the principle of divide and conquer to improve the complexity. This idea is due to V. Strassen [1969].

Let $A = (a_{\mu\nu})$ and $B = (b_{\mu\nu})$ be two real $n \times n$ matrices, and let $C = (c_{\mu\nu})$ be their product. We assume that $n = 2^k$ with $k \in \mathbb{N}$. This is no restriction, since every matrix can trivially be expanded to a larger one.

Lemma. *Let A and B be real $2^k \times 2^k$ matrices with $k \in \mathbb{N}$. Then the product $C = A \cdot B$ can be computed from $2^{k-1} \times 2^{k-1}$ matrices, using 7 matrix multiplications and 18 matrix additions.*

Proof. We decompose the matrices A, B and C as follows:

$$A = \begin{pmatrix} A_{11} & A_{12} \\ A_{21} & A_{22} \end{pmatrix}, \quad B = \begin{pmatrix} B_{11} & B_{12} \\ B_{21} & B_{22} \end{pmatrix}, \quad C = \begin{pmatrix} C_{11} & C_{12} \\ C_{21} & C_{22} \end{pmatrix}.$$

Here $A_{\mu\nu}$, $B_{\mu\nu}$ and $C_{\mu\nu}$ are matrices of size $2^{k-1} \times 2^{k-1}$. Then it is easy to see that

$$\begin{aligned} C_{11} &= M_1 + M_2 - M_4 + M_6, & C_{12} &= M_4 + M_5, \\ C_{21} &= M_6 + M_7, & C_{22} &= M_2 - M_3 + M_5 - M_7, \end{aligned}$$

where

$$\begin{aligned} M_1 &:= (A_{12} - A_{22})(B_{21} + B_{22}), & M_5 &:= A_{11}(B_{12} - B_{22}) \\ M_2 &:= (A_{11} + A_{22})(B_{11} + B_{22}), & M_6 &:= A_{22}(B_{21} - B_{11}), \\ M_3 &:= (A_{11} - A_{21})(B_{11} + B_{12}), & M_7 &:= (A_{21} + A_{22})B_{11}. \\ M_4 &:= (A_{11} + A_{12})B_{22}. \end{aligned}$$

This involves exactly 7 matrix multiplications and 18 matrix additions of matrices of size $2^{k-1} \times 2^{k-1}$, as asserted. □

Applying the principle of divide and conquer to the partition of the matrix multiplication problem given in this lemma, we get the

Theorem of Strassen. *If the multiplication of two real $2^k \times 2^k$ matrices is carried out according to the method described in the lemma, then the corresponding algorithm has complexity*

$$O(n^{\log_2 7})$$

as $n \to \infty$ with $n := 2^k$.

Proof. The number of multiplications required to multiply 7 matrices in $\mathbb{R}^{(\frac{n}{2},\frac{n}{2})}$ is

$$7 \cdot T_A^S(\frac{n}{2}).$$

The number of additions required to add 18 matrices in $\mathbb{R}^{(\frac{n}{2},\frac{n}{2})}$ is

$$18 \cdot (\frac{n}{2})^2.$$

Now the Lemma implies

$$T_A^S(n) \le 7 \cdot T_A^S(\frac{n}{2}) + 18\frac{n^2}{4}.$$

Moreover, $T_A^S(1) = 1$. Thus, the hypotheses of the Principle of Divide and Conquer are satisfied with $\alpha = 7$ and $r = 2$, and we get

$$T_A^S(n) = O(n^{\log_2 7})$$

as $n \to \infty$. □

In view of the fact that $\log_2 7$ is approximately 2.8, the improvement in complexity provided by the Strassen Algorithm is only of importance for large n. D. Coppersmith and S. Winograd [1986] recently gave a matrix multiplication algorithm with complexity order 2.388. As we have already noted, since the product C of two $n \times n$ matrices A and B has n^2 elements, it is impossible to construct an algorithm of order less than 2.

Remark. Our discussion of complexity is based on the assumption that we are working on a serial computer. For parallel computations, the definition of complexity must be appropriately modified. In that case, the speed of our algorithms can generally be further improved.

4.7 Problems. 1) Consider the following sorting method: To sort $2n$ numbers according to their size, divide them into two sets of size n, and sort these separately. Then recombine the sorted sets to get the final result. Show that repeatedly applying this method leads to a sorting method using $O(n \log_2 n)$ comparison operations.

2) Show that we can approximate the derivative of a three-times continuously differentiable function f by a difference quotient as follows:

a) $\frac{f(x+h)-f(x)}{h} = f'(x) + O(h)$;
b) $\frac{f(x+h)-f(x-h)}{2h} = f'(x) + O(h^2)$.

3) To multiply two complex numbers in the usual way requires 4 real multiplications. Find an analog of the Strassen Algorithm which gets by with 3 real multiplications.

4) a) Let A be a $2n \times 2n$ matrix, and let A_{ij} and C_{ij} be $n \times n$ matrices such that

$$A = \begin{bmatrix} A_{11} & A_{12} \\ A_{21} & A_{22} \end{bmatrix} \qquad A^{-1} = \begin{bmatrix} C_{11} & C_{12} \\ C_{21} & C_{22} \end{bmatrix}.$$

Show that the following algorithm produces the matrix A^{-1}:

$$M_1 := A_{11}^{-1} \qquad M_2 := A_{21} \cdot M_1 \qquad M_3 := M_1 \cdot A_{12} \qquad M_4 := A_{21} \cdot M_3$$

$$M_5 := M_4 - A_{22} \qquad M_6 := M_5^{-1} \qquad M_7 := M_3 \cdot C_{21}$$

$$C_{11} := M_1 - M_7 \qquad C_{12} := M_3 \cdot M_6 \qquad C_{21} := M_6 \cdot M_2 \qquad C_{22} := -M_6$$

Here we assume in advance that all inverses which appear exist.

b) By using the above method recursively on a $2^k \times 2^k$ matrix, we can define a "fast matrix inversion algorithm". Show that the number of arithmetic operations $T(2^k)$ of this fast method is given by

$$T(2^k) = 1.2 \cdot 7^{k+1} + 9.6 \cdot 2^k - 17 \cdot 4^k,$$

assuming that all required matrix multiplications are done with the fast matrix multiplication method of 4.6.
Hint: The fast matrix multiplication of two $2^k \times 2^k$ matrices requires a total of $7^{k+1} - 6 \cdot 4^k$ operations.

c) Show: $T(n) = O(n^{\log_2 7})$.

d) Inverting an $n \times n$ matrix using Gauss Elimination requires a total of $(2n^3 - 2n^2 + n)$ operations. Using a hand calculator, find out how large $n = 2^k$ must be before the fast matrix inversion method is really faster.

2
Linear Systems of Equations

Many problems in mathematics lead to linear systems of equations. In fact, in using computers to solve such problems, we frequently encounter very large linear systems. Thus, the development of efficient algorithms to solve such systems is of central importance in numerical analysis. We differentiate between two types of methods. *Direct methods* solve the problem in a finite number of steps, and so are not subject to method error, although, of course, the results can be very badly affected by roundoff error. *Indirect methods* seek to find the solution by iteration, and thus usually lead only to approximate solutions since the iteration has to be stopped at some point. Although in this case we have both method and roundoff errors, iterative methods have their advantages. In this chapter we will primarily discuss direct methods. Iterative methods for linear system of equations will be discussed in Chapter 8.

1. Gauss Elimination

Gauss developed an elimination method for solving linear systems in 1810 for use in solving certain problems in astronomy (see also Chap. 4, Sect. 6). His method remains one of the standard methods of numerical linear algebra, and is also discussed in most basic courses in linear algebra.

CARL FRIEDRICH GAUSS (1777–1855) influenced mathematics in the first half of the 19–th century more than any other mathematician. He is famous for both the breadth and depth of his work in every area of mathematics, including numerical analysis. We should be impressed not only by his many ideas, but also by the exceptional energy which he invested in carrying out huge calculations. His studies in geodesy, astronomy, and physics, of which the most important was probably his work with W. Weber on electro-magnetism (honored with a monument in the city of Göttingen), kept Gauss supplied with a constant stream of mathematical research problems. Conversely, he considered mathematics to be a part of human experience. For example, the impossibility of proving the parallel postulate of Euclidean Geometry forced him to conclude that non-Euclidean Geometry was equally valid, and that the question of which one truely described the structure of space could only be answered from human experience and experiment; see K. Reich ([1985], p. 62).

We devote the next several sections to an algorithmic description of the Gauss elimination method and its complexity.

1.1 Notation and Statement of the Problem. Throughout this chapter, whenever we talk about a vector in $\mathbb{C}^n$, we mean a column vector of the form

$$b = \begin{pmatrix} b_1 \\ \vdots \\ b_n \end{pmatrix}, \quad b_\nu \in \mathbb{C},\ 1 \le \nu \le n.$$

The transpose of b is the row vector $b^T = (b_1, \ldots, b_n)$. We denote the n unit vectors in $\mathbb{R}^n$ by $e^1, e^2, \ldots, e^n$. They are defined by $e^\nu_\mu = \delta_{\mu\nu}$ for $1 \le \mu, \nu \le n$, where $\delta_{\mu\nu}$ is the usual Kronecker delta symbol. We shall use the notations

$$A = (a_{\mu\nu}) \in \mathbb{C}^{(m,n)} \quad \text{and} \quad A^T = (a_{\nu\mu}) \in \mathbb{C}^{(n,m)}$$

for an $m \times n$ matrix of elements in $\mathbb{C}$, and its transpose. The unit matrix will be denoted by $I = (\delta_{\mu\nu})$.

Statement of the Problem. Consider the *linear system*

$$(*) \qquad Ax = b,$$

where $A \in \mathbb{C}^{(m,n)}$ is a matrix with $m \le n$, and the right-hand side $b \in \mathbb{C}^m$ is a given vector. Find a solution vector $x \in \mathbb{C}^n$. Clearly, we can always split the elements of A and the components of b into real and imaginary parts. It follows that every system of equations of this form can be rewritten as a system in $\mathbb{R}^{2n}$ of linear equations involving only real vectors and matrices.

1.2 The Elimination Method. The idea of the Gauss elimination method for solving the linear system of equations 1.1 is to choose appropriate combinations of the rows to force the elements below the diagonal of A to vanish. For the time being, we assume that each of the steps in the algorithm described in the following table can be carried out; i.e., no divisions by zero occur. We discuss this assumption in detail later.

row operation	matrix elements	$b^{(\mu)}$	$s^{(\mu)}$
$\mathbf{Z_1^{(1)}}$ **(1st row, step 1)**	$\mathbf{a_{11} \quad a_{12} \quad a_{13} \quad \cdots \quad a_{1n}}$	$\mathbf{b_1}$	$\mathbf{s_1}$
$Z_2^{(1)}$ (2nd row, step 1)	$a_{21} \quad a_{22} \quad a_{23} \quad \cdots \quad a_{2n}$	b_2	s_2
$\vdots$	$\vdots \quad \vdots \quad \vdots \quad \vdots$	$\vdots$	$\vdots$
$Z_m^{(1)}$ (m-th row, step 1)	$a_{m1} \quad a_{m2} \quad a_{m3} \quad \cdots \quad a_{mn}$	b_m	s_m
$\mathbf{Z_2^{(2)} := Z_2^{(1)} - \frac{a_{21}}{a_{11}} Z_1^{(1)}}$	$\mathbf{0 \quad a_{22}^{(2)} \quad a_{23}^{(2)} \quad \cdots \quad a_{2n}^{(2)}}$	$\mathbf{b_2^{(2)}}$	$\mathbf{s_2^{(2)}}$
$Z_3^{(2)} := Z_3^{(1)} - \frac{a_{31}}{a_{11}} Z_1^{(1)}$	$0 \quad a_{32}^{(2)} \quad a_{33}^{(2)} \quad \cdots \quad a_{3n}^{(2)}$	$b_3^{(2)}$	$s_3^{(2)}$
$\vdots \qquad \vdots$	$\vdots \quad \vdots \quad \vdots \quad \vdots$	$\vdots$	$\vdots$
$Z_m^{(2)} := Z_m^{(1)} - \frac{a_{m1}}{a_{11}} Z_1^{(1)}$	$0 \quad a_{m2}^{(2)} \quad a_{m3}^{(2)} \quad \cdots \quad a_{mn}^{(2)}$	$b_m^{(2)}$	$s_m^{(2)}$
$\vdots$	$\vdots$	$\vdots$	$\vdots$
$\mathbf{Z_m^{(m)} := Z_m^{(m-1)} - \frac{a_{m\,m-1}^{(m-1)}}{a_{m-1\,m-1}^{(m-1)}} Z_{m-1}^{(m-1)}}$	$\mathbf{0 \quad 0 \quad 0 \cdots 0 \; a_{mm}^{(m)} \quad \cdots \quad a_{mn}^{(m)}}$	$\mathbf{b_m^{(m)}}$	$\mathbf{s_m^{(m)}}$

This process leads to the following equivalent system of equations:

$$\begin{aligned} a_{11}x_1 + a_{12}x_2 + \cdots + a_{1m}x_m + \cdots + a_{1n}x_n &= b_1 \\ a_{22}^{(2)}x_2 + \cdots + a_{2m}^{(2)}x_m + \cdots + a_{2n}^{(2)}x_n &= b_2^{(2)} \\ \ddots \quad \vdots \qquad \vdots \qquad \vdots \qquad &\quad \vdots \\ a_{mm}^{(m)}x_m + \cdots + a_{mn}^{(m)}x_n &= b_m^{(m)}. \end{aligned}$$

If at least one of the coefficients $a_{m\nu}^{(m)}$ is nonzero, then the set of all solutions of this system of equations forms an affine space of dimension $(n-m)$. Every solution can be written as a sum of a solution of the inhomogeneous system and a linear combination of the basis vectors spanning the solution space of the homogeneous system of equations. For ease in computing a solution to the inhomogenous system, we may set $x_{m+1} = x_{m+2} = \cdots = x_n = 0$, and determine the remaining components of the solution vector x by solving the resulting system of equations. To find basis vectors spanning the solution space of the homgeneous system, we need only solve each of the systems

$$(x_{m+1}, x_{m+2}, \ldots, x_n)^T = e^{j-m} \in \mathbb{R}^{n-m}, \quad m+1 \le j \le n.$$

We now discuss some of the problems which can arise in carrying out this algorithm. First, we have to assure that no divisions by zero occur. We try to accomplish this by interchanging rows and columns so that at the μ-th step, $1 \le \mu \le m-1$, we have $a_{\mu\mu}^{(\mu)} \neq 0$. In exchanging columns, it is important to make sure that the corresponding components of the solution vector are correspondingly renumbered.

Special Case. If it is not possible to get $a_{\mu\mu}^{(\mu)} \neq 0$ at the μ-th step by interchanging rows and columns, then the Gauss algorithm stops at the $(\mu-1)$-st step. In this case, the last $(m-\mu+1)$ rows of the left-hand side of the system of equations are identically zero. Then there are two cases:

(a) There is some $\tilde{\mu}$ with $\mu \le \tilde{\mu} \le m$ such that $b_{\tilde{\mu}}^{(\mu)} \neq 0$;

(b) for all $\tilde{\mu}$ with $\mu \le \tilde{\mu} \le m$ we have $b_{\tilde{\mu}}^{(\mu)} = 0$.

In Case (a), the system of equations has no solution, while in Case (b), the solution space is of dimension $(m-\mu+1)$, and the general solution can be found as described above.

Remark. It is sometimes useful, especially when doing the calculations by hand, to keep track of the row sums

$$s_\mu^{(\ell)} := \sum_{\nu=\ell}^{n} a_{\mu\nu}^{(\ell)} + b_\mu^{(\ell)}, \quad 1 \le \ell \le m.$$

This provides a check on the arithmetic, since at each step,

$$s_\mu^{(\ell)} = \sum_{\nu=\ell-1}^{m} \left(a_{\mu\nu}^{(\ell-1)} - \frac{a_{\mu\ell-1}^{(\ell-1)}}{a_{\ell-1\,\ell-1}^{(\ell-1)}} a_{\ell-1\,\nu}^{(\ell-1)} \right) + \left(b_\mu^{(\ell-1)} - \frac{a_{\mu\ell-1}^{(\ell-1)}}{a_{\ell-1\,\ell-1}^{(\ell-1)}} b_{\ell-1}^{(\ell-1)} \right) =$$

$$= s_\mu^{(\ell-1)} - \frac{a_{\mu\ell-1}^{(\ell-1)}}{a_{\ell-1\,\ell-1}^{(\ell-1)}} s_{\ell-1}^{(\ell-1)}.$$

1.3 Triangular Decomposition by Gauss Elimination. Suppose A is an $n \times n$ nonsingular matrix. We now show that the elimination method discussed in the previous section leads to a decomposition of A into the product of two triangular matrices.

Consider the augmented matrix

$$(A|b) = \begin{pmatrix} a_{11} & \cdots & a_{1n} & b_1 \\ \vdots & & \vdots & \vdots \\ a_{n1} & \cdots & a_{nn} & b_n \end{pmatrix}$$

corresponding to the system of equations $(*)$ in 1.1. Then the first step of the Gauss elimination method is as follows:

(i) Find an index $r_1 \in \{1, 2, \ldots, n\}$ with $a_{r_1 1} \neq 0$.

(ii) Exchange the first row with the r_1-th row in the matrix $(A|b)$, and denote the resulting matrix by $(\hat{A}|\hat{b})$.

(iii) For each $\mu = 2, 3, \ldots, n$, subtract $\ell_{\mu 1} := \frac{\hat{a}_{\mu 1}}{\hat{a}_{11}}$ times the first row of the matrix $(\hat{A}|\hat{b})$ from the μ-th row. Then the result is a matrix $(A'|b')$ of the form

$$(A'|b') = \begin{pmatrix} a'_{11} & a'_{12} & \cdots & a'_{1n} & b'_1 \\ 0 & a'_{22} & \cdots & a'_{2n} & b'_2 \\ \vdots & \vdots & & \vdots & \vdots \\ 0 & a'_{n2} & \cdots & a'_{nn} & b'_n \end{pmatrix}.$$

The mapping $(A|b) \to (\hat{A}|\hat{b}) \to (A'|b')$ can be described in terms of matrix multiplications as follows: The row exchange operation in step (ii) corresponds to multiplying $(A|b)$ by a permutation matrix. In general, a

Permutation matrix is an $n \times n$ matrix $P_{\mu\nu}$ of the form

$$P_{\mu\nu} = \begin{pmatrix} 1 & & & & & & & & \\ & 1 & & & & & 0 & & \\ & & 0 & \cdots & \cdots & \cdots & 1 & & \\ & & \vdots & \ddots & & & \vdots & & \\ & & \vdots & & 1 & & \vdots & & \\ & & \vdots & & & \ddots & \vdots & & \\ & & 1 & \cdots & \cdots & \cdots & 0 & & \\ & & & & & & & 1 & \\ & 0 & & & & & & & \ddots & \\ & & & & & & & & & 1 \end{pmatrix}, \quad \begin{matrix} \leftarrow \mu\text{-th row} \\ \\ \leftarrow \nu\text{-th row} \end{matrix}$$

which differs from the identity matrix only in the μ-th and ν-th rows. Multiplication of a matrix by $P_{\mu\nu}$ exchanges the μ-th and ν-th rows in the matrix.

The effect of step (iii) can be described in terms of multiplication by a Frobenius matrix. A

Frobenius matrix is an $n \times n$ matrix G_μ of the form

$$G_\mu = \begin{pmatrix} 1 & & & & & \\ & \ddots & & & 0 & \\ & & 1 & & & \\ & & -\ell_{\mu+1\,\mu} & & & \\ & 0 & \vdots & & \ddots & \\ & & -\ell_{n\mu} & & & 1 \end{pmatrix}.$$

It differs from the identity matrix only in one column. Combining these facts, we can write the first step of Gauss elimination as follows:

$$(A'|b') = G_1(\hat{A}|\hat{b}) = G_1 P_{r_1 1}(A|b).$$

Remark. The matrices $P_{\mu\nu}$ and G_μ are nonsingular, and in fact, their inverses are given by

$$P_{\mu\nu}^{-1} = P_{\mu\nu}, \quad G_\mu^{-1} = \begin{pmatrix} 1 & & & & \\ & \ddots & & 0 & \\ & & 1 & & \\ & & \ell_{\mu+1\mu} & & \\ & 0 & \vdots & \ddots & \\ & & \ell_{n\mu} & & 1 \end{pmatrix}.$$

This claim is obvious for the permutation matrix. To verify it for the Frobenius matrix, we write G_μ in the form

$$G_\mu = I - m^{(\mu)}(e^\mu)^T,$$

where

$$m^{(\mu)} := (0, \ldots 0, \ \ell_{\mu+1\mu}, \ldots, \ell_{n\mu})^T.$$

Then the assertion follows from the string of identities

$$\begin{aligned} G_\mu \cdot (I + m^{(\mu)}(e^\mu)^T) &= (I - m^{(\mu)}(e^\mu)^T)(I + m^{(\mu)}(e^\mu)^T) = \\ &= I + m^{(\mu)}(e^\mu)^T - m^{(\mu)}(e^\mu)^T - m^{(\mu)}(e^\mu)^T m^{(\mu)}(e^\mu)^T = I. \end{aligned}$$

The entry $a_{r_1 1} =: \hat{a}_{11}$ which is defined in step (i), is called the *pivot element*, and the process of finding it is called the *pivot search*. For the algorithm described above, it is more accurate to refer to this as a *column pivot search*, since we are looking for the pivot element only in one column. From a theoretical standpoint, this always suffices for finding an element $a_{r_1 1} \neq 0$. However, from the standpoint of stability, it is preferable to search both columns and rows; i.e., to find r_1 such that

$$|a_{r_1 1}| = \max_{1 \leq \mu \leq n} |a_{\mu 1}|.$$

Moreover, the stability of the process can be further improved if prior to performing the maximum search, we *equilibrate* the rows of the matrix by multiplying them by factors so that the sum of the absolute values of the elements in each row is the same (cf. Problem 3 in 5.4).

After μ steps of the Gauss elimination process, we end up with a matrix of the form

$$(A^{(\mu)}|b^{(\mu)}) = \begin{pmatrix} A_{11}^{(\mu)} & A_{12}^{(\mu)} & b_1^{(\mu)} \\ 0 & A_{22}^{(\mu)} & b_2^{(\mu)} \end{pmatrix},$$

where $A_{12}^{(\mu)} \in \mathbb{R}^{(\mu, n-\mu)}$, $A_{22}^{(\mu)} \in \mathbb{R}^{(n-\mu, n-\mu)}$ and $b_1^{(\mu)} \in \mathbb{R}^\mu$, $b_2^{(\mu)} \in \mathbb{R}^{n-\mu}$. The matrix $A_{11}^{(\mu)} \in \mathbb{R}^{(\mu,\mu)}$ is an *upper triangular matrix*; i.e., all of its elements below the main diagonal are zero. In passing from $(A^{(\mu)}|b^{(\mu)})$ to $(A^{(\mu+1)}|b^{(\mu+1)})$ using the operation

$$(A^{(\mu+1)}|b^{(\mu+1)}) = G_\mu P_\mu (A^{(\mu)}|b^{(\mu)}), \quad P_\mu := P_{r_\mu \mu},$$

only the matrix $(A_{22}^{(\mu)}|b_2^{(\mu)}) \in \mathbb{R}^{(n-\mu,n-\mu+1)}$ is changed. Now after at most $(n-1)$ steps, we end up with an upper triangular matrix $R \in \mathbb{R}^{(n,n)}$ and a vector $c \in \mathbb{R}^n$, where

$$(R|c) = G_{n-1}P_{n-1}G_{n-2}P_{n-2}\cdots G_1P_1(A|b).$$

Here some of the factors $G_\nu P_\nu$ may be equal to the identity matrix.

In the μ-th elimination step, $(A^{(\mu)}|b^{(\mu)}) \to (A^{(\mu+1)}|b^{(\mu+1)})$, the elements below the main diagonal in the μ-th column are annihilated. Thus in carrying out Gauss elimination, we can store the numbers $\ell_{\nu\mu}$, $\mu+1 \le \nu \le n$ describing the matrix G_μ in these zeroed-out positions. Then, after the μ-th step, we have a matrix $T^{(\mu)} = (t_{\kappa\sigma}^{(\mu)})$ of the form

$$T^{(\mu)} = \begin{pmatrix} r_{11} & r_{12} & \cdots & r_{1\mu} & r_{1\mu+1} & \cdots & r_{1n} & c_1 \\ \lambda_{21} & r_{22} & \cdots & r_{2\mu} & r_{2\mu+1} & \cdots & r_{2n} & c_2 \\ \lambda_{31} & \lambda_{32} & \ddots & & \vdots & & \vdots & \\ \vdots & \vdots & & r_{\mu\mu} & r_{\mu\mu+1} & \cdots & r_{\mu n} & c_\mu \\ \vdots & \vdots & & \lambda_{\mu+1\mu} & a_{\mu+1\mu+1}^{(\mu+1)} & \cdots & a_{\mu+1n}^{(\mu+1)} & b_{\mu+1}^{(\mu+1)} \\ \vdots & \vdots & & \vdots & \vdots & & \vdots & \vdots \\ \lambda_{n1} & \lambda_{n2} & \cdots & \lambda_{n\mu} & a_{n\mu+1}^{(\mu+1)} & \cdots & a_{nn}^{(\mu+1)} & b_n^{(\mu+1)} \end{pmatrix},$$

where the entries $\lambda_{\kappa+1\kappa}, \lambda_{\kappa+2\kappa}, \ldots, \lambda_{n\kappa}$ in the κ-th column are some permutation of the numbers $\ell_{\kappa+1\kappa}, \ell_{\kappa+2\kappa}, \ldots, \ell_{n\kappa}$ appearing in the matrix G_κ. This process is called the *compact storage* method for Gauss elimination.

The mapping $T^{(\mu-1)} \to T^{(\mu)}$ can now be described as follows:

(i) Column pivot search: Find $r_\mu \in \{\mu, \mu+1, \ldots, n\}$ with $|t_{r_\mu\mu}^{(\mu-1)}| = \max_{\mu\le\kappa\le n} |t_{\kappa\mu}^{(\mu-1)}|$.

(ii) Permutation: Exchange the r_μ-th row with the μ-th row, and denote the entries of the new matrix $\hat{T}^{(\mu-1)}$ by $(\hat{t}_{\kappa\nu}^{(\mu-1)})$.

(iii) Elimination: Set $t_{\kappa\mu}^{(\mu)} := \hat{t}_{\kappa\mu}^{(\mu-1)}/\hat{t}_{\mu\mu}^{(\mu-1)}$, $\mu+1 \le \kappa \le n$, $t_{\kappa\rho}^{(\mu)} := \hat{t}_{\kappa\rho}^{(\mu-1)} - t_{\kappa\mu}^{(\mu)}\hat{t}_{\mu\rho}^{(\mu-1)}$, $\mu+1 \le \kappa \le n$, $\mu+1 \le \rho \le n$, and $t_{\kappa\rho}^{(\mu)} := \hat{t}_{\kappa\rho}^{(\mu-1)}$ otherwise.

Comment. We began with the assumption that the matrix A is nonsingular. In fact, this assumption can be dropped, since in step (i) of the algorithm, if $t_{\kappa\mu}^{(\mu-1)} = 0$ for all $\mu \le \kappa \le n$, then we know that A is singular. In practice $t_{\kappa\mu}^{(\mu-1)} = 0$ rarely happens because of roundoff, so it is common practice to declare the matrix A to be *numerically singular* as soon as $|t_{\kappa\mu}^{(\mu-1)}| < \varepsilon$ for all $\mu \le \kappa \le n$, where $\varepsilon > 0$ is a prescribed tolerance.

In step (iii), the elements $\ell_{\mu+1\mu}, \ldots, \ell_{n\mu}$ of the matrix G_μ are stored in their natural order. In later steps of the algorithm, $T^{(\kappa)} \to T^{(\kappa+1)}$ for $\mu \le \kappa \le n-2$, this order will be changed by the permutation step. When the elimination process is finished, the elements below the diagonal and in the μ-th column of $T^{(n-1)}$ are the last $(n-\mu)$ components of the vector

$$P_{r_{n-1}n-1} \cdot P_{r_{n-2}n-2} \cdots P_{r_{\mu+1}\mu+1} m^{(\mu)} =: \tilde{m}^{(\mu)}.$$

After these preliminaries, we can now prove the

Triangular Decomposition Theorem. *Suppose $T^{(n-1)} = (t_{\mu\nu})$ is the result of applying Gauss elimination with compact storage to the nonsingular matrix $A \in \mathbb{R}^{(n,n)}$. In addition, suppose $P := P_{r_{n-1}n-1} \cdots P_{r_2 2} P_{r_1 1}$ is the product of all the permutation matrices which were used, and that*

$$L = \begin{pmatrix} 1 & & & 0 \\ t_{21} & \ddots & & \\ \vdots & \ddots & \ddots & \\ t_{n1} & \cdots & t_{n\,n-1} & 1 \end{pmatrix} \quad \text{and} \quad R = \begin{pmatrix} t_{11} & \cdots & t_{1n} \\ & \ddots & \vdots \\ 0 & & t_{nn} \end{pmatrix}.$$

Then $P \cdot A = L \cdot R$ is a triangular decomposition of $P \cdot A$.

Proof. Using the representation of the matrix $(R|c)$, the Remark above, and the fact that for every vector $z \in \mathbb{R}^n$, $P_{r_\mu \mu}(I - z(e^\nu)^T)P_{r_\mu \mu} = = I - P_{r_\mu \mu} z (P_{r_\mu \mu} e^\nu)^T = I - (P_{r_\mu \mu} z)(e^\nu)^T$ for $\mu, r_\mu > \nu$, it follows that

$$\begin{aligned} R &= G_{n-1} P_{r_{n-1}n-1} G_{n-2} P_{r_{n-2}n-2} \cdots G_2 P_{r_2 2} G_1 P_{r_1 1} A = \\ &= G_{n-1} P_{r_{n-1}n-1} G_{n-2} P_{r_{n-1}n-1} P_{r_{n-1}n-1} P_{r_{n-2}n-2} G_{n-3} \cdots G_1 P_{r_1 1} A = \\ &= G_{n-1}(I - P_{r_{n-1}n-1} m^{(n-2)} (e^{n-2})^T) P_{r_{n-1}n-1} P_{r_{n-2}n-2} G_{n-3} \cdots G_1 P_{r_1 1} A \\ &= G_{n-1} \tilde{G}_{n-2} P_{r_{n-1}n-1} P_{r_{n-2}n-2} G_{n-3} \cdots G_1 P_{r_1 1} A = \cdots = \\ &= G_{n-1} \tilde{G}_{n-2} \tilde{G}_{n-3} \cdots \tilde{G}_1 P_{r_{n-1}n-1} P_{r_{n-2}n-2} \cdots P_{r_1 1} A, \end{aligned}$$

where here we have written $\tilde{G}_\nu := I - \tilde{m}^{(\nu)}(e^\nu)^T$. We have thus shown that

$$P \cdot A = \tilde{G}_1^{-1} \cdot \tilde{G}_2^{-1} \cdots \tilde{G}_{n-2}^{-1} \cdot \tilde{G}_{n-1}^{-1} \cdot R.$$

By the Remark, every $\tilde{G}_\nu^{-1}$ has the form

$$\tilde{G}_\nu^{-1} = I + \tilde{m}^{(\nu)}(e^\nu)^T,$$

and using induction, it is easy to show that

$$(I + \tilde{m}^{(1)}(e^1)^T)(I + \tilde{m}^{(2)}(e^2)^T) \cdots (I + \tilde{m}^{(n-1)}(e^{n-1})^T) = I + \sum_{\nu=1}^{n-1} \tilde{m}^{(\nu)}(e^\nu)^T.$$

Now the right-hand side of this equation is precisely the lower triangular matrix L described in the statement of the theorem, and we have proved that $P \cdot A = L \cdot R$. □

The above theorem takes account of the possibility that rows of the matrix A may have to be permuted in order to get a triangular decomposition. The following example shows that this is necessary in practice.

Example. The matrix $A = \begin{pmatrix} 0 & 1 \\ 1 & 0 \end{pmatrix}$ cannot be decomposed as $A = L \cdot R$ with L and R triangular matrices as in the Theorem. On the other hand, the matrix $P_{11} A = I$ trivially has such a decomposition.

Supplements. (i) Starting with the triangular decomposition $PA = LR$, we can solve the linear system of equations $Ax = b$ as follows: Set $c := Pb$ and solve $Lu = c$ and then $Rx = u$. The first of these two systems is lower triangular, and hence can easily be solved by successively computing the components u_ν, starting with u_1. Similarly, the second system is upper triangular, and the x_ν can be found starting with x_n.

(ii) To compute A^{-1}, we recommend first finding a triangular decomposition of A. Then the ν-th column x^ν of A^{-1} can be found by solving $Ax^\nu = e^\nu$. Thus to find A^{-1}, we have to solve a total of n systems of equations of the form $LRx^\nu = Pe^\nu$, $1 \leq \nu \leq n$. To accomplish this, we first solve $Lu^\nu = Pe^\nu$, and then $Rx^\nu = u^\nu$ for each ν. Both systems of equations have a triangular form, and hence are easy to solve.

(iii) Since $\det P = \pm 1$, it follows that $\det(A) = \pm(\det(L)) \cdot (\det(R))$. But $\det(R) = \prod_{\nu=1}^{n} t_{\nu\nu}$ and $\det(L) = 1$, and hence $\det(A) = \pm \prod_{\nu=1}^{n} t_{\nu\nu}$.

1.4 Some Special Matrices. The numerical solution of ordinary and partial differential equations by discretization frequently leads to linear systems of equations involving matrices whose elements are zero except in a band surrounding the main diagonal.

Definition. A matrix $A \in \mathbb{R}^{(n,n)}$ is called an (m, k)-*banded matrix* provided that all its elements $a_{\mu\nu}$ with indices $\nu - \mu > k$ and $\mu - \nu > m$ are zero.

Thus, in an (m, k)-banded matrix, nonzero elements appear only in the main diagonal, and in at most m subdiagonals and k superdiagonals. A $(1,1)$-banded matrix is called a *tridiagonal matrix*, while $(1, n-1)$- and $(n-1, 1)$-banded matrices are called upper and lower *Hessenberg matrices*, respectively.

If an (m, k)-banded matrix A has a triangular decomposition $A = L \cdot R$, then L is $(m, 0)$-banded and R is $(0, k)$-banded. If row exchanges are needed in order to decompose A, i.e., $P \cdot A = L \cdot R$, then R will be $(0, m + k)$-banded and L will be $(2m, 0)$-banded with at most $m + 1$ entries in each column. This observation is of practical importance, since it assures that in performing Gauss elimination on an (m, k)-banded matrix, which thus has at most $n(m+k+1)$ entries, we can obtain the decomposition using at most $n(m + 2k + 1)$ storage locations.

Since we will also be using tridiagonal matrices later in this book for the computation of quadratic splines (cf. 6.4.2), we now discuss them in more detail. Suppose we want to solve a tridiagonal system of equations $Ax = d$, $A \in \mathbb{C}^{(n,n)}$, $d \in \mathbb{C}^n$, where the matrix A has the form

$$A = \begin{pmatrix} a_1 & c_1 & & & \\ b_2 & \ddots & \ddots & & 0 \\ & \ddots & \ddots & \ddots & \\ & 0 & \ddots & \ddots & c_{n-1} \\ & & & b_n & a_n \end{pmatrix} =: \operatorname{tridiag}(b_\mu, a_\mu, c_\mu).$$

We can then prove the

Decomposition Theorem for Tridiagonal Matrices. *Suppose the elements of the matrix $A = \text{tridiag}(b_\mu, a_\mu, c_\mu)$ satisfy the following inequalities:*

$$(*) \qquad \begin{aligned} &|a_1| > |c_1| > 0, \\ &|a_\mu| \geq |b_\mu| + |c_\mu|, \; b_\mu \neq 0, c_\mu \neq 0, \; 2 \leq \mu \leq n-1, \\ &|a_n| \geq |b_n| > 0. \end{aligned}$$

Then:

(i) The quantities $\alpha_1 := a_1$, $\gamma_1 := c_1 \alpha_1^{-1}$ and

$$\alpha_\mu := a_\mu - b_\mu \gamma_{\mu-1} \;\text{ for }\; 2 \leq \mu \leq n, \quad \gamma_\mu := c_\mu \alpha_\mu^{-1} \;\text{ for }\; 2 \leq \mu \leq n-1$$

satisfy the inequalities

$$|\gamma_\mu| < 1, \; 1 \leq \mu \leq n-1; \quad 0 < |a_\mu| - |b_\mu| < |\alpha_\mu| < |a_\mu| + |b_\mu|, \; 2 \leq \mu \leq n.$$

(ii) A possesses the triangular decomposition $A = L \cdot R$ with

$$L := \text{tridiag}(b_\mu, \alpha_\mu, 0), \quad R := \text{tridiag}(0, 1, \gamma_\mu).$$

(iii) A is nonsingular.

Proof. (i) From $(*)$ it follows immediately that $|\gamma_1| = |c_1| \cdot |a_1|^{-1} < 1$. Now let $|\gamma_\nu| < 1$ for $\nu = 1, 2, \ldots, \mu - 1$. Then

$$|\gamma_\mu| = \left| \frac{c_\mu}{a_\mu - b_\mu \gamma_{\mu-1}} \right| \leq \frac{|c_\mu|}{\big||a_\mu| - |b_\mu|\,|\gamma_{\mu-1}|\big|} < \frac{|c_\mu|}{|a_\mu| - |b_\mu|} \leq 1.$$

Moreover,

$$\begin{aligned} |a_\mu| + |b_\mu| > |a_\mu| + |b_\mu|\,|\gamma_{\mu-1}| \geq |\alpha_\mu| \geq |a_\mu| - |b_\mu|\,|\gamma_{\mu-1}| > \\ > |a_\mu| - |b_\mu| \geq |c_\mu| > 0. \end{aligned}$$

(ii) We check that the decomposition $A = \text{tridiag}(b_\mu, \alpha_\mu, 0) \cdot \text{tridiag}(0, 1, \gamma_\mu)$ holds by multiplying out the factors:

$$\begin{aligned} a_{\mu\mu+1} &= \alpha_\mu \gamma_\mu = \alpha_\mu (c_\mu \alpha_\mu^{-1}) = c_\mu, \; 1 \leq \mu \leq n-1; \\ a_{\mu\mu} &= b_\mu \gamma_{\mu-1} + \alpha_\mu = b_\mu \gamma_{\mu-1} + (a_\mu - b_\mu \gamma_{\mu-1}) = a_\mu, \; 2 \leq \mu \leq n; \\ a_{\mu+1\mu} &= b_{\mu+1}, \; 1 \leq \mu \leq n-1, \quad a_{11} = \alpha_1 = a_1. \end{aligned}$$

(iii) The nonsingularity of A follows immediately from the fact that $\det(A) = \det(L)\det(R) = \prod_{\mu=1}^{n} \alpha_\mu \neq 0$. □

Remark. Tridiagonal matrices with the property $(*)$ are called *irreducibly diagonally dominant*. The theorem above can now be reformulated as follows: An irreducibly diagonally dominant matrix A possesses a triangular decomposition $A = L \cdot R$ where L is $(1,0)$-banded and R is $(0,1)$-banded, and where the main diagonal of R contains all ones.

In linear optimization (cf 9.3.6), the problem arises of solving a series of linear systems of equations, where at each step the corresponding matrix $\tilde{A}$ differs from the matrix A in the previous step in just one column. We now discuss efficient ways of handling such problems. Suppose A is a matrix of size $n \times n$ with triangular decomposition $A = L \cdot R$, and suppose the column vectors of A are a^μ, $1 \le \mu \le n$. In addition, suppose that the matrix $\tilde{A} = (a^1, a^2, \ldots, a^{\nu-1}, a^{\nu+1}, \ldots, a^{n-1}, \tilde{a})$ is obtained from A by dropping the ν-th column and adding a new last column. Since $L^{-1} \cdot A = R$, it follows that $L^{-1} \cdot \tilde{A}$ has the form

$$L^{-1} \cdot \tilde{A} = (L^{-1}a^1, L^{-1}a^2, \ldots, L^{-1}a^{\nu-1}, L^{-1}a^{\nu+1}, \ldots, L^{-1}a^n, L^{-1}\tilde{a}) =$$

$$= \begin{pmatrix}
r_{11} & r_{12} & \cdots & r_{1\nu-1} & r_{1\nu+1} & \cdots & r_{1n} & \tilde{r}_1 \\
 & r_{22} & \cdots & r_{2\nu-1} & r_{2\nu+1} & & r_{2n} & \tilde{r}_2 \\
 & & & \vdots & \vdots & & \vdots & \vdots \\
 & & & r_{\nu-1\nu-1} & \vdots & & \vdots & \vdots \\
 & & & & r_{\nu\nu+1} & & \vdots & \vdots \\
 & 0 & & & r_{\nu+1\nu+1} & \ddots & \vdots & \vdots \\
 & & & & & \ddots & r_{n-1n} & \tilde{r}_{n-1} \\
 & & & & & & r_{nn} & \tilde{r}_n
\end{pmatrix}.$$

In order to bring this matrix into triangular form, we now only need to carry out $(n-\nu)$ simple elimination steps. This saves a great deal of work as compared to doing a full Gauss elimination on $\tilde{A}$.

1.5 On Pivoting. In 1.3 we have introduced pivoting in order to assure that for nonsingular matrices, Gauss elimination does not involve a division by zero. The use of pivoting also has the advantage that it improves the numerical properties of the algorithm.

Example. The solution of the system of equations

$$\begin{pmatrix} 0.005 & 1 \\ 1 & 1 \end{pmatrix} \begin{pmatrix} x_1 \\ x_2 \end{pmatrix} = \begin{pmatrix} 0.5 \\ 1 \end{pmatrix},$$

rounded to three digits, is $Rd_3(x_1, x_2) = (0.503, 0.497)$. Now if we carry out Gauss elimination using two-digit floating-point arithmetic and choosing the pivot element to be $a_{11} = 0.005$, we get the system of equations

$$\begin{pmatrix} 0.005 & 1 \\ 0 & -200 \end{pmatrix} \begin{pmatrix} x_1 \\ x_2 \end{pmatrix} = \begin{pmatrix} 0.5 \\ -99 \end{pmatrix},$$

whose solution is $x_1 = 0.50$, $x_2 = 0$. Using column pivoting, we get

$$\begin{pmatrix} 1 & 1 \\ 0 & 1 \end{pmatrix} \begin{pmatrix} x_1 \\ x_2 \end{pmatrix} = \begin{pmatrix} 1 \\ 0.5 \end{pmatrix},$$

whose solution is $x_1 = 0.50$, $x_2 = 0.50$ which is the correct answer, rounded to two digits.

Column pivoting does not lead to better results in all cases. For example, suppose we multiply the first row of the above system of equations by 200:

$$\begin{pmatrix} 1 & 200 \\ 1 & 1 \end{pmatrix} \begin{pmatrix} x_1 \\ x_2 \end{pmatrix} = \begin{pmatrix} 100 \\ 1 \end{pmatrix}.$$

Then in the first step of Gauss elimination, the pivot element is $a_{11} = 1$, and we get the solution $x_1 = 0$, $x_2 = 0.5$.

In this example, the matrices have elements whose sizes are of different orders of magnitude (where the mantissa length is $t = 2$). In such cases, it is recommended to use *total pivoting*:

Find $r_\mu, s_\mu \in \{\mu, \mu+1, \ldots, n\}$ with $|a^{(\mu)}_{r_\mu s_\mu}| = \max_{\mu \le \kappa, \lambda \le n} |a^{(\mu)}_{\kappa\lambda}|$, and exchange the μ-th and r_μ-th rows and the μ-th and s_μ-th columns in $T^{(\mu-1)}$ to get the new matrix for the μ-th elimination step.

If we are solving a system of equations, then each column exchange in the Gauss elimination process requires that we also exchange the corresponding components of the solution vector. We have to keep track of these exchanges if in the end we want the components of the solution vector to be in their correct order. Since the computational effort involved in total pivoting is considerably higher than that required for column pivoting, it should only be applied when the order of magnitude of the matrix elements varies significantly.

There are some cases where, at least theoretically, pivoting is not required at all. For example, we have the

Theorem. *A nonsingular matrix $A \in \mathbb{C}^{(n,n)}$ possesses a triangular decomposition $A = L \cdot R$ if and only if all of its principal minors are nonzero.*

Proof. If no pivoting is required, then after j steps of Gauss elimination, we see that the principal minor $\det(A_{jj})$ of A satisfies

$$(*) \qquad \det(A_{jj}) = a_{11} \cdot a^{(2)}_{22} \cdot a^{(3)}_{33} \cdots a^{(j)}_{jj},$$

and thus,

$$a^{(j)}_{jj} = \frac{\det(A_{jj})}{\det(A_{j-1\,j-1})}.$$

This implies that all principal minors of A are nonzero. Conversely, if none of the principal minors of A vanishes, then we can argue from $(*)$ successively that $a_{11} \neq 0$, $a^{(2)}_{22} \neq 0, \ldots, a^{(n)}_{nn} \neq 0$. □

1.6 Complexity of Gauss Elimination. Not counting the work required to find the pivot, and counting only additions (together with subtractions) as well as multiplications (together with divisions), we find that in carrying out the step from $T^{(\mu-1)}$ to $T^{(\mu)}$ (cf. 1.3) in the computation of the triangular decomposition $P\cdot A = L\cdot R$ by Gauss elimination requires $((n-\mu)^2+(n-\mu))$ multiplications and $(n-\mu)^2$ additions. This gives a total of

$$\begin{aligned} &\textstyle\sum_{\mu=1}^{n-1}(n-\mu)^2 + \sum_{\mu=1}^{n-1}\mu = \frac{n^3}{3} - \frac{n}{3} \quad \text{multiplications} \\ \text{and}\quad &\textstyle\sum_{\mu=1}^{n-1}(n-\mu)^2 = \frac{n^3}{3} - \frac{n^2}{2} + \frac{n}{6} \quad \text{additions.} \end{aligned}$$

Thus, the complexity of computing the triangular decomposition using Gauss elimination is $O(n^3)$ as $n \to \infty$. In order to solve the system of equations $Ax = b$, we proceed as in Supplement 1.3. The computation of the solution vector u satisfying $Lu = Pb$ requires $\frac{1}{2}n(n-1)$ additions and multiplications. The work required to solve $Rx = u$ is $\frac{1}{2}n(n-1)$ additions and $\frac{1}{2}n(n+1)$ multiplications.

Summarizing, we have the

Proposition. To solve the system of equations $Ax = b$ using Gauss elimination requires

$$\begin{aligned} &\frac{1}{3}n^3 + n^2 - \frac{1}{3}n \quad \text{multiplications and divisions} \\ \text{and}\quad &\frac{1}{3}n^3 + \frac{n^2}{2} - \frac{5}{6}n \quad \text{additions and subtractions,} \end{aligned}$$

and so the complexity is $T_A^S(n) = O(n^3)$ as $n \to \infty$.

Since Gauss elimination can be interpreted as forming the LR-decomposition of the matrix A, where both L and R can be obtained by matrix multiplication, we can apply fast matrix multiplication methods to get a faster algorithm for solving a system of equations. Thus, in the sense of 1.4.6, Gauss elimination is not optimal (see also 1.4.7, Problem 4).

Remark. There is a variant of Gauss elimination called the *Gauss-Jordan method* which eliminates the need to solve the two systems $Lu = Pb$ and $Rx = u$. Without going into the details, we describe the first two elimination steps. To simplify the discussion, we assume that pivoting is not necessary. The first step of the Gauss-Jordan method is identical with the first step of Gauss elimination. In the second step, we eliminate not only the elements $a_{\mu 2}^{(2)}$, $2 < \mu \leq n$, but also $a_{12}^{(1)}$ by subtracting $a_{12}^{(1)}/a_{22}^{(2)}$ times the second row from the first row. After μ steps of the Gauss-Jordan method, we get a system of equations of the following form:

$$\begin{array}{llllll} a_{11}^{(1)}x_1 & & & +a_{1\mu+1}^{(\mu)}x_{\mu+1} & +\cdots+ a_{1n}^{(\mu)}x_n & = b_1^{(\mu)} \\ & a_{22}^{(2)}x_2 & & +a_{2\mu+1}^{(\mu)}x_{\mu+1} & +\cdots+ a_{2n}^{(\mu)}x_n & = b_2^{(\mu)} \\ & & \ddots & \vdots & \vdots \quad \vdots & \vdots \\ & & a_{\mu\mu}^{(\mu)}x_\mu & +a_{\mu\mu+1}^{(\mu)}x_{\mu+1} & +\cdots+ a_{\mu n}^{(\mu)}x_n & = b_\mu^{(\mu)} \\ & & & a_{\mu+1\mu+1}^{(\mu)}x_{\mu+1} & +\cdots+ a_{\mu+1n}^{(\mu)}x_n & = b_{\mu+1}^{(\mu)} \\ & & & \vdots & \vdots \quad \vdots & \vdots \\ & & & a_{n\mu+1}^{(\mu)}x_{\mu+1} & +\cdots+ a_{nn}^{(\mu)}x_n & = b_n^{(\mu)}. \end{array}$$

After completing all elimination steps, the components of the solution vector x can be simply computed as $x_\mu = b_\mu^{(n)}/a_{\mu\mu}^{(\mu)}$, $\mu = 1, 2, \ldots, n$. The complexity of the Gauss-Jordan method is $T_A^S(n) = O(n^3)$ as $n \to \infty$. Although its complexity is no better than that of Gauss elimination, the Gauss-Jordan method has some advantages when the computation is to be carried out on a computer capable of parallel operations on several processors simultaneously. We do not have space here, however, to discuss this point further.

1.7 Problems. 1) Find the LR-decomposition of the matrix

$$A := \begin{pmatrix} 1 & 2 & 3 & 4 \\ 1 & 4 & 9 & 16 \\ 1 & 8 & 27 & 64 \\ 1 & 16 & 81 & 256 \end{pmatrix},$$

and use this decomposition to solve the system of equations $Ax = b$ with right-hand side $b := (3, 1, -15, -107)^T$.

2)a) Let $\{a^1, a^2, \ldots, a^n\}$ be a basis for $\mathbb{R}^n$, and let $\{a^1, \ldots, \tilde{a}^k, \ldots, a^n\}$ be another basis which differs from the first only in that the vector a^k is replaced by the vector $\tilde{a}^k$. How can one find the coordinates of a prescribed vector with respect to the second basis, given the coordinates of $\tilde{a}^k$ with respect to the first basis?

b) Consider the following situation: Suppose we want to solve a linear system of equations, but after having computed the LR-decomposition, we discover that one column in the original matrix A is wrong. How can the decomposition nevertheless be used to find the correct solution? Formulate a corresponding algorithm, and apply it to the system of equations in Problem 1, where the first column of A is to be replaced by $(0, 0, 6, 36)^T$.

3)a) Let $a, b, c \in \mathbb{R}^n$ with $|a_\nu| \geq \sum_{\substack{\mu=1 \\ \mu\neq\nu}}^n |a_\mu|$, $a_\nu \neq 0$, $|b_\kappa| \geq \sum_{\substack{\mu=1 \\ \mu\neq\kappa}}^n |b_\mu|$ and $\nu \neq \kappa$. Suppose the vector c is defined by $c_\mu := b_\mu - \frac{b_\nu}{a_\nu} a_\mu$, $1 \leq \mu \leq n$. Show that $|c_\kappa| \geq \sum_{\substack{\mu=1 \\ \mu\neq\kappa}}^n |c_\mu|$.

b) If $|a_{\mu\mu}| \geq \sum_{\substack{\nu=1 \\ \nu\neq\mu}}^n |a_{\mu\nu}|$, then the matrix $A = (a_{\mu\nu})$ is called *weakly diagonally dominant.*

Prove that if A is a weakly diagonally dominant nonsingular matrix, then Gauss elimination without pivoting can be used to compute a decomposition of the form $L \cdot R = A$.

4) In general, is the inverse of a nonsingular band matrix a band matrix?

5) Write a computer program for Gauss elimination with complete pivoting. Test your program on the example

$$a_{\mu\nu} := 1/(\mu + \nu - 1),\ 1 \leq \mu, \nu \leq n, \quad b_\mu := 1/(n + \mu - 1),\ 1 \leq \mu \leq n.$$

2. The Cholesky Decomposition

For general nonsingular $n \times n$ matrices, pivoting is needed in order to construct the corresponding LR decomposition. Theorem 1.5 shows that for certain matrices, pivoting is not needed, but the criterion given there is difficult to use in practice because it requires too much computation. In this section, we show that LR decompositions of positive definite matrices can be computed without pivoting, and discuss a special method for constructing them.

2.1 Review of Positive Definite Matrices. In this section we review some useful properties of positive definite matrices; for proofs, see any book on linear algebra, e.g. G. Strang [1976].

Definition. A symmetric matrix $A \in \mathbb{R}^{(n,n)}$ is called *positive definite* provided $x^T Ax > 0$ for all vectors $x \in \mathbb{R}^n$ with $x \neq 0$. We call A *positive semidefinite* if $x^T Ax \geq 0$ for all $x \in \mathbb{R}^n$.

The positive definiteness of a matrix can be checked using the following

Criterion. The following two equivalent conditions

(i) there exists a nonsingular matrix W with $A = W^T W$,

and

(ii) all principal minors $\det A_{\mu\mu}$, $1 \leq \mu \leq n$, of A are positive,

are necessary and sufficient for the symmetric matrix $A \in \mathbb{R}^{(n,n)}$ to be positive definite.

Moreover, positive definite matrices have the following

Properties. Let $A \in \mathbb{R}^{(n,n)}$ be positive definite and symmetric. Then A^{-1} exists, is symmetric, and is positive definite. In addition, every principal submatrix $A_{\mu\mu}$ of A with $1 \leq \mu \leq n$ is symmetric and positive definite.

2.2 The Cholesky Decomposition. In view of 2.1, a symmetric matrix $A \in \mathbb{R}^{(n,n)}$ is positive definite if there exists a matrix $W \in \mathbb{R}^{(n,n)}$ with $A = W^T W$. We now show that W can be chosen to be triangular.

Theorem. *Let $A \in \mathbb{R}^{(n,n)}$ be symmetric and positive definite. Then there exists a triangular decomposition of the form $A = LL^T$ with a uniquely defined nonsingular lower triangular matrix $L = (\ell_{\mu\nu}) \in \mathbb{R}^{(n,n)}$ such that $\ell_{\mu\mu} > 0$, $1 \leq \mu \leq n$.*

Proof. We proceed by induction on n. For $n = 1$, $A = (a_{11})$ with $a_{11} > 0$, and so $L = L^T = (\sqrt{a_{11}})$.

Now let $A \in \mathbb{R}^{(n,n)}$ be symmetric and positive definite, and suppose the assertion holds for $n-1$. We partition the matrix A in the form

$$A = \begin{pmatrix} A_{n-1n-1} & b \\ b^T & a_{nn} \end{pmatrix}.$$

Since A_{n-1n-1} is a principal submatrix of a positive definite matrix, by Properties 2.1, it is itself positive definite. The element a_{nn} is positive and $b \in \mathbb{R}^{n-1}$. Now by the inductive hypothesis, there exists a unique nonsingular lower triangular matrix L_{n-1} such that $A_{n-1n-1} = L_{n-1} \cdot L_{n-1}^T$, with $\ell_{\mu\mu} > 0$ for $\mu = 1, 2, \ldots, n-1$. We now look for the desired matrix L in the form

$$(*) \qquad L = \begin{pmatrix} L_{n-1} & 0 \\ c^T & \alpha \end{pmatrix},$$

where the vector $c \in \mathbb{R}^{n-1}$ and the constant $\alpha > 0$ are to be determined from the equation

$$A = \begin{pmatrix} A_{n-1n-1} & b \\ b^T & a_{nn} \end{pmatrix} = \begin{pmatrix} L_{n-1} & 0 \\ c^T & \alpha \end{pmatrix} \begin{pmatrix} L_{n-1}^T & c \\ 0 & \alpha \end{pmatrix}$$

by comparing elements. This leads to $L_{n-1}c = b$ and $c^T c + \alpha^2 = a_{nn}$. Since L_{n-1} is nonsingular, it follows that $c = L_{n-1}^{-1} b$. Now $0 < \det(A) = = \alpha^2 \cdot (\det(L_{n-1}))^2$ implies that α^2 is positive and real. Hence, there exists a unique positive number α with $c^T c + \alpha^2 = a_{nn}$. □

The French major ANDRÉ-LOUIS CHOLESKY (1875–1918) was involved in geodesy and surveying from 1906 to 1909 during the international occupation of Crete, and later on also in North Africa. He developed the method named after him to compute solutions of least squares data fitting problems (cf. Chap. 5, Sect. 6). The factorization of a symmetric positive definite matrix A in the form $A = LL^T$ can, however, also be deduced from an earlier theorem of G. G. JACOBI, (cf. M. Koecher ([1985], p. 124).

We now derive formulae for computing the elements $\ell_{\mu\nu}$ of L by comparing elements in the equation

$$\begin{pmatrix} a_{11} & \cdots & a_{1n} \\ \vdots & \ddots & \vdots \\ a_{n1} & \cdots & a_{nn} \end{pmatrix} = \begin{pmatrix} \ell_{11} & & 0 \\ \vdots & \ddots & \\ \ell_{n1} & \cdots & \ell_{nn} \end{pmatrix} \begin{pmatrix} \ell_{11} & \cdots & \ell_{n1} \\ & \ddots & \vdots \\ 0 & & \ell_{nn} \end{pmatrix}.$$

This gives $a_{\nu\mu} = \sum_{\kappa=1}^{\mu} \ell_{\nu\kappa} \cdot \ell_{\mu\kappa}$, and in view of the symmetry of A, we need only consider indices ν with $\nu \geq \mu$. Proceeding one column at a time for $\mu = 1, 2, \ldots, n$, we get

$$\ell_{\mu\mu} = \left(a_{\mu\mu} - \sum_{\kappa=1}^{\mu-1} \ell_{\mu\kappa}^2 \right)^{1/2}, \quad \ell_{\nu\mu} = \frac{1}{\ell_{\mu\mu}} \left(a_{\nu\mu} - \sum_{\kappa=1}^{\mu-1} \ell_{\nu\kappa} \cdot \ell_{\mu\kappa} \right), \; \mu+1 \leq \nu \leq n.$$

Remarks. (i) The Cholesky decomposition $A = L \cdot L^T$ implies

$$\sum_{\kappa=1}^{\mu} \ell_{\mu\kappa}^2 \leq \max_{1 \leq \mu \leq n} |a_{\mu\mu}|$$

for all $1 \leq \mu \leq n$. It follows that all elements of the matrix L are bounded by $\max_{1\leq\mu\leq n} \sqrt{|a_{\mu\mu}|}$. Thus, the elements of the decomposition cannot grow too rapidly, a property which contributes to the stability of the method.

(ii) Since A is symmetric, we need only work with the elements above and on the main diagonal. The elements $\ell_{\mu\nu}$ with $\nu < \mu$ can be stored below the diagonal. The diagonal elements $\ell_{\mu\mu}$ have to be stored in an extra vector of length n.

(iii) The above algorithm for constructing the Cholesky decomposition simultaneously provides sufficient information to decide whether the matrix A is positive definite. We leave it to the reader to verify this, and to formulate an algorithm.

2.3 Complexity of the Cholesky Decomposition. For each μ, the computation of the elements $\ell_{\mu\nu}$ requires $\frac{1}{2}(n-\mu)(n-\mu+1)$ additions, $\frac{1}{2}(n-\mu)(n-\mu+1)$ multiplications, and $(n-\mu)$ divisions. Summing over μ, we see that a total of $\frac{1}{6}(n^3-n)$ additions and multiplications and $\frac{1}{6}(3n^2-3n)$ divisions are required. In addition, n square roots must be computed. For large n, we can ignore the square roots, and so the complexity of the Cholesky decomposition is

$$T_A^S(n) = \frac{1}{3}(n^3 + 3n^2 - n) = O(n^3)$$

as $n \to \infty$.

We see that in comparison with the LR decomposition obtained by Gauss elimination, the Cholesky decomposition requires approximately half as much work.

2.4 Problems. 1) Let $A \in \mathbb{R}^{(n,n)}$ be symmetric and positive definite. Show that for all $\mu \neq \nu$,

a) $|a_{\mu\nu}| < 0.5(a_{\mu\mu} + a_{\nu\nu})$, b) $|a_{\mu\nu}| < (a_{\mu\mu} \cdot a_{\nu\nu})^{1/2}$.

2) Let $A \in \mathbb{R}^{(n,n)}$ be symmetric and positive definite. Show that there is a unique decomposition of the form $A = SDS^T$, where S is a lower triangular matrix with $s_{\mu\mu} = 1$ for $1 \leq \mu \leq n$, and D is a diagonal matrix. Derive formulae similar to those for the Cholesky method for computing the elements of $S = (s_{\mu\nu})$ and $D = \text{diag}(d_\mu)$.

3) Let $A = (a_{\mu\nu})$ be a symmetric positive definite band matrix with band width m. Show that the matrix L in the Cholesky decomposition $A = L \cdot L^T$ has bandwidth m.

4) Write a computer program to solve a linear system of equations $Ax = b$ using the Cholesky method, and test it on the problems

$$a_{\mu\nu} = \frac{1 + (-1)^{\mu+\nu}}{\mu + \nu - 1}, \quad 1 \leq \mu, \nu \leq n,$$

$$b_\mu = \frac{(2n)!(1 - (-1)^{n+\mu})}{(n!)^2 \cdot (n + \mu)}, \quad 1 \leq \mu \leq n,$$

for $n = 5$ and $n = 10$. What does Gauss elimination give?

3. The QR Decomposition of Householder

In 1.3 we used Frobenius matrices to construct triangular decompositions $P \cdot A = L \cdot R$. In this section we show that a triangular decomposition of the form $A = Q \cdot R$ can also be obtained by application of appropriate orthogonal matrices to A. This method has the advantage that it is highly stable, and leads to a decomposition where Q is an orthogonal matrix, and R is upper triangular. To solve a linear system of equations $Ax = b$, we first compute $Q^T b =: u$ by a matrix multiplication, and then find x by solving the upper triangular system $Rx = u$.

3.1 Householder Matrices. Gauss elimination leads to an LR decomposition where the matrix L is the product of elementary matrices. To construct an orthogonal matrix Q giving a QR decomposition, we will also use products of elementary matrices.

Definition. A matrix $H \in \mathbb{R}^{(k,k)}$, $k \in \mathbb{Z}_+$, is called a *Householder matrix* provided that $H = I - 2hh^T$, where the vector $h \in \mathbb{R}^k$ is of the form $h = (0, \ldots, 0, h_\mu, \ldots, h_k)^T$ and has Euclidean length one. This means:
(i) there exists an index $\mu \in \{1, 2, \ldots, k\}$ with $h = (0, \ldots, 0, h_\mu, \ldots, h_k)^T$;
(ii) $\sum_{\kappa=\mu}^{k} h_\kappa^2 = 1$.

We denote the Euclidean length $(\sum_{\kappa=1}^{k} x_\kappa^2)^{1/2}$ of a vector $x \in \mathbb{R}^k$ by $\|x\|_2$.

The definition implies that a Householder matrix must have the form

$$H = \begin{pmatrix} 1 & & & & & \\ & \ddots & & & 0 & \\ & & 1 & & & \\ & & & 1 - 2h_\mu^2 & -2h_\mu h_{\mu+1} & \cdots & -2h_\mu h_k \\ & & & -2h_\mu h_{\mu+1} & 1 - 2h_{\mu+1}^2 & \cdots & -2h_{\mu+1} h_k \\ & 0 & & \vdots & & & \\ & & & -2h_\mu h_k & -2h_{\mu+1} h_k & \cdots & 1 - 2h_k^2 \end{pmatrix}$$

Clearly, H is symmetric, and since

$$H^2 = (I - 2hh^T)^2 = I - 4hh^T + 4hh^T hh^T = I,$$

it is also orthogonal.

Geometrically, the transformation H describes a reflection of the Euclidean space $\mathbb{R}^k$ relative to the hyperplane $H_{h,0} := \{z \in \mathbb{R}^k | h^T z = 0\}$.

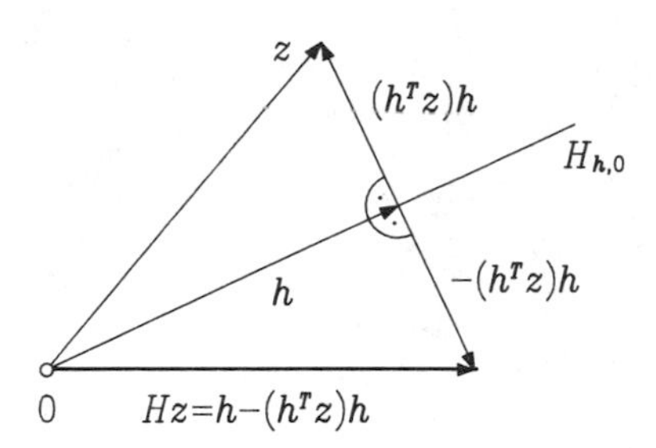

In particular, if we decompose the vector z into a component in the direction of h, and an orthogonal component as $z = (h^T z)h + (z - (h^T z)h)$, then it obviously follows that

$$Hz = (I - 2hh^T)z = (h^T z)h + (z - (h^T z)h) - 2hh^T(h^T z)h =$$
$$= -(h^T z)h + (z - (h^T z)h).$$

We now use Householder matrices to transform A stepwise into an upper triangular matrix.

3.2 The Basic Problem. In each step of the QR decomposition algorithm, we want to construct a reflection in $\mathbb{R}^k$ such that the vector $x \in \mathbb{R}^k$ is transformed into a multiple of the unit vector e^1 in $\mathbb{R}^k$. Thus, for given $x \in \mathbb{R}^k$, $x \neq 0$, we have to find a vector $h \in \mathbb{R}^k$ with $\|h\|_2 = 1$ such that $Hx = (I_k - 2hh^T)x = \sigma e^1$ for some $\sigma \in \mathbb{R}$.

Since H is orthogonal, σ is determined up to its sign by the equation $\|x\|_2 = \|Hx\|_2 = \|\sigma e^1\|_2 = |\sigma|$.

Now from $Hx = x - 2(hh^T)x = x - 2(h^T x)h = \sigma e^1$, it follows that h must be a multiple of the vector $x - \sigma e^1$. Coupled with the requirement that $\|h\|_2 = 1$, this implies that

$$h = \frac{x - \sigma e^1}{\|x - \sigma e^1\|_2},$$

where $\sigma \in \mathbb{R}$ must satisfy $|\sigma| = \|x\|_2$. Since we have already used all of the conditions which H is required to satisfy, the sign of σ remains undetermined. In order to prevent cancellation and improve the stability of the algorithm, we choose $\sigma := -\operatorname{sgn}(x_1) \cdot \|x\|_2$ and take $\operatorname{sgn}(x_1) = 1$ if $x_1 = 0$. To compute h, we note that

$$\|x - \sigma e^1\|_2^2 = \|x + \operatorname{sgn}(x_1) \cdot \|x\|_2 \cdot e^1\|_2^2 =$$
$$= \big|\, |x_1| + \|x\|_2 \,\big|^2 + \sum_{\mu=2}^{k} |x_\mu|^2 = 2\|x\|_2^2 + 2|x_1|\,\|x\|_2.$$

The following matrix H solves the basic problem:

(i) $H = I - \beta u \cdot u^T$,

(ii) $\beta := (\|x\|_2(|x_1| + \|x\|_2))^{-1}$,

(iii) $u := (\operatorname{sgn}(x_1)(|x_1| + \|x\|_2), x_2, \ldots, x_k)^T$.

We shall apply these kinds of matrices H to transform an arbitrary matrix $A \in \mathbb{R}^{(n,n)}$ into upper triangular form.

3.3 The Householder Algorithm. Let A be an arbitrary $n \times n$ matrix. Now set $A^{(0)} := A$, and, as described in 3.2, find an orthogonal matrix $H^{(1)}$ with $H^{(1)}(a^1)^{(0)} = \sigma e^1$, where $(a^1)^{(0)}$ is the vector in the first column of $A^{(0)}$. After $(\mu - 1)$ steps of this kind, we obtain a matrix $A^{(\mu-1)}$ of the form

$$A^{(\mu-1)} = \begin{pmatrix} B_{\mu-1} & C_{\mu-1} \\ 0 & \tilde{A}^{(\mu-1)} \end{pmatrix},$$

where $B_{\mu-1} \in \mathbb{R}^{(\mu-1,\mu-1)}$ is upper triangular, and $C_{\mu-1}$ and $\tilde{A}^{(\mu-1)}$ belong to $\mathbb{R}^{(n-\mu+1,n-\mu+1)}$. Then in the next step we find an orthogonal matrix $\tilde{H}^{(\mu)} \in \mathbb{R}^{(n-\mu+1,n-\mu+1)}$ such that $\tilde{H}^{(\mu)}(a^1)^{(\mu-1)} = \sigma e^1 \in \mathbb{R}^{n-\mu+1}$, where $(a^1)^{(\mu-1)}$ is the first column of the $(n-\mu+1) \times (n-\mu+1)$ matrix $\tilde{A}^{(\mu-1)}$. Now if we set

$$H^{(\mu-1)} := \begin{pmatrix} I_{\mu-1} & 0 \\ 0 & \tilde{H}^{(\mu-1)} \end{pmatrix} \in \mathbb{R}^{(n,n)},$$

then $H^{(\mu-1)}$ is symmetric and orthogonal, and $A^{(\mu)} := H^{(\mu-1)}A^{(\mu-1)}$ satisfies:

(i) $B_{\mu-1}$ and $C_{\mu-1}$ remain unchanged;

(ii) $a_{\nu\mu}^{(\mu)} = 0$ for $\nu > \mu$.

After a total of $(n-1)$ steps, we arrive at an upper triangular matrix $R := A^{(n-1)}$ and an orthogonal symmetric matrix $Q = (H^{(n-1)} \cdots H^{(1)})^{-1} = H^{(1)} \cdot H^{(2)} \cdots H^{(n-1)}$ with $A = Q \cdot R$.

Summarizing we have

The QR Decomposition Theorem. *Every real $n \times n$ matrix A can be decomposed into a product $A = Q \cdot R$ of an orthogonal matrix Q and an upper triangular matrix R.*

Remark. The QR Decomposition Theorem can be easily extended to both complex and nonsquare matrices. We leave it to the reader to make the necessary modifications.

The Householder Algorithm for solving a linear system of equations $Ax = b$ can now be described as follows:

Input: $n \in \mathbb{Z}_+$, $C := (A|b) =: (c_{\mu\nu}) \in \mathbb{R}^{(n,n+1)}$.

1. Initialization: $\mu := 1$.

2. Elimination step: $s := \left(\sum_{\kappa=\mu}^{n} c_{\kappa\mu}^2\right)^{1/2}$.

(i) If $s = 0$, stop since A is singular. Otherwise
$\beta := (s(|c_{\mu\mu}| + s))^{-1}$;
$u := (0, \ldots, 0, c_{\mu\mu} + \operatorname{sgn}(c_{\mu\mu})s, c_{\mu+1,\mu}, \ldots, c_{n\mu})^T$,
$\operatorname{sgn}(c_{\mu\mu}) = 1$, if $c_{\mu\mu} = 0$;
$H^{(\mu)} := I - \beta u u^T$,

(ii) $C := H^{(\mu)} \cdot C =: (c_{\mu\nu})$,

3. Check: If $\mu+1 \leq n-1$, set $\mu := \mu+1$ and go to step 2. Otherwise, stop.

3.4 Complexity of the QR Decomposition. In the μ-th elimination step, we first compute the quantity s using $(n - \mu + 1)$ multiplications and additions and one square root. The determination of the factor β requires one addition, one multiplication and one division. The operation $H^{(\mu)} \cdot C = C - \beta u u^T C$ in step (ii) of the algorithm for computing $u^T C$ requires exactly $(n-\mu+1)(n-\mu)$ multiplications and $(n-\mu+1)(n-\mu)+1$ additions, along with $(n-\mu+1)(n-\mu)$ multiplications and $(n-\mu)$ additions to compute the product $u \cdot (\beta \cdot u^T C)$. To this we have to add $(n-\mu+1)(n-\mu)$ additions for the formation of the difference $C - (\beta u u^T C)$. Thus, the μ-th step requires a total of

$$\begin{array}{rl} 2(n-\mu+1)^2 & \text{multiplications,} \\ (n-\mu+1)^2 + (n-\mu+1)(n-\mu) + 2 & \text{additions,} \\ 1 & \text{division,} \\ 1 & \text{square root.} \end{array}$$

Summing these numbers over the $(n-1)$ steps, it follows that the complexity of the QR decomposition is

$$T_A^S(n) = \frac{4}{3}n^3 + \frac{3}{2}n^2 + \frac{19}{6}n - 6 = O(n^3)$$

as $n \to \infty$, not counting the $(n-1)$ square roots.

In the next chapter, we show that the Householder decomposition will also be of use for the computation of eigenvalues.

3.5 Problems. 1) Show by example that the Householder method does not generally preserve the band structure of a matrix.

2) Write a computer program using the Householder method to solve the linear system of equations $Ax = b$, where $A \in \mathbb{R}^{(n,n)}$ and $b \in \mathbb{R}^n$. Test your program for the matrix $A = (a_{\mu\nu})$ with entries $a_{\mu\nu} = (\mu + \nu - 1)^{-1}$ for the following cases:

a) $n = 5$, $b = (1, 1, 1, 1)^T$

b) $n = 5, 8, 10$; $b = (b_1, \ldots, b_n)^T$, $b_\mu - \sum_{\nu=1}^{n} (\mu + \nu - 1)^{-1}$.

3) Show that the QR decomposition of a nonsingular matrix $A \in \mathbb{R}^{(n,n)}$ is unique up to the sign of the diagonal elements of R.

4) Suppose the columns of $A = (a_{\mu\nu}) \in \mathbb{R}^{(n,n)}$ are given by the vectors $a^1, a^2, \ldots, a^n \in \mathbb{R}^n$. By applying the QR decomposition, show that the following inequality of J. Hadamard holds: $|\det(A)| \leq \prod_{\nu=1}^{n}((a^\nu)^T a_\nu)^{1/2}$.

4. Vector Norms and Norms of Matrices

In this section we collect some definitions and results on norms of vectors and matrices which will be useful for analyzing the error in various methods for solving linear systems of equations. The material presented here can also be regarded as a preparation for Chapter 4, where norms will be discussed in the more general framework of function spaces.

4.1 Norms on Vector Spaces. Let X be a vector space over the field $\mathbb{K} := \mathbb{C}$ of complex or over the field $\mathbb{K} := \mathbb{R}$ of real numbers.

A *norm* is a mapping $\|\cdot\| : \mathrm{X} \to \mathbb{R}$, $x \to \|x\|$, which for all $x, y \in \mathrm{X}$ satisfies the following

Norm Conditions:

(i) $\|x\| = 0 \Leftrightarrow x = 0$;

(ii) $\|\alpha x\| = |\alpha|\ \|x\|$ for all $\alpha \in \mathbb{K}$; *Homogeneity*

(iii) $\|x + y\| \leq \|x\| + \|y\|$; *Triangle Inequality.*

The norm conditions (i)–(iii) imply the *definiteness* $\|x\| > 0$ for $x \neq 0$ of the norm, as well as the inequality

$$(*) \qquad |\,\|x\| - \|y\|\,| \leq \|x + y\|.$$

The pair $(\mathrm{X}, \|\cdot\|)$ is called a *normed vector space*; in this section we treat only the two finite dimensional vector spaces $\mathbb{C}^n$ and $\mathbb{R}^n$.

Example. Let $\mathrm{X} := \mathbb{C}^n$ and let $\|\cdot\| := \|\cdot\|_p$, where p is an integer satisfying $1 \leq p \leq \infty$. Here

$$\|x\|_p := (\sum_{\nu=1}^{n} |\, x_\nu\,|^p)^{\frac{1}{p}} \quad \text{for} \quad 1 \leq p < \infty$$

and

$$\|x\|_\infty := \max_{1 \leq \nu \leq n} |\, x_\nu\,|.$$

It is immediate that the norm conditions (i) and (ii) hold for all p, and that (iii) holds for $p = 1, \infty$. For $1 < p < \infty$, the triangle inequality (iii) is precisely the well-known

Minkowski Inequality

$$(\sum_{\nu=1}^{n} |\, x_\nu + y_\nu\,|^p)^{\frac{1}{p}} \leq (\sum_{\nu=1}^{n} |\, x_\nu\,|^p)^{\frac{1}{p}} + (\sum_{\nu=1}^{n} |\, y_\nu\,|^p)^{\frac{1}{p}}.$$

Proof: See e.g. D. Luenberger ([1969], p. 31). □

Continuity of the Norm. *The norm $\|x\|$ is a continuous function of the components $x_1, \ldots, x_n$ of the vector x.*

Proof: We prove the result assuming that the size of vectors $x \in \mathbb{R}^n$ is measured by the uniform norm $\|\cdot\|_\infty$, but the result holds in general by the equivalence of norms shown in 4.3 below. The inequality (*) above implies that if $z = (z_1, \ldots, z_n)$, then

$$|\,\|x + z\| - \|x\|\,| \leq \|z\|.$$

Let $\{e^1, \ldots, e^n\}$ be the canonical basis in X so that

$$z = \sum_1^n z_\nu e^\nu \quad \text{and} \quad \|e^\nu\| = 1 \quad \text{for} \quad 1 \leq \nu \leq n.$$

Then $\|z\| \leq \sum_1^n |z_\nu|\,\|e^\nu\| \leq n \max_{1 \leq \nu \leq n} |z_\nu|$. Thus if $\max_{1 \leq \nu \leq n} |z_\nu| \leq \frac{\varepsilon}{n}$, then it follows that $|\,\|x + z\| - \|x\|\,| \leq \varepsilon$, and the proof is complete. □

4.2 The Natural Norm of a Matrix. The set of $m \times n$ matrices with real or complex entries form a vector space $\mathbb{K}^{(m,n)}$ of dimension $m \cdot n$ over $\mathbb{R}$ or $\mathbb{C}$, respectively, and hence we can define various norms on these spaces. In this section we consider matrices as operators, and definine their norms in a general way.

An $m \times n$ matrix A defines a linear mapping of the n-dimensional linear space $(\mathrm{X}, \|\cdot\|_\mathrm{X})$ into the m-dimensional linear space $(\mathrm{Y}, \|\cdot\|_\mathrm{Y})$. Since the function $x \to \|Ax\|_\mathrm{Y}$ is continuous on the compact set $\{x \in \mathbb{C}^n \mid \|x\|_\mathrm{X} = 1\}$ relative to the norm $\|\cdot\|_\mathrm{X}$, it must assume its maximum there. It follows that

$$\|A\| := \sup_{x \in \mathbb{C}^n \setminus \{0\}} \frac{\|Ax\|_\mathrm{Y}}{\|x\|_\mathrm{X}} = \max_{\|x\|_\mathrm{X} = 1} \|Ax\|_\mathrm{Y}$$

is a finite number, and that

$$\|Ax\|_\mathrm{Y} \leq \|A\|\,\|x\|_\mathrm{X}.$$

From now on we consider only square $n \times n$ matrices, and assume that the same vector norm is used for both X and Y; i.e., $\|\cdot\|_\mathrm{X} = \|\cdot\|_\mathrm{Y} =: \|\cdot\|$. In this case we have the

Estimate

$$\|Ax\| \leq \|A\|\,\|x\|.$$

Claim. The mapping $A \to \|A\|$ defines a norm; i.e., it satisfies conditions (i)–(iii) in (2.1). Indeed, the homogeneity and the triangle inequality are obvious. The property $\|A\| = 0 \Leftrightarrow A = 0$ is also immediate since $A = 0$ trivially implies $\|A\| = 0$, while $\|Ax\| = 0$ for all $x \in \mathrm{X}$ implies that A is the zero matrix.

Since $\|A\|$ is defined in terms of the vector norm $\|\cdot\|$, we call it the *induced norm* or *natural norm* of the matrix A. Clearly, $\|I\| = 1$.

Comment. It is clear that $C := \|A\|$ is the smallest constant such that $\|Ax\| \leq C\|x\|$ for all $x \in X$. Indeed, equality holds whenever x is a vector for which $\|Ax\|$ takes on its maximum value.

Remark. For a natural norm defined on matrices in $\mathbb{K}^{(n,n)}$, we also have

$$\|A \cdot B\| \leq \|A\| \, \|B\|.$$

Indeed, in this case $\|ABx\| \leq \|A\| \, \|Bx\| \leq \|A\| \, \|B\| \, \|x\|$. In general, however, the inequality

$$\|ABx\| \leq \|AB\| \, \|x\|$$

is sharper, and in fact is best possible.

4.3 Special Norms of Matrices. In this section we present the most important natural matrix norms.

Definition. Let $A \in \mathbb{K}^{(n,n)}$, and let $\lambda_1, \lambda_2, \ldots, \lambda_n \in \mathbb{C}$ be the eigenvalues of A. Then

$$\rho(A) := \max_{1 \leq i \leq n} |\lambda_i|$$

is called the *spectral radius* of A.

The following theorem deals with matrix norms induced by the vector norms discussed in Example 4.1.

Theorem. *Let $\|\cdot\|_p$ be the norm of a matrix induced on the space of matrices $\mathbb{K}^{(n,n)}$ by the vector norm $\|\cdot\|_p$. Then*

$$(1) \qquad \|A\|_1 = \max_{1 \leq \nu \leq n} \sum_{\mu=1}^{n} |a_{\mu\nu}|,$$

$$(2) \qquad \|A\|_\infty = \max_{1 \leq \mu \leq n} \sum_{\nu=1}^{n} |a_{\mu\nu}|,$$

$$(3) \qquad \|A\|_2 = (\rho(\overline{A}^T A))^{\frac{1}{2}}.$$

These norms are called the maximum column-sum norm, the maximum row-sum norm, and the spectral norm, respectively.

Proof. We leave the proof of assertion (1) to the reader.

(2) It follows from Example 4.1 and Estimate 4.2 that

$$\|A\|_\infty \leq \max_{1 \leq \mu \leq n} \sum_{\nu=1}^{n} |a_{\mu\nu}|.$$

It remains to show that equality actually holds. To show this, suppose k is such that $\sum_{\nu=1}^{n} |a_{k\nu}| = \max_{1\le\mu\le n} \sum_{\nu=1}^{n} |a_{\mu\nu}|$. Now it is easy to see that the vector $\hat{x} \in \mathbb{K}^n$ with components

$$\hat{x}_\nu := \begin{cases} 1 & \text{if } a_{k\nu} = 0, \\ \frac{\overline{a}_{k\nu}}{|a_{k\nu}|} & \text{otherwise} \end{cases}$$

is such that $\|\hat{x}\|_\infty = 1$ and $\|A\hat{x}\|_\infty = \sum_{\nu=1}^{n} |a_{k\nu}|$.

(3) By 4.2, there exists $y \in \mathbb{K}^n$ with $\|y\|_2 = 1$ and $\|Ay\|_2 = \|A\|_2$ such that $\|A\|_2^2 = \overline{y}^T\overline{A}^T Ay$. Since $\overline{A}^T A$ is a Hermitian matrix, it has a complete orthogonal system of eigenvectors $\{x^1, x^2, \ldots, x^n\}$ with $(\overline{x}^\mu)^T x^\nu = \delta_{\mu\nu}$. Let $\lambda_1, \ldots, \lambda_n$ be the associated eigenvalues. Then $\overline{A}^T Ax^\mu = \lambda_\mu x^\mu$, and consequently $0 \le \|Ax^\mu\|_2^2 = (\overline{x}^\mu)^T \overline{A}^T Ax^\mu = \lambda_\mu$. This shows that the matrix $\overline{A}^T A$ is positive semidefinite.

Now writing y in the form $y = \sum_{\nu=1}^{n} \alpha_\nu x^\nu$, $\alpha_\nu \in \mathbb{K}$, we see that

$$1 = \|y\|_2^2 = (\sum_{\mu=1}^{n} \overline{\alpha}_\mu (\overline{x}^\mu)^T)(\sum_{\nu=1}^{n} \alpha_\nu x^\nu) = \sum_{\mu=1}^{n} |\alpha_\mu|^2,$$

and thus that

$$\begin{aligned}\|Ay\|_2^2 &= (\sum_{\mu=1}^{n} \overline{\alpha}_\mu (\overline{x}^\mu)^T)\overline{A}^T A(\sum_{\nu=1}^{n} \alpha_\nu x^\nu) = \\ &= (\sum_{\mu=1}^{n} \overline{\alpha}_\mu (\overline{x}^\mu)^T)(\sum_{\nu=1}^{n} \alpha_\nu \lambda_\nu x^\nu) = \sum_{\mu=1}^{n} \lambda_\mu |\alpha_\mu|^2 \le \\ &\le (\max_{1\le\mu\le n} \lambda_\mu) \sum_{\mu=1}^{n} |\alpha_\mu|^2 = \rho(\overline{A}^T A).\end{aligned}$$

Conversely, if λ_k is the largest eigenvalue of $\overline{A}^T A$, then

$$\|A\|_2^2 \ge \|Ax^k\|_2^2 = (\overline{x}^k)^T \overline{A}^T Ax^k = \lambda_k = \rho(\overline{A}^T A). \qquad \square$$

Equivalence of Norms. It can be shown that any two vector norms $\|\cdot\|_X$ and $\|\cdot\|_Y$ on a finite dimensional vector space X satisfy

$$m\|x\|_X \le \|x\|_Y \le M\|x\|_X$$

for all $x \in X$, where m and M are constants. In view of this fact, we say that all vector norms on a given finite dimensional space are *equivalent*. The proof can be accomplished by showing that any norm is equivalent to the norm $\|\cdot\|_\infty$. We leave the details to the reader. The equivalence of vector norms implies that all natural norms on a fixed space of matrices are also equivalent.

Norm Bounds. Since it is sometimes difficult to compute the natural norm of a matrix, it is often useful to be able to compute an upper bound for the norm. The spectral norm $\|A\|_2$ of a matrix is such an example, since it involves finding the largest eigenvalue of $\overline{A}^T A$. A "matrix norm" $\|A\|$ defined on $n \times n$ matrices is called *compatible* with the vector norm $\|x\|$ provided that it satisfies the conditions (i)–(iii) as well as the inequality $\|AB\| \leq \|A\|\,\|B\|$, and provided that $\|Ax\| \leq \|A\|\,\|x\|$ for all $x \in \mathbb{K}^n$. The natural norm provides the smallest possible constant in this inequality, and in this sense is minimal among the set of norms compatible with $\|x\|$.

Example. $\|A\|_F := \sqrt{\text{trace}(\overline{A}^T A)}$ is a matrix norm. Indeed, $\sqrt{\text{trace}(\overline{A}^T A)} = \left[\sum_{\mu,\kappa=1}^{n} |a_{\mu\kappa}|^2\right]^{1/2}$ can be regarded as the Euclidean norm $\|\cdot\|_2$ of a vector $y \in \mathbb{C}^{(n,n)}$ with components $a_{\mu\kappa}$, and so (i)–(iii) holds. Moreover,

$$\|AB\|_F^2 = \sum_{\nu,\kappa=1}^{n} \Big|\sum_{\mu=1}^{n} a_{\nu\mu} b_{\mu\kappa}\Big|^2 \leq \sum_{\nu,\kappa=1}^{n} \Big[\Big(\sum_{\mu=1}^{n} |a_{\nu\mu}|^2\Big)\Big(\sum_{\mu=1}^{n} |b_{\mu\kappa}|^2\Big)\Big] =$$

$$= \Big(\sum_{\mu,\nu=1}^{n} |a_{\nu\mu}|^2\Big)\Big(\sum_{\mu,\kappa=1}^{n} |b_{\mu\kappa}|^2\Big) = \|A\|_F^2 \|B\|_F^2.$$

Since we also have

$$\|Ax\|_2^2 = \sum_{\nu=1}^{n} \Big|\sum_{\kappa=1}^{n} a_{\nu\kappa} x_\kappa\Big|^2 \leq \sum_{\nu=1}^{n} \Big[\sum_{\kappa=1}^{n} |a_{\nu\kappa}|^2 \sum_{\kappa=1}^{n} |x_\kappa|^2\Big] = \|A\|_F^2 \|x\|_2^2,$$

it follows that $\|A\|_F$ is compatible with $\|x\|_2$. It provides a useful upper bound for the natural norm $\|A\|_2$, and is called the *Frobenius norm* or *Erhard Schmidt norm.*

4.4 Problems. 1) Determine the best possible constants m', M' and m'', M'' such that $m'\|x\|_1 \leq \|x\|_\infty \leq M'\|x\|_1$ and $m''\|x\|_\infty \leq \|x\|_2 \leq M''\|x\|_\infty$.

2) Where is the finite dimensionality of a vector space X needed in the proof of the equivalence of all norms defined on X?

3) Show that $\|A\| := n \max_{\mu,\nu} |a_{\mu\nu}|$ defines a matrix norm which is compatible with $\|\cdot\|_1$, $\|\cdot\|_2$ and $\|\cdot\|_\infty$.

4) Show that if A is a square matrix and $\|\cdot\|$ is a norm satisfying $\|A \cdot B\| \leq \|A\| \cdot \|B\|$, then $\rho(A) \leq \|A\|$. Compute $\rho(A)$ for the matrices $A := \begin{pmatrix} 1 & 3 & 1 \\ 2 & 0 & 1 \\ 1 & 3 & 2 \end{pmatrix}$ and $A := \begin{pmatrix} 1 & 3 & 1 \\ 2 & 0 & 1 \\ 1 & 1 & 2 \end{pmatrix}$, and compare with $\|A\|_1$, $\|A\|_\infty$ and $\|A\|_F$.

5. Error Bounds

In many practical situations, the matrix and right-hand side of a linear system of equations will be known only approximately. In such cases, the methods discussed above produce the solution of a problem which is near, but not exactly the same as, the desired problem. Moreover, even when the matrix and right-hand side are stored exactly in the computer, a numerical algorithm will usually not produce an exact solution; i.e., the defect will be nonzero. This suggests that we need a way to bound the error in the solution. In this section we discuss how this can be done.

5.1 Condition of a Matrix. Suppose $A \in \mathbb{C}^{(n,n)}$ and $b \in \mathbb{C}^n$. In this section we study the effect of changes in A and b on the solution of the linear system $Ax = b$. We begin with some preliminaries.

Suppose the matrix A is nonsingular, and that instead of $Ax = b$, we have the perturbed linear system of equations

$$A(x + \Delta x) = b + \Delta b$$

with $\Delta b \in \mathbb{C}^n$. Then we can write the error Δx in the solution vector x as $\Delta x = A^{-1}\Delta b$. Now choosing an arbitrary vector norm and the corresponding induced matrix norm, it follows that the error is bounded by

$$\|\Delta x\| \leq \|A^{-1}\| \cdot \|\Delta b\|.$$

Since $\|b\| \leq \|A\| \cdot \|x\|$, if $b \neq 0$, this gives the bound

$$\frac{\|\Delta x\|}{\|x\|} \leq \|A^{-1}\| \cdot \|A\| \frac{\|\Delta b\|}{\|b\|}$$

for the relative error $\|\Delta x\|/\|x\|$ in the solution of the original system of equations. The factor $\|A^{-1}\| \cdot \|A\|$ measures the sensitivity of the relative error of the solution x to a perturbation in the right-hand side b.

Definition. Let $A \in \mathbb{C}^{(n,n)}$ be nonsingular. Then $\text{cond}(A) := \|A^{-1}\| \cdot \|A\|$ is called the *condition number* of the matrix A.

If we use a natural matrix norm, then the fact that

$$1 = \|I\| = \|A^{-1} \cdot A\| \leq \|A^{-1}\| \cdot \|A\| = \text{cond}(A)$$

implies that the condition number of A is greater or equal to one. The condition number of a matrix A depends on the norm being used. When it is necessary to explicitly indicate which norm is being used, we will use subscripts; for example we write $\text{cond}_2(A)$ in the case where the spectral norm is being used.

In order to estimate the effect of changes in the matrix A on the solution of a linear system of equations $Ax = b$, we now prove the following

Lemma. *Suppose $A \in \mathbb{C}^{(n,n)}$, and suppose that for some natural matrix norm, $\|A\| < 1$. Then $(I+A)^{-1}$ exists, and*

$$\frac{1}{1+\|A\|} \leq \|(I+A)^{-1}\| \leq \frac{1}{1-\|A\|}.$$

Proof. For $x \in \mathbb{C}^n$, $x \neq 0$, we have

$$\|(I+A)x\| = \|x + Ax\| \geq \|x\| - \|Ax\| \geq (1 - \|A\|)\|x\|.$$

This implies that $(I+A)x = 0$ can only hold for $x = 0$, and so the matrix $(I+A)$ is nonsingular. Moroever, with $C := (I+A)^{-1}$, we have

$$1 = \|I\| = \|(I+A)C\| = \|C + AC\| \geq \|C\| - \|C\| \cdot \|A\|,$$

and analogously $1 \leq \|C\| + \|C\| \cdot \|A\|$. Combining these facts leads to the assertion. □

We can now apply the lemma to obtain the following result about matrices which are close to each other:

Perturbation Lemma. *Suppose $A, B \in \mathbb{C}^{(n,n)}$ are two matrices, and that A is nonsingular. In addition, suppose that for some real α and κ, $\|A^{-1}\| \leq \alpha$ and $\|A^{-1}\| \cdot \|B - A\| \leq \kappa < 1$, where the norm is any natural norm of a matrix. Then B is also nonsingular, and $\|B^{-1}\| \leq \frac{\alpha}{1-\kappa}$.*

Proof. Since $\|A^{-1}(B-A)\| \leq \|A^{-1}\| \cdot \|B - A\| < 1$, the Lemma implies that $(I + A^{-1}(B-A))^{-1}$ exists. This means that $A^{-1}B$ is also invertible, and thus so is B. Moreover, the Lemma gives the bound

$$\|B^{-1}\| \leq \|B^{-1}A\| \cdot \|A^{-1}\| \leq \frac{1}{1 - \|A^{-1}\| \cdot \|B - A\|} \cdot \alpha \leq \frac{\alpha}{1-\kappa}.$$ □

Remark. The assertion of this perturbation lemma also holds in the far more general framework of the perturbation theory of linear operators, and can be used there in the same way to establish error bounds.

5.2 An Error Bound for Perturbed Matrices. In this section we consider invertible matrices $A \in \mathbb{C}^{(n,n)}$. For such matrices we can compute their condition number, and in terms of it, establish the following

Theorem. *Suppose A and ΔA are matrices in $\mathbb{C}^{(n,n)}$, and that x and Δx are such that $Ax = b$ and $(A + \Delta A)(x + \Delta x) = b$ with $b \neq 0$. In addition, suppose that $\|\Delta A\| \cdot \|A^{-1}\| < 1$ holds for some induced norm of a matrix. Then the relative error satisfies*

$$\frac{\|\Delta x\|}{\|x\|} \leq \frac{\operatorname{cond}(A)}{1 - \operatorname{cond}(A)\frac{\|\Delta A\|}{\|A\|}} \frac{\|\Delta A\|}{\|A\|}.$$

Proof. Since $\|A^{-1}\| \cdot \|\Delta A\| < 1$, by the Perturbation Lemma 5.1, the inverse of $(A + \Delta A)$ exists, and can be bounded by

$$\|(A+\Delta A)^{-1}\| \leq \frac{\|A^{-1}\|}{1 - \|A^{-1}\|\,\|\Delta A\|}.$$

Moreover,

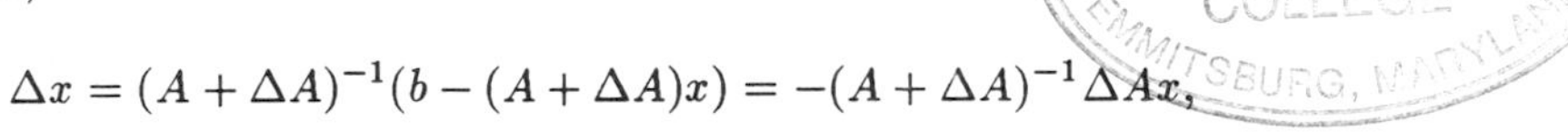

$$\Delta x = (A+\Delta A)^{-1}(b - (A+\Delta A)x) = -(A+\Delta A)^{-1}\Delta A x,$$

and so $\|\Delta x\| \leq \frac{\|A^{-1}\|}{1-\|A^{-1}\|\cdot\|\Delta A\|}\|\Delta A\|\|x\|$. This implies

$$\frac{\|\Delta x\|}{\|x\|} \leq \frac{\|A^{-1}\| \cdot \|A\|}{1 - \|A^{-1}\|\,\|A\|\frac{\|\Delta A\|}{\|A\|}} \cdot \frac{\|\Delta A\|}{\|A\|}. \qquad \square$$

Remark. In order to be able to use this bound in practice, we need an upper bound on $\|A^{-1}\|$. Since in general such a bound will not be known, or can only be computed with a great deal of effort, it will be useful to make the following

Observation. If either Gauss elimination or the Householder algorithm is used to solve the system of equations $Ax = b$, then we have corresponding factorizations $L \cdot R$ or $Q \cdot R$ of A for which the defect $D := L \cdot R - A$ or $D := Q \cdot R - A$ generally satisfies $\|D\| \ll 1$. Since L and R are triangular matrices and Q is an orthogonal matrix, we can easily find their inverses. But then the perturbation lemma gives the bound

$$\|A^{-1}\| \leq \frac{\alpha}{1-\kappa},$$

where α is such that $\|D\| \leq \frac{\kappa}{\alpha}$ and either $\|(L\cdot R)^{-1}\| \leq \alpha$ or $\|(Q\cdot R)^{-1}\| \leq \alpha$, respectively.

We recall that once we have an LR-decomposition of the matrix A, then to solve the system of equations $Ax = b$, we proceed in two steps. First we find the solution of $Ly = b$, and then the solution of $Rx = y$. Now in practice it can happen that one or both of the matrices L and R can be poorly-conditioned, even though A is well-conditioned in the sense that its condition number is close to one. This means that even for well-conditioned problems, the Gauss elimination process can give poor results. The situation is different for the QR-decomposition method of Householder. Suppose we work with the Euclidean vector norm and its associated induced matrix norm. Then because of the orthogonality of Q, $\|Qx\|_2 = \|x\|_2$ for all vectors $x \in \mathbb{C}^n$, and thus $\|Q\|_2 = 1$. Similarly, it follows that $\|Q^T\|_2 = \|Q^{-1}\|_2 = 1$, and thus the condition number of Q is one. On the other hand,

$$\|A\|_2 = \|Q^T QA\|_2 \leq \|QA\|_2 \leq \|A\|_2,$$

and so $\|A\|_2 = \|QA\|_2$. Similarly, $\|A^{-1}Q^{-1}\|_2 = \|A^{-1}\|_2$. We have established the

Proposition. The condition number of a matrix A remains unchanged if the QR-algorithm is applied; i.e.,

$$\text{cond}_2(A) = \text{cond}_2(QA).$$

This proposition and the theorem above account for the stability of the QR-algorithm.

5.3 Acceptability of Solutions. By using the techniques of backward error analysis, c.f. 1.3.3, we can derive an error bound without having to compute A^{-1}. Given $A = (a_{\mu\nu}) \in \mathbb{C}^{(n,n)}$ and $b = (b_\mu) \in \mathbb{C}^n$, suppose that $|A| := (|a_{\mu\nu}|)$ and $|b| := (|b_\mu|)$. Interpreting "$\leq$" componentwise, we obviously have $|A \cdot B| \leq |A| \cdot |B|$ and $|Ax| \leq |A| \cdot |x|$ for all $A, B \in \mathbb{C}^{(n,n)}$ and $x \in \mathbb{C}^n$. We now prove the

Theorem of Prager and Oettli. *Let $A_0, \Delta A \in \mathbb{C}^{(n,n)}$ and $b_0, \Delta b \in \mathbb{C}^n$ be given with $\Delta A \geq 0$ and $\Delta b \geq 0$. Suppose $\tilde{x} \in \mathbb{C}^n$ is a vector, and that $r(\tilde{x}) := b_0 - A_0\,\tilde{x}$ is the corresponding residual. Then the following two assertions are equivalent:*

(i) $|r(\tilde{x})| \leq \Delta A|\,\tilde{x}\,| + \Delta b$;

(ii) There exist $A \in \mathcal{A} := \{B \in \mathbb{C}^{(n,n)} \mid |B - A_0| \leq \Delta A\}$ *and* $b \in \mathcal{B} := \{c \in \mathbb{C}^n \mid |c - b_0| \leq \Delta b\}$ *with* $A\,\tilde{x} = b$.

Proof. Suppose first that $A \in \mathcal{A}$ and $b \in \mathcal{B}$ are such that $A\,\tilde{x} = b$. Then $\delta A := A - A_0$ and $\delta b := b - b_0$ satisfy the estimates $|\delta A| \leq \Delta A$ and $|\delta b| \leq \Delta b$. Moreover,

$$\begin{aligned}|r(\tilde{x})| = |b_0 - A_0\,\tilde{x}\,| = |b - \delta b - (A - \delta A)\,\tilde{x}\,| = |-\delta b + \delta A\,\tilde{x}\,| \leq \\ \leq \Delta A|\,\tilde{x}\,| + \Delta b.\end{aligned}$$

Conversely, suppose $|r(\tilde{x})| \leq \Delta A|\,\tilde{x}\,| + \Delta b$. We now show how to construct $A \in \mathcal{A}$ and $b \in \mathcal{B}$ satsifying $A\,\tilde{x} = b$. Using the abbreviations $r := r(\tilde{x})$, $s := \Delta b + \Delta A|\,\tilde{x}\,|$, let

$$\delta a_{\mu\nu} := \begin{cases} 0, & \text{if } s_\mu = 0 \\ r_\mu \Delta a_{\mu\nu} \cdot \text{sgn}(\tilde{x}_\nu) s_\mu^{-1}, & \text{if } s_\mu \neq 0 \end{cases}$$

and

$$\delta b_\mu := \begin{cases} 0, & \text{if } s_\mu = 0 \\ -r_\mu \Delta b_\mu s_\mu^{-1}, & \text{if } s_\mu \neq 0 \end{cases},$$

where (see 2.3.2) the "sgn" of a (possibly) complex number $\tilde{x}_\nu$ is defined by

$$\text{sgn}(\tilde{x}_\nu) = \begin{cases} 1 & \text{if } \tilde{x}_\nu = 0 \\ \frac{\tilde{x}_\nu}{|\tilde{x}_\nu|} & \text{otherwise.} \end{cases}$$

Since by hypothesis $|r| \leq s$, it follows that $r_\mu / s_\mu \leq 1$ for $1 \leq \mu \leq n$. We conclude that $A := A_0 + \delta A \in \mathcal{A}$ and $b := b_0 + \delta b \in \mathcal{B}$. Let $A_0 =: (a_{0\mu\nu})$ and $b_0 =: (b_{0\mu})$. We now show that $A\tilde{x} = b$. If $s_\mu \neq 0$, then

$$r_\mu = b_{0\mu} - \sum_{\nu=1}^{n} a_{0\mu\nu}\,\tilde{x}_\nu = b_\mu - \delta b_\mu - \sum_{\nu=1}^{n} a_{\mu\nu}\,\tilde{x}_\nu + \sum_{\nu=1}^{n} \delta a_{\mu\nu}\,\tilde{x}_\nu =$$

$$= b_\mu - \sum_{\nu=1}^{n} a_{\mu\nu}\,\tilde{x}_\nu + (\Delta b_\mu + \sum_{\nu=1}^{n} \Delta a_{\mu\nu} |\,\tilde{x}_\nu\,|) s_\mu^{-1} r_\mu = b_\mu - \sum_{\nu=1}^{n} a_{\mu\nu}\,\tilde{x}_\nu + r_\mu .$$

On the other hand, if $s_\mu = 0$, then

$$0 = r_\mu = b_{0\mu} - \sum_{\nu=1}^{n} a_{0\mu\nu}\,\tilde{x}_\nu = b_\mu - \sum_{\nu=1}^{n} a_{\mu\nu}\,\tilde{x}_\nu\,.$$

This shows $A\tilde{x} = b$ and the proof is complete. □

This theorem provides a means of judging the suitability of a vector $\tilde{x} \in \mathbb{C}^n$ as a solution of a linear system of equations, based on the size of the residual $|r(\tilde{x})|$. For example, if all components of A_0 and b_0 have the same relative accuracy ε, i.e.,

$$\Delta A = \varepsilon |A_0| \quad \text{and} \quad \Delta b = \varepsilon |b_0|,$$

then condition (i) of the theorem holds when $|b_0 - A_0\,\tilde{x}\,| \leq \varepsilon(|b_0| + |A_0|\,|\,\tilde{x}\,|)$. Then statement (ii) of the theorem can be interpreted as asserting that $\tilde{x}$ is the exact solution of a "neighboring" linear system of equations.

Example. Suppose

$$A_0 = \begin{pmatrix} 2 & -1 \\ 1 & 1 \end{pmatrix}, \quad b_0 = \begin{pmatrix} 1 \\ 2 \end{pmatrix}, \quad \tilde{x} = \begin{pmatrix} 0.95 \\ 1.05 \end{pmatrix}.$$

Then

$$|A_0\,\tilde{x} - b_0| = \begin{pmatrix} 0.15 \\ 0 \end{pmatrix} \quad \text{and} \quad |b_0| + |A_0|\,|\,\tilde{x}\,| = \begin{pmatrix} 3.95 \\ 4 \end{pmatrix}.$$

Now the above observation tells us that $\tilde{x}$ is acceptable as a solution of the linear system whenever the relative accuracy ε of $|A_0|$ and $|b_0|$ is at least 0.038. This says that if A_0 and b_0 have a relative accuracy of 4 %, then $\tilde{x}$ is a reasonable solution of $A_0 x = b_0$.

5.4 Problems. 1) Determine the condition number of the matrix $\begin{pmatrix} 1 & 2 \\ 3 & 7 \end{pmatrix}$ with respect to the matrix norm induced by the vector norms $\|\cdot\|_1$, $\|\cdot\|_2$ and $\|\cdot\|_\infty$.

2) Consider the following $n \times n$ matrices with $n \geq 3$ and $a \geq 0$:

$$A := \begin{pmatrix} a+1 & \cdots & a+1 & a \\ \vdots & \cdot^{\cdot^{\cdot}} & \cdot^{\cdot^{\cdot}} & \vdots \\ a+1 & a & \cdots & a \\ a & a & \cdots & a-1 \end{pmatrix},$$

$$B := \begin{pmatrix} -a & 0 & \cdots & 0 & 1 & a \\ 0 & & & 1 & -1 & 0 \\ \vdots & 0 & \cdot^{\cdot^{\cdot}} & \cdot^{\cdot^{\cdot}} & & \vdots \\ 0 & \cdot^{\cdot^{\cdot}} & \cdot^{\cdot^{\cdot}} & 0 & & \vdots \\ 1 & -1 & & & & 0 \\ a & 0 & \cdots & & 0 & -(a+1) \end{pmatrix}.$$

a) Show $B = A^{-1}$.

b) Compute the condition number of A with respect to $\|\cdot\|_\infty$.

c) Let $Ax = b$ and $A(x + \Delta x) = b + \Delta b$ be two linear systems of equations. By making an appropriate choice of x and Δb, show the sharpness of the estimate

$$\frac{\|\Delta x\|_\infty}{\|x\|_\infty} \leq \operatorname{cond}_\infty(A) \cdot \frac{\|\Delta b\|_\infty}{\|b\|_\infty},$$

where $\operatorname{cond}_\infty(A) = \|A\|_\infty \cdot \|A^{-1}\|_\infty$.

3) A matrix with the property that the sums of the absolute values of the entries in each row are equal is called *row-equilibrated*. Every nonsingular matrix can be transformed into a row-equilibrated matrix by multiplying by a diagonal matrix. Show that if A is a row-equilibrated matrix, then

$$\operatorname{cond}_\infty(A) \leq \operatorname{cond}_\infty(DA)$$

for every nonsingular diagonal matrix D; i.e., equilibration improves the condition of the matrix.

4) Generalize Theorem 5.2 by showing that the solutions of the linear systems $Ax = b$ and $(A + \Delta A)(x + \Delta x) = b + \Delta b$ satisfy

$$\frac{\|\Delta x\|}{\|x\|} \leq \frac{\operatorname{cond}(A)}{1 - \operatorname{cond}(A)\frac{\|\Delta A\|}{\|A\|}} \left(\frac{\|\Delta A\|}{\|A\|} + \frac{\|\Delta b\|}{\|b\|} \right).$$

5) On the basis of the Theorem of Prager and Oettli, which of the approximate solutions $\tilde{x} = (1.1, 0.9)^T$, $\tilde{y} = (1.5, 0.6)^T$ or $\tilde{z} = (0, 2)^T$ is an acceptable solution of the system

$$\begin{pmatrix} 1 & 2 \\ 3 & 7 \end{pmatrix} \begin{pmatrix} x_1 \\ x_2 \end{pmatrix} = \begin{pmatrix} 3 \\ 10 \end{pmatrix},$$

assuming that the relative error in the data (matrix and right-hand side) is of size

a) 2.5% ; b) 10% ; c) 20% ?

6. Ill-Conditioned Problems

Linear systems of equations can often be very poorly conditioned, in which case the methods discussed above usually produce unsatisfactory results. A well known example of a poorly-conditioned matrix is the *Hilbert matrix*

$$A = (a_{\mu\nu}), \quad a_{\mu\nu} = \frac{1}{\mu+\nu-1}, \quad 1 \le \mu, \nu \le n.$$

To illustrate what can happen, we consider solving the linear system of equations $Ax = b$, $b_\mu = \sum_{\kappa=1}^{n} \frac{1}{\kappa+\mu-1}$, whose exact solution is $x = (1, 1, \ldots, 1)^T$. The following table shows the relative errors obtained with the methods of Gauss, Cholesky, and Householder.

Method	Relative error ($n = 8$)	Relative error ($n = 10$)
Gauss	0.406	3.39
Gauss with equilibration	0.0915	3.55
Cholesky	0.421	A is numerically not positive definite
Householder	0.208	91.7
Householder with equilibration	0.0560	5.44

We now discuss a method which produces significantly better results for ill-conditioned problems.

6.1 The Singular-Value Decomposition of a Matrix. In this section we make use of the concept of an eigenvalue of a matrix; see any book on linear algebra (eg. G. Strang [1976]). The numerical treatment of matrix eigenvalue problems will be treated in the following chapter.

Definition. Let A be a real $m \times n$ matrix. A decomposition of the form

$$A = U\Sigma V^T,$$

where $U \in \mathbb{R}^{(m,m)}$ and $V \in \mathbb{R}^{(n,n)}$ are orthogonal matrices and the $m \times n$ matrix $\Sigma = (\sigma_\mu \delta_{\mu\nu})$ is a diagonal matrix, is called a *singular-value decomposition* of A.

If $U = (u^1, u^2, \ldots, u^m)$ and $V = (v^1, v^2, \ldots, v^n)$ with $u^\nu \in \mathbb{R}^m$ and $v^\nu \in \mathbb{R}^n$, then it immediately follows that the decomposition $A = U\Sigma V^T$ can also be written in the form

$$Av^\nu = \sigma_\nu u^\nu, \quad 1 \le \nu \le m,$$
$$Av^\nu = 0, \quad m+1 \le \nu \le n,$$

when $m < n$, and in the form

$$Av^\nu = \sigma_\nu u^\nu, \quad 1 \le \nu \le n,$$

when $m \ge n$. Since the choice of the signs of the components of the vectors u^ν and v^ν is free, we may assume that the numbers σ_ν are nonnegative. The positive numbers σ_ν are called the *singular values* of A.

Remark. Since $A^T = V\Sigma^T U^T$, if A has a singular-value decomposition, then so does A^T, and the singular values are the same. In addition, we have

$$A^T A = V\Sigma^T \Sigma V^T,$$

so that $A^T A$ can be transformed to diagonal form using V. Similarly, $AA^T = U\Sigma\Sigma^T U^T$, and AA^T can be transformed to diagonal form using U. It follows that the squares of the singular values of A are eigenvalues of both $A^T A$ and AA^T, and that complete systems of orthonormal eigenvectors are given by $v^1, v^2, \ldots, v^n$ and $u^1, u^2, \ldots, u^m$, respectively.

This remark suggests a way to give a constructive proof of the existence of a singular-value decomposition of an arbitrary matrix A.

Lemma. *Let $\lambda > 0$ be an eigenvalue of the matrix $A^T A$, $A \in \mathbb{R}^{(m,n)}$. Then λ is also an eigenvalue of AA^T with the same multiplicity.*

Proof. If $x \in \mathbb{R}^n$, $x \ne 0$, is an eigenvector of the matrix $A^T A$ corresponding to the eigenvalue $\lambda > 0$, then since $A^T Ax = \lambda x$, it follows that $Ax \ne 0$. Now $AA^T Ax = \lambda Ax$ implies that Ax is an eigenvector of the matrix AA^T corresponding to the eigenvalue λ.

Now let $v^1, v^2, \ldots, v^k \in \mathbb{R}^n$ be an orthonormal basis for the eigenspace of the matrix $A^T A$ corresponding to the eigenvalue λ. As we just showed, this means $Av^\nu \ne 0$ for all $\nu = 1, 2, \ldots, k$, and Av^ν is an eigenvector of the matrix AA^T corresponding to the eigenvalue λ. Moreover, $(Av^\nu)^T(Av^\kappa) = v^{\nu T} A^T Av^\kappa = \lambda v^{\nu T} v^\kappa = \lambda\delta_{\nu\kappa}$. Thus, the vectors Av^ν, $1 \le \nu \le k$, are orthogonal, and hence linearly independent. It follows that the dimension of the eigenspace of AA^T corresponding to the eigenvalue λ is at least as large as the dimension of the eigenspace of $A^T A$ corresponding to the same eigenvalue λ. Now by symmetry, the same chain of arguments is valid starting with AA^T, and we conclude that the dimensions of the two eigenspaces corresponding to the eigenvalue λ are the same. This means that the multiplicities of λ are also the same. □

For completeness, we establish the following well-known

Proposition. For an arbitrary $m \times n$ matrix,

$$\text{rank}\,(A)=\text{rank}\,(A^T)=\text{rank}\,(AA^T)=\text{rank}\,(A^TA).$$

Proof. One of the basic results of linear algebra is the fact that the row rank of a matrix is equal to its column rank. Moreover, the rank of a matrix $B \in \mathbb{R}^{(r,s)}$ is related to the dimension of its kernel by

$$\text{rank}(B) = s - \dim(\text{kernel } B),$$

see e.g. G. Strang [1976]. We apply this formula to the matrices $B := A^TA$ and $B := A$ in $\mathbb{R}^{(m,n)}$ to get

$$(*) \qquad \text{rank}(A) = \text{rank}(A^TA) + \dim(\text{kernel}(A^TA)) - \dim(\text{kernel}(A)).$$

But kernel$(A) \subset$ kernel(A^TA), and the condition that $A^TAx = 0$ implies that $x^TA^TAx = \|Ax\|_2^2 = 0$, and hence $Ax = 0$. It follows that kernel$(A^TA) \subset$ kernel(A). Combining these facts with $(*)$, we are led to rank$(A) =$ rank(A^TA). □

Since A^TA is a positive semidefinite matrix, there exist an orthogonal matrix $V \in \mathbb{R}^{(n,n)}$ and a diagonal matrix $L = (\lambda_\mu \delta_{\mu\nu}) \in \mathbb{R}^{(n,n)}$ with eigenvalues $\lambda_1 \geq \cdots \geq \lambda_n \geq 0$ such that

$$(**) \qquad A^TA = VLV^T.$$

Analogously, there exist an orthogonal matrix $U \in \mathbb{R}^{(m,m)}$ and a diagonal matrix $\tilde{L} = (\tilde{\lambda}_\mu \delta_{\mu\nu}) \in \mathbb{R}^{(m,m)}$, $\tilde{\lambda}_1 \geq \tilde{\lambda}_2 \geq \cdots \geq \tilde{\lambda}_m \geq 0$, with

$$AA^T = U\tilde{L}U^T.$$

Corollary. Let $r :=$ rank(A). Then $\lambda_\mu = \tilde{\lambda}_\mu$ for $\mu = 1, 2, \ldots, r$ and $\lambda_{r+1} = \cdots = \lambda_n = \tilde{\lambda}_{r+1} = \cdots = \tilde{\lambda}_m = 0$.

This fact follows immediately from the Lemma and the Proposition.

We now formulate these results as the following

Existence Theorem for a Singular-Value Decomposition. *Suppose $A \in \mathbb{R}^{(m,n)}$ with* rank$(A) = r$. *Also suppose $\lambda_1 \geq \lambda_2 \geq \cdots \geq \lambda_r > 0$ and $\lambda_{r+1} = \cdots = \lambda_n = 0$ are the eigenvalues of A^TA and that $v^1, v^2, \ldots, v^n$ is a corresponding orthonormal system of eigenvectors. Then $u^\nu := \frac{1}{\sigma_\nu} Av^\nu$ with $\sigma_\nu := +\sqrt{\lambda_\nu}$, $1 \leq \nu \leq r$ is an orthonormal system of eigenvectors for*

AA^T corresponding to the eigenvalues $\lambda_1, \lambda_2, \ldots, \lambda_r$, and this system can be extended to an orthonormal system $u^1, u^2, \ldots, u^m$ of eigenvectors of the matrix AA^T. Moreover, if we set $V = (v^1, v^2, \ldots, v^n)$, $U = (u^1, u^2, \ldots, u^m)$ and $\Sigma = (\sigma_\mu \delta_{\mu\nu}) \in \mathbb{R}^{(m,n)}$ with $\sigma_\mu := +\sqrt{\lambda_\mu}$ for $\mu = 1, 2, \ldots, r$ and $\sigma_{r+1} = \sigma_{r+2} = \cdots = \sigma_{\min(m,n)} = 0$, then A and A^T admit the singular-value decompositions

$$A = U\Sigma V^T \quad \text{and} \quad A^T = V\Sigma^T U^T,$$

with the r singular values $\sigma_1 \geq \sigma_2 \geq \cdots \geq \sigma_r > 0$.

Proof. Since $AA^T u^\nu = \frac{1}{\sigma_\nu} AA^T A v^\nu = \lambda_\nu \frac{1}{\sigma_\nu} A v^\nu$ and

$$u^{\mu T} u^\nu = \frac{1}{\sigma_\mu \sigma_\nu} (Av^\mu)^T (Av^\nu) = \frac{1}{\sigma_\mu \sigma_\nu} v^{\mu T} A^T A v^\nu = \frac{\sigma^\nu}{\sigma_\mu} v^{\mu T} v^\nu = \delta_{\mu\nu},$$

$u^1, \ldots, u^r$ are orthonormal eigenvectors corresponding to the eigenvalues $\lambda_1, \ldots, \lambda_r$ of the matrix AA^T. We can extend these to a complete system of orthonormal eigenvectors $u^1, u^2, \ldots, u^m$. Now by the definition of the vector u^ν, it follows that

$$Av^\nu = \sigma_\nu u^\nu, \quad 1 \leq \nu \leq r.$$

Moreover, in the proof of the Proposition it was shown that kernel(A) is the same as kernel($A^T A$), and thus

$$Av^\nu = 0, \quad r + 1 \leq \nu \leq n.$$

This is equivalent to the asserted singular-value decomposition . □

Remark. The diagonal matrix Σ in a singular-value decomposition is unique. Because of possible multiplicities of the eigenvalues of AA^T, this is not true for the transformation matrices U and V. If A is a symmetric $n \times n$ matrix, then its singular values are $\sigma_\mu = |\kappa_\mu|$, where κ_μ is the μ-th eigenvalue of A.

We now apply the singular-value decomposition to the problem of solving poorly-conditioned linear systems of equations.

6.2 Pseudo-Normal Solutions of Linear Systems of Equations. We now return to our original problem, the solution of a poorly-conditioned linear system of equations $Ax = b$. Instead of solving this system of equations, it turns out to be useful to replace it by the following

Minimization Problem. Let $A \in \mathbb{R}^{(m,n)}$ and $b \in \mathbb{R}^m$. Find a vector $\tilde{x} \in \mathbb{R}^n$ such that

$$\|A\tilde{x} - b\|_2 = \inf_{x \in \mathbb{R}^n} \|Ax - b\|_2.$$

In this formulation, we have extended the original problem with $m = n$ to the cases $m > n$ (an over-determined system) and $m < n$ (an under-determined system of equations). In the theorem below we will show that this minimization problem is always solvable. The singular-value decomposition $A = U\Sigma V^T$ provides a means of finding all solutions $\tilde{x}$ explicitly. Using the fact that U is an orthogonal matrix, and setting $z := V^T x$ and $d := U^T b$, we get

$$\|Ax - b\|_2^2 = \|U^T(Ax - b)\|_2^2 = \|\Sigma V^T x - U^T b\|_2^2 = \|\Sigma z - d\|_2^2.$$

We can now read off the solutions of the minimization problem:

$$\tilde{z}_\mu = \frac{1}{\sigma_\mu} d_\mu \quad \text{for } \mu = 1, 2, \ldots, r \text{ and}$$
$$\tilde{z}_\mu \in \mathbb{R} \quad \text{for } \mu = r+1, \ldots, n.$$

Every solution $\tilde{x}$ of the minimization problem can thus be represented in the form

$$\tilde{x} = \sum_{\mu=1}^{r} \frac{1}{\sigma_\mu} d_\mu v^\mu + \sum_{\mu=r+1}^{n} \tilde{z}_\mu v^\mu.$$

By construction, the last $n - r$ columns of the matrix V span the kernel of the mapping $A^T A$. In addition, we have kernel$(A^T A)$ = kernel(A), a fact which we have already used several times (cf. the proof of Proposition 6.1). Now the solution set L of the minimization problem can be expressed as

$$(*) \qquad \mathrm{L} = \overline{x} + \mathrm{kernel}(A) \quad \text{with } \overline{x} := \sum_{\mu=1}^{r} \frac{1}{\sigma_\mu} d_\mu v^\mu.$$

The set L is in general not a singleton, and hence it makes sense to look for some particular solution. This leads us to the

Definition. A vector $x^+ \in \mathbb{R}^n$ is called a *pseudo-normal solution* of the minimization problem, and of the corresponding linear system of equations $Ax = b$, provided that $\|x^+\|_2 \leq \|x\|_2$ for all $x \in \mathrm{L}$.

Proposition. The vector $\overline{x} := \sum_{\mu=1}^{r} \frac{1}{\sigma_\mu} d_\mu v^\mu$ is a pseudo-normal solution of the minimization problem.

Proof. From the formula $(*)$ and the orthonormality of the vectors v^μ, it follows that for every vector $\tilde{x} = \overline{x} + \sum_{\mu=r+1}^{n} z_\mu v^\mu \in \mathrm{L}$, we have

$$\|\tilde{x}\|_2^2 = \|\overline{x} + \sum_{\mu=r+1}^{n} z_\mu v^\mu\|_2^2 = \|\overline{x}\|_2^2 + \sum_{\mu=r+1}^{n} |z_\mu|^2 \cdot \|v^\mu\|_2^2 \geq \|\overline{x}\|_2^2. \qquad \square$$

We have now established the existence of a pseudo-normal solution of the form $x^+ = \sum_{\mu=1}^{r} \frac{1}{\sigma_\mu} d_\mu v^\mu$. We now prove the

Theorem on Uniqueness and Characterization of Pseudo-Normal Solutions. *There exists exactly one pseudo-normal solution* x^+ *of the minimization problem. It is characterized by* $x^+ \in \mathrm{L} \cap (\mathrm{kernel}(A))^\perp$, *where* $(\mathrm{kernel}(A))^\perp$ *is the orthogonal complement of* $\mathrm{kernel}(A)$ *in* $\mathbb{R}^n$.

Proof. The existence of $x^+ = \sum_{\mu=1}^r \frac{1}{\sigma_\mu} d_\mu v^\mu$ and its uniqueness follow from the inequality established in the proof of the Proposition. Because of the orthogonality of the vectors v^μ, it follows that $x^+ \in (\mathrm{kernel}(A))^\perp$. □

The pseudo-normal solution x^+ of the minimization problem is the solution with minimal Euclidean norm. If the system of equations $Ax = b$, $A \in \mathbb{R}^{(n,n)}$, has a unique solution, then x^+ reduces to $A^{-1}b$. Hence, for general $A \in \mathbb{R}^{(m,n)}$, the concept of the pseudo-normal solution provides a way of defining a generalized inverse for A.

6.3 The Pseudo-Inverse of a Matrix. Given a matrix $A \in \mathbb{R}^{(m,n)}$, by Theorem 6.2 on the uniqueness and characterization of pseudo-normal solutions, for every vector $b \in \mathbb{R}^m$, there is precisely one corresponding vector $x^+ \in \mathbb{R}^n$ with the property that it solves the Minimization Problem 6.2, and has the smallest Euclidean norm among all solutions. It follows from the construction of $x^+ = \sum_{\mu=1}^r \frac{1}{\sigma_\mu} d_\mu v^\mu = \sum_{\mu=1}^r \frac{1}{\sigma_\mu}(U^T b)_\mu v^\mu$ that the mapping $b \to x^+$ is linear, and thus there must be some matrix $A^+ \in \mathbb{R}^{(n,m)}$ with $A^+ b = x^+$.

Definition. The uniquely defined matrix $A^+ \in \mathbb{R}^{(n,m)}$ with $A^+ b = x^+$ is called the *pseudo-inverse* or *Moore-Penrose inverse* of the matrix A in $\mathbb{R}^{(m,n)}$.

The idea of the pseudo-inverse was first introduced in 1903 by I. FREDHOLM in connection with integral equations. For matrices, the definition is due to E. H. MOORE, who presented the concept of an inverse for a general $m \times n$ matrix in a lecture at a meeting of the American Mathematical Society in 1920. His idea was essentially forgotten for many years. In 1955, R. PENROSE independently discovered how to define generalized inverses for arbitrary matrices. Since then, the subject has enjoyed an explosive development. The Moore-Penrose inverse of linear operators has applications in functional analysis, numerical analysis, and mathematical statistics. For a survey of the current state-of-the-art, see e.g. A. Ben-Israel and T. N. E. Greville [1974].

Frequently, the pseudo-inverse of a matrix is defined axiomatically by certain properties. We have developed it in a constructive way, so we now pause to establish these properties.

Theorem. *Suppose* $A \in \mathbb{R}^{(m,n)}$. *Then:*

(i) There exists exactly one matrix $B \in \mathbb{R}^{(n,m)}$ *with*

$$AB = (AB)^T, \; BA = (BA)^T, \; ABA = A, \; BAB = B.$$

(ii) The matrix B *is the pseudo-inverse* A^+, *and* A^+A *is the orthogonal projection of* $\mathbb{R}^n$ *onto* $(\mathrm{kernel}(A))^\perp$; AA^+ *is the orthogonal projection of* $\mathbb{R}^m$ *onto the image of* A.

Proof. We first prove (i). Let $A = U\Sigma V^T$ be the singular-value decomposition of A. Let $B := V\tilde{\Sigma}U^T$ with $\tilde{\Sigma} := (\tau_\mu \cdot \delta_{\mu\nu}) \in \mathbb{R}^{(n,m)}$, and let

$$\tau_\mu := \begin{cases} \sigma_\mu^{-1} & \text{if } \sigma_\mu \neq 0 \\ 0 & \text{if } \sigma_\mu = 0 \end{cases}.$$

Then the matrix product $\Sigma \cdot \tilde{\Sigma}$ has the form

$$\Sigma \cdot \tilde{\Sigma} = \begin{pmatrix} 1 & & & & & \\ & \ddots & & & 0 & \\ & & 1 & & & \\ & & & 0 & & \\ & 0 & & & \ddots & \\ & & & & & 0 \end{pmatrix}.$$

It follows immediately that

$$AB = U\Sigma V^T V\tilde{\Sigma}U^T = U\Sigma\tilde{\Sigma}U^T = (U\Sigma\tilde{\Sigma}U^T)^T = (AB)^T.$$

Analogously, $BA = (BA)^T$, and we have $ABA = U\Sigma V^T V\tilde{\Sigma}U^T U\Sigma V^T = U\Sigma V^T = A$. The proof of the identity $BAB = B$ is similar.

To prove the uniqueness of the matrix B, we assume that there exists another matrix C with the same properties, and show that this leads to a contradiction. If C exists, then

$$\begin{aligned} B = BAB &= BB^T A^T C^T A^T = BB^T A^T AC = \\ &= BAA^T C^T C = A^T C^T C = CAC = C \end{aligned}$$

(ii) Now let $b \in \mathbb{R}^m$. From $Bb = V\tilde{\Sigma}U^T b = \sum_{\mu=1}^r \frac{1}{\sigma_\mu}(U^T b)_\mu v^\mu = A^+ b$ we deduce that the matrix B coincides with A^+, and thus

$$A^+ = V\tilde{\Sigma}U^T.$$

After a short calculation, we are led to the identity $\tilde{\Sigma} = \Sigma^+$, and so A^+ can be written as

$$A^+ = V\Sigma^+ U^T.$$

It remains to show that $P := A^+A$ and $\overline{P} = AA^+$ are orthogonal projections onto $(\text{kernel}(A))^\perp$ and $\text{image}(A)$, respectively. From (i) it follows that $P^T = P$ and $P^2 = (A^+AA^+)A = A^+A = P$ and $\overline{P}^T = \overline{P}$ and $\overline{P}^2 = A(A^+AA^+) = AA^+ = \overline{P}$. Thus, P and $\overline{P}$ are orthogonal projections.

Since P is an orthogonal projection, $\text{image}(P) = \text{kernel}(P))^\perp$ (see e.g. W. H. Greub ([1967], p. 214). Also $\text{kernel}(A^+A) \supset \text{kernel}(A)$, and conversely, since $AA^+A = A$, $\text{kernel}(A) = \text{kernel}(AA^+A) \supset \text{kernel}(A^+A)$. This implies $\text{image}(A^+A) = (\text{kernel}(A^+A))^\perp = (\text{kernel}(A))^\perp$. Similarly, we have image$(AA^+) \subset$ image(A) and image$(A) =$ image$(AA^+A) \subset$ image(AA^+). From this it follows that image$(AA^+) =$ image(A). □

Corollary. $(A^+)^+ = A$ and $(A^+)^T = (A^T)^+$.

Proof. In the proof of the previous theorem we showed that $A^+ = U\Sigma^+V^T$. Now since $(\Sigma^+)^+ = \Sigma$ and $(\Sigma^+)^T = (\Sigma^T)^+$, the assertion immediately follows. □

In this respect, the pseudo-inverse A^+ of a matrix $A \in \mathbb{R}^{(m,n)}$ has the same properties as the usual inverse A^{-1} of a nonsingular matrix $A \in \mathbb{R}^{(n,n)}$. On the other hand, we have the following

Difference. For general $A \in \mathbb{R}^{(m,n)}$, $B \in \mathbb{R}^{(n,p)}$, we have $(AB)^+ \neq B^+A^+$.

Example. Given $A = B = \begin{pmatrix} 1 & 1 \\ 0 & 0 \end{pmatrix}$, find A^+. The eigenvalues of the matrix A^TA are $\lambda_1 = 2$ and $\lambda_2 = 0$, and so its singular value is $\sigma_1 = \sqrt{2}$. An orthonormal system of eigenvectors for the matrix A^TA is given by $v^1 = \frac{\sqrt{2}}{2}(1,1)^T$, $v_2 = \frac{\sqrt{2}}{2}(1,-1)^T$. This gives $u^1 = \frac{1}{\sqrt{2}}\begin{pmatrix} 1 & 1 \\ 0 & 0 \end{pmatrix}\begin{pmatrix} \frac{\sqrt{2}}{2} \\ \frac{\sqrt{2}}{2} \end{pmatrix} = \begin{pmatrix} 1 \\ 0 \end{pmatrix}$. For u^2 we choose $u^2 = (0,1)^T$. It follows that A has the singular-value decomposition

$$A = \begin{pmatrix} 1 & 0 \\ 0 & 1 \end{pmatrix}\begin{pmatrix} \sqrt{2} & 0 \\ 0 & 0 \end{pmatrix}\begin{pmatrix} \frac{\sqrt{2}}{2} & \frac{\sqrt{2}}{2} \\ \frac{\sqrt{2}}{2} & -\frac{\sqrt{2}}{2} \end{pmatrix},$$

and the formula $A^+ = V\Sigma^+U^T$ becomes

$$A^+ = \frac{\sqrt{2}}{2}\begin{pmatrix} 1 & 1 \\ 1 & -1 \end{pmatrix}\begin{pmatrix} \frac{1}{\sqrt{2}} & 0 \\ 0 & 0 \end{pmatrix}\begin{pmatrix} 1 & 0 \\ 0 & 1 \end{pmatrix} = \frac{1}{2}\begin{pmatrix} 1 & 0 \\ 1 & 0 \end{pmatrix}.$$

Thus, $(A^+)^2 = \frac{1}{4}\begin{pmatrix} 1 & 0 \\ 1 & 0 \end{pmatrix}$. On the other hand, $A^2 = A$, and thus $(A^2)^+ = = A^+ = \frac{1}{2}\begin{pmatrix} 1 & 0 \\ 1 & 0 \end{pmatrix}$. This shows that in this case, $(AB)^+ \neq B^+A^+$.

In the next section we show how the concepts of singular-value decomposition and pseudo-inverse can be used to describe the condition of a general matrix $A \in \mathbb{R}^{(m,n)}$.

6.4 More on Linear Systems of Equations. We now return to the problem of solving a linear system of equations of the form $Ax = b$, where $A \in \mathbb{R}^{(m,n)}$ and $b \in \mathbb{R}^m$. The pseudo-normal solution of this system is given by $x^+ = A^+b$. We now assume that the right-hand side of the linear system has been perturbed by a vector $\Delta b \in \mathbb{R}^m$, giving us the system $A(x^+ + \Delta x) = b + \Delta b$ to solve. Then $x^+ + \Delta x = A^+(b + \Delta b)$, and hence the error is $\Delta x = A^+\Delta b$. Now

$$A^+(A^+)^T = V\Sigma^+U^TU(\Sigma^+)^TV^T = V\operatorname{diag}(\sigma_1^{-2},\ldots,\sigma_r^{-2},0,\ldots,0)V^T.$$

This equation implies that the spectral radius of $A^+(A^+)^T$ is given by $\rho(A^+(A^+)^T) = \sigma_r^{-2}$. By Theorem 4.3(3), it follows that $\|A^+\|_2 = \sigma_r^{-1}$, and so the error Δx satisfies

$$\|\Delta x\|_2 \leq \|A^+\|_2 \|\Delta b\|_2 = \sigma_r^{-1} \|\Delta b\|_2.$$

In addition, the pseudo-normal solution x^+ satisfies the inequality

$$\|x^+\|_2^2 = \sum_{\mu=1}^{r} \sigma_\mu^{-2} d_\mu^2 \geq \sigma_1^{-2} \sum_{\mu=1}^{r} d_\mu^2 = \sigma_1^{-2} \| \sum_{\mu=1}^{r} d_\mu v^\mu \|_2^2.$$

We recall that by the definition of d (cf. 6.2), $\sum_{\mu=1}^{r} d_\mu v^\mu$ is the projection of b onto image(A). For the relative error we therefore have

$$(*) \qquad \frac{\|\Delta x\|_2}{\|x^+\|_2} \leq \frac{\sigma_1}{\sigma_r} \frac{\|\Delta b\|_2}{\|P_{image\,(A)} b\|_2},$$

where $P_{image\,(A)}$ denotes the projection onto image(A). The error bound $(*)$ suggests the following

Definition. Suppose $A \in \mathbb{R}^{(m,n)}$ is a matrix with singular-value decomposition $A = U\Sigma V^T$. Then $\text{cond}_2(A) := \frac{\sigma_1}{\sigma_r}$ is called the *condition* of A.

In 5.1 we introduced the condition of a nonsingular $n \times n$ matrix as $\text{cond}(A) = \|A^{-1}\| \cdot \|A\|$. The definition above gives the same number if A is nonsingular, since $\|A\|_2 = (\rho(A^T A))^{1/2} = \sigma_1$ and $\|A^{-1}\|_2 = \|A^+\|_2 = \sigma_r^{-1}$. Thus, it is a natural extension of the concept of condition of a nonsingular matrix.

Remark. The problem of minimizing the expression $f(x) := \frac{1}{2}\|Ax - b\|_2^2$ over $x \in \mathbb{R}^n$ can also be solved by looking for a solution of the necessary conditions $\frac{\partial}{\partial x_\mu} f(x) = 0$, $1 \leq \mu \leq n$. The leads to the linear system of equations $A^T A x = A^T b$, the so-called *normal equations* (cf. 4.6.1). Since $\text{cond}_2(A^T A) = \text{cond}_2(A^2)$, the normal equations are in general more poorly-conditioned than the minimization problem.

6.5 Improving the Condition and Regularization of a Linear System of Equations. The above Definition 6.4 of the condition of a matrix $A \in \mathbb{R}^{(m,n)}$ suggests how to construct approximation problems related to $\|Ax - b\|_2 \stackrel{!}{=} \min$ which are better conditioned. We proceed as follows: Determine a singular-value decomposition $A = U\Sigma V^T$ of A, and set

$$(*) \qquad \Sigma_\tau^+ := (\eta_\mu \delta_{\mu\nu}), \quad \eta_\mu := \begin{cases} \sigma_\mu^{-1} & \text{if } \sigma_\mu \geq \tau \\ 0 & \text{otherwise} \end{cases}.$$

Here $\tau > 0$ is a parameter to be chosen appropriately. The passage from Σ^+ to Σ_τ^+ defined by $(*)$ involves eliminating the small singular values σ_μ. Now instead of taking the pseudo-normal solution $x^+ = A^+ b$, we consider the approximation $x_\tau^+ = A_\tau^+ b$, where $A_\tau^+ := V\Sigma_\tau^+ U^T$. By Definition 6.4, this approximation problem is better conditioned than the original. The matrix A_τ^+ is called the *effective pseudo-inverse* of A.

Remark. By the properties of the pseudo-inverse $B = A^+$ given in Theorem 6.3, it follows that A_τ^+ satisfies $A_\tau^+ A = (A_\tau^+ A)^T$, $AA_\tau^+ = (AA_\tau^+)^T$, and $A_\tau^+ AA_\tau^+ = A_\tau^+$. On the other hand,

$$\|AA_\tau^+ A - A\|_2 = \|U\Sigma V^T V\Sigma_\tau^+ U^T U\Sigma V^T - U\Sigma V^T\|_2 = \|U(\Sigma_\tau - \Sigma)V^T\|_2 \le \tau$$

with $\Sigma_\tau = (\tilde{\eta}_\mu \delta_{\mu\nu})$, $\tilde{\eta}_\mu := \begin{cases} \sigma_\mu & \text{if } \sigma_\mu \ge \tau \\ 0 & \text{otherwise} \end{cases}$.

The removal of small singular values is referred to as a *regularization* of the problem. This process improves the condition, but at the cost of some accuracy; i.e., some method error is introduced. There are several possibilities for regularizing an ill-conditioned problem. The best known method is due to A. N. Tichonov [1963]. It corresponds to dampening the influence of small singular values.

ANDREI NIKOLAIEVICH TICHONOV (born 1906) is Professor of Mathematics and Geophysics at the Moscow State University, and is a corresponding member of the Academy of Sciences of the U.S.S.R. He has made important contributions in topology, mathematical physics, and geophysics. One of his best-known theorems is in general topology: "The topological product of an arbitrary number of compact spaces is compact." In 1966 he received the Lenin prize for his papers on regularization of ill-posed problems. The theory and practice of ill-posed problems is discussed in detail in the book of B. Hofmann [1986].

To describe the principle of Tichonov regularization, we consider the linear system of equations $Ax = b$, and assume that the true right-hand side b is unknown. The problem is to solve $Ax = \tilde{b}$ for a modified right-hand side $\tilde{b}$, where it is known that $\tilde{b}$ lies in a δ-neighborhood of b, i.e., $\|b - \tilde{b}\|_2 \le \delta$. We may assume $\|\tilde{b}\|_2 > \delta$, since otherwise $b = 0$ is an admissable right-hand side, and the zero vector $x = 0$ would be a reasonable solution. We now replace the original problem by the following

Minimization Problem with a Constraint. Suppose $A \in \mathbb{R}^{(m,n)}$ and $\tilde{b} \in \mathbb{R}^m$. Let $\mathrm{M} := \{x \in \mathbb{R}^n \,|\, \|Ax - \tilde{b}\|_2 \le \delta\}$. Find a vector $\tilde{x} \in \mathbb{R}^n$ such that

$$\|\tilde{x}\|_2 = \inf\{\|x\|_2 \;\big|\; x \in \mathrm{M}\}.$$

Remark. Since $\|Ax - \tilde{b}\|_2 \le \delta$ for all $x \in \mathrm{M}$, where M is compact and strictly convex, it follows that this minimization problem has a unique solution $\tilde{x}$ (cf. also Chap. 4, Section 3). Moreover, the vector $\tilde{x}$ lies on the boundary of the constraint set; i.e., $\|A\tilde{x} - \tilde{b}\|_2 = \delta$. Indeed, if $\tilde{\delta} := \|A\tilde{x} - \tilde{b}\|_2 < \delta$, then it would follow that the vector $x_\kappa := (1-\kappa)\tilde{x}$ satisfies

$$\|Ax_\kappa - \tilde{b}\|_2 = \|A\tilde{x} - \tilde{b} - \kappa A\tilde{x}\|_2 \le \|A\tilde{x} - \tilde{b}\|_2 + \kappa\|A\|_2\|\tilde{x}\|_2 \le \delta,$$

where $\kappa := \min\{1, \frac{\delta-\tilde{\delta}}{\|A\|_2\,\|\tilde{x}\|_2}\}$ and $\|x_\kappa\|_2 = (1-\kappa)\|\tilde{x}\|_2 < \|\tilde{x}\|_2$. This contradicts the minimal property of $\tilde{x}$.

In view of the above, we can replace the minimization problem with an inequality constraint by the following equivalent

Minimization Problem with Equality Constraint. Find a vector $\tilde{x}$ in $\mathbb{R}^n$ such that

$$\|\tilde{x}\|_2 = \inf\{\|x\|_2 \mid \|Ax - \tilde{b}\|_2 = \delta\}.$$

From analysis it is known that this kind of problem can be solved using the Lagrange function

$$L(x, \lambda) := \|x\|_2^2 + \lambda(\|Ax - \tilde{b}\|_2^2 - \delta^2).$$

The number $\lambda \in \mathbb{R}_+$ is a Lagrange multiplier. A necessary condition for a solution of this minimization problem is that

$$\frac{1}{2}\text{grad}_x L(x, \lambda) = x + \lambda A^T(Ax - \tilde{b}) = 0,$$
$$\|Ax - \tilde{b}\|_2 = \delta.$$

We now set $\alpha := \lambda^{-1}$, and rewrite the linear system of equations as

$$A^T Ax + \alpha Ix = A^T\tilde{b}.$$

These equations are necessary (and also sufficient) for x to minimize

$$\|Ax - \tilde{b}\|_2^2 + \alpha\|x\|_2^2.$$

This formulation is called the *Tichonov regularization* of the system of equations $Ax = \tilde{b}$. The number $\alpha > 0$ is called the *regularization parameter.*

Connection with Singular Values. If we define $\overline{A} := \binom{A}{\alpha^{1/2}I}$ and $\overline{b} := \binom{\tilde{b}}{0}$, then the Tichonov regularization can also be expressed as the problem of minimizing

$$\|\overline{A}x - \overline{b}\|_2^2.$$

This problem can be solved using the singular-value decomposition of $\overline{A}$. If σ_μ are the singular values of A, then since $\overline{A}^T\overline{A} = A^TA + \alpha I$, the singular values of $\overline{A}$ are $\sqrt{\sigma_\mu^2 + \alpha}$. Thus, the condition of the Tichonov regularization is given by $\sqrt{(\sigma_1^2 + \alpha)(\sigma_r^2 + \alpha)^{-1}}$. The Tichonov regularization generally improves the condition of a problem. Indeed, the singular values are increased by a positive amount depending on the regularization parameter α. The determination of an optimal value for the regularization parameter α is generally not easy, however.

For the purposes of comparison, we consider again the problem of solving the system of equations $Ax = b$, where A is the Hilbert matrix discussed at the beginning of this section. The following table shows that both Tichonov regularization and the use of the singular-value decomposition with removal of small singular values produce much better results.

Method	Relative error ($n = 8$)	Relative error ($n = 10$)
Tichonov-Cholesky	$5.59 \cdot 10^{-3}$ ($\alpha = 4 \cdot 10^{-8}$)	0.0115 ($\alpha = 10^{-7}$)
Tichonov-Householder	$4.78 \cdot 10^{-5}$ ($\alpha = 6 \cdot 10^{-15}$)	$3.83 \cdot 10^{-4}$ ($\alpha = 6 \cdot 10^{-13}$)
Singular-value decomposition	$2 \cdot 10^{-4}$ ($\tau = 10^{-8}$)	$3.81 \cdot 10^{-4}$ ($\tau = 10^{-8}$)

6.6 Problems. 1) Compute a singular-value decomposition for the matrix

$$A = \begin{pmatrix} 1 & 1 \\ \sqrt{2} & 0 \\ 0 & \sqrt{2} \end{pmatrix}.$$

2) Let $A = (a_{11} a_{12}) \in \mathbb{R}^{(1,2)}$. Show that $A^+ = (a_{11}^2 + a_{12}^2)^{-1} \binom{a_{11}}{a_{12}}$.

3) (i) Let $A \in \mathbb{R}^{(m,n)}$. Show:

$$A^+ = (A^T A)^+ A^T = A^T (AA^T)^+.$$

(ii) A matrix $A \in \mathbb{R}^{(n,n)}$ is called *normal* provided that $AA^T = A^T A$. Show that for a normal matrix A, the pseudo-inverse A^+ is also normal.

(iii) Show: If A is a normal matrix, then $(A^2)^+ = (A^+)^2$.

4) Suppose $A \in \mathbb{R}^{(m,n)}$ is such that $\text{cond}_2(A) = \frac{\sigma_1}{\sigma_r}$ as in Definition 6.4. Show:

$$\text{cond}_2(A) = \|A\|_2 \cdot \|A^+\|_2.$$

5) Let $x_\alpha^\delta \in \mathbb{R}^n$ be a solution of the following Tichonov regularization problem: Minimize

$$\|Ax - \tilde{b}\|_2^2 + \alpha \|x\|_2^2.$$

Let $D(\alpha; \tilde{b}) := \|Ax_\alpha^\delta - \tilde{b}\|_2$ be the *residual* for the approximate solution x_α^δ. Show: If $\|b - \tilde{b}\|_2 \le \delta < \|\tilde{b}\|_2$, then the mapping $\alpha \to D(\alpha; \tilde{b})$ is continuous, strictly monotone increasing, and $\delta \in$ image $(D(\cdot; \tilde{b}))$.

6) Why is $\alpha_\delta > 0$ with $\delta = D(\alpha_\delta; \tilde{b})$ a convenient regularization parameter? (This way of choosing α is called the *residual method*).

3
Eigenvalues

In Chapter 2 we have seen that in order to compute a singular-value decomposition of a matrix A, we need to have the eigenvalues of $A^T A$. This process was illustrated in Example 2.6.3, where because of the small size of the problem, we were able to find the necessary eigenvalues by hand calculation. For larger problems, however, this is no longer possible, and we need to use a computer to find eigenvalues. Such problems arise, for example, in the study of oscillations, where the eigenfrequences are to be determined by discretizing the associated differential equation. In this chapter we discuss various methods for computing eigenvalues of matrices.

Let $A \in \mathbb{C}^{(n,n)}$ be an arbitrary square matrix. We are interested in the following

Eigenvalue Problem. Find a number $\lambda \in \mathbb{C}$ and a vector $x \in \mathbb{C}^n$, $x \neq 0$, such that the *eigenvalue equation*

$$Ax = \lambda x$$

is satisfied.

The number λ is called an *eigenvalue*, and the vector x an *eigenvector* of the matrix A corresponding to the eigenvalue λ. Eigenvalues and eigenvectors are described in detail in any book on linear algebra (see e.g. G. Strang [1976]), and so we do not present the theory beyond what is needed to formulate and understand algorithms.

Let $\lambda \in \mathbb{C}$ be an eigenvalue of the matrix A. Then it is well known that the space $\mathrm{E}(\lambda) := \{x \in \mathbb{C}^n \mid Ax = \lambda x\}$ is a linear subspace of $\mathbb{C}^n$. It is called the *eigenspace* corresponding to the eigenvalue λ. By the dimension formula for homomorphisms, its dimension is

$$d(\lambda) = n - \text{rank } (A - \lambda I).$$

Thus, $\lambda \in \mathbb{C}$ is an eigenvalue of A precisely when $d(\lambda) > 0$. The number $d(\lambda)$ is called the *geometric multiplicity* of the eigenvalue λ. On the other hand, $d(\lambda) > 0$ is also equivalent to the condition that the matrix $(A - \lambda I)$

be singular. This means that λ is an eigenvalue of A if and only if it is a zero of the *characteristic polynomial*

$$p(\lambda) := \det(A - \lambda I).$$

If λ is a zero of the characteristic polynomial of multiplicity $v(\lambda)$, then we say that the eigenvalue λ has ***algebraic multiplicity*** $v(\lambda)$. It is easy to check that

$$1 \le d(\lambda) \le v(\lambda) \le n.$$

If for each eigenvalue of the matrix $A \in \mathbb{C}^{(n,n)}$, its geometric and algebraic multiplicities are the same, then the corresponding eigenvectors form a basis for $\mathbb{C}^n$. In this case we say that A has a *complete system of eigenvectors.* Matrices which possess a complete system of eigenvectors are *diagonalizable.* A diagonalizable matrix A can be transformed into a diagonal matrix $T^{-1}AT$ whose diagonal elements are the eigenvalues of A, where the columns of the transformation matrix T are the eigenvectors of A.

Because it guarantees the existence of an expansion of an arbitrary vector in $\mathbb{C}^n$ in terms of the eigenvectors of A, the diagonalizability of a matrix A is clearly an important property as regards numerical methods for computing eigenvalues. The class of diagonalizable matrices includes the *normal matrices* which are characterized by the property $A\overline{A}^T = \overline{A}^T A$. Thus all Hermitian matrices are diagonalizable. This is a useful observation since it is easy to check whether a given matrix is normal or even Hermitian.

For the numerical computation of the eigenvalues of a matrix, it is generally not advisable to try to find the zeros of the characteristic polynomial. This is because the coefficients of p are in general only given approximately, whereas the zeros of p, especially the multiple zeros, are very sensitive to the values of the coefficients so that very imprecise results will be obtained. For more on this phenomenon, see the book of H. R. Schwarz [1989]. The methods discussed in the remainder of this chapter make no use of the characteristic polynomial.

1. Reduction to Tridiagonal or Hessenberg Form

Given a matrix $A \in \mathbb{C}^{(n,n)}$, we want to find $\lambda \in \mathbb{C}$ and $x \in \mathbb{C}^n$, $x \neq 0$ solving the eigenvalue equation $Ax = \lambda x$. Our approach will be to apply nonsingular transformations to the eigenvalue equation with the aim of simplifying the problem. Let $T \in \mathbb{C}^{(n,n)}$ be a nonsingular matrix. Setting $y := T^{-1}x$, we have

$$T^{-1}ATy = T^{-1}Ax = \lambda T^{-1}x = \lambda y,$$

which asserts that $\lambda \in \mathbb{C}$ is also an eigenvalue of the transformed matrix $T^{-1}AT$ with associated eigenvector $y = T^{-1}x$. The methods in the following sections are based on the idea of applying a finite sequence of such similarity transformations to the matrix A in order to produce a matrix B whose eigenvalues are simple to compute.

1.1 The Householder Method. The method of Householder (see also 2.3.2) is based on using similarity transformations $A_\mu := T_\mu^{-1} A_{\mu-1} T_\mu$, where $T_\mu = T_\mu^{-1} := I - \beta_\mu u^\mu (u^\mu)^T$ are orthogonal Householder matrices. We restrict our discussion to the case of symmetric matrices $A \in \mathbb{R}^{(n,n)}$. The analysis is similar for Hermitian matrices $A \in \mathbb{C}^{(n,n)}$; see J. Stoer and R. Bulirsch [1983].

The QR decomposition of a matrix A transforms A into an upper triangular matrix R by applying $(n-1)$ Householder transformations. This amounts to multiplying A on the left by $Q := T_{n-1} \cdot T_{n-2} \cdots T_1$. We now consider the similarity transformation where A is multiplied on both the left and right by Q. In general, for an arbitrary symmetric matrix, we cannot expect that this will lead to a diagonal matrix. We now show, however, that the transformed matrix is always tridiagonal.

We begin with $A_0 := (a_{\mu\nu}^{(0)}) = A$ and $T_0 = I$. Suppose that after the $(\kappa-1)$-st step, we have a matrix $A_{\kappa-1} := (a_{\mu\nu}^{(\kappa-1)})$ of the form

$$A_{\kappa-1} = \begin{pmatrix} D_{\kappa-1} & c & 0 \\ c^T & \delta_\kappa & a_\kappa^T \\ 0 & a_\kappa & \tilde{A}_{\kappa-1} \end{pmatrix},$$

where

$$\begin{pmatrix} D_{\kappa-1} & c \\ c^T & \delta_\kappa \end{pmatrix} = \begin{pmatrix} \delta_1 & \gamma_2 & & & 0 \\ \gamma_2 & \delta_2 & \ddots & 0 & \vdots \\ & \ddots & \ddots & \gamma_{\kappa-1} & 0 \\ & 0 & \gamma_{\kappa-1} & \delta_{\kappa-1} & \gamma_\kappa \\ 0 & \cdots & 0 & \gamma_\kappa & \delta_\kappa \end{pmatrix} \quad \text{and} \quad a_\kappa = \begin{pmatrix} a_{\kappa+1\,\kappa} \\ a_{\kappa+2\,\kappa} \\ \vdots \\ a_{n\kappa} \end{pmatrix}.$$

By 2.1.3 there exists an $(n-\kappa)\times(n-\kappa)$ Householder matrix $\tilde{T}_\kappa$ with

$$\tilde{T}_\kappa a_\kappa = \sigma\, e^1, \quad \sigma \in \mathbb{R}.$$

By 2.3.2(i)–(iii), the matrix $\tilde{T}_\kappa$ has the form $\tilde{T}_\kappa = I - \beta u u^T$ with

(i) $\beta = (\|a_\kappa\|_2(|a_{\kappa+1\kappa}| + \|a_\kappa\|_2))^{-1}$,
(ii) $u := (\operatorname{sgn}(a_{\kappa+1\kappa})(|a_{\kappa+1\kappa}| + \|a_\kappa\|_2), a_{\kappa+2\kappa}, \ldots, a_{n\kappa})^T$.

We now use the orthogonal matrix

$$T_\kappa := \begin{pmatrix} I_{\kappa-1} & & 0 \\ & 1 & \\ 0 & & \tilde{T}_\kappa \end{pmatrix}$$

to perform a similarity transformation. The result is

$$A_\kappa := T_\kappa^{-1} A_{\kappa-1} T_\kappa = T_\kappa A_{\kappa-1} T_\kappa = \begin{pmatrix} D_{\kappa-1} & c & 0 \\ c^T & \delta_\kappa & \sigma(e^1)^T \\ 0 & \sigma e^1 & \tilde{T}_\kappa \tilde{A}_{\kappa-1} \tilde{T}_\kappa \end{pmatrix}.$$

Writing $\gamma_{\kappa+1} := \sigma = -\operatorname{sgn}(a_{\kappa+1\kappa})\|a_\kappa\|_2$, where sgn(0):=1, we see that A_κ has the form

$$A_\kappa = \begin{pmatrix} \delta_1 & \gamma_2 & & & & & \\ \gamma_2 & \delta_2 & \ddots & & & 0 & \\ & \ddots & \ddots & \gamma_{\kappa-1} & & & \\ & & \gamma_{\kappa-1} & \delta_{\kappa-1} & \gamma_\kappa & & \\ & 0 & & \gamma_\kappa & \delta_\kappa & \gamma_{\kappa+1} & \\ & & & & \gamma_{\kappa+1} & & \\ & & & & & & \tilde{T}_\kappa \tilde{A}_{\kappa-1} \tilde{T}_\kappa \end{pmatrix}.$$

It remains to find a convenient way of computing

$$\begin{aligned} \tilde{T}_\kappa \tilde{A}_{\kappa-1} \tilde{T}_\kappa &= (I - \beta u u^T) \tilde{A}_{\kappa-1} (I - \beta u u^T) = \\ &= \tilde{A}_{\kappa-1} - \beta \tilde{A}_{\kappa-1} u u^T - \beta u u^T \tilde{A}_{\kappa-1} + \beta u u^T \tilde{A}_{\kappa-1} u u^T. \end{aligned}$$

Introducing the vectors

$$p := \beta \tilde{A}_{\kappa-1} u, \quad q := p - \frac{\beta}{2}(p^T u) u$$

in $\mathbb{R}^{n-\kappa}$, we get

$$\begin{aligned} \tilde{T}_\kappa \tilde{A}_{\kappa-1} \tilde{T}_\kappa &= \tilde{A}_{\kappa-1} - p u^T - u p^T + \beta (u p^T)(u u^T) = \\ &= \tilde{A}_{\kappa-1} - (p - \frac{\beta}{2}(p^T u) u) u^T - u (p - \frac{\beta}{2}(p^T u) u)^T = \\ &= \tilde{A}_{\kappa-1} - q u^T - u q^T. \end{aligned}$$

This equation completes the description of the κ-th similarity transformation $A_\kappa := T_\kappa^{-1} A_{\kappa-1} T_\kappa$. After $(n-2)$ steps we end up with a symmetric tridiagonal matrix A_{n-2}.

Remark. Instead of using orthogonal similarity transformations based on Householder matrices, it is also possible to use Frobenius matrices as in the LR decomposition of a matrix in order to transform A to tridiagonal form. Because of their stability properties, orthogonal transformations are, however, preferable. In addition to the Householder matrices, there are other orthogonal matrices such that the corresponding similarity transformations will tridiagonalize A; see 2.1.

In this section we have assumed that A is a symmetric matrix. The approach discussed here can be applied even if we drop this assumption. In this case, however, we no longer get a tridiagonal matrix, but instead end up with a matrix whose elements $a_{\mu\nu}$ are zero in general only for $\mu \ge \nu + 2$. These are the Hessenberg-matrices mentioned earlier in 2.1.4.

In the following sections we discuss the problem of computing the eigenvalues of tridiagonal and Hessenberg matrices.

1.2 Computation of the Eigenvalues of Tridiagonal Matrices. Let D be a real symmetric $n \times n$ tridiagonal matrix of the form

$$D = \begin{pmatrix} \alpha_1 & \beta_2 & & 0 \\ \beta_2 & \alpha_2 & \ddots & \\ & \ddots & \ddots & \beta_n \\ 0 & & \beta_n & \alpha_n \end{pmatrix}.$$

The eigenvalues of the matrix D are the zeros of the characteristic polynomial $p(\lambda) = \det(D - \lambda I)$. It is known that for a symmetric matrix D, the polynomial p has only real zeros. To compute these zeros, we can use e.g. the Newton method (cf. 8.2.1). We now derive recurrence formulae for computing the values of p and p' at an arbitrary λ which are needed to apply the Newton method. Let

$$p_\mu(\lambda) := \det(D_\mu - \lambda I) = \begin{vmatrix} (\alpha_1 - \lambda) & \beta_2 & & & 0 \\ \beta_2 & (\alpha_2 - \lambda) & \ddots & & \\ & \ddots & \ddots & \ddots & \beta_\mu \\ & 0 & & \beta_\mu & (\alpha_\mu - \lambda) \end{vmatrix}.$$

Expanding this determinant about the last column, we get

$$(*) \qquad p_\mu(\lambda) = (\alpha_\mu - \lambda)p_{\mu-1}(\lambda) - \beta_\mu^2 p_{\mu-2}(\lambda),$$

for $2 \le \mu \le n$. Setting $p_0(\lambda) := 1$ and $p_1(\lambda) = \alpha_1 - \lambda$, we can now use this recursion to compute the value of the characteristic polynomial $p(\lambda) = p_n(\lambda)$ at any point $\lambda \in \mathbb{R}$. Differentiating $(*)$, we see that the derivative $p'(\lambda)$ can be computed recursively using

$$(**) \qquad \begin{aligned} p'_\mu(\lambda) &= -p_{\mu-1}(\lambda) + (\alpha_\mu - \lambda)p'_{\mu-1}(\lambda) - \beta_\mu^2 p'_{\mu-2}(\lambda); \\ p'_0(\lambda) &= 0, \quad p'_1(\lambda) = -1, \end{aligned}$$

where $p'(\lambda) = p'_n(\lambda)$.

We hasten to point out that in computing $p(\lambda)$ and $p'(\lambda)$ for given λ using the formulae $(*)$ and $(**)$, we never deal with the coefficients of p. To apply the Newton method, we need some reasonable starting values for the iteration. Generally, these can be taken to be the eigenvalues of the matrix

$$\hat{D} := \begin{pmatrix} \hat{\alpha} & \hat{\beta} & & 0 \\ \hat{\beta} & \hat{\alpha} & \ddots & \\ & \ddots & \ddots & \hat{\beta} \\ 0 & & \hat{\beta} & \hat{\alpha} \end{pmatrix},$$

where

$$\hat{\alpha} := \frac{1}{n}\sum_{\mu=1}^{n} \alpha_\mu \quad \text{and} \quad \hat{\beta} := \frac{1}{n-1}\sum_{\mu=2}^{n} \beta_\mu.$$

Explicit formulae for these eigenvalues (and their corresponding eigenvectors) are given in the following

Theorem. *Let D be a real $n \times n$ tridiagonal matrix of the form*

$$D = \begin{pmatrix} b & c & & 0 \\ a & \ddots & \ddots & \\ & \ddots & \ddots & c \\ 0 & & a & b \end{pmatrix}$$

with $a \cdot c > 0$. Then the eigenvalues of D are

$$\lambda_\mu = b + 2\sqrt{ac}\,\operatorname{sgn}(a)\cos\left(\frac{\mu\pi}{n+1}\right), \quad 1 \le \mu \le n,$$

with corresponding eigenvectors $x^\mu \in \mathbb{R}^n$ whose components are

$$x^\mu_\nu = \left(\frac{a}{c}\right)^{\frac{\nu-1}{2}} \sin\left(\frac{\mu\pi\nu}{n+1}\right), \quad 1 \le \mu \le n,\ 1 \le \nu \le n.$$

Proof. Let x^μ be the vector in $\mathbb{R}^n$ with components as given in the statement of the theorem. Then the ν-th component of Dx^μ is given by

$$\begin{aligned}(Ax^\mu)_\nu &= b\left(\frac{a}{c}\right)^{\frac{\nu-1}{2}} \sin\left(\frac{\mu\pi\nu}{n+1}\right) + c\left(\frac{a}{c}\right)^{\frac{\nu}{2}} \left[\sin\left(\frac{\mu\pi(\nu-1)}{n+1}\right) + \sin\left(\frac{\mu\pi(\nu+1)}{n+1}\right)\right] \\ &= b\left(\frac{a}{c}\right)^{\frac{\nu-1}{2}} \sin\left(\frac{\mu\pi\nu}{n+1}\right) + 2\operatorname{sgn}(c)\sqrt{ac}\left(\frac{a}{c}\right)^{\frac{\nu-1}{2}} \sin\left(\frac{\mu\pi\nu}{n+1}\right)\cos\left(\frac{\mu\pi}{n+1}\right) \\ &= \left(b + 2\sqrt{ac}\,\operatorname{sgn}(a)\cos\left(\frac{\mu\pi}{n+1}\right)\right)x^\mu_\nu = \lambda_\mu x^\mu_\nu.\end{aligned}$$

This shows that λ_μ is an eigenvalue of D with associated eigenvector x^μ, as asserted. □

Remark. In discretizing boundary value problems for ordinary differential equations, we frequently get the special case where $\operatorname{sgn}(a) = -1$ and $a = c$. In this case the theorem says that the eigenvalues are

$$\lambda_\mu = b - 2|a|\cos\left(\frac{\mu\pi}{n+1}\right).$$

1.3 Computation of the Eigenvalues of Hessenberg Matrices. We have seen in 1.1 that for nonsymmetric matrices, orthogonal similarity transformations based on Householder matrices can be used to reduce A to a matrix of the form

$$B = \begin{pmatrix} b_{11} & b_{12} & \cdots & b_{1n} \\ b_{21} & b_{22} & \cdots & b_{2n} \\ 0 & \ddots & \ddots & \\ & & b_{n-1n} & b_{nn} \end{pmatrix}.$$

Such matrices are called Hessenberg matrices (after K. Hessenberg [1941]).

We now show how to compute the values of the characteristic polynomial $p(\lambda) = \det(B - \lambda I)$ and its derivative at a given point λ. To this end, for fixed λ we consider the following linear system of equations depending on the parameter α:

$$(*) \quad \begin{aligned} (b_{11} - \lambda)x_1(\lambda) + \quad b_{12}x_2(\lambda) + \cdots + \quad & b_{1n}x_n(\lambda) = \alpha \\ b_{21}x_1(\lambda) + (b_{22} - \lambda)x_2(\lambda) + \cdots + \quad & b_{2n}x_n(\lambda) = 0 \\ \ddots \qquad \ddots \qquad & \quad \vdots \qquad \vdots \\ b_{nn-1}x_{n-1}(\lambda) + (b_{nn} - \lambda)& x_n(\lambda) = 0. \end{aligned}$$

If λ is not an eigenvalue of B, then for every α, $(*)$ has a unique solution $x(\lambda;\alpha) = (x_1(\lambda;\alpha), \ldots, x_n(\lambda;\alpha))^T$. Using Cramer's rule, we find that the n-th component of the solution vector is equal to

$$x_n(\lambda; a) = (-1)^{n+1}\alpha \cdot b_{21} \cdot b_{32} \cdots b_{nn-1} \cdot (\det(B - \lambda I))^{-1}.$$

The system of equations $(*)$ can also be regarded as an under-determined system with the unknowns $x_1(\lambda), x_2(\lambda), \cdots, x_n(\lambda), \alpha(\lambda)$. Thus by the above formula, whenever $b_{21} \cdot b_{32} \cdots b_{n\,n-1} \neq 0$, as soon as we have computed one of the unknowns, then all the others are uniquely determined. Setting $x_n(\lambda;\alpha) = 1$, we get

$$p(\lambda) = (-1)^{n+1}\alpha(\lambda)b_{21} \cdot b_{32} \cdots b_{nn-1}.$$

Here the factor $\alpha(\lambda)$ is also uniquely determined for every fixed λ; it can be computed from the system of equations $(*)$, starting with $x_n(\lambda) = 1$ and working upwards from the bottom. To compute

$$p'(\lambda) = (-1)^{n+1}\alpha'(\lambda)b_{21} \cdot b_{32} \cdots b_{nn-1},$$

we need to find $\alpha'(\lambda)$. Differentiating the system of equations $(*)$ with respect to λ, it follows that

$$(**) \quad \begin{aligned} (b_{11} - \lambda)x_1'(\lambda) - x_1(\lambda) \quad & + b_{12}x_2'(\lambda) + \quad \cdots + \\ & + b_{1n}x_n'(\lambda) \quad = \alpha'(\lambda) \\ b_{21}x_1'(\lambda) \quad & + (b_{22} - \lambda)x_2'(\lambda) - x_2(\lambda) + \cdots + \\ & + b_{2n}x_n'(\lambda) \quad = 0 \\ \vdots \qquad & \qquad \vdots \qquad\qquad \vdots \\ b_{nn-1}x_{n-1}'(\lambda) + (b_{nn} - \lambda)&x_n'(\lambda) - x_n(\lambda) = 0. \end{aligned}$$

Since $x_n(\lambda) = 1$, using the fact that $x_{n-1}(\lambda), \ldots, x_1(\lambda)$ have already been computed from $(*)$, we can now start at the bottom of $(**)$ and successively determine $x'_{n-1}(\lambda), x'_{n-2}(\lambda), \ldots, x'_1(\lambda)$ using the n-th through the second equations. The value of $\alpha'(\lambda)$ then follows from the first equation.

Since we can compute $p(\lambda)$ and $p'(\lambda)$ for every value of λ, we can now apply the Newton method to compute the zeros of p. The choice of starting values can be problematical, however. To get starting values, we may use the methods for estimating eigenvalues to be discussed later.

1.4 Problems. 1) Show that a symmetric matrix can be transformed to tridiagonal form using an LR decomposition involving Frobenius and permutation matrices. Show that if the matrix is nonsymmetric, then the method leads to a Hessenberg matrix.

2) Compute the complexity of the algorithm for transforming a matrix $A \in \mathbb{R}^{(n,n)}$ to Hessenberg form using Householder matrices.

3) Show that using a similarity transformation with a diagonal matrix D, every Hessenberg matrix can be transformed to a matrix whose elements below the main diagonal are either zero or one.

4) Show that the components $x_\mu(\lambda)$, $1 \le \mu \le n$, of the solution vector $x(\lambda)$ in 1.3 are polynomials in λ of degree $n - \mu$.

5) Explain how the method in 1.3 for computing the eigenvalues of a Hessenberg matrix $B = (b_{\mu\nu})$ has to be modified when the assumption $b_{21} \cdot b_{32} \cdots b_{n\,n-1} \neq 0$ is dropped.

6) Write a computer program to compute all eigenvalues for the eigenvalue problem $Ax = \lambda x$ where the μ-th equation is given by

$$e^{-2\mu h}\Big(\big(-\frac{1}{h^2} - \frac{1}{2h}\big)x_{\mu-1} + \frac{2}{h^2}x_\mu + \big(-\frac{1}{h^2} + \frac{1}{2h}\big)x_{\mu+1}\Big) = \lambda x_\mu$$

with $h := 1/n + 1$, $x_0 := 0$, $x_{n+1} := 0$. Use the Newton method, where starting values are computed according to Theorem 1.2. Run the program for $n = 4$ and $n = 9$.

2. The Jacobi Rotation and Eigenvalue Estimates

Using similarity transformations based on Householder matrices, we can transform any matrix $A \in \mathbb{R}^{(n,n)}$ in a finite number of steps to either a tridiagonal or Hessenberg matrix. In 1.2 and 1.3 we have given fast algorithms based on the Newton method for computing the eigenvalues of matrices of these special types. In this section we study a method which directly computes the eigenvalues of certain matrices A, but requires an infinite number of iteration steps.

2.1 The Jacobi Method. Let A be a real symmetric $n \times n$ matrix. Then, it is well known that A has only real eigenvalues, and that there exist orthogonal matrices which can be used to transform A to a diagonal matrix containing the eigenvalues of A on the diagonal. Our aim now is to show how to transform A to diagonal form using an infinite sequence of orthogonal similarity transformations.

Definition. The $n \times n$ matrix

$$\Omega_{\mu\nu}(\varphi) := \begin{pmatrix} 1 & & & & & & & & & \\ & \ddots & & & & & 0 & & & \\ & & 1 & & & & & & & \\ & & & \cos\varphi & & \cdots & -\sin\varphi & & & \\ & & & & 1 & & & & & \\ & & & & & \ddots & & & & \\ & & & & & & 1 & & & \\ & & & \sin\varphi & & & \cos\varphi & & & \\ & & 0 & & & & & 1 & & \\ & & & & & & & & \ddots & \\ & & & & & & & & & 1 \end{pmatrix}, \quad \begin{matrix} \\ \\ \\ \leftarrow \text{row } \mu \\ \\ \\ \\ \leftarrow \text{row } \nu \\ \\ \\ \\ \end{matrix}$$

with $|\varphi| \leq \pi$ is called a *Jacobi rotation*.

Clearly, applying the matrix $\Omega_{\mu\nu}(\varphi)$ to a vector in the plane results in a vector which is rotated by an angle φ. The idea of the Jacobi method is to apply an infinite sequence of such Jacobi rotations to A so that the nondiagonal elements of the sequence of transformed matrices converge to zero.

CARL GUSTAV JACOBI (1804–1851), whose name comes up several times in this book, worked in Königsberg and Berlin. His numerous publications deal with just about every aspect of real and complex analysis, with number theory, and with mechanics. In numerical analysis, his contributions to the treatment of linear systems of equations and to numerical integration are especially influencial. Jacobi's interest in systems of equations was aroused by his study of the papers of Gauss on the method of least squares.

We restrict our attention here to the *classical Jacobi method.* In the first step of this method, we find a nondiagonal element of largest absolute value. Since $A_0 := A = (a_{\mu\nu})$ is assumed to be symmetric, in looking for this element it suffices to consider only those elements $a_{\mu\nu}$ with $\mu < \nu$. We denote the desired matrix element by $a_{\mu(0)\nu(0)}$. Now we choose φ so that if we apply the associated Jacobi-Rotation $\Omega_{\mu(0)\nu(0)}(\varphi)$ to A, we get a matrix $A_1 := \Omega^{-1}_{\mu(0)\nu(0)}(\varphi) A \Omega_{\mu(0)\nu(0)}(\varphi)$ with $a^{(1)}_{\mu(0)\nu(0)} = 0$. Since $\Omega_{\mu(0)\nu(0)}(\varphi)$ is an orthogonal matrix, $A_1 = \Omega^T_{\mu(0)\nu(0)}(\varphi) A_0 \Omega_{\mu(0)\nu(0)}(\varphi)$. We note that A_1 and A_0 differ only in the ν-th and μ-th columns and rows.

Since $A = A_0$ is symmetric,

$$\begin{pmatrix} \cos\varphi & \sin\varphi \\ -\sin\varphi & \cos\varphi \end{pmatrix}\begin{pmatrix} a_{\mu(0)\mu(0)} & a_{\mu(0)\nu(0)} \\ a_{\mu(0)\nu(0)} & a_{\nu(0)\nu(0)} \end{pmatrix}\begin{pmatrix} \cos\varphi & -\sin\varphi \\ \sin\varphi & \cos\varphi \end{pmatrix} = \begin{pmatrix} a^{(1)}_{\mu(0)\mu(0)} & a^{(1)}_{\mu(0)\nu(0)} \\ a^{(1)}_{\mu(0)\nu(0)} & a^{(1)}_{\nu(0)\nu(0)} \end{pmatrix}.$$

To compute the angle φ, we multiply out to get

$$\begin{aligned} a^{(1)}_{\mu(0)\nu(0)} &= \\ &= -a_{\mu(0)\mu(0)}\sin\varphi\cos\varphi - a_{\mu(0)\nu(0)}\sin^2\varphi + \\ &\qquad + a_{\mu(0)\nu(0)}\cos^2\varphi + a_{\nu(0)\nu(0)}\sin\varphi\cos\varphi = \\ &= (a_{\nu(0)\nu(0)} - a_{\mu(0)\mu(0)})\sin\varphi\cos\varphi + a_{\mu(0)\nu(0)}(\cos^2\varphi - \sin^2\varphi) = \\ &= \frac{1}{2}(a_{\nu(0)\nu(0)} - a_{\mu(0)\mu(0)})\sin 2\varphi + a_{\mu(0)\nu(0)}\cos 2\varphi. \end{aligned}$$

The requirement $a^{(1)}_{\mu(0)\nu(0)} = 0$ leads to the formula

$$\tan 2\varphi = \frac{2a_{\mu(0)\nu(0)}}{a_{\mu(0)\mu(0)} - a_{\nu(0)\nu(0)}}, \quad |\varphi| \le \frac{\pi}{4}.$$

This process can now be repeated, and in general, at step κ we choose the angle φ so that the element $a^{(\kappa)}_{\mu(\kappa-1)\nu(\kappa-1)}$ is zeroed out. This leads to

$$(*) \qquad \tan 2\varphi = \frac{2a^{(\kappa-1)}_{\mu(\kappa-1)\nu(\kappa-1)}}{a^{(\kappa-1)}_{\mu(\kappa-1)\mu(\kappa-1)} - a^{(\kappa-1)}_{\nu(\kappa-1)\nu(\kappa-1)}}, \quad |\varphi| \le \frac{\pi}{4}.$$

Remarks. 1) In carrying out a Jacobi rotation, it is not actually necessary to compute the angle φ. In fact, we need only find the numbers $c := \cos\varphi$ and $s := \sin\varphi$. We can get these by rewriting formula $(*)$ using trigonometric identities. Indeed, setting

$$\tau := \frac{|a^{(\kappa-1)}_{\mu(\kappa-1)\mu(\kappa-1)} - a^{(\kappa-1)}_{\nu(\kappa-1)\nu(\kappa-1)}|}{((a^{(\kappa-1)}_{\mu(\kappa-1)\mu(\kappa-1)} - a^{(\kappa-1)}_{\nu(\kappa-1)\nu(\kappa-1)})^2 + 4(a^{(\kappa-1)}_{\mu(\kappa-1)\nu(\kappa-1)})^2)^{1/2}}$$

and $\sigma := \operatorname{sgn}((a^{(\kappa-1)}_{\mu(\kappa-1)\mu(\kappa-1)} - a^{(\kappa-1)}_{\nu(\kappa-1)\nu(\kappa-1)})a^{(\kappa-1)}_{\mu(\kappa-1)\nu(\kappa-1)})$, it follows that $c = \left(\frac{1+\tau}{2}\right)^{1/2}$ and $s = \sigma\cdot\left(\frac{1-\tau}{2}\right)^{1/2}$.

2) In deriving formula $(*)$ and in the computation of the quantities c and s, we made no use of the assumption that $a^{(\kappa-1)}_{\mu(\kappa-1)\nu(\kappa-1)}$ is the nondiagonal element of the matrix $A_{\kappa-1}$ with largest absolute value. The method can be applied whenever $a^{(\kappa-1)}_{\mu(\kappa-1)\nu(\kappa-1)}) \neq 0$.

Although in each step of the method some matrix element is transformed to zero, in general it will happen that this same element is changed again in the following step. In this connection we have the following

Theorem. *Starting with $A_0 := A$, the classical Jacobi method leads to a sequence of matrices $A_{\kappa+1} = \Omega^T_{\mu(\kappa)\nu(\kappa)}(\varphi) \cdot A_\kappa \cdot \Omega_{\mu(\kappa)\nu(\kappa)}(\varphi)$ which converge elementwise to a diagonal matrix whose diagonal entries are the eigenvalues of A.*

Proof. Since A is a symmetric matrix, there exists an orthogonal matrix C and a diagonal matrix

$$D = \begin{pmatrix} \lambda_1 & & & 0 \\ & \lambda_2 & & \\ & & \ddots & \\ & 0 & & \lambda_n \end{pmatrix}$$

with $A = C^T DC$. The trace of a matrix is invariant under similarity transformations, and thus

$$\sum_{\mu=1}^{n} \sum_{\nu=1}^{n} a^2_{\mu\nu} = \text{trace } (A^T A) = \text{trace } (C^T DCC^T DC) =$$

$$= \text{trace } (C^T D^2 C) = \text{trace } (D^2) = \sum_{\mu=1}^{n} \lambda^2_\mu .$$

Setting $N(A) := 2 \sum_{\mu=1}^{n} \sum_{\substack{\nu=1 \\ \nu>\mu}}^{n} a^2_{\mu\nu}$, it follows that

$$\sum_{\mu=1}^{n} \lambda^2_\mu = \sum_{\mu=1}^{n} a^2_{\mu\mu} + N(A).$$

Applying this observation to the passage from $A_{\kappa-1}$ to A_κ, and using the fact that the similarity transformation $\Omega_{\mu(\kappa-1)\nu(\kappa-1)}(\varphi)$ alters only the elements in the μ-th and ν-th rows and columns, we get

$$N(A_\kappa) - N(A_{\kappa-1}) = \sum_{\mu=1}^{n} (a^{(\kappa-1)}_{\mu\mu})^2 - \sum_{\mu=1}^{n} (a^{(\kappa)}_{\mu\mu})^2 =$$

$$= (a^{(\kappa-1)}_{\mu(\kappa-1)\mu(\kappa-1)})^2 + (a^{(\kappa-1)}_{\nu(\kappa-1)\nu(\kappa-1)})^2 -$$

$$- (a^{(\kappa)}_{\mu(\kappa-1)\mu(\kappa-1)})^2 - (a^{(\kappa)}_{\nu(\kappa-1)\nu(\kappa-1)})^2 .$$

On the other hand, using

$$\begin{pmatrix} a^{(\kappa)}_{\mu(\kappa-1)\mu(\kappa-1)} & a^{(\kappa)}_{\mu(\kappa-1)\nu(\kappa-1)} \\ a^{(\kappa)}_{\mu(\kappa-1)\nu(\kappa-1)} & a^{(\kappa)}_{\nu(\kappa-1)\nu(\kappa-1)} \end{pmatrix} =$$

$$= \begin{pmatrix} \cos\varphi & \sin\varphi \\ -\sin\varphi & \cos\varphi \end{pmatrix} \begin{pmatrix} a^{(\kappa-1)}_{\mu(\kappa-1)\mu(\kappa-1)} & a^{(\kappa-1)}_{\mu(\kappa-1)\nu(\kappa-1)} \\ a^{(\kappa-1)}_{\mu(\kappa-1)\nu(\kappa-1)} & a^{(\kappa-1)}_{\nu(\kappa-1)\nu(\kappa-1)} \end{pmatrix} \begin{pmatrix} \cos\varphi & -\sin\varphi \\ \sin\varphi & \cos\varphi \end{pmatrix}$$

and the fact that the trace and determinant of a matrix are invariant under similarity transformations, it follows that

$$(a^{(\kappa)}_{\mu(\kappa-1)\mu(\kappa-1)})^2 + (a^{(\kappa)}_{\nu(\kappa-1)\nu(\kappa-1)})^2 + 2(a^{(\kappa)}_{\mu(\kappa-1)\nu(\kappa-1)})^2 =$$
$$= (a^{(\kappa-1)}_{\mu(\kappa-1)\mu(\kappa-1)})^2 + (a^{(\kappa-1)}_{\nu(\kappa-1)\nu(\kappa-1)})^2 + 2(a^{(\kappa-1)}_{\mu(\kappa-1)\nu(\kappa-1)})^2.$$

This implies the identity

$$N(A_\kappa) = N(A_{\kappa-1}) - 2((a^{(\kappa-1)}_{\mu(\kappa-1)\nu(\kappa-1)})^2 - (a^{(\kappa)}_{\mu(\kappa-1)\nu(\kappa-1)})^2) =$$
$$= N(A_{\kappa-1}) - 2(a^{(\kappa-1)}_{\mu(\kappa-1)\nu(\kappa-1)})^2.$$

Now for the classical Jacobi method we have the estimate

$$|a^{(\kappa-1)}_{\mu(\kappa-1)\nu(\kappa-1)}|^2 \geq \frac{N(A_{\kappa-1})}{n(n-1)},$$

and hence

$$N(A_\kappa) \leq N(A_{\kappa-1})(1 - \frac{2}{n(n-1)}), \quad n \geq 2.$$

This implies that all nondiagonal elements of the sequence (A_κ) converge to zero as $\kappa \to \infty$. □

Remark. There are several variants of the classical Jacobi method. For example, since it is very expensive to find the element of the matrix with largest absolute value, one approach is to successively zero the elements in columns 2 through n of the first row, then the elements in columns 3 through n of the second row, etc., and then to repeat the entire process until the nondiagonal elements have absolute value smaller than a prescribed $\varepsilon > 0$. This method is called the *cyclic Jacobi method.*

In using Jacobi methods in practice, we have to stop after a finite number of iterations. To decide when to stop the iteration, we need some way of estimating how accurately the diagonal elements of the current matrix approximate the true eigenvalues of A. We study this question in the following section.

2.2 Estimating Eigenvalues. Let $A = (a_{\mu\nu})$ be an arbitrary $n \times n$ matrix with real or complex elements. The equation $Ax = \lambda x$ can be written out as

$$(\lambda - a_{\mu\mu})x_\mu = \sum_{\substack{\kappa=1 \\ \kappa\neq\mu}}^{n} a_{\mu\kappa}x_\kappa, \quad 1 \leq \mu \leq n.$$

Now suppose μ is an index such that $|x_\mu| = \|x\|_\infty$. Then we immediately get the following simple bound on the eigenvalues λ found by S. A. Gerschgorin in 1931:

$$|\lambda - a_{\mu\mu}| \leq \sum_{\substack{\kappa=1 \\ \kappa\neq\mu}}^{n} |a_{\mu\kappa}|, \quad 1 \leq \mu \leq n.$$

We have established the

Gerschgorin Theorem. *All of the eigenvalues λ of an $n \times n$ matrix $A = (a_{\mu\nu})$ must lie in the union of the Gerschgorin disks*

$$\mathrm{K}_\mu := \{z \in \mathbb{C} \mid |z - a_{\mu\mu}| \le r_\mu\}$$

whose radii are $r_\mu := \sum_{\substack{\kappa=1 \\ \kappa \ne \mu}}^{n} |a_{\mu\kappa}|$.

If A is a diagonal matrix, then the Gerschgorin disks shrink to their center point which is precisely the diagonal element. Starting with this observation and noting that the roots of the characteristic polynomial of A depend continuously on its coefficients, we are led to the following

Refinement of the Gerschgorin Theorem. *If k Gerschgorin disks form a simply connected point set* G *which is disjoint from the remaining Gerschgorin disks, then* G *contains exactly k eigenvalues of the matrix A.*

Example. Consider the matrix

$$A := \begin{pmatrix} 1 + 0.5i & 0.5 & 0.1 \\ 0.3 & 1 - 0.5i & 0.5 \\ 0.4 & 0 & -0.5 \end{pmatrix}.$$

It is easy to compute that the Gerschgorin radii are $r_1 = 0.6$, $r_2 = 0.8$ and $r_3 = 0.4$. The following sketch shows the Gerschgorin disks in the complex plane. All three eigenvalues of A lie in their union. The point sets enclosed by the darker lines are dealt with in the following remark.

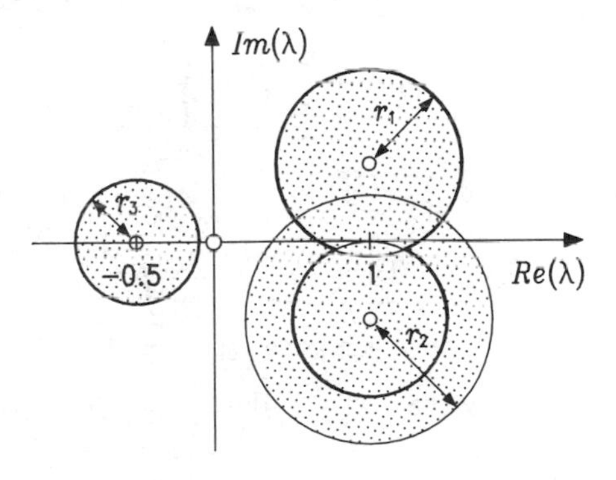

Remark. Since the transpose of a matrix A possesses the same eigenvalues as A, it follows that all eigenvalues must also lie in the union of the disks

$$\mathrm{K}'_\mu := \{z \in \mathbb{C} \mid |z - a_{\mu\mu}| \le r'_\mu\},$$

where $r'_\mu := \sum_{\substack{\kappa=1 \\ \kappa\neq\mu}}^{n} |a_{\kappa\mu}|$, and hence in the smaller set

$$\left(\bigcup_{\mu=1}^{n} \mathrm{K}_\mu\right) \cap \left(\bigcup_{\mu=1}^{n} \mathrm{K}'_\mu\right).$$

Another way to improve our bounds on the eigenvalues is to apply similarity transformations to A with the aim of reducing the size of the Gerschgorin disks.

Of the numerous other ways of bounding eigenvalues, we now discuss a simple approach based on the residual in the eigenvalue equation.

Theorem. *Let $A \in \mathbb{C}^{(n,n)}$ be a Hermitian matrix whose eigenvalues are given by $\lambda_1, \lambda_2, \ldots, \lambda_n$. Suppose that $\lambda \in \mathbb{R}$ and the vector $x \in \mathbb{C}^n$, $x \neq 0$ are such that $d := Ax - \lambda x$. Then*

$$\min_{1\leq\mu\leq n} |\lambda - \lambda_\mu| \leq \frac{\|d\|_2}{\|x\|_2}.$$

Proof. Since A is Hermitian, there exists an orthonormal basis $x^1, x^2, \ldots, x^n$ for $\mathbb{C}^n$ consisting of the eigenvectors of A. Let $x = \sum_{\mu=1}^{n} \alpha_\mu x^\mu$ be the expansion of x in terms of these basis elements. Now the difference is given by $d = \sum_{\mu=1}^{n} \alpha_\mu \cdot (\lambda_\mu - \lambda) x^\mu$, and the desired bound follows from

$$\|d\|_2^2 = \sum_{\mu=1}^{n} |\alpha_\mu|^2 |\lambda_\mu - \lambda|^2 \geq \min_\mu |\lambda_\mu - \lambda|^2 \cdot \|x\|_2^2. \qquad \square$$

Example. The matrix

$$A = \begin{pmatrix} 6 & 4 & 3 \\ 4 & 6 & 3 \\ 3 & 3 & 7 \end{pmatrix}$$

has the eigenvalues $\lambda_1 = 13$, $\lambda_2 = 4$, $\lambda_3 = 2$. To apply the theorem, we choose $x = (0.9, 1, 1.1)^T$ and $\lambda = 12$. Then $d = \begin{pmatrix} 12.7 \\ 12.9 \\ 13.4 \end{pmatrix} - \begin{pmatrix} 10.8 \\ 12 \\ 13.2 \end{pmatrix}$, and we obtain $\min_{1\leq\mu\leq 3} |\lambda - \lambda_\mu| \leq 1.22$ which is a good estimate of the largest eigenvalue.

It is also easy to show that $|\lambda| \leq \|A\|$; see also the extremal property of the Rayleigh-Quotient 3.3.

2.3 Problems. 1) Show that by applying a finite number of Jacobi rotations, we can transform any matrix $A \in \mathbb{R}^{(n,n)}$ into a Hessenberg matrix (or if A is symmetric, to a tridiagonal matrix).

2) Show that the cyclic Jacobi method converges as long as in each step of the process, we operate on an element whose absolute value exceeds $N(A)/2n^2$.

3) Write a computer program for the classical Jacobi method, and use it to find the eigenvalues of the matrix

$$\begin{pmatrix} n & n-1 & n-2 & \cdots & 2 & 1 \\ n-1 & n-1 & n-2 & \cdots & 2 & 1 \\ n-2 & n-2 & n-2 & \cdots & 2 & 1 \\ \vdots & & & & & \\ 2 & 2 & 2 & \cdots & 2 & 1 \\ 1 & 1 & 1 & \cdots & 1 & 1 \end{pmatrix}$$

for $n = 12$.

Hint: The eigenvalues are $\lambda_\mu = \frac{1}{2}(1 - \cos\frac{(2\mu-1)\pi}{2n+1})^{-1}$.

4) Estimate the eigenvalues of the matrix

$$\begin{pmatrix} 4.2 & 0.65 & 3.2 \\ 0.65 & 6.4 & 1.6 \\ 3.2 & 1.6 & 4.8 \end{pmatrix}$$

using the method of Gerschgorin.

5) Prove the following assertion: For every eigenvalue λ_μ of a matrix $A \in \mathbb{C}^{(n,n)}$ and for every matrix $B \in \mathbb{C}^{(n,n)}$, either $\det(\lambda_\mu I - B) = 0$, or $\lambda_\mu \in T := \{\lambda \in \mathbb{C} \mid \|(\lambda I - B)^{-1} \cdot (A - B)\| \geq 1\}$.

6) Apply Problem 5 to prove the Gerschgorin Theorem. *Hint:* The matrix B must be chosen appropriately.

7) Prove that if $A = (a_{\mu\nu})$ is Hermitian, then corresponding to every diagonal element $a_{\mu\mu}$, there exists an eigenvalue λ of the matrix A satisfying

$$|\lambda - a_{\mu\mu}| \leq \Big(\sum_{\substack{\nu=1 \\ \nu\neq\mu}}^{n} |a_{\mu\nu}|^2\Big)^{1/2}.$$

3. The Power Method

The methods presented above for computing the eigenvalues of a matrix are designed to find all of them. In many practical cases, however, one is not interested in finding all of the eigenvalues, but only some of them, for example, the one with largest or smallest absolute value. In such cases, it would be useful to have a method which finds these eigenvalues in the simplest possible way. Throughout this section we assume that the matrix A is diagonalizable.

3.1 An Iterative Method. Let $A \in \mathbb{C}^{(n,n)}$ be a diagonalizable matrix. Starting with an arbitrary initial vector $z^{(0)} \in \mathbb{C}^n$, we carry out the iteration

$$z^{(\kappa)} := Az^{(\kappa-1)}, \quad \kappa = 1, 2, \ldots,$$

which can be rewritten as

$$z^{(\kappa)} = A^{\kappa} z^{(0)}.$$

Suppose the eigenvalues are arranged in decreasing order

$$|\lambda_1| \geq |\lambda_2| \geq \cdots \geq |\lambda_n|,$$

and that

$$z^{(0)} = \alpha_1 x^1 + \alpha_2 x^2 + \cdots + \alpha_n x^n$$

is an expansion of $z^{(0)}$ with respect to a basis $\{x^1, \ldots, x^n\}$ of eigenvectors of the matrix A. Then the iteration produces the vectors

$$\begin{aligned} z^{(\kappa)} &= \alpha_1 \lambda_1^{\kappa} x^1 + \alpha_2 \lambda_2^{\kappa} x^2 + \cdots + \alpha_n \lambda_n^{\kappa} x^n = \\ &= \lambda_1^{\kappa} \Big[\alpha_1 x^1 + \alpha_2 \Big(\frac{\lambda_2}{\lambda_1}\Big)^{\kappa} x^2 + \cdots + \alpha_n \Big(\frac{\lambda_n}{\lambda_1}\Big)^{\kappa} x^n\Big]. \end{aligned}$$

We now assume that $z^{(0)}$ is chosen so that $\alpha_1 \neq 0$. We now discuss the two cases

(i) $|\lambda_1| > |\lambda_2|$

and

(ii) $|\lambda_1| = \cdots = |\lambda_m|$ with $|\lambda_m| > |\lambda_{m+1}|$ in case $m < n$.

In case (i) we note that

$$\lim_{\kappa \to \infty} \frac{1}{\lambda_1^{\kappa}} z^{(\kappa)} = \alpha_1 x^1,$$

and thus the quotients $q_\nu^{(\kappa)} := \frac{z_\nu^{(\kappa)}}{z_\nu^{(\kappa-1)}}$, $z_\nu^{(\kappa-1)} \neq 0$, satisfy

$$(*) \qquad \lim_{\kappa \to \infty} q_\nu^{(\kappa)} = \lambda_1, \quad \text{provided } x_\nu^1 \neq 0.$$

The sequence $(q^{(\kappa)})$ with $q^{(\kappa)} := \frac{\|z^{(\kappa)}\|}{\|z^{(\kappa-1)}\|}$ has a better rate of convergence, but only gives

$$(**) \qquad \lim_{\kappa \to \infty} q^{(\kappa)} = |\lambda_1|.$$

In the typical case where $A \in \mathbb{R}^{(n,n)}$, the assumption $|\lambda_1| > |\lambda_2|$ implies that λ_1 is real. Then if $z^{(0)}$ is chosen to be real, the entire iteration involves only real values.

Practical Hint. It is useful to normalize $z^{(\kappa)}$ after every step of the iteration. This will avoid large changes in the sizes of the iterates.

The assumption $\alpha_1 \neq 0$ cannot be checked in advance, since x^1 is not known before beginning the computation. On the other hand, an arbitrary initial vector $z^{(0)}$ will in general have a component in the direction of x^1, and even if this is not the case, in the course of the calculation, roundoff error will eventually introduce a component in this direction, and so convergence to λ_1 will be assured.

(ii) In the case where $|\lambda_1| = \cdots = |\lambda_m|$, the convergence behavior varies. For example, if $\lambda_1 = \cdots = \lambda_m$, then we have the same situation as in case (i), and the sequences $(q_\nu^{(\kappa)})$ and $(q^{(\kappa)})$ satisfy $(*)$ and $(**)$.

If, however, $\lambda_2 = -\lambda_1$, $m = 2$, for example, then

$$z^{(2\rho)} = \lambda_1^{2\rho}[\alpha_1 x^1 + \alpha_2 x^2 + y^{(2\rho)}],$$
$$z^{(2\rho+1)} = \lambda_1^{2\rho+1}[\alpha_1 x^1 - \alpha_2 x^2 + y^{(2\rho+1)}]$$

with $\lim_{\kappa\to\infty} y^{(\kappa)} = 0$. In this case

$$\lim_{\kappa\to\infty} \frac{z_\nu^{(\kappa+2)}}{z_\nu^{(\kappa)}} = \lambda_1^2, \quad \text{provided } x_\nu^1 \neq 0,$$

and

$$\lim_{\kappa\to\infty} \frac{\|z^{(\kappa+2)}\|}{\|z^{(\kappa)}\|} = |\lambda_1|^2.$$

That $\lambda_2 = -\lambda_1$ leads to a special case can be seen already from the form of the sequence $(z^{(\kappa)})$, since in this case after normalization it splits into two convergent subsequences.

We do not discuss the other special cases which arise in Case (ii), since they are all similar to the one just treated; see Problem 1.

The power method is frequently referred to as *Von Mises Iteration*. RICHARD EDLER VON MISES (1883–1953) suggested this method, and investigated it along with other numerical methods for solving linear systems of equations (R. v. Mises and H. Pollaczek-Geiringer [1929]). V. Mises worked in Vienna, Brno, Strassburg, Dresden, Berlin, Istanbul and at Harvard University. His broad mathematical interests ranged from mechanics to calculation of probabilities. Among other things, he contributed to the analysis of aircraft wings and to boundary layer theory, and even worked as a builder of aircraft.

3.2 Computation of Eigenvectors and Further Eigenvalues. By the discussion of Case (i) in 3.1, we see that if $|\lambda_1| > |\lambda_2|$, then the normalized iterates $z^{(\kappa)}/\|z^{(\kappa)}\|$ converge to the normalized eigenvector x^1. In Case (ii), the iterates converge to a linear combination of the eigenvectors corresponding to the eigenvalues involved, which can then be used to find the individual eigenvectors; see Problem 2.

If we are looking for the eigenvalue of a diagonalizable matrix A which is of smallest rather than largest absolute value, then we may apply the iteration

$$z^{(\kappa)} = A^{-1} z^{(\kappa-1)},$$

which can also be written in the form

$$Az^{(\kappa)} = z^{(\kappa-1)}.$$

This latter form of the iteration avoids the computation of the inverse A^{-1}, but does require the solution of a system of equations at each step of the iteration. It depends on the nature of A which of the two methods is more appropriate.

In order to compute additional eigenvalues, we have to modify the matrix A. This can be accomplished by *deflation*, whereby A is transformed to a matrix for which the eigenvalue λ_1 is replaced by an eigenvalue zero, and the remaining eigenvalues $\lambda_2, \ldots, \lambda_n$ remain unchanged. Another possibility is the *reduction* of A, whereby A is used to generate an $(n-1) \times (n-1)$ matrix with eigenvalues $\lambda_2, \ldots, \lambda_n$. Both transformations make use of the eigenvalue λ_1 and the eigenvector x^1. The accuracy to which these have been determined controls the numerical usefulness of these methods. For details, see e.g. A. S. Householder [1964]. In this regard we remark that the power method is primarily used to compute the eigenvalues of largest and smallest absolute values, and that if all eigenvalues are needed, then one of the methods of Sections 1 and 2 should be applied.

3.3 The Rayleigh Quotient. Let A be a Hermitian matrix; i.e., $A = \overline{A}^T$. In this section we show how the eigenvalues of A can be characterized by the extremal property of the *Rayleigh Quotients* $\frac{\overline{x}^T Ax}{\|x\|_2^2}$.

For every matrix $A \in \mathbb{C}^{(n,n)}$, the equation $Ax^\mu = \lambda_\mu x^\mu$ for an eigenvalue λ_μ and associated eigenvector x^μ, $\|x^\mu\|_2 = 1$ implies that $\lambda_\mu = (\overline{x}^\mu)^T Ax^\mu$. Now if A is Hermitian, then the quadratic form $\overline{x}^T Ax$ is always a real number since $\overline{x}^T Ax = \overline{x}^T \overline{A}^T x = x^T \overline{A}\,\overline{x} = \overline{\overline{x}^T Ax}$ for all $x \in \mathbb{C}^n$.

Suppose $\lambda_1 \geq \cdots \geq \lambda_n$ are the eigenvalues and that $\{x^1, \ldots, x^n\}$ is the associated orthonormal system of eigenvectors of the Hermitian matrix A. Then the normalized vector $x \in \mathbb{C}^n$ can be written in the form

$$x = \alpha_1 x^1 + \cdots + \alpha_n x^n \quad \text{with } |\alpha_1|^2 + \cdots + |\alpha_n|^2 = 1,$$

and so

$$\overline{x}^T Ax = \overline{\left(\sum_{\nu=1}^{n} \alpha_\nu x^\nu\right)}^T \left(\sum_{\nu=1}^{n} \alpha_\nu \lambda_\nu x^\nu\right) = \sum_{\nu=1}^{n} |a_\nu|^2 \lambda_\nu \leq \lambda_1.$$

Now observing that $(\overline{x}^1)^T Ax^1 = \lambda_1$, we get the

Extremal Property of Rayleigh Quotients

$$\lambda_1 = \max_{\|x\|_2=1} \overline{x}^T Ax.$$

Analogously,

$$\lambda_n = \min_{\|x\|_2=1} \overline{x}^T Ax.$$

The other eigenvalues of a Hermitian matrix are also extreme values of Rayleigh quotients. In particular, it can be shown that for $1 \le k \le n-2$,

$$\lambda_{k+1} = \max_{\|x\|_2=1} \overline{x}^T Ax$$

$$\text{under the side conditions } \overline{x}^T x^\nu = 0 \text{ for } 1 \le \nu \le k.$$

The proof can be carried out using the method of Lagrange multipliers, and is left to the reader (Problem 3).

In using the power method to compute the eigenvalue of largest absolute value of a Hermitian matrix, the extremal property of Rayleigh quotients provides some important additional information. In the notation of Section 3.3, either λ_1 or λ_n is the eigenvalue of largest absolute value. We denote it by λ^*, and suppose the associated eigenvector is x^*. Then the remainder term of the Taylor expansion $\overline{x}^T Ax$ about the extremal point x^* leads to

$$\overline{x}^T Ax = \lambda^* + O(\|x - x^*\|_2^2)$$

for all vectors $x \in \mathrm{U} := \{x \in \mathbb{C}^n \mid \|x - x^*\|_2 < \delta\}$.

This implies that the sequence of Rayleigh quotients $((\overline{z^{(\kappa)}})^T Az^{(\kappa)})$ normalized so that $\|z^{(\kappa)}\|_2 = 1$ converges *quadratically* to the extremal value $\lambda^* = (\overline{x}^*)^T Ax^*$. Moreover, this sequence provides the correct sign of the eigenvalue of largest absolute value, as well as a sequence of lower bounds for λ^* (when $(\lambda^* > 0)$) or upper bounds (when $(\lambda^* < 0)$).

3.4 Problems. 1) Let $A \in \mathbb{R}^{(n,n)}$, and suppose λ_1 and λ_2 are a conjugate pair of complex eigenvalues with $\lambda_2 = \overline{\lambda_1}$ and $|\lambda_1| > |\lambda_3|$. Study the behavior of the sequence $(z^{(\kappa)})$, $z^{(0)} \in \mathbb{R}^n$, and show that after a sufficient number of iteration steps, up to the numerical accuracy of the computations, the linear dependence relation $\gamma_0 z^{(\kappa)} + \gamma_1 z^{(\kappa+1)} + z^{(\kappa+2)} = 0$ holds, and thus the eigenvalues λ_1 and $\overline{\lambda}_1$ can be found by computing γ_0 and γ_1 and solving the equation $\gamma_0 + \gamma_1\lambda + \lambda^2 = 0$.

2) Let $A \in \mathbb{C}^{(n,n)}$, $\lambda_2 = -\lambda_1$, and let $|\lambda_2| > |\lambda_3|$. Show that in the power method, the eigenvectors x^1 and x^2 satisfy

$$\frac{x^1}{\|x^1\|} = \lim_{\rho\to\infty} \frac{\lambda_1^{2\rho} z^{(2\rho)} + z^{(2\rho+1)}}{\|\lambda_1^{2\rho} z^{(2\rho)} + z^{(2\rho+1)}\|}$$

and

$$\frac{x^2}{\|x^2\|} = \lim_{\rho\to\infty} \frac{\lambda_1^{2\rho} z^{(2\rho)} - z^{(2\rho+1)}}{\|\lambda_1^{2\rho} z^{(2\rho)} - z^{(2\rho+1)}\|}.$$

3) Prove that for a Hermitian matrix, the eigenvalues $\lambda_2, \ldots, \lambda_{n-1}$ can be obtained as limits of the Rayleigh Quotients 3.3 subject to appropriate

side conditions. Restrict yourself to real matrices, and find the necessary conditions for a relative extreme value using the method of Lagrange multipliers.

4) Since $(n+1)$ vectors in $\mathbb{C}^n$ are always linearly dependent, the iterates $z^{(0)}, \ldots, z^{(n)}$ corresponding to a matrix $A \in \mathbb{C}^{(n,n)}$ always satisfy an equation of the form

$$\alpha_0 z^{(0)} + \cdots + \alpha_{n-1} z^{(n-1)} = -z^{(n)}.$$

Show:

a) The coefficients $\alpha_0, \ldots, \alpha_{n-1}$ are the coefficients of the characteristic equation

$$p(\lambda) = \det(A - \lambda I) = (-1)^n(\alpha_0 + \alpha_1\lambda + \cdots + \lambda^n) = 0.$$

Hint: Use the identity $p(A) = 0$ (Theorem of Cayley-Hamilton).

b) If all eigenvalue are simple, then $\det(z^{(0)}, \ldots, z^{(n-1)}) \neq 0$, and so the coefficients $\alpha_0, \ldots, \alpha_{n-1}$ can be computed.

c) Discuss the case of multiple eigenvalues. The corresponding method for computing the characteristic equation of a matrix is called the *Krylov method.*

4. The QR Algorithm

The QR Decomposition Theorem 2.3.3 provides the basis for still another method for iteratively computing all of the eigenvalues of a real matrix. Given such a matrix $A \in \mathbb{R}^{(n,n)}$, there exist an orthogonal matrix Q and an upper triangular matrix R with $A = Q \cdot R$. We use this fact to construct a sequence of matrices $(A_\kappa)_{\kappa \in \mathbb{N}}$ by

$$\begin{aligned} A_0 &:= A, \\ A_{\kappa+1} &:= R_\kappa \cdot Q_\kappa, \quad \kappa \in \mathbb{N}, \end{aligned}$$

where the orthogonal matrix Q_κ and the upper triangular matrix R_κ are defined as in 2.3.3 so that $A_\kappa = Q_\kappa \cdot R_\kappa$. It can be shown that under appropriate conditions, this sequence converges to a limit matrix from which all of the eigenvalues of the matrix A can be read off. The method in this form is called the *QR algorithm*, and is due to J. G. F. Francis [1961]. An analogous method, the *LR algorithm*, was developed earlier by H. Rutishauser [1958], based on the LR decomposition of a matrix (cf. 2.1.3). We will concentrate primarily on the convergence question for the QR algorithm because of its wider range of applicability.

4.1 Convergence of the QR Algorithm. In preparation for the proof of a convergence theorem, we first establish the following

Proposition. *Let $D = \operatorname{diag}(d_{\mu\mu}) \in \mathbb{R}^{(n,n)}$ be a diagonal matrix with*

$$|d_{\mu\mu}| > |d_{\mu+1\mu+1}| > 0$$

for all $1 \leq \mu \leq n-1$. In addition, let $L = (\ell_{\mu\nu})$ be a $n \times n$ matrix of the form

$$L = \begin{pmatrix} 1 & & & \\ & & 0 & \\ \ell_{21} & 1 & & \\ \vdots & & \ddots & \\ \ell_{n1} & \cdots & \ell_{nn-1} & 1 \end{pmatrix},$$

and let $\hat{L}_\kappa := (\ell_{\mu\nu} \cdot \frac{d^\kappa_{\mu\mu}}{d^\kappa_{\nu\nu}}) \in \mathbb{R}^{(n,n)}$. Then $D^\kappa L = \hat{L}_\kappa D^\kappa$, and $\hat{L}_\kappa$ converges to the identity matrix as $\kappa \to \infty$.

Proof. The identity $D^\kappa L = \hat{L}_\kappa D^\kappa$ is trivial for $\kappa = 0$. Assuming that it holds for $\imath \in \mathbb{N}$, we see that

$$\begin{aligned} D^{\imath+1}L = D(D^\imath L) = D(\hat{L}_\imath D^\imath) = (d_{\mu\mu} \cdot \ell_{\mu\nu} \frac{d^\imath_{\mu\mu}}{d^\imath_{\nu\nu}} d^\imath_{\nu\nu}) = \\ = (\ell_{\mu\nu} \frac{d^{\imath+1}_{\mu\mu}}{d^{\imath+1}_{\nu\nu}} d^{\imath+1}_{\nu\nu}) = \hat{L}_{\imath+1} D^{\imath+1}, \end{aligned}$$

which establishes the result by induction. The convergence of $\hat{L_\kappa} \to I$ as $\kappa \to \infty$ follows immediately from the special form of L and the properties of the matrix elements $d_{\mu\mu}$. □

We now formulate the convergence theorem for the QR method in a special case. For results on convergence in the general case, see the book of H. R. Schwarz [1989].

Theorem. *Let A be a real $n \times n$ matrix with eigenvalues $\lambda_1, \lambda_2, \ldots, \lambda_n$ with $|\lambda_1| > |\lambda_2| > \cdots > |\lambda_n| > 0$. Suppose the associated eigenvectors are $x^1, x^2, \ldots, x^n$, and that the matrix T^{-1} with $T := (x^1, x^2, \ldots, x^n)$ has an LR decomposition. Then the matrix sequence (Q_κ) from the QR algorithm converges to a diagonal matrix. Moreover, the sequence (A_κ) contains a convergent subsequence which converges to an upper triangular matrix whose diagonal elements $r_{\mu\mu}$ are the eigenvalues λ_μ, $1 \leq \mu \leq n$.*

Proof. By the construction of the matrix sequence (A_κ), we have

$$(*) \qquad \begin{aligned} A_{\kappa+1} &= Q_\kappa^{-1} A_\kappa Q_\kappa = Q_\kappa^{-1} Q_{\kappa-1}^{-1} A_{\kappa-1} Q_{\kappa-1} Q_\kappa = \cdots = \\ &= (Q_0 \cdot Q_1 \cdots Q_\kappa)^{-1} A_0 (Q_0 \cdot Q_1 \cdots Q_\kappa) = Q_{0\kappa}^{-1} A_0 Q_{0\kappa}, \end{aligned}$$

where for convenience we use the abbreviation $Q_{0\kappa} = Q_0 \cdot Q_1 \cdots Q_\kappa$. This implies that $A_{\kappa+1}$ and A_0 are similar matrices, and consequently have the same eigenvalues. Moreover, the identity

$$\begin{aligned} A^\kappa &= A_0^\kappa = A_0^{\kappa-1} Q_0 R_0 = A^{\kappa-2} Q_0 A_1 R_0 = A^{\kappa-2} Q_0 Q_1 R_1 R_0 = \cdots = \\ &= Q_0 Q_1 \cdots Q_{\kappa-1} R_{\kappa-1} R_{\kappa-2} \cdots R_0 \end{aligned}$$

implies that the QR algorithm produces a QR decomposition

$$(**) \qquad A^\kappa = Q_{0\kappa-1} \cdot R_{\kappa-1\,0}$$

of the powers A^κ of A with $R_{\kappa-1\,0} := R_{\kappa-1} \cdots R_0$. By Problem 3 in 2.3.5, this decomposition is unique provided that we assure that the diagonal elements of the matrices R_μ, $0 \le \mu \le \kappa - 1$, are positive.

Since the eigenvector x^ν corresponding to the eigenvalue λ_ν of the matrix A is also an eigenvector corresponding to the eigenvalue λ_ν^κ of the matrix A^κ, we also know that A^κ satisfies the formula

$$A^\kappa = TD^\kappa T^{-1}$$

with $D := \operatorname{diag}(\lambda_\mu)$.

By assumption, the matrix T^{-1} has an LR decomposition

$$T^{-1} = L \cdot R.$$

We can always arrange that the diagonal elements of R are positive. Substituting in the above, we get

$$A^\kappa = TD^\kappa LR.$$

In the proposition we have shown that there exists a matrix $\hat{L}_\kappa$ such that $D^\kappa \cdot L = \hat{L}_\kappa \cdot D^\kappa$ and $\hat{L}_\kappa \to I$ as $\kappa \to \infty$. This leads to

$$A^\kappa = T\hat{L}_\kappa D^\kappa R,$$

and using the QR decomposition $T = \hat{Q} \cdot \hat{R}$, to

$$A^\kappa = \hat{Q}\hat{R}\hat{L}_\kappa D^\kappa R.$$

Here we have again chosen the QR decomposition of T so that the diagonal elements of $\hat{R}$ are positive.

Now replacing $\hat{R}\hat{L}_\kappa$ by the QR decomposition $\hat{R}\hat{L}_\kappa = \check{Q}_\kappa \check{R}_\kappa$, where $\check{R}_\kappa$ has positive diagonal elements, it follows that the product $\check{Q}_\kappa \cdot \check{R}_\kappa$ converges to $\hat{R}$, and $\check{Q}_\kappa$ converges to the identity matrix as $\kappa \to \infty$, since $\hat{L}_\kappa \to I$ and the QR decomposition of $\hat{R}$ is unique. Thus

$$A^\kappa = \hat{Q}\check{Q}_\kappa \check{R}_\kappa D^\kappa R$$

and

$$\lim_{\kappa \to \infty} \check{Q}_\kappa = I.$$

From this we conclude that in addition to $(**)$, A^κ also has another QR decomposition

$$A^\kappa = (\hat{Q}\check{Q}_\kappa \Sigma^\kappa)(\Sigma^\kappa \check{R}_\kappa D^\kappa R)$$

with $\Sigma^\kappa := \text{diag}\,(\text{sgn}\lambda_\mu^\kappa)$.

It is easy to see that the diagonal elements of the matrix $\Sigma^\kappa \check{R}_\kappa D^\kappa R$ are all positive. Because of the uniqueness of such a QR decomposition, comparing with (**), we find that

$$Q_{0\kappa-1} = \hat{Q}\check{Q}_\kappa \Sigma^\kappa,$$

and substituting this in (*) leads to

$$\begin{aligned} A_{\kappa+1} &= \Sigma^{\kappa+1}\check{Q}_{\kappa+1}^{-1}\hat{Q}^{-1}A\hat{Q}\check{Q}_{\kappa+1}\Sigma^{\kappa+1} = \\ &= \Sigma^{\kappa+1}\check{Q}_{\kappa+1}^{-1}\hat{R}T^{-1}AT\hat{R}^{-1}\check{Q}_{\kappa+1}\Sigma^{\kappa+1} = \\ &= \Sigma^{\kappa+1}\check{Q}_{\kappa+1}^{-1}\hat{R}D\hat{R}^{-1}\check{Q}_{\kappa+1}\Sigma^{\kappa+1}. \end{aligned}$$

Since $\check{Q}_\kappa \to I$, we conclude that

$$\lim_{\kappa\to\infty} \Sigma^{\kappa+1}A_{\kappa+1}\Sigma^{\kappa+1} = \hat{R}D\hat{R}^{-1}.$$

Now looking at the diagonal elements only, we obtain the asserted convergence

$$\lim_{\kappa\to\infty} a_{\mu\mu}^{(\kappa+1)} = \lambda_\mu,$$

for $1 \le \mu \le n$. In addition, the identity

$$Q_\kappa = Q_{0\kappa-1}^{-1} \cdot Q_{0\kappa},$$

together with $\lim_{\kappa\to\infty} \Sigma^\kappa Q_\kappa \Sigma^{\kappa+1} = \lim_{\kappa\to\infty} \check{Q}_\kappa^{-1} \cdot \check{Q}_{\kappa+1} = I$ implies that

$$\lim_{\kappa\to\infty} Q_\kappa = \text{diag}\,(\frac{\lambda_\mu}{|\lambda_\mu|}),$$

and the theorem is proved. □

This proof is due to J. H. Wilkinson [1965].

In applying the QR algorithm in practice, it is better to first transform the matrix A to Hessenberg or tridiagonal form (cf. 1.1 and 1.3), and then apply the QR algorithm. This saves considerably on computing time since the QR decomposition of such matrices requires only a few operations. In addition, all QR transformations A_κ of a Hessenberg matrix are again of Hessenberg form, while all A_κ are symmetric tridiagonal matrices whenever A is a symmetric tridiagonal matrix.

Remark. A more precise analysis of the convergence of the QR algorithm shows the relationship of this method with the vector iteration methods in 3.1. For details, see J. Stoer and R. Bulirsch [1983]. The rate at which the subdiagonal elements of $A_{\mu+1\mu}^{(\kappa)}$, $1 \le \mu \le n$, converge to zero depends on how fast the quotients $|\frac{\lambda_{\mu+1}}{\lambda_\mu}|^\kappa$ go to zero as $\kappa \to \infty$. In some cases, we may have to perform a large number of QR transformations. The method can be accelerated by introducing an appropriate *spectral shift* at each step. For example, if σ is a good approximation to the eigenvalue $\lambda_{\mu+1}$, then $|\lambda_{\mu+1}-\sigma| \ll |\lambda_\mu - \sigma|$, and the quotients $|\lambda_{\mu+1}-\sigma|^\kappa/|\lambda_\mu-\sigma|^\kappa$ go to zero quickly as $\kappa \to \infty$. The corresponding modified *QR algorithm with shift* has the form

$$A_0 := A,$$
$$A_{\kappa+1} := R_\kappa \cdot Q_\kappa + \sigma_\kappa I, \quad \kappa \in \mathbb{N},$$

where we now start with the QR decomposition $A_\kappa - \sigma_\kappa I = Q_\kappa \cdot R_\kappa$.

We do not have space here for a detailed analysis of the shifting process.

4.2 Remarks on the LR Algorithm. Suppose now that A is a real $n \times n$ matrix with an LR decomposition which can be constructed without first multiplying by a permutation matrix (cf. 2.1.3). In analogy with Section 4.1, we define the *LR algorithm* by

$$A_0 := A,$$
$$A_{\kappa+1} := R_\kappa \cdot L_\kappa, \quad \kappa \in \mathbb{N},$$

using the LR decomposition $A_\kappa = L_\kappa \cdot R_\kappa$.

If A is a positive definite symmetric matrix, then it has a Cholesky decomposition, in which case R_κ can be chosen to be L_κ^T. We now analyse the convergence of the LR algorithm in this case.

Theorem. *Suppose $A \in \mathbb{R}^{(n,n)}$ is positive definite and symmetric with eigenvalues $\lambda_1 > \lambda_2 > \cdots > \lambda_n > 0$. Then the matrix sequence (A_κ) generated by the LR algorithm using the Cholesky decomposition satisfies*

$$\lim_{\kappa\to\infty} A_\kappa = \operatorname{diag}(\lambda_\mu).$$

Proof. Since $A_{\kappa+1} = L_\kappa^{-1} A_\kappa L_\kappa$, all of the matrices $A_\kappa = (a_{\nu\mu}^{(\kappa)})$ in the sequence (A_κ) are similar to each other, and so have the same eigenvalues. Moreover, they are positive definite, and hence have positive diagonal elements $a_{\mu\mu}^{(\kappa)}$, which implies that the traces

$$s_\rho^{(\kappa)} = \sum_{\iota=1}^{\rho} a_{\iota\iota}^{(\kappa)}$$

of the principal submatrices of A_κ satisfy the inequalities

$$0 < s_1^{(\kappa)} < s_2^{(\kappa)} < \cdots < s_n^{(\kappa)}.$$

Since $s := \operatorname{trace} A = \operatorname{trace} A_\kappa = s_n^{(\kappa)}$, $s_n^{(\kappa)}$ does not depend on κ. From the Cholesky decomposition $A_\kappa = L_\kappa \cdot L_\kappa^T$ with $L_\kappa = (\ell_{\mu\nu}^{(\kappa)})$, we note that

$$a_{\mu\mu}^{(\kappa)} = \sum_{\iota=1}^{\mu} (\ell_{\mu\iota}^{(\kappa)})^2,$$

while $A_{\kappa+1} = L_\kappa^T \cdot L_\kappa$ implies

$$a_{\nu\nu}^{(\kappa+1)} = \sum_{\imath=\nu}^{n} (\ell_{\imath\nu}^{(\kappa)})^2.$$

Summing both equations, we get

$$s_\rho^{(\kappa)} = \sum_{\mu=1}^{\rho} a_{\mu\mu}^{(\kappa)} = \sum_{\mu=1}^{\rho}\sum_{\imath=1}^{\mu} (\ell_{\mu\imath}^{(\kappa)})^2 = \sum_{\imath=1}^{\rho}\sum_{\mu=\imath}^{\rho} (\ell_{\mu\imath}^{(\kappa)})^2, \quad s_\rho^{(\kappa+1)} = \sum_{\imath=1}^{\rho}\sum_{\mu=\imath}^{n} (\ell_{\mu\imath}^{(\kappa)})^2,$$

and thus

$$s_\rho^{(\kappa+1)} - s_\rho^{(\kappa)} = \sum_{\imath=1}^{\rho}\sum_{\mu=\rho+1}^{n} (\ell_{\mu\imath}^{(\kappa)})^2 \geq 0.$$

It follows that the sequence $(s_\rho^{(\kappa)})$ is monotone increasing and bounded above by s. This implies that it converges, and so $\lim_{\kappa\to\infty} \ell_{\mu\imath}^{(\kappa)} = 0$ for $1 \leq \imath \leq \rho$ and $\rho < \mu \leq n$. The index ρ was chosen arbitrarily subject to the restriction $1 \leq \rho \leq n$, and hence

$$\lim_{\kappa\to\infty} \ell_{\mu\imath}^{(\kappa)} = 0 \quad \text{for} \quad 1 \leq \imath < \mu \leq n.$$

The diagonal elements $\ell_{\mu\mu}^{(\kappa)}$ converge also as $\kappa \to \infty$, since $\lim_{\kappa\to\infty} s_\rho^{(\kappa)}$ implies the existence of

$$\lim_{\kappa\to\infty} (s_{\rho+1}^{(\kappa)} - s_\rho^{(\kappa)}) = \lim_{\kappa\to\infty} (a_{\rho+1\,\rho+1}^{(\kappa)})^2.$$

Summarizing, we have now shown that

$$\lim_{\kappa\to\infty} A_\kappa = \lim_{\kappa\to\infty} L_\kappa \cdot L_\kappa^T$$

exists and is a diagonal matrix. It follows that

$$\lim_{\kappa\to\infty} A_\kappa = \operatorname{diag}(\lambda_\mu). \qquad \square$$

Under the special assumptions of this theorem, we can now extend the above considerations to the following result on the

Speed of Convergence. Suppose A is a real positive definite and symmetric $n \times n$ matrix with all simple eigenvalues. Then the elements $a_{\mu\nu}^{(\kappa)}$ of the matrix A_κ exhibit the following asymptotic behavior as $\nu > \mu$:

$$a_{\mu\nu}^{(\kappa)} = O\Big(\big(\frac{\lambda_\nu}{\lambda_\mu}\big)\Big)^{\frac{\kappa}{2}} \quad \text{as } \kappa \to \infty.$$

Here we have numbered the eigenvalues so that $\lambda_1 > \lambda_2 > \cdots > \lambda_n$.

To see why this asymptotic behavior holds, suppose that the matrix A_κ is already nearly diagonal; i.e., that the nondiagonal elements have small absolute value as compared with one, while the elements on the main diagonal are already good approximations $\tilde{\lambda}_\mu$ to the eigenvalues λ_μ:

$$A_\kappa := \begin{pmatrix} \tilde{\lambda}_1 & & & \\ & \tilde{\lambda}_2 & & \varepsilon_{\mu\nu} \\ \varepsilon_{\mu\nu} & & \ddots & \\ & & & \tilde{\lambda}_n \end{pmatrix}, \; |\varepsilon_{\mu\nu}| \ll 1 \text{ and } |\tilde{\lambda}_\mu - \lambda_\mu| \ll 1.$$

Now using the formulae for the matrix elements $\ell_{\mu\nu}$ of the Cholesky Decomposition 2.2.2, and ignoring the quadratic terms in $\varepsilon_{\mu\nu}$, we get the approximations

$$\tilde{\ell}_{\mu\mu} = \sqrt{\tilde{\lambda}_\mu}, \quad 1 \le \mu \le n \text{ and } \tilde{\ell}_{\nu\mu} = \tilde{\ell}_{\mu\nu} = \frac{1}{\sqrt{\tilde{\lambda}_\mu}} \varepsilon_{\mu\nu}, \quad 1 < \mu < \nu \le n,$$

of the elements $\ell_{\mu\nu}$. In the next step of the LR algorithm, we get the approximation

$$\tilde{A}_{\kappa+1} := \begin{pmatrix} \tilde{\lambda}_1 & & & \\ & \tilde{\lambda}_2 & & \varepsilon_{\mu\nu}\sqrt{\frac{\tilde{\lambda}_\nu}{\tilde{\lambda}_\mu}} \\ \varepsilon_{\mu\nu}\sqrt{\frac{\tilde{\lambda}_\nu}{\tilde{\lambda}_\mu}} & & \ddots & \\ & & & \tilde{\lambda}_n \end{pmatrix}$$

of $A_{\kappa+1}$, where here again terms of second order in $\varepsilon_{\mu\nu}$ are ignored in forming the product $\tilde{L}_\kappa^T \cdot \tilde{L}_\kappa$. Thus at each step, we are multiplying the nondiagonal elements by the factor $\left(\frac{\tilde{\lambda}_\nu}{\tilde{\lambda}_\mu}\right)^{1/2}$, and our asymptotic assertion follows. With some additional technicalities, this argument can be made precise.

Analogous to the discussion in Remark 4.1, we may use shifting in order to accelerate the convergence. We do not have space for the details here.

4.3 Problems. 1) Show that the QR transform A_κ of a Hessenberg matrix A is again a Hessenberg matric, and that the QR transform of a symmetric tridiagonal matrix is a symmetric tridiagonal matrix.

2) Prove that the QR decomposition of a Hessenberg matrix or a symmetric tridiagonal matrix $A \in \mathbb{R}^{(n,n)}$ can be accomplished by using $(n-1)$ rotation matrices.

3) Write a computer program for the QR algorithm (and the LR algorithm), and find the eigenvalues of the matrix A in Example 8.4.3. How does the shifting method affect the rate of convergence? How many steps are needed if all we want to find is the spectral radius?

4) Prove that if A is a symmetric band matrix, then the transformed matrix obtained by the LR method using Cholesky is again a band matrix with the same band width.

5) In analogy with Theorem 4.1, show the following: If A is a matrix with eigenvalues λ_μ satisfying

$$|\lambda_1| > |\lambda_2| > \cdots > |\lambda_n| > 0,$$

then the matrix sequence (A_κ) produced by the LR algorithm converges to an upper triangular matrix, and that T^{-1} as well as all of the matrices $T = (x^1, x^2, \dots, x^n)$ formed from the associated eigenvectors have LR decompositions.

4

Approximation

In Chapters 2 and 3 we have discussed methods of numerical linear algebra. In this chapter we turn to another central question in numerical analysis, the approximation of mathematical objects. The methods presented here have a wide variety of applications.

1. Preliminaries

Linear spaces provide an appropriate framework for the study of approximation methods, particularly since they allow us to work with operators, and to apply some of the methods of functional analysis. In this section we review some of the concepts and results which we will need in this book. This material can be found in most elementary books, but for some of the results we include short proofs. In addition, this section also contains some comments and isolated results without proofs which we felt should be included for completeness, but which we will not use explicitly.

1.1 Normed Linear Spaces. We use the notation $(\mathrm{V}, \|\cdot\|)$ for a linear space V of arbitrary dimension over either the field $\mathbb{K} := \mathbb{C}$ or the field $\mathbb{K} := \mathbb{R}$, with norm $\|\cdot\|$; cf. 2.4.1. In the case where the elements of the linear space are functions of one or more variables, we denote them by $f, g, \ldots$ or $\varphi, \psi, \ldots$. If $f \in \mathrm{V}$, $f \neq 0$, then the element $\frac{f}{\|f\|}$ has norm one. An element of norm one is called *normed.*

Metric. The function $d(f,g) := \|f - g\|$ provides a metric for the normed linear space $(\mathrm{V}, \|\cdot\|)$. Indeed, d is a mapping $d : \mathrm{V} \times \mathrm{V} \to [0, \infty)$, and using the properties of a norm 2.4.1, we see that d satisfies the defining properties of a metric. In particular, for all $f, g, h \in \mathrm{V}$, we have

$$\begin{aligned}
&d(f,g) = 0 \Leftrightarrow f = g && \text{by 2.4.1(i)},\\
&d(f,g) = d(g,f) && \text{by 2.4.1(ii)},\\
&d(f,g) \leq d(f,h) + d(h,g) && \text{by 2.4.1(iii)}.
\end{aligned}$$

Example. One of the standard examples of a normed infinite dimensional linear space is given by the space $(\mathrm{C}[a,b], \|\cdot\|_\infty)$ of all continuous real-valued functions on a closed interval $[a, b]$, endowed with the norm $\|f\|_\infty := \max_{x\in[a,b]} |f(x)|$. This is the so-called *Chebyshev norm.* Here the underlying field is the set $\mathbb{R}$ of real numbers. Interpreting the addition of two functions $f, g \in C[a, b]$ pointwise, we see that $C[a, b]$ is a linear space, and that $\|\cdot\|_\infty$ is a corresponding norm.

Strict Norms. A norm is called a ***strict norm*** provided that it has the property that equality holds in the triangle inequality only if the two elements involved are linearly dependent. Such norms are defined by the property that if $f, g \in V$, $f \neq 0$, $g \neq 0$ are such that

$$\|f + g\| = \|f\| + \|g\|,$$

then there exists a number $\lambda \in \mathbb{C}$ with $g = \lambda f$. In this case we can show that in fact $\lambda \in \mathbb{R}$ and $\lambda \geq 0$. Indeed, if $\|f + g\| = \|f + \lambda f\| = \|f\| + \|\lambda f\|$, it follows from $\|f + \lambda f\| = |1 + \lambda| \, \|f\|$ and $\|f\| + \|\lambda f\| = (1 + |\lambda|)\|f\|$ that $|1 + \lambda| = 1 + |\lambda|$, and thus that $\lambda = |\lambda|$.

The norm $\|\cdot\|_2$ on $\mathbb{C}^n$ is a strict norm, since it is easy to see that in this case equality in the triangle inequality can hold only if it also holds in the Cauchy inequality $|\sum_1^n x_\nu \overline{y}_\nu| \leq \|x\|_2 \|y\|_2$. This, however, can only happen if x and y are linearly dependent.

On the other hand, the linear space $(C[a, b], \|\cdot\|_\infty)$ is not strictly normed. This follows from the fact that the functions $f(x) := 1$ and $g(x) := x$ on $[a, b] := [0, 1]$ satisfy $\|f + g\|_\infty = \|f\|_\infty + \|g\|_\infty$, but are nevertheless linearly independent.

1.2 Banach Spaces. If a linear space $(V, \|\cdot\|)$ has the property that every Cauchy sequence of elements is convergent to an element of V, then we say that V is *complete*, and call it a *Banach space.*

STEFAN BANACH (1892–1945) worked in Kracow and Lvov, Poland. He was a member of an important group of mathematicians who where working in Lvov around 1930, including S. Mazur, H. Steinhaus, J. Schauder and S. Ulam. It is said that the group's favorite meeting place was the "Scottish Café", where they often wrote their problems on the marble table tops. A very significant part of modern functional analysis arose out of the work of this group. Among other results of particular importance for numerical analysis is the famous *Banach Fixed Point Theorem*, also called the *Contraction Theorem*, in which the contraction principle for general operators is formulated.

$(C[a, b], \|\cdot\|_\infty)$ is a Banach space, since the elements of $C[a, b]$ are continuous functions, and convergence with respect to the Chebyshev norm corresponds to uniform convergence. It is well known that in this norm every Cauchy sequence of continuous functions converges to a continuous function, i.e., an element of $C[a, b]$. This shows that the linear space $C[a, b]$ is complete.

Being a finite dimensional linear space, $(\mathbb{C}^n, \|\cdot\|_2)$ is complete. This follows since convergence in $\mathbb{C}^n$ implies componentwise convergence. Thus for any Cauchy sequence in $\mathbb{C}^n$, each of its n components is a Cauchy sequence in $\mathbb{C}$, and hence must converge to a number in $\mathbb{C}$.

The Spaces $\mathrm{C}_m(\mathrm{G})$. In addition to the usual finite dimensional linear spaces $\mathbb{C}^n$ and $\mathbb{R}^n$, the linear spaces of continuous and continuously differentiable functions play an important role in numerical analysis. We now discuss these spaces in more detail.

Let G be a bounded set in $\mathbb{R}^n$, and let $\overline{\mathrm{G}}$ be the the closure of G. We write C(G) for the linear space of all continuous real-valued functions on G.

A multi-index γ is an n-tuple of natural numbers $\gamma = (\gamma_1, \ldots, \gamma_n)$. Corresponding to γ, we write $|\gamma| := \sum_1^n \gamma_\nu$, and define the partial derivative of order γ of a function f depending on the variables $x = (x_1, \ldots, x_n)$ by

$$D^\gamma f := \frac{\partial^{|\gamma|} f}{\partial x_1^{\gamma_1} \cdots \partial x_n^{\gamma_n}}.$$

We write $\mathrm{C}_m(\mathrm{G})$ for the linear space of all functions defined on G which are continuous, and all of whose derivatives $D^\gamma f$ with $|\gamma| \leq m$ are also continuous. The space $\mathrm{C}_m(\overline{\mathrm{G}})$ is defined similarly. This space is a Banach space (cf. Problem 3) under the norm

$$\|f\|_\infty := \sum_{|\gamma| \leq m} \max_{x \in \overline{\mathrm{G}}} |D^\gamma f(x)|.$$

As an example of the above, the space $\mathrm{C}_m(a, b)$ is the linear space of all functions on (a, b) which are m-times continuously differentiable. We usually write $\mathrm{C}(a, b)$ for $\mathrm{C}_0(a, b)$. The Banach space of m-times continuously differentiable functions on the closed interval $[a, b]$ endowed with the Chebyshev norm is denoted by $(\mathrm{C}_m[a, b], \|\cdot\|_\infty)$. Here the derivatives at a and at b are understood to be the right- and left-hand derivatives, respectively.

1.3 Hilbert Spaces and Pre-Hilbert Spaces. Normed linear spaces whose norm is induced by an inner product have several additional important properties which we now describe in detail.

A mapping $\langle \cdot, \cdot \rangle : \mathrm{V} \times \mathrm{V} \to \mathbb{C}$ is called an *inner product* provided that the following properties hold for all $f, g, h \in \mathrm{V}$ and $\alpha \in \mathbb{C}$:

$$\begin{aligned}
&\langle f + g, h \rangle = \langle f, h \rangle + \langle g, h \rangle && \text{linearity,} \\
&\langle \alpha f, g \rangle \;= \alpha \langle f, g \rangle && \text{homogeneity,} \\
&\langle f, g \rangle \;= \overline{\langle g, f \rangle} && \text{symmetry,} \\
&\langle f, f \rangle \;> 0 \;\text{ for } f \neq 0 && \text{positivity.}
\end{aligned}$$

In view of these properties, $\|f\| := \langle f, f \rangle^{\frac{1}{2}}$ defines a norm on V. Indeed, conditions 2.4.1(i) and 2.4.1(ii) for a norm are immediate. To check the triangle inequality 2.4.1(iii), we need the following

Schwarz Inequality. For all $f, g \in \mathrm{V}$,

$$|\langle f, g\rangle| \leq \|f\| \, \|g\|.$$

Proof. Since the estimate clearly holds if either $f := 0$ or $g := 0$, we can assume that $f \neq 0$ and $g \neq 0$. For all $\lambda \in \mathbb{C}$, we have $\langle \lambda f + g, \lambda f + g\rangle \geq 0$, and thus

$$|\lambda|^2 \langle f, f\rangle + \overline{\lambda}\langle g, f\rangle + \lambda\langle f, g\rangle + \langle g, g\rangle \geq 0.$$

Now choosing $\lambda := -\frac{\langle g,f\rangle}{\langle f,f\rangle}$, we have $\overline{\lambda} = -\frac{\overline{\langle g,f\rangle}}{\langle f,f\rangle}$ and $|\lambda|^2 = \frac{|\langle g,f\rangle|^2}{\langle f,f\rangle^2}$, and thus that

$$|\langle f, g\rangle|^2 \leq \langle f, f\rangle\langle g, g\rangle. \qquad \square$$

A normed linear space whose norm is induced by an inner product is called a *pre-Hilbert space*. We claim now that pre-Hilbert spaces are always strictly normed linear spaces. Indeed, since

$$\begin{aligned} \langle f+g, f+g\rangle &= \|f\|^2 + \|g\|^2 + \langle f, g\rangle + \langle g, f\rangle \\ &\leq \|f\|^2 + \|g\|^2 + 2|\langle f, g\rangle|, \\ \|f+g\|^2 &\leq (\|f\| + \|g\|)^2, \end{aligned}$$

equality can hold in the triangle inequality only if equality also holds in the Schwarz inequality. But this can happen only if $\langle \lambda f + g, \lambda f + g\rangle = 0$, which in turn implies $\lambda f + g = 0$; that is, f and g are linearly dependent and $\langle f, g\rangle = \langle g, f\rangle = |\langle f, g\rangle|$.

The space $(\mathbb{C}^n, \|\cdot\|_2)$ provides a simple example of a pre-Hilbert space, since the Euclidean norm $\|\cdot\|_2$ is induced by the inner product $\langle x, y\rangle := = \sum_1^n x_\nu \cdot \overline{y}_\nu$, defined for any two vectors $x, y \in \mathbb{C}^n$.

Another important example of a pre-Hilbert space is provided by the space $(\mathrm{C}[a,b], \|\cdot\|_2)$, with norm $\|f\| = [\int_a^b f^2(x)dx]^{\frac{1}{2}}$. This norm is induced by the inner product $\langle f, g\rangle := \int_a^b f(x)g(x)dx$. This space can be generalized by introducing a weight function $w : (a,b) \to \mathbb{R}$, $w(x) > 0$ for $x \in (a,b)$ with $0 < \int_a^b w(x)dx < \infty$ into the inner product, which then becomes $\langle f, g\rangle := \int_a^b w(x)f(x)g(x)dx$. It is easy to check that this is an admissible inner product, and that the corresponding norm is $\|f\| = [\int_a^b w(x)f^2(x)dx]^{\frac{1}{2}}$. When dealing with linear spaces whose elements are complex-valued functions on an interval $[a,b]$, then because of the symmetry condition, we need to modify the inner product $\langle f, g\rangle$ to

$$\langle f, g\rangle := \int_a^b f(x)\overline{g}(x)dx.$$

We have already shown in 1.2 that the space $(\mathbb{C}^n, \|\cdot\|_2)$ is complete. A pre-Hilbert space with this property is called a *Hilbert space*.

The situation for the linear space $(C[a,b], \|\cdot\|_2)$ is, however, different. It is easy to see that this space is not complete, since not every Cauchy sequence of continuous functions which is convergent in the $\|\cdot\|_2$ norm necessarily converges to a continuous function (cf Problem 5). To get a Hilbert space containing $C[a,b]$ and using the norm $\|\cdot\|_2$, we have to replace $C[a,b]$ by the larger space $L^2[a,b]$ of square integrable functions (in the sense of Lebesgue).

DAVID HILBERT (1862–1943) grew up in Königsberg in East Prussia, and from 1895, worked in Göttingen. He was one of the greatest mathematicians of his time. His papers, which covered a wide spectrum from number theory to physics, were fundamental for the development of both pure and applied mathematics in this century. In the obituary "David Hilbert and His Mathematical Work", Bull. Amer. Math. Soc. 50, 612–654 (1944), H. Weyl (1885–1955), another of the most important mathematicians of this century wrote: "A great master of mathematics passed away when David Hilbert died in Göttingen on February 14th, 1943, at the age of eighty-one. In retrospect it seems to us that the era of mathematics upon which he impressed the seal of his spirit and which is now sinking below the horizon achieved a more perfect balance than prevailed before and after, between the mastering of single concrete problems and the formation of general abstract concepts ...". Hilbert's work on integral equations, which is especially noteworthy as a mathematical model for physical processes, led to the concept which we now refer to as Hilbert space. The book of C. Reid [1973] contains a detailed biography of Hilbert.

1.4 The Spaces $L^p[a,b]$. For completeness, we also introduce the linear space $L^p[a,b]$ for given $1 \le p < \infty$. This space consists of all real functions f for which $|f|^p$ is Lebesgue integrable, endowed with the norm

$$\|f\|_p := [\int_a^b |f(x)|^p dx]^{\frac{1}{p}}.$$

We immediately see that the norm conditions 2.4.1(i) and 2.4.1(ii) hold. The triangle inequality 2.4.1(iii) is referred to as the

Minkowski Inequality

$$\|f+g\|_p \le \|f\|_p + \|g\|_p.$$

For the case of Riemann integrals, see e.g. D. Luenberger ([1969], p. 33). The result also holds, however, for the Lebesgue integral.

The following inequality involving p-norms is also important:

Hölder Inequality. For all $p, q > 1$ with $\frac{1}{p} + \frac{1}{q} = 1$,

$$|\langle f, g\rangle| \le \|f\|_p \|g\|_q.$$

This inequality also holds both for the Riemann integral; cf. D. Luenberger ([1969], p. 32), as well as for the Lebesgue integral. For $p = q = 2$ it reduces to the Schwarz inequality.

All of the spaces $L^p[a,b]$ are Banach spaces, but among these, only the space $L^2[a,b]$ is a Hilbert space. If we take $p = \infty$ and restrict our attention to $C[a,b]$, then the norm $\|\cdot\|_p$ reduces to the Chebyshev norm, and we obtain the Banach space $(C[a,b], \|\cdot\|_\infty)$ with $\|f\|_\infty = \max_{x\in[a,b]} |f(x)|$.

In addition to the cases $p = 2$ and $p = \infty$, the case $p = 1$ is also of some interest in numerical analysis. The normed linear space $(C[a,b], \|\cdot\|_1)$ is of particular interest. This space is certainly not complete, since the limit of a convergent sequence with respect to the norm $\|\cdot\|_1$ is not necessarily continuous (Problem 5).

Of the normed linear spaces of the type $C_m(G)$ and $L^p[a,b]$ defined above, in this book we will use only the Banach spaces $(C_m(\overline{G}), \|\cdot\|_\infty)$, the pre-Hilbert space $(C[a,b], \|\cdot\|_2)$, the Hilbert space $L^2[a,b]$, and the non-complete normed linear space $(C[a,b], \|\cdot\|_1)$.

1.5 Linear Operators. We now consider mappings of a linear space into another linear space, or into itself; see Definition 2.4.2. Let X and Y be linear spaces, and let Q be a mapping which uniquely associates the elements of one subset $D \subset X$ with elements of another subset $W \subset Y$. Then we call Q an *operator*, D its *domain* and W its *image*. We write $Q : D \to W$.

If D is a linear subspace of X, then we call Q a *linear operator* provided that

$$Q(\alpha f + \beta g) = \alpha Qf + \beta Qg$$

for all $\alpha, \beta \in \mathbb{K}$ and for all $f, g \in D$.

Example 1. Let $f \in C[a,b]$. Then the definite integral $Jf := \int_a^b w(x)f(x)dx$ with the weight function w can be thought of as a linear operator J. The operator J maps $C[a,b]$ into $\mathbb{R}$.

A linear operator (such as the one in Example 1) whose image lies in $\mathbb{R}$ or $\mathbb{C}$ is called a *linear functional.*

Example 2. The matrix $A := (a_{\mu\nu})_{\substack{\mu=1,\ldots,m \\ \nu=1,\ldots,n}}$, $a_{\mu\nu} \in \mathbb{C}$ maps the linear space $\mathbb{C}^n$ into $\mathbb{C}^m$, and is, of course, also a linear operator.

Bounded Linear Operators. The linear operator L is called *bounded* provided that there exists a number $K \in \mathbb{R}$ such that for all elements $x \in D$,

$$\|Lx\| \le K\|x\|.$$

This concept of boundedness of an operator is the natural generalization to operators of the concept of Lipschitz-boundedness of functions. Indeed, on the one hand, $\|L(x-y)\| = \|Lx - Ly\| \le K\|x-y\|$ holds for any bounded operator L. Conversely, if L is Lipschitz-bounded; i.e., $\|Lx - Ly\| \le$ $\le K\|x-y\|$, then taking $y := 0$ and using the fact that $L0 = 0$ for any linear operator, we conclude that the operator L is bounded.

We can now introduce the *norm* of a bounded linear operator.

Definition. We define the norm of a bounded linear operator L to be the number $\|L\| := \inf\{K \in \mathbb{R} \mid \|Lx\| \leq K\|x\| \text{ for all } x \in \mathrm{D}\}$. With this definition, we have

$$\|Lx\| \leq \|L\|\,\|x\|.$$

Fact. $\|L\| = \sup_{0\neq x\in \mathrm{D}} \frac{\|Lx\|}{\|x\|}$. To see this, note that $\frac{\|Lx\|}{\|x\|} \leq \|L\|$ for all $x \in \mathrm{D}, x \neq 0$, and in particular $\sup_{0\neq x\in \mathrm{D}} \frac{\|Lx\|}{\|x\|} =: M \leq \|L\|$. But, on the other hand $\|Lx\| = \frac{\|Lx\|}{\|x\|}\|x\| \leq M\|x\|$ for $0 \neq x \in \mathrm{D}$, and thus $\|L\| \leq M$. This shows that $M \leq \|L\| \leq M$, and the assertion is established. □

The norm of L can also be written in the form $\|L\| = \sup_{\|x\|=1} \|Lx\|$. It is easy to show that the mapping $\|L\|$ satisfies the properties of a norm. Moreover, the *product* $(L_1L_2)x := L_1(L_2x)$ of two linear operators L_1 and L_2 satisfies the inequality

$$\|L_1L_2\| \leq \|L_1\|\,\|L_2\|,$$

since $\|(L_1L_2)x\| \leq \|L_1\|\,\|L_2x\| \leq \|L_1\|\,\|L_2\|\,\|x\|$.

Application. We consider once again the two examples of linear operators given above.

Example 1. The integral operator $J : \mathrm{C}[a,b] \to \mathbb{R}$ on the space $(\mathrm{C}[a,b], \|\cdot\|_\infty)$ is a bounded linear operator since

$$|Jf| = |\int_a^b w(x)f(x)dx| \leq \int_a^b w(x)dx\|f\|_\infty \quad \text{for } w(x) > 0 \text{ in } \ (a,b),$$

and thus $\|J\| = \sup_{\|f\|_\infty=1} |Jf| \leq \int_a^b w(x)dx$. Since J is a mapping into $\mathbb{R}$, it is in fact a bounded linear functional.

We also note that for the element $f^* := 1$, the estimate $\sup_{\|f\|_\infty=1} |Jf| \geq$ $\geq |Jf^*| = \int_a^b w(x)dx$ holds, and thus $\|J\| \geq \int_a^b w(x)dx$. Combining these facts, we conclude that the norm of J is given by $\|J\| = \int_a^b w(x)dx$.

Example 2. In view of the results in 2.4.2, it follows that every finite-dimensional matrix is a bounded linear operator. Various matrix norms were calculated in 2.4.3.

1.6 Problems. 1) Show that the mapping

$$a : \mathrm{C}_1[0,1] \to \mathbb{R},\ a(f) := (\int_0^1 |f'(x)|^2 w(x)dx)^{\frac{1}{2}} + \sup_{x\in[0,1]} |f(x)|$$

defines a norm on $\mathrm{C}_1[0,1]$. Is this norm strict if $w(x) := 1$?

2) Let $\|\cdot\|_a$ and $\|\cdot\|_b$ be norms on the linear space V and suppose that $\|\cdot\|_a$ is strict. Show that the norm $\|v\| := \|v\|_a + \|v\|_b$ on V is also strict.

3) Show that the mapping

$$a : \mathrm{C}_m(\overline{\mathrm{G}}) \to \mathbb{R},\ a(f) := \sum_{|\gamma|\leq m} \max_{x\in\overline{\mathrm{G}}} |D^\gamma f(x)|$$

defines a norm on the linear space $C_m(\overline{G})$, and that $C_m(\overline{G})$ equipped with this norm is a Banach space.

4) Let $(V, \|\cdot\|)$ be a normed linear space over $\mathbb{R}$. Show that the norm $\|\cdot\|$ is induced by an inner product $\langle\cdot,\cdot\rangle$ if and only if the "parallelogram law"

$$\|f+g\|^2 + \|f-g\|^2 = 2(\|f\|^2 + \|g\|^2)$$

holds for all $f, g \in V$. Note that in $(\mathbb{R}^2, \|\cdot\|_2)$, the parallelogram law with $\langle x, y\rangle = 0$ reduces to the Pythagorean Theorem.

Hint: Assume $\langle f, g\rangle := \frac{1}{4}(\|f+g\|^2 - \|f-g\|^2)$.

5) By investigating the convergence of the sequence $(f_n)_{n\in\mathbb{Z}_+}$ defined on $[a,b] := [-1,+1]$ by

$$f_n(x) := \begin{cases} -1 & \text{for } x \in [-1, -\frac{1}{n}] \\ nx & \text{for } x \in [-\frac{1}{n}, +\frac{1}{n}] \\ 1 & \text{for } x \in [\frac{1}{n}, 1], \end{cases}$$

show that the linear space $C[a,b]$ is not complete with respect to either the norm $\|\cdot\|_2$ or the norm $\|\cdot\|_1$.

6) Show that the mapping $Ff := \sum_1^n \alpha_\nu f(x_\nu)$, $\alpha_\nu \in \mathbb{R}$ defined for functions $f \in C[a,b]$ is a bounded linear functional on the normed linear space $(C[a,b], \|\cdot\|_\infty)$, and that $\|F\| = \sum_1^n |\alpha_\nu|$.

2. The Approximation Theorems of Weierstrass

We begin our discussion of approximation theory with the classical problem of approximating a function. A more general approximation problem will be treated later in this chapter. In this section we shall present several approximation theorems of Weierstrass which show how to approximate an arbitrary continuous function by simple functions.

2.1 Approximation by Polynomials. It is known from calculus that an analytic function f can be written as a power series

$$f(x) = a_0 + a_1 x + \cdots + a_n x^n + \cdots$$

which uniformly converges to the function f inside a certain convergence interval.

Consider now the sequence $(\sigma_n)_{n\in\mathbb{N}}$ of partial sums of the power series defined by

$$\sigma_n(x) = a_0 + a_1 x + \cdots + a_n x^n.$$

Then it is clear that for every $\varepsilon > 0$, there exists a number $N(\varepsilon) \in \mathbb{N}$ such that $\|f - \sigma_n\|_\infty < \varepsilon$ for every $n > N$. In other words, for any given interval, there always exists a polynomial which uniformly approximates the analytic function arbitrarily well.

It is now natural to ask whether a similar assertion still holds if we assume only that f is continuous. In general, such an approximation cannot be in the form of a power series, since as is well known, power series represent functions which are infinitely differentiable, whereas certainly not every continuous function f has derivatives.

We answer this question in the following section by establishing the classical approximation theorem of Weierstrass. Although we shall later discuss a more general theorem of Korovkin, it is worthwhile to first present the original Weierstrass Theorem with a direct proof. Indeed, in this way we can formulate the theorem in a simple instructive way, and moreover, we can give a constructive proof due to S. N. BERNSTEIN in 1912 which serves to motivate the later results of P. P. KOROVKIN.

KARL WEIERSTRASS (1815–1897) established his approximation theorems in the paper "Über die analytische Darstellbarkeit sogenannter willkürlicher Funktionen reeller Argumente" (Sitzg. ber. Kgl. Preuss. Akad. d. Wiss. Berlin 1885, pp. 663–639, 789–805). He gave non-constructive proofs of his theorems. Weierstrass became famous primarily for his fundamental results in analysis. He is considered to be one of the founders of modern function theory; the starting point of his work is the power series. In addition, Weierstrass fully understood the great importance of mathematics for applications to problems in physics and astronomy. For this reason he gave mathematics a leading position,"since through it alone can one obtain a truely satisfactory understanding of nature". (Quote from I. Runge ([1949], p. 29)).

Because of its potential applications, we now present S. N. Bernstein's constructive proof of the approximation theorem for continuous functions. The so-called Bernstein polynomials which appear in the proof came originally from probability theory.

Before proceeding, we mention that there are a series of alternative proofs of these approximation theorems, for example by E. LANDAU (1908), H. LEBESGUE (1908), and others. We also mention a generalization to topological spaces due to M. H. STONE (1948).

2.2 The Approximation Theorem for Continuous Functions. In this section we prove that every continuous function on a given finite closed interval can be uniformly approximated arbitrarily well by a polynomial. This means that the polynomials are dense in the space $\mathrm{C}[a, b]$ of continuous functions.

Let P_n denote the $(n+1)$-dimensional linear space of all polynomials of maximal degree n over the field $\mathbb{R}$, defined by

$$\mathrm{P}_n := \{p \in \mathrm{C}(-\infty, +\infty) \mid p(x) = \sum_{\nu=0}^{n} a_\nu x^\nu\}.$$

Approximation Theorem of Weierstrass. *Let $-\infty < a < b < +\infty$, and suppose that $f \in C[a,b]$ is an arbitrary continuous function. Then for every $\varepsilon > 0$, there exists an $n \in \mathbb{N}$ and a polynomial $p \in P_n$ such that $\|f - p\|_\infty < \varepsilon$.*

Proof. Since every interval $[a,b]$ can be mapped onto $[0,1]$ by a linear transformation, we may restrict our attention to the case $[a,b] := [0,1]$. To establish the theorem, we shall show that the sequence $(B_n f)$ of *Bernstein-Polynomials*

$$(B_n f)(x) := \sum_{\nu=0}^{n} f\left(\frac{\nu}{n}\right)\binom{n}{\nu} x^\nu (1-x)^{n-\nu}, \quad n = 1, 2, \cdots,$$

converges uniformly to f on $[0,1]$.

First we note that $(B_n f)(0) = f(0)$ and $(B_n f)(1) = f(1)$ for all n. Now

$$1 = [x + (1-x)]^n = \sum_{\nu=0}^{n} \binom{n}{\nu} x^\nu (1-x)^{n-\nu} =: \sum_{\nu=0}^{n} q_{n\nu}(x)$$

implies

$$(*) \qquad f(x) - (B_n f)(x) = \sum_{\nu=0}^{n} \left[f(x) - f\left(\frac{\nu}{n}\right)\right] q_{n\nu}(x),$$

and thus

$$|f(x) - (B_n f)(x)| \le \sum_{\nu=0}^{n} \left| f(x) - f\left(\frac{\nu}{n}\right) \right| q_{n\nu}(x)$$

for all $x \in [0,1]$.

By the (uniform) continuity of f, for every $\varepsilon > 0$ there exists a number δ, not depending on x, such that $|f(x) - f(\frac{\nu}{n})| < \frac{\varepsilon}{2}$ for all points x with $|x - \frac{\nu}{n}| < \delta$.

For every $x \in [0,1]$, consider the two sets

$$N' := \left\{\nu \in \{0,1,\ldots,n\} : \left| x - \frac{\nu}{n} \right| < \delta\right\}$$

$$N'' := \left\{\nu \in \{0,1,\ldots,n\} : \left| x - \frac{\nu}{n} \right| \ge \delta\right\},$$

and split the sum into two parts $\sum_{\nu=0}^{n} = \sum_{\nu \in N'} + \sum_{\nu \in N''}$. Then the first sum satisfies

$$\sum_{\nu \in N'} \left| f(x) - f\left(\frac{\nu}{n}\right) \right| q_{n\nu}(x) \le \frac{\varepsilon}{2} \sum_{\nu \in N'} q_{n\nu}(x) \le \frac{\varepsilon}{2} \sum_{\nu=0}^{n} q_{n\nu}(x) = \frac{\varepsilon}{2}.$$

Moreover, with $M := \max_{x \in [0,1]} |f(x)|$ we also have

$$\sum_{\nu \in N''} \left| f(x) - f\left(\frac{\nu}{n}\right) \right| q_{n\nu}(x) \le \sum_{\nu \in N''} \left| f(x) - f\left(\frac{\nu}{n}\right) \right| q_{n\nu}(x) \frac{(x - \frac{\nu}{n})^2}{\delta^2} \le$$

$$\le \frac{2M}{\delta^2} \sum_{\nu=0}^{n} q_{n\nu}(x) \left(x - \frac{\nu}{n}\right)^2.$$

Since $(x - \frac{\nu}{n})^2 = x^2 - 2x\frac{\nu}{n} + (\frac{\nu}{n})^2$, the last sum can be separated into the following three parts:

$$(1) \sum_{\nu=0}^{n} \binom{n}{\nu} x^{\nu}(1-x)^{n-\nu} = 1;$$

$$(2) \sum_{\nu=0}^{n} \binom{n}{\nu} x^{\nu}(1-x)^{n-\nu} \frac{\nu}{n} = x \sum_{\nu=1}^{n} \binom{n-1}{\nu-1} x^{\nu-1}(1-x)^{(n-1)-(\nu-1)} = x;$$

$$(3) \sum_{\nu=0}^{n} \binom{n}{\nu} x^{\nu}(1-x)^{n-\nu} \left(\frac{\nu}{n}\right)^2 =$$

$$= \frac{x}{n} \sum_{\nu=1}^{n} (\nu - 1) \binom{n-1}{\nu-1} x^{\nu-1}(1-x)^{(n-1)-(\nu-1)} + \frac{x}{n} =$$

$$= \frac{x^2}{n}(n-1) \sum_{\nu=2}^{n} \binom{n-2}{\nu-2} x^{\nu-2}(1-x)^{(n-2)-(\nu-2)} + \frac{x}{n} = x^2(1 - \frac{1}{n}) + \frac{x}{n} =$$

$$= x^2 + \frac{x}{n}(1-x).$$

Thus, for all $x \in [0,1]$,

$$(**) \qquad \sum_{\nu=0}^{n} q_{n\nu}(x)\left(x - \frac{\nu}{n}\right)^2 = x^2 \cdot 1 - 2x \cdot x + x^2 + \frac{x(1-x)}{n} \le \frac{1}{4n}$$

and

$$\sum_{\nu \in N''} \left| f(x) - f\left(\frac{\nu}{n}\right) \right| q_{n\nu}(x) \le \frac{2M}{\delta^2} \frac{1}{4n} < \frac{\varepsilon}{2},$$

provided that we choose $n > \frac{M}{\delta^2 \varepsilon}$. Combining these facts, we obtain the estimate

$$|f(x) - (B_n f)(x)| < \frac{\varepsilon}{2} + \frac{\varepsilon}{2} = \varepsilon$$

for all $x \in [0,1]$, which establishes the uniform convergence of the sequence $(B_n f)$. □

Remark. We can now give an answer to the question raised in 2.1. Every analytic function can be expanded in a power series, while every continuous function can be represented as an expansion in terms of polynomials as follows:

$$f(x) = (B_1 f)(x) + [(B_2 f)(x) - (B_1 f)(x)] + \cdots + [(B_n f)(x) - (B_{n-1} f)(x)] + \cdots.$$

This series converges uniformly, but in general cannot be rearranged into a power series.

2.3 The Korovkin Approach. Examining the proof given in the previous section once again, we note that the estimation of the sums (1) – (3) is the essential part of the proof of the convergence of the sum (*). Indeed, it is clear that the convergence essentially depends on being able to establish the uniform convergence of the sums (1), (2) and (3) for the functions $e_1(x) := 1$, $e_2(x) := x$, and $e_3(x) := x^2$. This suggests that the convergence of the sequence of Bernstein polynomials to an arbitrary continuous function is already determined by the way in which the Bernstein polynomials behave for the three elements e_1, e_2, $e_3 \in \mathrm{C}[a,b]$.

This conjecture turns out to be correct. In 1953, P. P. Korovkin established a general approximation theorem which contains this assertion. His proof depends in an essential way on the concept of a

Monotone Linear Operator. Let $f, g \in \mathrm{C}(\mathrm{I})$ be given functions such that $f \leq g$, where this notation means that $f(x) \leq g(x)$ for all $x \in \mathrm{I}$. Then a linear operator $L : \mathrm{C}(\mathrm{I}) \to \mathrm{C}(\mathrm{I})$ is called *monotone* provided that $Lf \leq Lg$. This property is equivalent to the property of *positivity*, i. e., $f \geq 0$ implies $Lf \geq 0$. In 2.4 we shall exploit the fact that the Bernstein operators defined there are positive.

Korovkin investigated sequences $(L_n)_{n \in \mathbb{N}}$ of positive linear operators $L_n : \mathrm{C}(\mathrm{I}) \to \mathrm{C}(\mathrm{I})$ for $\mathrm{I} := [0,1]$ which map continuous functions $f \in \mathrm{C}(\mathrm{I})$ to polynomials, as well as similar operators which map a continuous and 2π-periodic function $f \in \mathrm{C}_{2\pi}(\mathrm{I})$ with $I := [-\pi, \pi]$ to trigonometric polynomials of maximal degree n. He showed that for every $f \in \mathrm{C}([0,1])$, the sequence $(L_n f)$ converges uniformly to f provided that uniform convergence holds for the three functions $e_1(x) := 1$, $e_2(x) := x$, $e_3(x) := x^2$, and that the same holds for every $f \in \mathrm{C}_{2\pi}([-\pi, \pi])$, provided that it holds for each of the three functions $e_1(x) := 1$, $e_2(x) := \sin(x)$, $e_3(x) := \cos(x)$.

Korovkin's proofs of these two facts are similar, but not exactly the same. Here we present a unified and generalized version of the proof due to E. Schäfer [1989]. This proof can, in fact, be further simplified if one is interested only in the two special cases of continuous functions mentioned above.

Consider the linear space $(C(I), \|\cdot\|_\infty)$. Let $Q := \{f_1, \dots, f_k\}$, $Q \subset C(I)$, and let $e_1 \in \operatorname{span}(Q)$. We call the set Q a *test set* provided that there exists a function $p \in C(I \times I)$ such that $p(t,x) := \sum_{\kappa=1}^{k} a_\kappa(t) f_\kappa(x)$ with $a_\kappa \in C(I)$ for $1 \le \kappa \le k$, $p(t,x) \ge 0$ for all $(t,x) \in I \times I$, and $p(t,t) = 0$ for all $t \in I$.

We denote by $Z(g) := \{(t,x) \in I \times I \mid g(t,x) = 0\}$ the zero set of a function $g \in C(I \times I)$, and write $d_f(t,x) := f(x) - f(t)$ for the "difference function" associated with a given $f \in C(I)$. We now have the following

Theorem. *Let $(L_n)_{n \in \mathbb{N}}$, $L_n : C(I) \to C(I)$, be a sequence of positive linear operators, and let Q be a test set with associated function p. Suppose that for every element $f \in Q$, $\lim_{n\to\infty} \|L_n f - f\|_\infty = 0$. Then it follows that $\lim_{n\to\infty} \|L_n f - f\|_\infty = 0$ for every element $f \in C(I)$ which satisfies the condition $Z(p) \subset Z(d_f)$.*

Proof. In part (a) of the proof we show that for $\lim_{n\to\infty} \|f - L_n f\|_\infty = 0$, it suffices to establish that $\lim_{n\to\infty} \max_{t\in I} |(L_n d_f(t,\cdot))(t)| = 0$. The proof that $\lim_{n\to\infty} \max_{t\in I} |(L_n d_f(t,\cdot))(t)| = 0$ for all elements $f \in C(I)$ such that $Z(p) \subset Z(d_f)$ is presented in part (b).

(a) $d_f(t,\cdot) = f - f(t)e_1$ satisfies $f - L_n f = f - f(t)L_n e_1 - L_n d_f(t,\cdot)$. From this it follows that

$$|f(t) - (L_n f)(t)| \le \|f\|_\infty \|e_1 - L_n e_1\|_\infty + \max_{t\in I} |(L_n d_f(t,\cdot))(t)|,$$

uniformly for all $t \in I$. Since $e_1 \in \operatorname{span}(Q)$, we get $\lim_{n\to\infty} \|e_1 - L_n e_1\|_\infty = 0$, and thus $\lim_{n\to\infty} \max_{t\in I} |(L_n d_f(t,\cdot))(t)| = 0$ gives $\lim_{n\to\infty} \|f - L_n f\|_\infty = 0$.

(b) The difference function depends continuously on the variables x and t. Hence, for every $\varepsilon > 0$, there exists an open neighborhood Ω of $Z(d_f)$, where $|d_f(t,x)| < \varepsilon$ for all $(t,x) \in \Omega$. The diagonal set D defined by $D := \{(t,x) \in I \times I \mid t = x\}$ is thus surely a subset of $Z(d_f)$. By the assumption that $Z(p) \subset Z(d_f)$, it follows that $p(t,x) > 0$ in the complement $\Omega' := I \times I \setminus \Omega$.

Ω' is closed and hence compact, which assures that the minimum $0 < m := \min_{(t,x)\in\Omega'} p(t,x)$ exists. Thus

$$|d_f(t,x)| \le \|d_f\|_\infty \frac{p(t,x)}{m} \quad \text{for} \quad (t,x) \in \Omega',$$

and we have

$$|d_f(t,x)| \le \frac{\|d_f\|_\infty}{m} p(t,x) + \varepsilon \quad \text{for} \quad (t,x) \in I \times I.$$

Applying the positive operator L_n with respect to x for fixed t, it follows that

$$|(L_n d_f(t,\cdot))(t)| \le \frac{\|d_f\|_\infty}{m} (L_n p(t,\cdot))(t) + \varepsilon (L_n e_1)(t) \le$$

$$\le \frac{\|d_f\|_\infty}{m} \max_{t\in I} (L_n p(t,\cdot))(t) + \varepsilon \|L_n e_1\|_\infty.$$

Since $p(t,t) = 0$ for all $t \in \mathrm{I}$, we can write

$$(L_n p(t,\cdot))(t) = \sum_{\kappa=1}^{k} a_\kappa(t)[(L_n f_\kappa)(t) - f_\kappa(t)].$$

The convergence of the sequence (L_n) on span(Q) thus implies that

$$\lim_{n\to\infty} \max_{t\in\mathrm{I}} (L_n p(t,\cdot))(t) = 0.$$

Since $\|L_n e_1\|_\infty$ is uniformly bounded in n, we finally arrive at the assertion

$$\lim_{n\to\infty} \max_{t\in\mathrm{I}} |(L_n d_f(t,\cdot))(t)| = 0. \qquad \square$$

2.4 Applications of Theorem 2.3. In this section we apply Theorem 2.3 to obtain the classical approximation theorems of Weierstrass. Although we have already established the approximation theorem for continuous functions in 2.2, here we reprove it by showing how it follows from Theorem 2.3.

In order to apply Theorem 2.3, we must find an appropriate test set and a sequence of positive linear operators which converges on this test set. We begin by establishing the Approximation Theorem 2.2 with the help of

Bernstein-Operators. The Bernstein polynomial $B_n f$ introduced in the proof of Theorem 2.2 defines a mapping of the space of continuous functions into the linear subspace of polynomials P_n. Considering B_n as an operator $B_n : \mathrm{C(I)} \to \mathrm{C(I)}$, it is easy to see that it is linear und monotone. First, from the definition

$$(B_n f)(x) = \sum_{\nu=0}^{n} f\left(\frac{\nu}{n}\right)\binom{n}{\nu} x^\nu (1-x)^{n-\nu},$$

it follows immediately that $B_n(\alpha f + \beta g) = \alpha B_n f + \beta B_n g$, and thus that B_n is linear. Since $f \geq 0$ implies $B_n f \geq 0$, it follows that B_n is positive, or equivalently, monotone.

A natural choice for a set of test functions Q is the set $\{f_1, f_2, f_3\}$ with $f_1(x) := e_1(x) = 1$, $f_2(x) := e_2(x) = x$, $f_3(x) := e_3(x) = x^2$, with corresponding p defined by $p(x,t) := (t-x)^2 = t^2 - 2tx + x^2$. The condition $\mathrm{Z}(p) \subset \mathrm{Z}(d_f)$ holds for every $f \in \mathrm{C(I)}$ since $p(x,t) = 0$ if and only if $x = t$.

Our choice of the elements e_1, e_2, e_3 for the set Q is motivated by the fact that in the proof of Theorem 2.2, we established the fact that $\lim_{n\to\infty} \|B_n e_\kappa - e_\kappa\|_\infty = 0$ for $\kappa = 1, 2, 3$. This together with Theorem 2.3 implies that $\lim_{n\to\infty} \|B_n f - f\|_\infty = 0$ for all elements $f \in \mathrm{C(I)}$. We have obtained the Approximation Theorem 2.2 as an application of Theorem 2.3.

Periodic Functions. A natural way to approximate a 2π-periodic function as a linear combination of given elements is to use the Fourier series expansion in terms of trigonometric functions. It is known, however, that the sequence $(S_n f)_{n\in\mathbb{N}}$ of Fourier sums

$$(S_n f)(x) = \frac{a_0}{2} + \sum_{\nu=1}^{n} [a_\nu \cos(\nu x) + b_\nu \sin(\nu x)]$$

with

$$a_\nu = \frac{1}{\pi}\int_{-\pi}^{+\pi} f(x)\cos(\nu x)dx \quad \text{for } \nu = 0,\cdots,n,$$
$$b_\nu = \frac{1}{\pi}\int_{-\pi}^{+\pi} f(x)\sin(\nu x)dx \quad \text{for } \nu = 1,\cdots,n$$

does not converge uniformly to f for every function $f \in C_{2\pi}[-\pi,+\pi]$, and indeed that sometimes we don't even have pointwise convergence.

This lack of convergence can be overcome by considering *Cesáro summation* as introduced by E. CESÁRO (1859–1906). The n-th Cesáro sum is defined to be the arithmetic mean of the first n terms of the sequence $S_0 f, \ldots, S_{n-1} f$; i.e.,

$$F_n f := \frac{S_0 f + \cdots + S_{n-1} f}{n}.$$

It will be convenient to write $(F_n f)(x)$ as an integral. We start with an integral representation of the Fourier sum

$$(S_j f)(x) = \frac{1}{2\pi}\int_{-\pi}^{+\pi} f(t)\frac{\sin((2j+1)\frac{t-x}{2})}{\sin\frac{t-x}{2}}dt$$

in terms of the Dirichlet kernel (cf. e. g. P. Davis [1963], Chap. XII). Applying

$$\sin((j+\frac{1}{2})u)\sin\frac{u}{2} = \frac{1}{2}[\cos(ju) - \cos((j+1)u)],$$

we obtain

$$\sum_{j=0}^{n-1}\sin((j+\frac{1}{2})u)\sin\frac{u}{2} = \frac{1}{2}\sum_{j=0}^{n-1}[\cos(ju) - \cos((j+1)u)] =$$

$$= \frac{1}{2}[1-\cos(nu)] = \sin^2\frac{nu}{2}.$$

It follows that

$$(F_n f)(x) = \frac{1}{2\pi n}\int_{-\pi}^{+\pi} f(t)\left[\sum_{j=0}^{n-1}\frac{\sin((2j+1)\frac{t-x}{2})}{\sin\frac{t-x}{2}}\right]dt =$$

$$= \frac{1}{2\pi n}\int_{-\pi}^{+\pi} f(t)\frac{\sin^2(\frac{n(t-x)}{2})}{\sin^2(\frac{t-x}{2})}dt.$$

The operator $F_n : C_{2\pi}[-\pi,+\pi] \to C_{2\pi}[-\pi,+\pi]$ is called the *Fejér operator*, after L. FEJÉR (1880–1959). One immediately sees that F_n is linear and positive (or equivalently monotone).

An appropriate test set for the application of Theorem 2.3 is given by $f_1(x) := 1$, $f_2(x) := \cos(x)$, $f_3(x) := \sin(x)$ with associated function $p(t,x) := 1 - \cos(t-x) = 1 - \cos(t)\cos(x) - \sin(t)\sin(x)$. The zero set $Z(p)$ is now $Z(p) = D \cup \{(-\pi,+\pi), (+\pi,-\pi)\}$, where D is the diagonal set defined in the proof in 2.3. Now if f is a function in $C_{2\pi}[-\pi,+\pi]$, then by the periodicity, $\{(-\pi,+\pi), (+\pi,-\pi)\} \subset Z(d_f)$. Since $D \subset Z(d_f)$, it follows that $Z(p) \subset Z(d_f)$.

It remains to show that $\lim_{n\to\infty} \|F_n f_\kappa - f_\kappa\|_\infty = 0$ for $\kappa = 1,2,3$. This follows immediately from the three identities $(F_n f_1)(x) = 1$ for $n \geq 0$, $(F_n f_2)(x) = \frac{n-1}{n}\cos(x)$, and $(F_n f_3)(x) = \frac{n-1}{n}\sin(x)$ for $n \geq 1$.

We have established

The Weierstrass Approximation Theorem for Periodic Functions. *Every continuous periodic function can be uniformly approximated arbitrarily well by trigonometric polynomials.*

Functions of Several Variables. Let f be a continuous function of m variables $x_1,\ldots,x_m \in [0,1]$. As a direct generalization of the univariate case, we define the multivariate Bernstein polynomial by

$$(B_{n_1\ldots n_m} f)(x_1,\ldots,x_m) :=$$
$$= \sum_{\nu_1=0}^{n_1} \cdots \sum_{\nu_m=0}^{n_m} f\left(\frac{\nu_1}{n_1},\ldots,\frac{\nu_m}{n_m}\right) \cdot q_{n_1\nu_1}(x_1)\cdots q_{n_m\nu_m}(x_m).$$

The associated operator $B_{n_1\cdots n_m}$ is again linear and monotone. To apply Theorem 2.3 we need a test set. Here we may take

$$p(t_1,\ldots,t_m,x_1,\ldots,x_m) := \sum_{\mu=1}^{m} (t_\mu - x_\mu)^2,$$

and define the test set to consist of the functions $f_1(x_1,\ldots,x_m) = 1$, $f_\kappa(x_1,\ldots,x_m) = x_{\kappa-1}$, $\kappa = 2,\ldots,m+1$ and $f_{m+2}(x_1,\ldots,x_m) = \sum_{\mu=1}^m x_\mu^2$.

In the same way as in the proof of Theorem 2.2, we can now show that the sequence $(B_{n_1\cdots n_m} f_\kappa)$ for $\kappa = 1,\ldots,m+2$ converges uniformly to f_κ provided that $\min_{1\leq\mu\leq m} n_\mu \to \infty$. Thus, it follows that the Weierstrass Approximation Theorem 2.2 also holds for continuous functions of several variables. This result also appeared already in K. Weierstrass [1885].

2.5 Approximation Error. The fundamental question of whether a continuous function can be approximated by polynomials has been answered by the Approximation Theorem 2.2 of Weierstrass. There remains, however, the question of how usable the Bernstein polynomials are. We cannot, of course, expect that the same convergence rates will hold for all continuous functions. The class of continuous functions contains a wide variety of functions whose special properties influence the convergence rate.

In order to differentiate between various degrees of continuity, we now introduce the *modulus of continuity* of a function f defined by

$$\omega_f(\delta) := \sup_{\substack{|x'-x''|\le\delta \\ x',x''\in[0,1]}} |f(x') - f(x'')|.$$

We want to estimate the approximation error $|f(x) - (B_n f)(x)|$ in terms of this modulus of continuity.

Given x' and x'', let $\lambda = \lambda(x', x''; \delta)$ be the largest integer less than or equal to $\frac{|x'-x''|}{\delta}$. Then since $\omega_f(\delta_1) \le \omega_f(\delta_2)$ for $\delta_1 \le \delta_2$, it follows that

$$|f(x') - f(x'')| \le \omega_f(|x' - x''|) \le \omega_f((\lambda + 1)\delta).$$

Since $\omega_f(\mu\delta) \le \mu\omega_f(\delta)$ for $\mu \in \mathbb{N}$, we obtain

$$|f(x') - f(x'')| \le (\lambda + 1)\omega_f(\delta).$$

Now let $\mathrm{N}^* := \{\nu \in \{0, \ldots, n\} \mid \lambda(x, \frac{\nu}{n}\delta) \ge 1\}$. Then, starting as in the proof of Theorem 2.2, we have the estimate

$$|f(x)-(B_n f)(x)| \le \sum_{\nu=0}^{n} |f(x)-f(\frac{\nu}{n})| q_{n\nu}(x) \le \omega_f(\delta) \sum_{\nu=0}^{n} (1+\lambda(x, \frac{\nu}{n}; \delta)) q_{n\nu}(x).$$

Since $\lambda(x, \frac{\nu}{n}; \delta) = 0$ for all values $\nu \notin \mathrm{N}^*$, it also follows that

$$\begin{aligned}
|f(x) - (B_n f)(x)| &\le \omega_f(\delta)(1 + \sum_{\nu\in\mathrm{N}^*} \lambda(x, \frac{\nu}{n}; \delta) q_{n\nu}(x)) \le \\
&\le \omega_f(\delta)(1 + \delta^{-1} \sum_{\nu\in\mathrm{N}^*} |x - \frac{\nu}{n}| q_{n\nu}(x)) \le \\
&\le \omega_f(\delta)(1 + \delta^{-2} \sum_{\nu=0}^{n} (x - \frac{\nu}{n})^2 q_{n\nu}(x)) \le \\
&\le \omega_f(\delta)(1 + \frac{1}{4n\delta^2}) \ \text{ because of } (**) \text{ in 2.2 .}
\end{aligned}$$

Choosing $\delta := \frac{1}{\sqrt{n}}$, we obtain the following uniform error bound for all values $x \in [0, 1]$:

Error Bound

$$|f(x) - (B_n f)(x)| \le \frac{5}{4}\omega_f(\frac{1}{\sqrt{n}}).$$

Remark. Suppose, for example, that a function $f \in C[0,1]$ is such that $\omega_f(\delta) \le K\delta^\alpha$. A function with this property is called Hölder continuous if $0 < \alpha < 1$ and is called Lipschitz bounded if $\alpha := 1$. Then the error bound becomes

$$|f(x) - (B_n f)(x)| \le \frac{5}{4} K n^{-\frac{\alpha}{2}}.$$

Comment. Depending on the behavior of the modulus of continuity, the expression in the error bound given above can converge to 0 arbitrarily slowly. On the other hand, for smoother functions the convergence of the sequence $(B_n f)$ to f can be very fast. We shall encounter this type of situation frequently in the sequel.

In fact, Bernstein polynomials are of limited practical use in approximating continuous functions (see, however, the remark in Problem 4 below). The convergence of the sequence $(B_n f)$ is in general relatively slow, and we will develop more powerful methods later. The results of this section (in particular the theorems of Weierstrass and their proofs) are valuable, however, as building blocks for a theory of approximation. Now that we have some approximation methods, it is natural to ask if it is possible somehow to construct a *best approximation*, assuming that we can define an appropriate measure of goodness of an approximation. The definition of such measures, their connection with normed linear spaces, and the development of general approximation results as well as practical methods for the calculation of best approximations will be discussed in detail in Sections 3 – 6 of this chapter.

2.6 Problems. 1) Let $f \in C[a,b]$, $0 \le \varepsilon_1 < \varepsilon_2$. Show that there always exists a polynomial p such that $\|f - p\|_\infty \le \varepsilon_2$ and $f(x) - p(x) \ge \varepsilon_1$ for all $x \in [a,b]$. Explain the case $\varepsilon_1 = 0$.

2) Show: a) Every sequence in $C[a,b]$ which converges with respect to the norm $\|\cdot\|_\infty$ also converges with respect to $\|\cdot\|_1$.

b) The converse of the assertion a) is false.

3) Let $f : [0,1] \to \mathbb{R}$, $f(x) := x^3$. Show: a) $B_n f$ is a polynomial of degree 3 for all $n \ge 3$.

b) $\lim_{n\to\infty} \max_{x\in[0,1]} |f(x) - (B_n f)(x)| = 0$.

4) Show that a function $f : [0,1] \to \mathbb{R}$ and its associated Bernstein polynomial $(B_n f)(x) = \sum_{\nu=0}^{n} f\left(\frac{\nu}{n}\right)\binom{n}{\nu} x^\nu (1-x)^{n-\nu}$ have the following properties:

a) if f is monotone, then so is $B_n f$ in the same sense,

b) if f is convex (resp. concave), then so is $B_n f$.

Remark. Although the Bernstein polynomial $B_n f$ is in general not a good uniform approximation of f for small n, it does preserve global geometric properties of f. This is the reason why the Bernstein polynomials are useful for applications in geometric modelling.

5) Show by the construction of a counter-example, that the operator defined in 2.4 on periodic functions by the Dirichlet kernel is not monotone.

6) Let $f : [a, b] \to \mathbb{R}$.

a) Show: f is uniformly continuous on $[a, b]$ if and only if the modulus of continuity satisfies $\lim_{\delta \to 0} \omega_f(\delta) = 0$.

b) Compute $\omega_f(\delta)$ for $f(x) := \sqrt{x}$, $[a, b] := [0, 1]$.

c) Using b), find $N \in \mathbb{N}$, so that for all $n \geq N$, $|(B_n\sqrt{\cdot})(x) - \sqrt{x}| \leq 10^{-2}$.

7) Let $f \in C[0, 1]$ be Lipschitz bounded; i.e., $\omega_f(\delta) \leq K\delta$. Show directly that the factor $\frac{5}{4}$ in the estimate 2.5 can be improved to $\frac{1}{2}$.

8) Let $f : [0, 1] \times [0, 1] \to \mathbb{R}$ and $f(0, 0) = f(0, 1) = f(1, 0) = f(1, 1) = 0$, $f(0, \frac{1}{2}) = f(1, \frac{1}{2}) = f(\frac{1}{2}, 0) = f(\frac{1}{2}, 1) = 1$, $f(\frac{1}{2}, \frac{1}{2}) = \lambda \geq 2$. Study and sketch the surface corresponding to the Bernstein polynomial $B_{22}f$ in two variables. How does this surface vary with λ?

3. The General Approximation Problem

The concept of approximation plays an essential role in mathematics. This is particularly true in applied mathematics, and indeed approximation methods of various kinds are the main objects of interest in numerical analysis.

We want to present a sufficiently general formulation of the general approximation problem to include a wide variety of special cases of interest. We shall formulate the general problem in a normed linear space since the metric defined by the norm will provide a way of measuring the quality of the approximation.

3.1 Best Approximations. Let $(V, \|\cdot\|)$ be a normed linear space, and let $T \subset V$ be an arbitrary subset. Given an element $v \in V$, we consider $u \in T$ to be a good approximation of v if the distance $\|v - u\|$ between the elements is small. We call $\tilde{u} \in T$ a *best approximation* or *proximum* provided that $\|v - \tilde{u}\| \leq \|v - u\|$ for every element $u \in T$.

The following two examples show that in some cases a best approximation exists, while in others it does not.

Example 1. Let $V := \mathbb{R}^2$, $\|\cdot\| := \|\cdot\|_2$, and let $T := \{x \in V \mid \|x\| \leq 1\}$. It is easy to see from elementary geometrical considerations (cf. the sketch on the next page) that for every element $y \in V$, there exists a corresponding best approximation $\tilde{x} \in T$.

Example 2. Let $T := \{u \in V \mid u(x) = e^{\beta x}, \beta > 0\}$ be a subset of the space $(C[0, 1], \|\cdot\|_\infty)$, and let v be the constant function $v(x) := \frac{1}{2}$. Consider the problem of finding a best approximation $\tilde{u} \in T$. We are seeking $\tilde{u}(x) = e^{\tilde{\beta} x}$ such that the value of $\max_{x \in [0,1]} |\frac{1}{2} - e^{\beta x}|$ is minimal over all choices of $\beta > 0$. Now $\max_{x \in [0,1]} |\frac{1}{2} - e^{\beta x}| = e^{\beta} - \frac{1}{2}$, and since $\inf_{\beta > 0}(e^{\beta} - \frac{1}{2}) = \frac{1}{2}$ is not taken on by any element in T, it follows that this approximation problem has no solution.

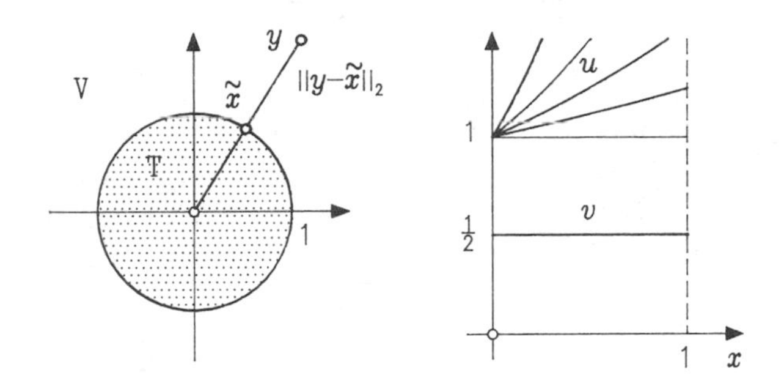

Definition of a Best Approximation. Let T be a subset of a normed linear space $(V, \|\cdot\|)$, and let $v \in V$. A $\tilde{u} \in T$ is called a *best approximation* of v provided that $\|v - \tilde{u}\| = \inf_{u \in T} \|v - u\|$. We call $E_T(v) := \inf_{u \in T} \|v - u\|$ the *minimal deviation* of the element v from the subset T.

Remark. The trivial case $v \in T$ is not excluded. In this case there always exists a best approximation; namely, $\tilde{u} = v$ since then $\|v - \tilde{u}\| = 0$.

3.2 Existence of a Best Approximation. The essential difference between the two examples presented above is the fact that the subset T chosen in the first example is a compact subset of V, while in the second example, it is not. We now explore this observation further.

Minimizing Sequences. A sequence of elements $(u_\nu)_{\nu \in \mathbb{N}}$ from $T \subset V$ is called a *minimizing sequence* for $v \in V$ provided $\lim_{\nu \to \infty} \|v - u_\nu\| = E_T(v)$. By the definition of the distance $E_T(v)$, it is clear that for every non-empty subset T and every element $v \in V$, there always exists a minimizing sequence. But for a minimizing sequence we only require that the norm $\|v - u_\nu\|$ converges, which for arbitrary subsets T is not enough to conclude that (u_ν) converges to an element of T or even to an element of V. However, the following lemma always holds for minimizing sequences.

Lemma. *Let $v \in V$ and let u^* be a cluster point of a minimizing sequence. If u^* lies in T, then it is a best approximation of v in T.*

Proof. Let (u_ν) be a minimizing sequence; i.e. $\lim_{\nu \to \infty} \|v - u_\nu\| = E_T(v)$, and suppose that the subsequence $(u_{\mu(\nu)})$ converges to the element $u^* \in T$. Then since $\lim_{\mu \to \infty} \|v - u_\mu\| = E_T(v)$ and $\lim_{\mu \to \infty} \|u_\mu - u^*\| = 0$, it follows from $\|v - u^*\| \leq \|v - u_\mu\| + \|u_\mu - u^*\|$ for all μ, that $\|v - u^*\| \leq E_T(v)$. Moreover, for the distance we have $E_T(v) \leq \|v - u\|$ for all $u \in T$. We conclude that $\|v - u^*\| = E_T(v)$ and thus that u^* is a best approximation. □

Theorem. *Let $T \subset V$ be a compact subset. Then for every $v \in V$ there exists a best approximation $\tilde{u} \in T$.*

Proof. Let $(u_\nu)_{\nu\in\mathbb{N}}$ be a minimizing sequence in T for $v \in$ V. Since T is compact, this minimizing sequence contains a convergent subsequence. By the lemma this subsequence converges to a best approximation $\tilde{u} \in$ T. □

3.3 Uniqueness of a Best Approximation. In addition to the question of the existence of a best approximation, it is also of interest to discuss its uniqueness. The best approximation in Example 1 of 3.1 is obviously unique. Suppose we modify this example to consider best approximation from the set

$$\hat{\mathrm{T}} := \mathrm{T} \setminus \mathrm{T}^*, \quad \mathrm{T}^* := \{x \in \mathrm{V} \mid \|x\| \le 1 \text{ with } x_1 > 0, x_2 > 0\}.$$

Then clearly both of the points $(0,1)$ and $(1,0)$ are best approximations of $(1,1)$ from $\hat{\mathrm{T}}$.

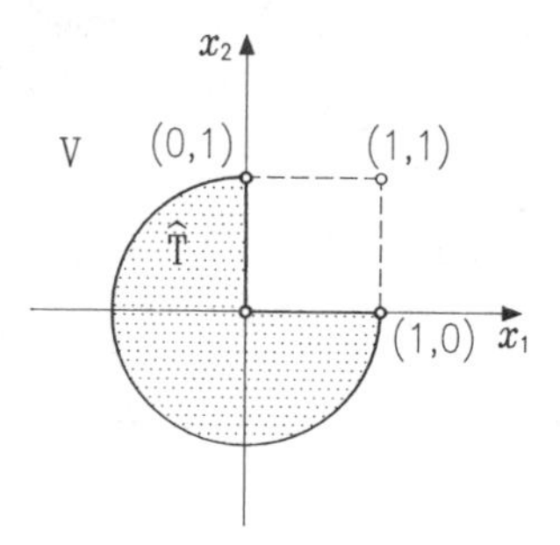

The essential property which led to the uniqueness of the best approximation in Example 1 of 3.1 is

Convexity. A subset T ⊂ V is called *convex* provided that for any two arbitrary elements u_1 and u_2 of T, all elements of the set $\{\lambda u_1 + (1-\lambda)u_2$ for $0 < \lambda < 1\}$ also lie in T. If all elements of this set with $u_1 \neq u_2$ are interior points of T, then we say that *T* is *strictly convex.*

Remark. The convexity of a subset T implies that all points on the line joining u_1 and u_2 belong to T. Strong convexity means that the boundary of T does not contain any straight line segments.

We have the following

Uniqueness Assertion. *Let* T *be a compact and strictly convex subset of a normed linear space* V. *Then for every* $v \in$ V, *there exists exactly one best approximation of* v *in* T.

Proof. Let $\tilde{u}_1$ and $\tilde{u}_2$, $\tilde{u}_1 \neq \tilde{u}_2$, be two best approximations of $v \in$ V in T. Then $\|\frac{1}{2}(\tilde{u}_1 + \tilde{u}_2) - v\| \le \frac{1}{2}\|\tilde{u}_1 - v\| + \frac{1}{2}\|\tilde{u}_2 - v\| \Rightarrow \|\frac{1}{2}(\tilde{u}_1 + \tilde{u}_2) - v\| \le$

$\leq E_{\mathrm{T}}(v) \Rightarrow \|\frac{1}{2}(\tilde{u}_1+\tilde{u}_2)-v\| = E_{\mathrm{T}}(v)$. Since T is strictly convex, there exist numbers $\lambda \in (0,1)$ such that $\tilde{u} := \frac{1}{2}(\tilde{u}_1 + \tilde{u}_2) + \lambda[v - \frac{1}{2}(\tilde{u}_1 + \tilde{u}_2)]$ lies in T. If $\hat{\lambda} > 0$ is one of these values, then

$$\|\tilde{u}-v\| = \|\frac{1}{2}(1-\hat{\lambda})(\tilde{u}_1+\tilde{u}_2)-(1-\hat{\lambda})v\| = (1-\hat{\lambda})E_{\mathrm{T}}(v) \Rightarrow \|\tilde{u}-v\| < E_{\mathrm{T}}(v).$$

This contradicts the assumption that $\tilde{u}_1 \neq \tilde{u}_2$, and the uniqueness is established. □

3.4 Linear Approximation. For applications, the special case where the space T := U is a finite dimensional linear subspace of V is particularly important. Suppose that U := span$(u_1, u_2, \ldots, u_n)$. The problem of finding a best approximation $\tilde{u} \in$ U of a given element $v \in$ V then reduces to finding a linear combination $\tilde{u} = \tilde{\alpha}_1 u_1 + \cdots + \tilde{\alpha}_n u_n$ such that the distance given by $d(\alpha) := \|v - (\alpha_1 u_1 + \cdots + \alpha_n u_n)\|$ is minimal among all linear combinations of the form $u = \alpha_1 u_1 + \cdots + \alpha_n u_n$.

In the trivial case where $v \in$ U, this approximation problem reduces to the problem of finding a representation of $\tilde{u} = v$ in terms of the basis $(u_1, u_2, \ldots, u_n)$ of U. This case, which is characterized by $d(\tilde{\alpha}) = 0$, is not excluded here, but will also be discussed in more detail in Chapter 5.

We now restrict our attention to the case where $v \notin$ U. In this case we cannot directly apply Theorem 3.2 since the assumption of compactness is not always satisfied for a finite dimensional linear subspace.

We now show, however, that in studying minimizing sequences in U corresponding to a given $v \in$ V, it suffices to work with a bounded subset of U.

Lemma. *Every minimizing sequence in* U *is bounded.*

Proof. Let $(u_\nu)_{\nu\in\mathbb{N}}$ be a minimizing sequence in U corresponding to $v \in$ V. Then

$$E_{\mathrm{U}}(v) \leq \|v - u_\nu\| \leq E_{\mathrm{U}}(v) + 1$$

for all $\nu \geq N$. Thus $\|u_\nu\| \leq \|v - u_\nu\| + \|v\| \leq E_{\mathrm{U}}(v) + 1 + \|v\| =: K_1$ for $\nu \geq N$. Now let $K_2 \geq \|u_\nu\|$ for all $\nu < N$ and let $K := \max\{K_1, K_2\}$. Then $\|u_\nu\| \leq K$ for all $\nu \in \mathbb{N}$. □

We are now in a position to establish the following important result on the existence of a best approximation.

Fundamental Theorem of Approximation Theory in Normed Linear Spaces. *If* U *is a finite dimensional linear subspace of a normed linear space* V*, then for every element* $v \in$ V*, there exists at least one best approximation* $\tilde{u} \in$ U.

Proof. By the lemma, every minimizing sequence for $v \in$ V is bounded, and therefore possesses a cluster point u^*. Since U is closed, this cluster point must lie in U, and by Lemma 3.2 is thus a best approximation $\tilde{u}$. □

Remark. In this theorem it is essential that the linear space U has finite dimension. Indeed, it is easy to see from the approximation theorem of Weierstrass that the finite dimensionality cannot be dispensed with. The importance of this fundamental theorem, and hence also its name, can be justified as follows. By the theorem, for any given element in a normed linear space, such as a function given in a complicated closed form or given pointwise by calculation or as the result of experimental measurements, and given a finite set of approximating elements, there always exists a "best possible" approximating linear combination of these elements.

We continue our study of best approximation from a finite dimensional linear subspace in the following section.

3.5 Uniqueness in Finite Dimensional Linear Subspaces. The following result answers the question of uniqueness of best approximations in the case of finite dimensional linear subspaces.

Uniqueness. *Let V be strictly normed. Then the best approximation of $v \in$ V from an arbitrary finite dimensional linear subspace U is always unique.*

Proof. If $v \in$ U, then clearly $\tilde{u} = v$ is always uniquely determined, no matter what normed linear space we are working in. Thus we may assume that $v \notin$ U. Suppose now that $\tilde{u}_1$ and $\tilde{u}_2$ are both best approximations. Then as in 3.3,

$$\|v - \frac{1}{2}(\tilde{u}_1 + \tilde{u}_2)\| \leq \frac{1}{2}\|v - \tilde{u}_1\| + \frac{1}{2}\|v - \tilde{u}_2\| = E_{\mathrm{U}}(v), \text{ and thus}$$

$$\|(v - \tilde{u}_1) + (v - \tilde{u}_2)\| = \|v - \tilde{u}_1\| + \|v - \tilde{u}_2\|.$$

Since the norm $\|\cdot\|$ is strict, it follows that

$$v - \tilde{u}_1 \lambda(v - \tilde{u}_2) \quad \text{and} \quad (1 - \lambda)v = \tilde{u}_1 - \lambda\tilde{u}_2.$$

Since $v \notin$ U, these equations can hold only for $\lambda = 1$, and we conclude that $\tilde{u}_1 = \tilde{u}_2$. Thus, uniqueness of the best approximation is established. □

If we drop the assumption that V is strictly normed, we can still get the inequality in the above proof; i.e., if $\tilde{u}_1$ and $\tilde{u}_2$ are best approximations, then so is $\frac{1}{2}(\tilde{u}_1 + \tilde{u}_2)$. In fact, if this is the case, then every element $\lambda\tilde{u}_1 + (1-\lambda)\tilde{u}_2$ with $\lambda \in [0, 1]$ is a best approximation. This implies the following

Remark. In a normed linear space V, the best approximation of an element $v \in$ V from a finite dimensional linear subspace is either unique, or there exist infinitely many best approximations.

Example 1. Let V := C$[a, b]$, $\|\cdot\| := \|\cdot\|_2$. The norm $\|\cdot\|_2$ is a strict norm. Indeed, for any norm obtained from an inner product, we have the Schwarz inequality $|\langle v_1, v_2\rangle| \le \|v_1\|\,\|v_2\|$, and (cf. 1.3) equality holds precisely when v_1 and v_2 are linearly dependent. The same holds for the triangle inequality. Thus the problem of finding a best approximation $\tilde{u} \in \mathrm{U}$ of $v \in \mathrm{V}$ always has a unique solution.

Example 2. Let $\mathrm{V} := \mathbb{R}^3$, $\|\cdot\| := \|\cdot\|_\infty$. This linear space is not strictly normed. Indeed, consider the elements $x := (1,0,0) \in \mathrm{V}$ and $y := (1,1,0) \in \mathrm{V}$. Then $\|x\|_\infty = \|y\|_\infty = 1$ and $\|x+y\|_\infty = 2$. It follows that $\|x+y\|_\infty = \|x\|_\infty + \|y\|_\infty$ even though x and y are not linearly dependent.

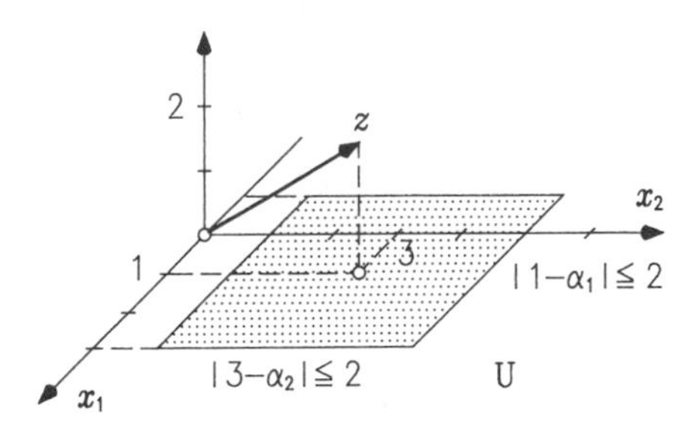

If $\mathrm{U} \subset \mathrm{V}$ and $z \notin \mathrm{U}$, then it can in fact happen that z has infinitely many best approximations. Consider, for example, the problem of approximating $z := (1,3,2)$ in the plane $\mathrm{U} := \mathrm{span}(x^1, x^2)$ with $x^1 := (1,0,0)$, $x^2 := (0,1,0)$. Then

$$\|z - \tilde{z}\|_\infty = \min_{\alpha_1,\alpha_2 \in \mathbb{R}} \|z - (\alpha_1 x^1 + \alpha_2 x^2)\|_\infty = 2.$$

The minimum is taken on for all values α_1, α_2 with $|1-\alpha_1| \le 2$ and $|3-\alpha_2| \le 2$.

In Example 2 we saw that the Chebyshev norm in the space $\mathbb{R}^3$ is not strict. By 1.1, the same holds for the linear space of continuous functions equipped with the Chebyshev norm. Thus, in this case we cannot establish uniqueness on the basis of properties of the norm.

The same two functions f and g used in 1.1 to show that the space $(\mathrm{C}[0,1], \|\cdot\|_\infty)$ is not strictly normed serve to verify that the linear space $(\mathrm{C}[0,1], \|\cdot\|_1)$ is also not strictly normed. Again, $\|f+g\|_1 = \|f\|_1 + \|g\|_1$, without f and g being linearly dependent.

On the other hand, it is precisely the space $(\mathrm{C}[a,b], \|\cdot\|_\infty)$ which is especially important for the approximation of functions, since using the Chebyshev norm gives a measure of the largest pointwise deviation of a best approximation of a given function, and therefore provides a useful numerical error bound.

The treatment of Example 1 shows that in every pre-Hilbert space V, the best approximation of an arbitrary element $v \in \mathrm{V}$ from a finite dimensional linear subspace is always uniquely determined. This fact goes back to the properties of the Schwarz inequality. We also mention that the linear space $\mathrm{V} := \mathbb{C}^n$ equipped with the norm $\|\cdot\|_p$ is a strictly normed linear space if

$1 < p < \infty$, cf. Example 2.4.1. In this case, the triangle inequality for the norm reduces to the Minkowski inequality 2.4.1 where equality holds only if the corresponding elements are linearly dependent. The same assertion holds for the linear spaces $L^p[a,b]$, and in particular also for the space $C[a,b]$ equipped with the norm $\|\cdot\|_p$, as long as $1 < p < \infty$. We have already observed above that the situation is different for $p = 1$ and $p = \infty$.

The existence of a strict norm for a linear space is a sufficient condition that for every finite dimensional linear subspace, every element of the space has a unique best approximation. On the other hand, there also exist finite dimensional linear subspaces of linear spaces which are not strictly normed from which best approximations are always uniquely determined; i.e., being strictly normed is not a necessary condition for uniqueness. The case of $(C[a,b], \|\cdot\|_\infty)$ is particularly important, and will be treated in detail in §4.

We now give an example to show that there are non-strictly normed linear spaces and finite-dimensional linear subspaces where nonuniqueness always occurs.

Example. Let V be an arbitrary non-strictly normed linear space over $\mathbb{R}$. We now construct a finite dimensional linear subspace $U \subset V$ and an element $v \in V$ such that v has more than one best approximation from U.

a) Since V is not strictly normed, there exist two linearly independent elements v_1^* and v_2^* with $0 < \|v_1^*\| \leq \|v_2^*\|$ such that equality holds in the triangle inequality; i.e., $\|v_1^* + v_2^*\| = \|v_1^*\| + \|v_2^*\|$. The same also holds for the normed elements $v_1 := \frac{v_1^*}{\|v_1^*\|}$ and $v_2 := \frac{v_2^*}{\|v_2^*\|}$. Indeed,

$$\|v_1 + v_2\| = \|\frac{v_1^*}{\|v_1^*\|} + \frac{v_2^*}{\|v_2^*\|}\| = \|(\frac{v_1^*}{\|v_1^*\|} + \frac{v_2^*}{\|v_1^*\|}) - (\frac{v_2^*}{\|v_1^*\|} - \frac{v_2^*}{\|v_2^*\|})\| \geq$$
$$\geq \frac{1}{\|v_1^*\|}\|v_1^* + v_2^*\| - |\frac{1}{\|v_1^*\|} - \frac{1}{\|v_2^*\|}|\|v_2^*\| = \frac{1}{\|v_1^*\|}(\|v_1^*\| + \|v_2^*\|) -$$
$$-(\frac{1}{\|v_1^*\|} - \frac{1}{\|v_2^*\|})\|v_2^*\| = 2,$$

and thus $\|v_1+v_2\| \geq 2$. Together with the inequality $\|v_1+v_2\| \leq \|v_1\|+\|v_2\| = 2$, this implies $\|v_1 + v_2\| = \|v_1\| + \|v_2\|$.

Consider now the finite dimensional linear subspace $U := \text{span}(v_1 - v_2)$ which consists of all elements of the form $u(\lambda) := \lambda(v_1 - v_2)$, $\lambda \in \mathbb{R}$. Suppose we want to approximate the element $w := -v_2 \notin U$ from U. We now show that both $u(0) = 0$ and $u(1) = v_1 - v_2$ are best approximations; i.e., $\|w - u(0)\| = \|w - u(1)\| \leq \|w - u(\lambda)\|$ for all $\lambda \in \mathbb{R}$.

Let $d(\lambda) := u(\lambda) - w = \lambda v_1 + (1-\lambda)v_2$. Since $d(0) = v_2$ and $d(1) = v_1$, we have $\|d(0)\| = \|d(1)\| = 1$. To show that $\|d(\lambda)\| \geq 1$ for all values of λ, we distinguish several cases.

1) $\lambda < 0$: In this case we write $v_2 = \frac{-\lambda}{1-2\lambda}(v_1 + v_2) + \frac{1}{1-2\lambda}[\lambda v_1 + (1-\lambda)v_2]$, and use

$$\|v_2\| \leq \frac{-\lambda}{1-2\lambda}(\|v_1\| + \|v_2\|) + \frac{1}{1-2\lambda}\|d(\lambda)\|,$$
$$\|d(\lambda)\| \geq (1-2\lambda)\|v_2\| + \lambda(\|v_1\| + \|v_2\|) = 1.$$

Similar arguments can be applied in the following cases:

2) $0 < \lambda < \frac{1}{2}$: $\quad v_1 + v_2 = \frac{1-2\lambda}{1-\lambda} v_1 + \frac{1}{1-\lambda} d(\lambda)$;
3) $\lambda = \frac{1}{2}$: $\quad \frac{1}{2}(v_1 + v_2) = d(\frac{1}{2})$;
4) $\frac{1}{2} < \lambda < 1$: $\quad v_1 + v_2 = \frac{2\lambda-1}{\lambda} v_2 + \frac{1}{\lambda} d(\lambda)$;
5) $1 < \lambda$: $\quad v_1 = \frac{\lambda-1}{2\lambda-1}(v_1 + v_2) + \frac{1}{2\lambda-1} d(\lambda)$.

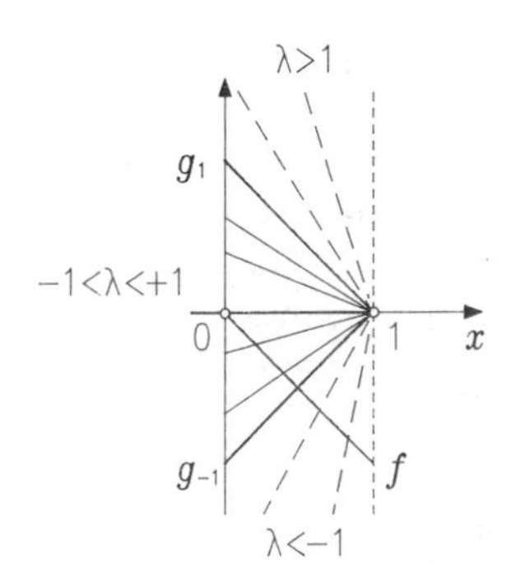

b) We now illustrate this example in the concrete case of the non-strictly normed linear space $(C[0,1], \|\cdot\|_\infty)$. Let $v_1(x) := 1$, $v_2(x) := x$ for all $x \in [0,1]$. In this case U is the pencil of lines $g_\lambda(x) := \lambda(1-x)$. Consider now the problem of best approximation of the function $f(x) := -x$ from U. Then $d_\lambda(x) := \lambda + (1-\lambda)x$, and thus $\|d(\lambda)\| = 1$ for $\lambda \in [-1,+1]$ and $\|d(\lambda)\| > 1$ for all other values of λ.

This shows that in addition to g_0 and g_1, all elements of the form g_λ for $\lambda \in [0,1]$ are also best approximations; cf. Remark 3.5.

3.6 Problems. 1) Consider the subset $T := \{u \in C[0,1] \mid u(0) = 0\}$ of the normed linear space $(C[0,1], \|\cdot\|_\infty)$. Show that the sequence $(u_\nu)_{\nu \in \mathbb{N}}$, $u_\nu(t) := t^\nu$, is a minimizing sequence corresponding to the function $v(t) := 1$, but that it does not converge to an element in T.

2) a) Show that in $(\mathbb{R}^2, \|\cdot\|_2)$ every element $x = (x_1, x_2)$ has a unique best approximation from the closed lower half plane.

b) Let $T := \{u \in V \mid \|u\| \leq 1\}$ be the closed unit ball in the normed linear space $(V, \|\cdot\|)$. Show that for every $v \in V$, the element $\tilde{u} := \begin{cases} v, & \text{if } v \in T \\ \frac{v}{\|v\|}, & \text{if } v \notin T \end{cases}$ provides a best approximation of v from T.

3) Show that the set of all polynomials with nonnegative coefficients is convex.

4) Sketch the unit ball $\|x\| = 1$, $x \in \mathbb{R}^2$ for the norms $\|\cdot\|_1$, $\|\cdot\|_2$ and $\|\cdot\|_\infty$. What properties of the norm can you deduce from the convexity and strict convexity of the unit ball?

5) Determine whether uniqueness always holds for best approximation from finite dimensional linear subspaces of the following normed linear spaces:

a) $V := C_2[0,1]$, $\|f\| := \left(\int_0^1 |f''(x)|^2 dx\right)^{\frac{1}{2}} + |f(0)| + |f(1)|$;

b) $V := C_n[0,1]$, $\|f\| := \left(\sum_0^n \int_0^1 |f^{(\nu)}(x)|^2 dx\right)^{\frac{1}{2}}$, $n \in \mathbb{N}$;
c) $V := \{(x_\nu)_{\nu\in\mathbb{N}} \mid x_1 = 0, \sum_1^\infty |x_{\nu+1} - x_\nu| < \infty\}$, $\|x\| := \sum_1^\infty |x_{\nu+1} - x_\nu|$.
6) Let $V := P_1$, equipped with the norm $\|p\| = |p(0)| + |p(1)|$. Determine the set of all best approximations of the polynomial $p(x) := x$ from $U := P_0$.

4. Uniform Approximation

The problem of approximating a continuous function by a finite linear combination of given functions can be approached in various ways. For the purpose of representing arbitrary continuous functions by elementary functions, e.g. polynomials, it is natural to use the maximal deviation of the approximation as a measure of the quality of the approximation. In this case the appropriate normed linear space is $C[a,b]$, equipped with the Chebyshev norm $\|\cdot\|_\infty$. We refer to approximation methods based on this norm as *uniform approximation* since the Chebyshev norm provides a uniform bound on the deviation throughout the entire interval.

PAFNUTII LVOVITSCH CHEBYSHEV (1821–1894) worked mainly in St. Petersburg, now called Leningrad. He was a universal mathematician whose papers are still influential in various areas of mathematics. These include number theory, probability theory, the theory of orthogonal functions, and theoretical mechanics. Chebyshev is considered to be a pathbreaker in constructive function theory, which includes the theory of uniform approximation. He was the first to formulate and prove the basic Alternation Theorem 4.3.

For a given $f \in C[a,b]$ and a given linear subspace U, the existence of a best approximation $\tilde{f}$ of f has already been established in Theorem 3.4 above. We were not able to give any general result on uniqueness since $(C[a,b], \|\cdot\|_\infty)$ is not strictly normed. However, as we shall see below, uniqueness can be shown for special subspaces.

4.1 Approximation by Polynomials. We begin by studying approximation of a continuous function by polynomials of degree at most $n-1$. This corresponds to the choice of the subspace $U := P_{n-1} = \mathrm{span}(g_1, \ldots, g_n)$ with $g_j(x) := x^{j-1}$, $1 \le j \le n$. We first present a criterion which can be used to check if a given polynomial is a best approximation.

Theorem. *Let $g \in P_{n-1}$, $f \in C[a,b]$ and $\rho := \|f-g\|_\infty$. Suppose there exist $(n+1)$ points $a \le x_1 < x_2 \cdots < x_{n+1} \le b$ such that $(f-g)$ assumes its maximial absolute value ρ at these points with alternating signs; that is, $|f(x_\nu) - g(x_\nu)| = \rho$ for $1 \le \nu \le n+1$ and $f(x_{\nu+1}) - g(x_{\nu+1}) = -[f(x_\nu) - g(x_\nu)]$ for $1 \le \nu \le n$. Then g is a best approximation of f.*

Proof. To prove this result, we first derive another characterization of a best approximation. Given a polynomial $p^* \in P_{n-1}$, let M be the set of points where the difference $d^* := f - p^*$ takes on the extreme values $\pm\|f-p^*\|_\infty$:

$$M := \{x \in [a,b] \mid |f(x) - p^*(x)| = \|f - p^*\|_\infty\}.$$

If p^* is not a best approximation, then there is a best approximation $\tilde{f}$ which can be written in the form $\tilde{f} = p^* + p$, $p \neq 0$ for some appropriate $p \in \mathrm{P}_{n-1}$. Then for all $x \in \mathrm{M}$, we have

$$|f(x) - (p^*(x) + p(x))| < |f(x) - p^*(x)|,$$

or equivalently $|d^*(x) - p(x)| < |d^*(x)|$. This can only happen if for these values of x, the sign of $p(x)$ and the sign of $d^*(x)$ are the same; that is, if $[f(x) - p^*(x)]p(x) > 0$ for $x \in \mathrm{M}$. Reversing this chain of implications, it follows that p^* is a best approximation whenever there is *no* polynomial $p \in \mathrm{P}_{n-1}$ satisfying this condition.

We return to the proof of the theorem. Suppose $g \in \mathrm{P}_{n-1}$ is such that $|f(x_\nu) - g(x_\nu)| = \rho$ and $f(x_{\nu+1}) - g(x_{\nu+1}) = -[f(x_\nu) - g(x_\nu)]$ at $(n+1)$ points

$$a \leq x_1 < x_2 < \cdots < x_{n+1} \leq b.$$

Then we claim that the conditions $[f(x_\nu) - g(x_\nu)]p(x_\nu) > 0$ for $1 \leq \nu \leq n+1$ cannot hold for any polynomial $p \in \mathrm{P}_{n-1}$. Indeed, if they did hold, then p would have to have at least n sign changes in $[a, b]$, and hence also n zeros there. But by the Fundamental Theorem of Algebra, this is impossible. We conclude that g is a best approximation, and the proof is complete. □

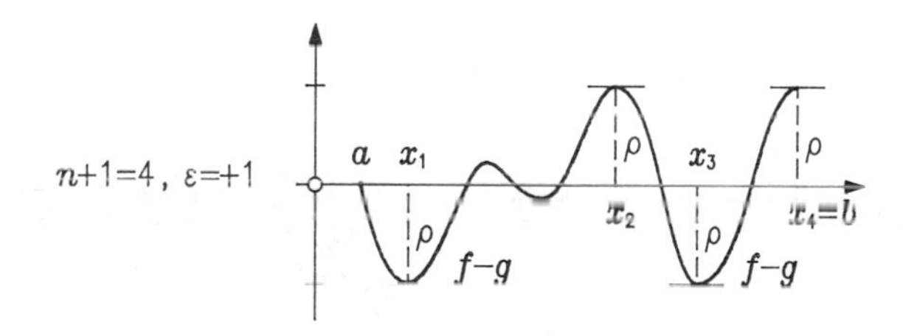

Remark. Suppose that the function $f \in \mathrm{C}[a, b]$ is given only at $m \geq n+1$ points $x_1 < x_2 < \cdots < x_m$, and that we want to find a polynomial of degree at most $n-1$ which approximates f best on these m points with respect to the Chebyshev norm. Then the same theorem can be established with $\rho := \max_{1 \leq \mu \leq m} |f(x_\mu) - g(x_\mu)|$. The proof of this variant of the theorem can be taken over word for word.

Comment. This theorem asserts only that g is a best approximation when there are *at least* $(n+1)$ points satisfying the assumption of the theorem. In general, there can be more points where the maximal deviation is achieved. For example, suppose we want to approximate the function $f(x) := \sin(3x)$ in $\mathrm{C}[0, 2\pi]$ by a polynomial in P_{n-1}. It follows immediately from the theorem that if $n - 1 \leq 4$, then the polynomial $g = 0$ is a best approximation of f.

Indeed, in this case the difference $(f - g)$ alternates between its maximal absolute value at six points, whereas the theorem only requires $n+1$ points. On the other hand, for $n - 1 = 5$ we have $n + 1 = 7$, and $g = 0$ no longer satisfies the hypothesis of the theorem. Indeed, in this case $g = 0$ is not a best approximation from P_5. This is not immediately clear, but we shall show in 4.3 below that the condition presented in this theorem is not only sufficient, but is also necessary for a polynomial g to be a best approximation.

4.2 Haar Spaces. The only property of the subspace P_{n-1} used in establishing Theorem 4.1 was the fact that the polynomials satisfy the Fundamental Theorem of Algebra. In fact, we only used the weaker property that any polynomial of degree $(n-1)$ has *at most* $(n-1)$ distinct zeros in $[a, b]$. This property is shared by a larger class of functions.

Definition. Suppose that $g_1, \ldots, g_n \in \mathrm{C}[a, b]$ are n linearly independent functions such that every element $g \in \operatorname{span}(g_1, \ldots, g_n)$, $g \neq 0$ in $[a, b]$, has *at most* $(n-1)$ distinct zeros in $[a, b]$. Then we say that $\mathrm{U} := \operatorname{span}(g_1, \ldots, g_n)$ is a *Haar Space.*

This concept is named after the Austrian-Hungarian mathematician ALFRED HAAR (1885 – 1933), who is well known for his papers in functional analysis. He completed his habilitation in 1910 at Göttingen, and from 1912 to 1920 taught in Cluj, which at the time belonged to Hungary. After Cluj became Rumanian, he moved to Szeged. In Szeged he established a mathematical center together with Friedrich Riesz (1880 – 1956). Many important contributions to modern functional analysis stem from this center.

Chebyshev Systems. A basis $\{g_1, \ldots, g_n\}$ for a Haar space is called a *Chebyshev System.* We have already seen that $\{1, x, \ldots, x^{n-1}\}$ is a Chebyshev system. Two other interesting examples are $\{1, e^x, \ldots, e^{(n-1)x}\}$, $x \in \mathbb{R}$ and $\{1, \sin(x), \ldots, \sin(mx), \cos(x), \ldots, \cos(mx)\}$, $x \in [0, 2\pi)$.

The system of exponentials can be shown to be a Chebyshev system by using the transformation $t := e^x$. The system of sines and cosines can be treated by passing to complex-valued functions via the equation

$$\sum_{\mu=0}^{m} (\alpha_\mu \sin(\mu x) + \beta_\mu \cos(\mu x)) = \sum_{|\mu| \leq m} \gamma_\mu e^{i\mu x} = e^{-imx} q(e^{ix}),$$

where q is a suitable polynomial in e^{ix} of degree at most $2m$ (which thus can have at most $2m = n - 1$ zeros). By the periodicity of the trigonometric functions, it is clear that they also form a Chebyshev system on any interval of the form $[a, b]$ with $0 < b - a < 2\pi$.

It is clear that Theorem 4.1 holds for any Haar space. It provides a sufficient condition for an element $g \in U$ to be a best approximation from U of a prescribed f.

4.3 The Alternation Theorem. Theorem 4.1 provides a tool for checking when a function is a best approximation. In this section we show that it can be extended to provide a necessary and sufficient condition for best approximations. To this end we introduce the following

Definition. A set of $(n+1)$ points $a \leq x_1 < \cdots < x_{n+1} \leq b$ is called an *alternant* for $f \in C[a,b]$ and $g \in \text{span}(g_1, \ldots, g_n)$ provided that the difference $d := f - g$ satisfies $\text{sgn}\, d(x_\nu) = \varepsilon(-1)^\nu$ with $\varepsilon \in \{-1, +1\}$, $1 \leq \nu \leq n+1$.

We now formulate the extension of Theorem 4.1. The extended version also holds for Haar subspaces, but for convenience we state and prove the theorem only for the important special case where $U := P_{n-1}$.

Alternation Theorem. *The polynomial $g \in P_{n-1}$ is a best approximation of the function $f \in C[a,b]$ if and only if there exists an alternant $a \leq x_1 < \cdots < x_{n+1} \leq b$ with the property $f(x_\nu) - g(x_\nu)| = \|f - g\|_\infty$ for $\nu = 1, \ldots, n+1$.*

Proof. The sufficiency part of the theorem is precisely the content of Theorem 4.1. To prove the necessity assertion, we now show, motivated by the proof of Theorem 4.1, that if $p^* \in P_{n-1}$ is such that there exists a polynomial $p \in P_{n-1}$ with $d^*(x)p(x) = [f(x) - p^*(x)]p(x) > 0$ for all $x \in M$, then p^* can be modified to produce a better approximation.

We assume that the polynomial p satisfies $|p(x)| \leq 1$ for all $x \in [a,b]$, and shall show that there exists $\theta > 0$ such that $\max_{x \in [a,b]} |d^*(x) - \theta p(x)| < \max_{x \in [a,b]} |d^*(x)|$.

Consider the set M' of all x such that $d^*(x)p(x) \leq 0$. This set is closed, and since M and M' are disjoint, we see that $d := \max_{x \in M'} |d^*(x)|$ must satisfy the inequality $d < \max_{x \in M} |d^*(x)|$. If M' is empty, then we take $d := 0$.

Now let $\theta := \frac{1}{2}[\max_{x \in [a,b]} |d^*(x)| - d]$, and let $\xi \in [a,b]$ be chosen such that $|d^*(\xi) - \theta p(\xi)| = \max_{x \in [a,b]} |d^*(x) - \theta p(x)|$. If $\xi \in M'$, then we have the estimate

$$\max_{x \in [a,b]} |d^*(x) - \theta p(x)| \leq |d^*(\xi)| + |\theta p(\xi)| \leq d + \theta$$

$$= \frac{1}{2}[\max_{x \in [a,b]} |d^*(x)| + d] < \max_{x \in [a,b]} |d^*(x)|.$$

On the other hand, if $\xi \notin M'$, then since $d^*(\xi)$ and $p(\xi)$ have the same sign, we get

$$|d^*(\xi) - \theta p(\xi)| < \max[|d^*(\xi)|, |\theta p(\xi)|].$$

In either case $p^* + \theta p$ is a better approximation of f than p^*.

Now suppose no alternant exists. Then there are at most $k \leq n$ points ξ_ν, such that $|d(\xi_\nu)| = \|d\|_\infty$ and $\text{sgn}\, d(\xi_\nu) = \varepsilon(-1)^\nu$ for $\nu = 1, \ldots, k$. But then we can always find a polynomial p such that $[f(\xi_\nu) - g(\xi_\nu)]p(\xi_\nu) > 0$ for $\nu = 1, \ldots, k$. Indeed, we may choose a polynomial that has the simple zeros $\xi'_1, \ldots, \xi'_{k-1}$ with $\xi_\kappa < \xi'_\kappa < \xi_{\kappa+1}$, $1 \le \kappa \le k-1$, and no other zeros in $[a, b]$. □

Remark. As was the case for Theorem 4.1, the alternation theorem is also valid for a function given only at a discrete set of points. The proof proceeds exactly as above, except that now the alternant must satisfy $|f(x_\nu) - g(x_\nu)| = = \rho := \max_{1 \le \mu \le m} |f(x_\mu) - g(x_\mu)|$.

Extension. The proof given above for polynomials generalizes immediately to Haar spaces provided that we can show that for a general Haar space $U = \operatorname{span}(g_1, \ldots, g_n)$, there exists $p \in U$ such that $[f(\xi_\nu) - g(\xi_\nu)]p(\xi_\nu) > 0$ for $\nu = 1, \ldots, k$ whenever $k \le n$. This follows, for example, from Theorem 5.1.1 below on interpolation from Haar spaces.

4.4 Uniqueness. The Alternation Theorem 4.3 completely characterizes when a function in a Haar subspace is a best approximation of a given continuous function. The theorem can also be used to establish the uniqueness of the best approximation. We prove the following

Uniqueness Theorem. *Let* U := span$(g_1, \ldots, g_n)$ *be a Haar subspace of* C$[a, b]$. *Then for any* $f \in$ C$[a, b]$, *there is a unique best approximation* $\tilde{f} \in$ U.

Proof. Suppose both h_1 and h_2 are best approximations of f from U. By Remark 3.4, $\frac{1}{2}(h_1 + h_2)$ is also a best approximation. But then by the alternation theorem, there exists an alternant $a \le x_1 < x_2 < \cdots < x_{n+1} \le b$ such that

$$f(x_\nu) - \frac{1}{2}[h_1(x_\nu) + h_2(x_\nu)] = \varepsilon(-1)^\nu \rho \text{ with } \varepsilon = \pm 1 \text{ for } \nu = 1, \ldots, n+1.$$

But then

$$\frac{1}{2}[f(x_\nu) - h_1(x_\nu)] + \frac{1}{2}[f(x_\nu) - h_2(x_\nu)] = \varepsilon(-1)^\nu \rho.$$

Now since $|f(x_\nu) - h_j(x_\nu)| \le \rho$, $(j = 1, 2)$, it follows that $f(x_\nu) - h_1(x_\nu) = = f(x_\nu) - h_2(x_\nu)$, thus $h_1(x_\nu) = h_2(x_\nu)$ for $\nu = 1, \ldots, n+1$, and therefore $h_1 = h_2$, since U is a Haar space. □

4.5 An Error Bound. In certain simple cases, Theorem 4.1 can be used to explicitly construct the best approximation of a given continuous function f. For example, suppose f is in the set C$_2[a, b] \subset$ C$[a, b]$ of functions whose second derivative does not change sign, and that we want to approximate f by a linear polynomial. In this case a three point alternant is given by $a = x_1 < x_2 < x_3 = b$, where x_2 is chosen so that $f'(x_2) = \frac{f(b)-f(a)}{b-a}$. Then the linear polynomial

$$\tilde{p}(x) = \frac{f(b) - f(a)}{b - a}(x - \frac{a + x_2}{2}) + \frac{1}{2}[f(a) + f(x_2)]$$

is the best approximation.

In general, it will not be so easy to find a polynomial to which Theorem 4.1 can be applied. Thus, it is useful to have an estimate for how good an approximation is, given an alternant for it. The following result of C. de la Vallée-Poussin (1866–1962) provides such an estimate.

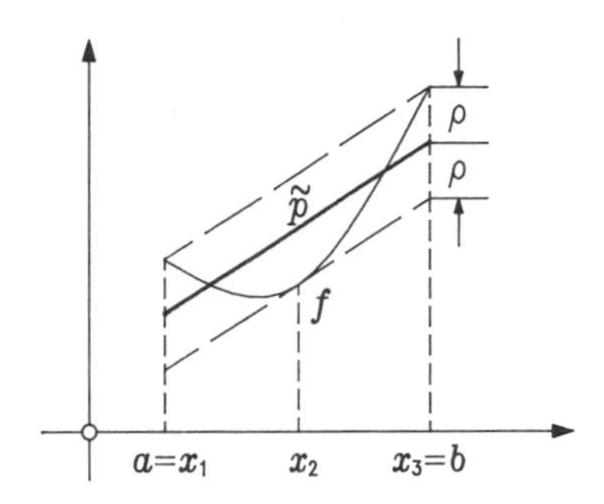

Error Bound. *Let* $\mathrm{U} := \operatorname{span}(g_1, \ldots, g_n)$ *be a Haar subspace of* $\mathrm{C}[a,b]$. *Given* $f \in \mathrm{C}[a,b]$ *and* $g \in \mathrm{U}$, *let* $x_1, \ldots, x_{n+1}$ *be a corresponding alternant. Then the minimal deviation* $E_{\mathrm{U}}(f) = \|f - \tilde{f}\|_\infty$ *satisfies* $\delta \leq E_{\mathrm{U}}(f) \leq \Delta$ *with* $\delta := \min_{1 \leq \nu \leq n+1} |d(x_\nu)|$ *and* $\Delta := \max_{x \in [a,b]} |d(x)|$.

Proof. The right-hand inequality is obvious. To establish the left-hand inequality, we show that assuming $E_{\mathrm{U}}(f) < \delta$ leads to a contradiction.

Indeed, suppose that

$$\max_{1 \leq \nu \leq n+1} |f(x_\nu) - \tilde{f}(x_\nu)| \leq \|f - \tilde{f}\|_\infty < \min_{1 < \nu < n+1} |f(x_\nu) - g(x_\nu)|.$$

But then the function $\tilde{f} - g = (f - g) - (f - \tilde{f})$ must satisfy

$$\operatorname{sgn}[\tilde{f}(x_\nu) - g(x_\nu)] = \operatorname{sgn}[f(x_\nu) - g(x_\nu)] = \varepsilon(-1)^\nu$$

for $\nu = 1, \ldots, n+1$. This implies that $\tilde{f} - g \in U$ has at least one zero in each of the n subintervals $(x_\nu, x_{\nu+1})$, $1 \leq \nu \leq n$. From this it follows that $g = \tilde{f}$, which is a contradiction. □

This result allows us to compute upper and lower bounds for the minimal deviation whenever we have an approximant g and a corresponding alternant.

4.6 Computation of the Best Approximation. Theorem 4.1 provides a basis for designing a method for the computation of best approximations of continuous functions. The method works in general for Chebyshev systems, but for convenience, we present the details only for the most important practical case, approximation by polynomials.

The Exchange Method of Remez. Let $f \in C[a,b]$, and suppose we want to compute a best approximation $\tilde{p} \in P_{n-1}$.

The method starts with the selection of $(n+1)$ points

$$a \leq x_1^{(0)} < x_2^{(0)} < \cdots < x_{n+1}^{(0)} \leq b$$

forming an alternant for $f - p^{(0)}$. In the first step of the method we determine an associated first approximation $p^{(0)} \in P_{n-1}$.

Step 1: We determine $p^{(0)} \in P_{n-1}$ from the conditions that $\{x_\nu^{(0)}\}_{\nu=1}^{n+1}$ be an alternant for $f - p^{(0)}$ and that the value $|\rho^{(0)}|$ of the deviation is the same at every alternant point.

These conditions can be written as

$$(f - p^{(0)})(x_\nu^{(0)}) = (-1)^{\nu-1}\rho^{(0)}, \quad 1 \leq \nu \leq n+1,$$

and writing $p^{(0)}(x) =: \alpha_0^{(0)} + \alpha_1^{(0)}\, x + \cdots + \alpha_{n-1}^{(0)}\, x^{n-1}$, we obtain the linear system of equations

$$(-1)^{\nu-1}\rho^{(0)} + \alpha_0^{(0)} + \alpha_1^{(0)}\, x_\nu^{(0)} + \cdots + \alpha_{n-1}^{(0)}(x_\nu^{(0)})^{n-1} = f(x_\nu^{(0)}),$$

$1 \leq \nu \leq n+1$, for the unknowns $\rho^{(0)}, \alpha_0^{(0)}, \ldots, \alpha_{n-1}^{(0)}$.

This system has a unique solution since the determinant of the matrix $A^{(0)}$ of the system is positive. Indeed,

$$\det(A^{(0)}) := \begin{vmatrix} 1 & 1 & x_1^{(0)} & \cdots & (x_1^{(0)})^{n-1} \\ -1 & 1 & x_2^{(0)} & & (x_2^{(0)})^{n-1} \\ \vdots & \vdots & & & \vdots \\ (-1)^n & 1 & x_{n+1}^{(0)} & \cdots & (x_{n+1}^{(0)})^{n-1} \end{vmatrix} = \sum_{\lambda=1}^{n+1} \det(A_\lambda^{(0)}),$$

where

$$\det(A_\lambda^{(0)}) := \begin{vmatrix} 1 & x_1^{(0)} & \cdots & (x_1^{(0)})^{n-1} \\ \vdots & & & \\ 1 & x_{\lambda-1}^{(0)} & \cdots & (x_{\lambda-1}^{(0)})^{n-1} \\ 1 & x_{\lambda+1}^{(0)} & \cdots & (x_{\lambda+1}^{(0)})^{n-1} \\ \vdots & & & \\ 1 & x_{n+1}^{(0)} & \cdots & (x_{n+1}^{(0)})^{n-1} \end{vmatrix} = \prod_{\mu>\nu} (x_\mu^{(0)} - x_\nu^{(0)}),$$

$(\mu, \nu = 1, \ldots, \lambda-1, \lambda+1, \ldots, n+1)$, $1 \leq \lambda \leq n+1$. These are Vandermonde determinants, and are positive since $(x_\mu^{(0)} - x_\nu^{(0)}) > 0$ for $\mu > \nu$.

Suppose now that $\|f - p^{(0)}\|_\infty = |f(\xi^{(1)}) - p^{(0)}(\xi^{(1)})|$ for some $\xi^{(1)}$ in $[a, b]$. If $\xi^{(1)} \in \{x_1^{(0)}, \ldots, x_{n+1}^{(0)}\}$, then $\|f - p^{(0)}\|_\infty = |f(x_\nu^{(0)}) - p^{(0)}(x_\nu^{(0)})|$ for all $1 \leq \nu \leq n+1$ with alternating signs. In this case $p^{(0)} =: \tilde{p}$ is already the best approximation. In the case where $\xi^{(1)} \notin \{x_1^{(0)}, \ldots, x_{n+1}^{(0)}\}$, we replace one of the points $x_1^{(0)}, \ldots, x_{n+1}^{(0)}$ by $\xi^{(1)}$, following the rules given in the table below. The exchange assures that the remaining n points in $\{x_1^{(0)}, \ldots, x_{n+1}^{(0)}\}$ together with $\xi^{(1)}$ result in an $(n+1)$-tuple $x_1^{(1)} < \cdots < x_{n+1}^{(1)}$ which forms a new alternant for $f - p^{(0)}$. The absolute value of the deviation at the alternant point $\xi^{(1)}$ is given by $\|f - p^{(0)}\|_\infty > \delta^{(0)} := |\rho^{(0)}|$ while the absolute value of the deviation at the other n alternant points is $\delta^{(0)}$.

The following table gives the exchange rules to be used to find the $(j+1)$-st alternant $\{x_1^{(j+1)}, x_2^{(j+1)}, \ldots, x_{n+1}^{(j+1)}\}$ from the j-th:

$\xi^{(j+1)} \in$	$\operatorname{sgn}[f - p^{(j)}](\xi^{(j+1)}) =$	Replace $\xi^{(j+1)}$ by
$[a, x_1^{(j)})$	$+\operatorname{sgn}[f - p^{(j)}](x_1^{(j)})$	$x_1^{(j)}$
	$-\operatorname{sgn}[f - p^{(j)}](x_1^{(j)})$	$x_{n+1}^{(j)}$
$(x_\nu^{(j)}, x_{\nu+1}^{(j)})$	$+\operatorname{sgn}[f - p^{(j)}](x_\nu^{(j)})$	$x_\nu^{(j)}$
$\nu = 1, \ldots, n$	$-\operatorname{sgn}[f - p^{(j)}](x_\nu^{(j)})$	$x_{\nu+1}^{(j)}$
$(x_{n+1}^{(j)}, b]$	$+\operatorname{sgn}[f - p^{(j)}](x_{n+1}^{(j)})$	$x_{n+1}^{(j)}$
	$-\operatorname{sgn}[f - p^{(j)}](x_{n+1}^{(j)})$	$x_1^{(j)}$

Step 2: The second step of the method involves the construction of the polynomial $p^{(1)} \in \mathrm{P}_{n-1}$ such that $\{x_1^{(1)}, \ldots, x_{n+1}^{(1)}\}$ is an alternant for $f - p^{(1)}$ with a common value of $\delta^{(1)} := |\rho^{(1)}|$ for the deviations at each of the alternant points. We construct $p^{(1)}$ by solving the linear system of equations

$$(*) \quad (-1)^{\nu-1}\rho^{(1)} + \alpha_0^{(1)} + \cdots + \alpha_{n-1}^{(1)}(x_\nu^{(1)})^{n-1} = f(x_\nu^{(1)}), \quad 1 \leq \nu \leq n+1.$$

For later use, we denote the matrix of this system by $A^{(1)}$.

We claim that $\delta^{(1)} > \delta^{(0)}$. To see this, subtract $p^{(0)}(x_\nu^{(1)})$ from both sides of $(*)$ for $1 \leq \nu \leq n-1$. This leads to the system

$$(-1)^{\nu-1}\rho^{(1)} + (\alpha_0^{(1)} - \alpha_0^{(0)}) + \cdots + (\alpha_{n-1}^{(1)} - \alpha_{n-1}^{(0)})(x_\nu^{(1)})^{n-1} = (f - p^{(0)})(x_\nu^{(1)}),$$

$1 \leq \nu \leq n+1$. Applying the Cramer rule, we obtain

$$\rho^{(1)} = \left[\sum_{\lambda=1}^{n+1} \det(A_\lambda^{(1)})\right]^{-1} \sum_{\lambda=1}^{n+1} (-1)^{\lambda+1} \det(A_\lambda^{(1)})(f - p^{(0)})(x_\lambda^{(1)})$$

where $\det(A_\lambda^{(1)})$ are the corresponding subdeterminants. In view of the sign change properties of $f - p^{(0)}$, it follows that

$$\delta^{(1)} = \left[\sum_{\lambda=1}^{n+1} \det(A_\lambda^{(1)})\right]^{-1} \sum_{\lambda=1}^{n+1} \det(A_\lambda^{(1)})|(f-p^{(0)})(x_\lambda^{(1)})|.$$

This is a weighted average, and since we assumed $\delta^{(0)} < \|f - p^{(0)}\|_\infty$, we get $\delta^{(1)} > \delta^{(0)}$.

Iteration: The above steps are repeated until the best approximation $\tilde{p}$ is approximated to a sufficient accuracy. A complete discussion of the convergence of the exchange method can be found in the book of G. Meinardus [1967]. The problem of convergence does not arise in the case where the best approximation is to be determined on the discrete set x_ν, $1 \le \nu \le m$, with $m \ge n+1$. In this case there are only $\binom{m}{n+1}$ ways to choose $(n+1)$-tuples of points $\{x_1^{(j)}, x_2^{(j)}, \ldots, x_{n+1}^{(j)}\}$, while by the monotonicity $\delta^{(j)} < \delta^{(j+1)}$, and so the same $(n+1)$-tuple cannot appear twice.

Example. We illustrate the Remez algorithm with a simple example. Consider the problem of computing the best approximation of $f(x) := x^2$ for $x \in [0,1]$ from the space of linear polynomials P_1. As a starting alternant we choose $\{x_1^{(0)}, x_2^{(0)}, x_3^{(0)}\} = \{0, \frac{1}{3}, 1\}$.
Step 1: We determine $p^{(0)}$ from the equations

$$\begin{aligned} \rho^{(0)} + \alpha_0^{(0)} \qquad\qquad &= 0 \\ -\rho^{(0)} + \alpha_0^{(0)} + \alpha_1^{(0)}\tfrac{1}{3} &= \tfrac{1}{9} \\ \rho^{(0)} + \alpha_0^{(0)} + \alpha_1^{(0)} \quad &= 1. \end{aligned}$$

The solution gives $\alpha_0^{(0)} = -\frac{1}{9}$, $\alpha_1^{(0)} = 1$ and $\rho^{(0)} = \frac{1}{9}$ so that $p^{(0)}(x) = -\frac{1}{9} + x$. This is the best approximation on the set $\{0, \frac{1}{3}, 1\}$, and satisfies

$$\|f - p^{(0)}\|_\infty = \max_{x\in[0,1]} |x^2 - x + \frac{1}{9}| = \frac{5}{36} > \frac{1}{9}.$$

This value is assumed for $\xi^{(1)} = \frac{1}{2}$. Hence we replace the alternant point $x_2^{(0)}$ by $\xi^{(1)}$ so that the new alternant for $p^{(1)}$ is $\{x_1^{(1)}, x_2^{(1)}, x_3^{(1)}\} = \{0, \frac{1}{2}, 1\}$.
Step 2: $p^{(1)}$ and $\rho^{(1)}$ are computed from

$$\begin{aligned} \rho^{(1)} + \alpha_0^{(1)} \qquad\qquad &= 0 \\ -\rho^{(1)} + \alpha_0^{(1)} + \alpha_1^{(1)}\tfrac{1}{2} &= \tfrac{1}{4} \\ \rho^{(1)} + \alpha_0^{(1)} + \alpha_1^{(1)} \quad &= 1. \end{aligned}$$

This gives $\alpha_0^{(1)} = -\frac{1}{8}$, $\alpha_1^{(1)} = 1$ and $\rho^{(1)} = \frac{1}{8}$. The corresponding polynomial is $p^{(1)}(x) = -\frac{1}{8} + x$, and $\|f - p^{(1)}\|_\infty = \max_{x\in[0,1]} |x^2 - x + \frac{1}{8}| = \frac{1}{8}$. Since this value is assumed at all three points $x_1^{(1)} = 0$, $x_2^{(1)} = \frac{1}{2}$ and $x_3^{(1)} = 1$, it follows that $p^{(1)}$ is the best approximation, and the algorithm stops.

In general, we cannot expect that the algorithm will lead to the exact solution in a finite number of steps as was the case for this simple example. In practice it is common to terminate the iteration when the bounds $\delta^{(k)}$ and $\|f - p^{(k)}\|_\infty$ at the k-th step are sufficiently close to each other.

4.7 Chebyshev Polynomials of the First Kind. We now consider the problem of finding a best possible uniform approximation to the monomial $f(x) := x^n$ on $[-1, +1]$ by a polynomial from P_{n-1}, $(n = 1, 2, \ldots)$. We next show that the solution of this problem can be found using the Alternation Theorem.

The problem is to find the unique polynomial $\tilde{p} \in P_{n-1}$ satisfying

$$\max_{x\in[-1,+1]} |x^n - (\tilde{\alpha}_{n-1}x^{n-1} + \cdots + \tilde{\alpha}_0)| =$$
$$= \min_{\alpha\in\mathbb{R}^n} \max_{x\in[-1,+1]} |x^n - (\alpha_{n-1}x^{n-1} + \cdots + \alpha_0)|.$$

Solution: For $n = 1$,

$$\min_{\alpha_0\in\mathbb{R}} \max_{x\in[-1,+1]} |x - \alpha_0| = \min_{\alpha_0\in\mathbb{R}} \max(|1 - \alpha_0|, |-1 - \alpha_0|) = 1,$$

and thus $\tilde{\alpha}_0 = 0$. It follows that $\tilde{p} = 0$ is the best approximation from P_0.

For $n = 2$, the solution can be found from the construction 4.4. In this case the best approximation $\tilde{p} \in P_1$ of $f(x) := x^2$ on $[-1, +1]$ is $\tilde{p}(x) = \frac{1}{2}$ since the difference $d(x) = x^2 - \frac{1}{2}$ satisfies $d(-1) = -d(0) = d(1) = \frac{1}{2}$, and thus the points $\{-1, 0, 1\}$ form an alternant with maximal deviation.

We claim that, in general, the solution is given by the polynomial $\tilde{p}(x) = = x^n - \hat{T}_n(x)$ with $\hat{T}_n(x) := \frac{1}{2^{n-1}} T_n(x)$, $T_n(x) := \cos(n \arccos(x))$. The proof proceeds as follows:

1) $\tilde{p} \in P_{n-1}$. For $n = 1$ we directly compute $T_1(x) = \cos(\arccos(x)) = x$ and $\hat{T}_1(x) = x$ with $\tilde{p}(x) = 0$. For $n > 1$, we employ the substitution $\theta := \arccos(x)$ (or equivalently $x = \cos(\theta)$) which gives a mapping $\theta : [-1, +1] \to [-\pi, 0]$. Then $T_n(x(\theta)) = \cos(n\theta)$.

Now the trigonometric identity $\cos((n+1)\theta) + \cos((n-1)\theta) = 2\cos(\theta)\cos(n\theta)$ implies the recursion relation $T_{n+1}(x) = 2xT_n(x) - T_{n-1}(x)$ for $n \in \mathbb{Z}_+$. Starting with $T_0(x) = 1$, we obtain

$$T_2(x) = 2x^2 - 1, \quad T_3(x) = 4x^3 - 3x, \quad \ldots, \quad T_n(x) = 2^{n-1}x^n - \cdots \quad \text{etc.}$$

The polynomials $\hat{T}_n$ are normalized so that the leading coefficient is 1, and thus $\tilde{p}(x) = x^n - \hat{T}_n(x)$ is a polynomial in P_{n-1}.

2) $\tilde{p} \in P_{n-1}$ is a best approximation. At the $n\theta_\nu := -(n - \nu + 1)\pi$, $1 \le \nu \le n + 1$, we have $T_n(x(\theta_\nu)) = \cos(n\theta_\nu) = (-1)^{n-\nu+1}$. It follows that the points $x_\nu := \cos(-\frac{n-\nu+1}{n}\pi) = \cos((1 - \frac{\nu-1}{n})\pi)$ form an

alternant for $d(x) := \hat{T}_n(x) = x^n - \tilde{p}(x)$. Moreover, since $|\hat{T}_n(x_\nu)| = \frac{1}{2^{n-1}} = \|d\|_\infty$, the maximal deviation is assumed at these points; that is, $d(x_\nu) = \varepsilon(-1)^\nu \|d\|_\infty$ with $\varepsilon = \pm 1$ for $\nu = 1, \ldots, n+1$.

Clearly, the polynomial T_n has n simple zeros in the interval $(-1, +1)$ located at the points $x_\nu = \cos \frac{2\nu - 1}{2n}\pi$, $1 \leq \nu \leq n$.

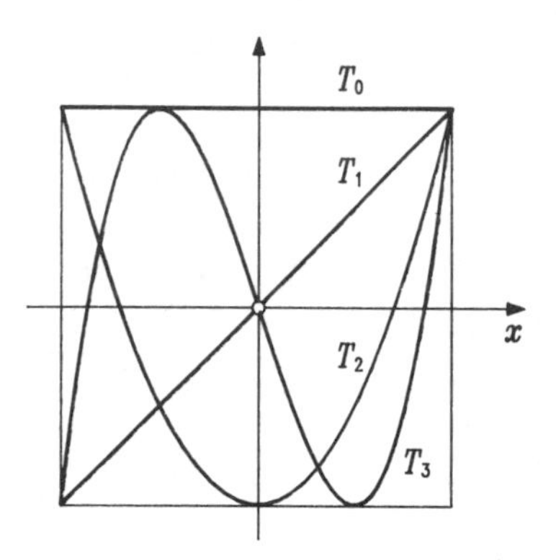

The polynomials $T_n(x) = \cos(n \arccos(x))$ are called *Chebyshev polynomials of the first kind.* They are defined for all $n \geq 0$.

The approximation problem discussed in this section can be reinterpreted as follows: Find a polynomial of degree n, with leading coefficient one, whose maximum in $[-1, +1]$ is minimal. This is equivalent to finding a polynomial in the subset

$$\hat{\mathrm{P}}_n := \{p \in \mathrm{P}_n \mid p(x) = x^n + \alpha_{n-1}x^{n-1} + \cdots + \alpha_0\}$$

which best approximates the function $f = 0$ on $[-1, +1]$. The solution of this problem is given by $\tilde{p}(x) = x^n - \hat{T}_n(x)$. This says that the polynomial $\hat{T}_n(x) = x^n - \tilde{p}(x)$ is the unique polynomial in $\hat{\mathrm{P}}_n$ of minimal norm; i.e., $\|\hat{T}_n\|_\infty \leq \|p\|_\infty$ for all $p \in \hat{\mathrm{P}}_n$.

This reformulation of the approximation problem of this section is a relatively simple example of a nonlinear approximation problem. Indeed, the subset $\hat{\mathrm{P}}_n$ is not a linear space, although it is an affine subspace of a linear space. We have been able to derive a remarkable minimal property of the Chebyshev polynomial of the first kind by considering an appropriate linear approximation problem.

4.8 Expansions in Chebyshev Polynomials. Using the fact that the Chebyshev polynomials of the first kind were defined in terms of trigonometric functions, it is easy to show that they form an orthogonal system of functions with respect to the weight function $w(x) := \frac{1}{\sqrt{1-x^2}}$. In fact,

$$\int_{-1}^{+1} T_k(x)T_\ell(x)\frac{dx}{\sqrt{1-x^2}} = \int_0^\pi \cos(k\theta)\cos(\ell\theta)\frac{\sin\theta}{\sin\theta}d\theta = 0 \quad \text{for} \quad k \neq \ell$$

and

$$\int_{-1}^{+1} T_k^2(x)\frac{dx}{\sqrt{1-x^2}} = \begin{cases} \pi & \text{for } k=0 \\ \frac{\pi}{2} & \text{for } k \neq 0. \end{cases}$$

It is known from a theorem in analysis that every function $f \in C[a,b]$ can be expanded in terms of a complete orthogonal system of functions. The partial sums of such a Fourier series expansion provide approximations to f which converge with respect to the norm $\|f\| := [\int_a^b f^2(x)w(x)dx]^{\frac{1}{2}}$ associated with the weight function w. In 5.5-5.8 we will discuss this fact further, especially for the case where the norm is $\|\cdot\|_2$. Here we proceed directly, and for given f, define approximations of f in terms of Chebyshev polynomials $T_0, T_1, \ldots$ by

$$\tilde{f}_n(x) = \frac{c_0}{2} + \sum_{k=1}^{n} c_k T_k(x),$$

where

$$c_k = \frac{2}{\pi}\int_{-1}^{+1} f(x)T_k(x)\frac{dx}{\sqrt{1-x^2}}, \qquad k \in \mathbb{N},$$

or equivalently,

$$c_k = \frac{2}{\pi}\int_0^{\pi} f(\cos\theta)\cos(k\theta)d\theta = \frac{1}{\pi}\int_{-\pi}^{\pi} f(\cos\theta)\cos(k\theta)d\theta.$$

Under appropriate hypotheses, this sequence of approximations can even be shown to converge to f with respect to the uniform norm $\|\cdot\|_\infty$. Uniform convergence is especially interesting for the following reason. Suppose that a function $f \in C[a,b]$ can be expanded in a uniformly convergent series in terms of a system of polynomials $\{\psi_0, \psi_1, \ldots\}$ which are normed so that $|\psi_k(x)| \leq 1$ in $[a,b]$. Then

$$|f(x) - \tilde{f}_n(x)| = |\sum_{k=n+1}^{\infty} c_k\psi_k(x)| \leq \sum_{k=n+1}^{\infty} |c_k|,$$

and thus if the coefficients c_k for $k \geq n+1$ are negligibly small, then $\tilde{f}_n$ provides a good approximation to the best approximating polynomial $\tilde{p} \in P_n$ of f with respect to the Chebyshev norm. The following theorem shows that this situation persists for the Chebyshev expansion, provided we restrict f to lie in $C_2[-1,+1]$.

Expansion Theorem. *Let $f \in C_2[-1,+1]$. Then the expansion of f in terms of Chebyshev polynomials T_k of the first kind for $x \in [-1,+1]$ converge uniformly on this interval to f. Moreover, the corresponding coefficients satisfy*

$$|c_k| \leq \frac{A}{k^2},$$

where the constant A depends only on f.

Proof. From the formula for the coefficients given above, it follows, after integrating by parts twice and writing $\varphi(\theta) := f(\cos\theta)$, that

$$c_k = -\frac{2}{\pi k}\int_0^\pi \frac{d\varphi}{d\theta}\sin(k\theta)d\theta = \frac{2}{\pi k^2}\frac{d\varphi}{d\theta}\cos(k\theta)\Big|_0^\pi - \frac{2}{\pi k^2}\int_0^\pi \frac{d^2\varphi}{d\theta^2}\cos(k\theta)d\theta.$$

This immediately implies the estimate $|c_k| \le \frac{A}{k^2}$. In addition, it follows that there exists a function $g \in \mathrm{C}[-1,+1]$ with $\lim_{n\to\infty}\|\tilde{f}_n - g\|_\infty = 0$. Since $\lim_{n\to\infty}\|\tilde{f}_n - f\| = 0$ while

$$\|\tilde{f}_n - g\| = [\int_{-1}^{+1}(\tilde{f}_n(x)-g(x))^2\frac{dx}{\sqrt{1-x^2}}]^{\frac{1}{2}} \le \|\tilde{f}_n - g\|_\infty\left(\int_{-1}^{+1}\frac{dx}{\sqrt{1-x^2}}\right)^{\frac{1}{2}},$$

the inequality

$$\|f-g\| \le \|f-\tilde{f}_n\| + \|\tilde{f}_n - g\|$$

implies $f = g$ and the assertion is established. □

Practical Consequence. Given a function $f \in \mathrm{C}_2[-1,+1]$, we can obtain a good approximation to the best approximation $\tilde{p} \in \mathrm{P}_n$ by taking the partial sum $\tilde{f}_n = \sum_0^n c_k T_k$. This method is particularly applicable when the coefficients c_k are simple to compute.

Example. Consider the approximation of the function $f(x) := \sqrt{1-x^2}$ on $[-1,+1]$ by partial sums of the Chebyshev polynomial expansion of f. In this case

$$c_k = \frac{2}{\pi}\int_0^\pi \cos(kt)\sin t\,dt = \begin{cases}\frac{4}{\pi}\frac{1}{1-k^2} & \text{for } k = 2\kappa\\ 0 & \text{for } k = 2\kappa+1\end{cases}, \quad \kappa \in \mathbb{N}.$$

This leads to the approximations

$$\tilde{f}_0(x) = \frac{2}{\pi}, \quad \tilde{f}_2(x) = \frac{2}{3\pi}(5-4x^2),$$
$$\tilde{f}_4(x) = \frac{2}{15\pi}(23-4x^2-16x^4), \quad \text{etc.}$$

We note that in this example the bound for $|c_k|$ given in the Expansion Theorem above is valid, even though here f is only two-times continuously differentiable on $(-1,+1)$.

In practice it is not usually possible to determine the coefficients c_k of the Chebyshev expansion by explicit integration, as was the case in this simple example. Generally, it will be necessary to use numerical quadrature formula to calculate these coefficients. An example is given in Problem 7 in 7.4.4.

4.9 Convergence of Best Approximations. Given a function $f \in \mathrm{C}[a,b]$ and the sequence $(\tilde{p}_n)$ of best approximating polynomials, where $\tilde{p}_n \in \mathrm{P}_n$ is best in the Chebyshev sense, it is natural to ask whether this sequence converges to f. This question can be answered with the help of the Weierstrass Approximation Theorem 2.2. Indeed, let $(p_n)_{n\in\mathbb{N}}$ be some sequence of polynomials $p_n \in \mathrm{P}_n$ such that $\lim_{n\to\infty}\|f-p_n\|_\infty = 0$. Then since $\|f-\tilde{p}_n\|_\infty \le \|f-p_n\|_\infty$ for all $n \in \mathbb{N}$, it follows immediately from $\lim_{n\to\infty}\|f-p_n\|_\infty = 0$ that $\lim_{n\to\infty}\|f-\tilde{p}_n\|_\infty = 0$. We have established the

Convergence Theorem. *Let $f \in \mathbb{C}[a,b]$. Then the sequence $(\tilde{p}_n)_{n\in\mathbb{N}}$ of best approximations $\tilde{p}_n \in \mathrm{P}_n$ with respect to the norm $\|\cdot\|_\infty$ converges uniformly to f.*

4.10 Nonlinear Approximation. In this section we discuss an approximation problem involving a nonlinear subset of $(\mathrm{C}[a,b], \|\cdot\|_\infty)$ which is especially important; namely, approximation by rational functions. We shall restrict our discussion primarily to the problem of the existence of a best approximation.

Let $\mathrm{R}_{n,m}[a,b]$ be the set of continuous rational functions on the interval $[a,b]$ of the form $r(x) := \frac{p(x)}{q(x)}$, where $p \in \mathrm{P}_n$, $q \in \mathrm{P}_m$, $\|q\|_\infty = 1$ and $q(x) > 0$ for $x \in [a,b]$. In addition, suppose that p and q have no common factors, so that they have no common zeros anywhere on $\mathbb{C}$. The following theorem settles the problem of existence of a best approximation $\tilde{r} \in \mathrm{R}_{n,m}[a,b]$.

Theorem. *Let $f \in \mathrm{C}[a,b]$. Then in the set $\mathrm{R}_{n,m}[a,b]$ of continuous rational functions, there always exists a best approximation $\tilde{r}$ of f.*

Proof. Let $(r_\nu)_{\nu\in\mathbb{N}}$ be a minimizing sequence for f in $\mathrm{R}_{n,m}$, and suppose $r_\nu = \frac{p_\nu}{q_\nu}$, where $p_\nu \in \mathrm{P}_n$ and $q_\nu \in \mathrm{P}_m$ have no common factors. $\|q_\nu\|_\infty = 1$ implies that (q_ν) is a bounded sequence in P_m, and thus contains a convergent subsequence $(q_{\nu(\kappa)})$. Since P_m is finite dimensional, as $\kappa \to \infty$ this subsequence converges to some $q^* \in \mathrm{P}_m$, $\|q^*\|_\infty = 1$.

By Lemma 3.4, the minimizing sequence (r_μ), $\mu := \nu(\kappa)$, is itself bounded. Now $\frac{|p_\mu(x)|}{|q_\mu(x)|} \le C$ for $x \in [a,b]$ implies that $\|p_\mu\|_\infty \le C$, which again gives the existence of a convergent subsequence $(p_{\mu(\kappa)})$ converging to some $p^* \in \mathrm{P}_n$. Clearly, $|p^*(x)| \le C|q^*(x)|$, and thus if $x_1, \cdots, x_k$ are zeros of q^*, $k \le m$, then they are also zeros of p^*. Thus the common factors of $\frac{p^*}{q^*}$ can be cancelled to produce a rational function $\frac{\hat{p}}{\hat{q}} \in \mathrm{R}_{n,m}$ with $\hat{q}(x) > 0$ for $x \in [a,b]$ and

$$|f(x) - \frac{\hat{p}(x)}{\hat{q}(x)}| = |f(x) - \frac{p^*(x)}{q^*(x)}| \le |f(x) - \frac{p_{\mu(\kappa)}(x)}{q_{\mu(\kappa)}(x)}| + |\frac{p_{\mu(\kappa)}(x)}{q_{\mu(\kappa)}(x)} - \frac{p^*(x)}{q^*(x)}|$$

$$\Rightarrow \|f - \frac{p^*}{q^*}\|_\infty \le \|f - \frac{p_{\mu(\kappa)}}{q_{\mu(\kappa)}}\|_\infty + \|\frac{p_{\mu(\kappa)}}{q_{\mu(\kappa)}} - \frac{p^*}{q^*}\|_\infty.$$

Since $\lim_{\kappa\to\infty} \|f - \frac{p_{\mu(\kappa)}}{q_{\mu(\kappa)}}\|_\infty = E_{\mathrm{R}_{n,m}}(f)$ and $\lim_{\kappa\to\infty} \|\frac{p_{\mu(\kappa)}}{q_{\mu(\kappa)}} - \frac{p^*}{q^*}\|_\infty = 0$, we conclude that $\|f - \frac{p^*}{q^*}\|_\infty \le E_{\mathrm{R}_{n,m}}(f)$. Now since $\frac{p^*}{q^*} \in \mathrm{R}_{n,m}[a,b]$, we know that $E_{\mathrm{R}_{n,m}}(f) \le \|f - \frac{p^*}{q^*}\|_\infty$, and so $\|f - \frac{p^*}{q^*}\|_\infty = E_{\mathrm{R}_{n,m}}(f)$; i.e., $\frac{p^*}{q^*}$ is a best approximation of f in $\mathrm{R}_{n,m}[a,b]$. □

It is beyond the scope of this book to develop fully further properties of approximation by rational functions, but we complete this section by mentioning a result on uniqueness and an outline of how best approximations can be computed.

Uniqueness Theorem. *Every $f \in C[a,b]$ has a unique best approximation $\tilde{r} \in R_{n,m}[a,b]$.*

For a proof of this result, see e.g. the book of G. A. Watson [1980].

Computation of the Best Approximation. There is also an alternation theorem for approximation by rational functions, and thus the best approximation $\tilde{r} \in R_{n,m}[a,b]$ can be computed with an appropriate exchange method. For details of such a method, which is an extension of the Remez algorithm to rational function, see e.g. G. Meinardus [1967].

4.11 Remarks on Approximation in $(C[a,b], \|\cdot\|_1)$. Occasionally the problem arises of approximating a continuous function with respect to the norm $\|\cdot\|_1$. For example, if we are looking for a polynomial approximation, then the problem is to find a polynomial $\tilde{p}$ which minimizes $\int_a^b |f(x)-p(x)|dx$ over $p \in P_n$.

By the Fundamental Theorem of Approximation Theory in Normed Linear Spaces 3.4, we know that there always exists a best approximation $\tilde{p} \in P_n$. The uniqueness result 3.5 cannot, however, be applied here, since as we have already seen in Example 2 of 3.5, the linear space $(C[a,b], \|\cdot\|_1)$ is not strictly normed. One can, nevertheless, show that, as in the case of Chebyshev approximation, if we are approximating from a Haar subspace, then there is a unique best approximation. A proof can be found in the book of G. A. Watson [1980].

Approximation in the norm $\|\cdot\|_1$ is of interest in those situations where it is desired that the best approximation not be sensitive to local changes in f. In particular, it can be shown that a best approximation $\tilde{p}$ of f remains best if the value $f(x)$ is modified, as long as the sign of $(f(x) - \tilde{p}(x))$ does not change. As a consequence, it follows that the characterization of best approximations in this case involves properties of the function $\mathrm{sgn}(f - \tilde{p})$. We illustrate this situation in the simplest possible case of approximation by a constant $p \in P_0$.

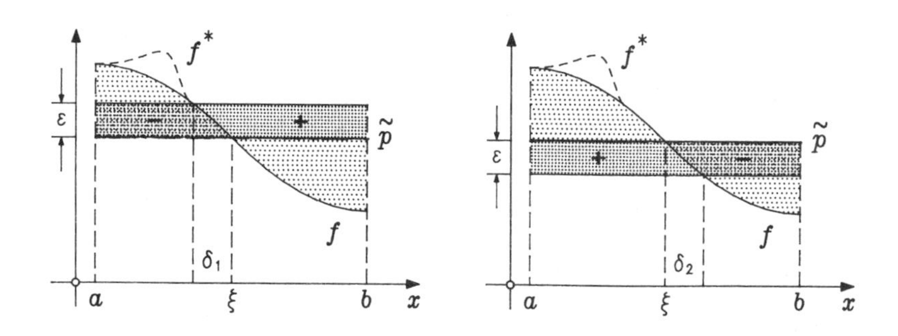

Suppose f is a strictly monotone decreasing continuous function on $[a,b]$. We claim that $\tilde{p} = f(\frac{a+b}{2})$ is the best approximation in P_0. Indeed, consider $p = f(\xi)$ with $\xi \in [a,b]$. If we move p upwards by an amount ε, then the linear part of the change in $\|f-p\|_1$ is $-(\xi-a)\varepsilon + (b-\xi)\varepsilon = = [\frac{a+b}{2} - \xi]2\varepsilon$, where (see the sketch below), $\delta(\varepsilon)$ is the change in ε. Similarly, if we move p downwards by an amount ε, then the linear part of the change is $(\xi-a)\varepsilon - (b-\xi)\varepsilon = [\xi - \frac{a+b}{2}]2\varepsilon$. It follows that if $\xi \neq \frac{a+b}{2}$, then the value of $\|f-p\|_1$ can be reduced, and hence the best approximation corresponds to $\xi = \frac{a+b}{2}$.

It is clear in this example that modifying the function f (for example to a function f^* as shown in the figure) does not affect our analysis, and thus $\tilde{p}$ is not changed as long as the function $\operatorname{sgn}(f-\tilde{p})$ is not affected.

4.12 Problems. 1) Let $f \in C_{n+1}[a,b]$ be such that $f^{(n+1)}(x) \neq 0$ for all $x \in [a,b]$. Let $g_j(x) := x^{j-1}$, $1 \leq j \leq n+1$. Show: $\{g_1, \ldots, g_{n+1}, f\}$ is a Chebyshev system.

2) Let $g_1, \ldots, g_n \in C[a,b]$. Show that the following assertions are equivalent:

a) $\{g_1, \ldots, g_n\}$ forms a Chebyshev system.

b) for any n distinct points $x_1, \ldots, x_n \in [a,b]$, $\det(g_\mu(x_\nu))^n_{\mu,\nu=1} \neq 0$.

3) Determine the best approximation in $(C[0,1], \|\cdot\|_\infty)$ from P_1 of the following functions:

a) $f(x) := \cos(2\pi x) + x$; b) $f(x) := e^x$;

c) $f(x) := \min(5x - 2x^2, 22(1-x)^2)$.

4) Find the best approximation $\tilde{p} \in P_2$ of $f(x) := x|x|$ in $(C[0,1], \|\cdot\|_\infty)$.

a) Carry out three steps of the Remez algorithm with the starting alternant $x_1^{(0)} := -\frac{1}{2}$, $x_2^{(0)} := 0$, $x_3^{(0)} := \frac{1}{2}$, $x_4^{(0)} := 1$.

b) After 3 steps, how far are you from the minimal distance (in percent)?

c) Compute $\tilde{p}$ in each step, and tabulate the convergence behavior of $\delta^{(j)}$, $\alpha_0^{(j)}$, $\alpha_1^{(j)}$ and $\alpha_2^{(j)}$ for $j = 1,2,3$.

5) Find the best approximation of a polynomial $p \in P_n$ from P_{n-1} in $(C[-1,+1], \|\cdot\|_\infty)$.

6) Continue the computations of Example 4.8 by finding the bounds in 4.5 for the approximations $\tilde{f}_2$ and $\tilde{f}_4$ in order to judge the quality of the approximations. Draw sketches comparing these approximations with the corresponding polynomials of the same degree obtained by truncating the Taylor expansions about $x = 0$.

Compare the computed approximation with the polynomial of second degree obtained by truncating the Taylor expansion of f about $x = 0$.

7) Compute and compare:

a) the best approximations of $f(x) := \alpha x + \beta$, $(\alpha, \beta \in \mathbb{R})$ from P_0 in $(C[a,b], \|\cdot\|_\infty)$ and in $(C[a,b], \|\cdot\|_1)$.

b) the best approximations of $f(x) := e^x$ from P_0 in $(C[0,1], \|\cdot\|_\infty)$ and in $(C[0,1], \|\cdot\|_1)$.

8) Let $\sum_{\nu=1}^n a_{\mu\nu} x_\nu = b_\mu$, $1 \leq \mu \leq m$ and $m > n$, be an over-determined linear system.

a) By formulating it as a Chebyshev approximation problem, show that the problem of minimizing $\max_\mu |\sum_{\nu=1}^n a_{\mu\nu}x_\nu - b_\mu|$ over all $x_1, \ldots, x_n$ has a solution.

b) Show also that the solution $\tilde{x}_1, \ldots, \tilde{x}_n$ is unique if and only if the matrix $A := (a_{\mu\nu})_{\substack{\mu=1,\ldots,m \\ \nu=1,\ldots,n}}$ is of full rank.

9) Problem 8a) can be treated graphically in the case $\nu = 1$. Find the solution of Problem 8a) for $A := \begin{pmatrix} 1 \\ 2 \\ -1 \end{pmatrix}$, $b := \begin{pmatrix} 2 \\ 1 \\ 1 \end{pmatrix}$ with the help of a sketch.

5. Approximation in Pre-Hilbert Spaces

In addition to approximation based on the Chebyshev norm, as treated in Section 4 above, for applications it is also important to consider approximation with respect to the norm $\|\cdot\|_2$. While uniform approximation involves making the largest deviation as small as possible, approximation with respect to the norm $\|\cdot\|_2$ is a smoothing process whereby the overall error is important. In this section we consider approximation from a finite dimensional linear subspace.

5.1 Characterization of the Best Approximation. Let V be a linear space, and for any two elements f and g in V, let $\langle f, g\rangle$ denote their inner product. Let $\|f\| := \langle f, f\rangle^{1/2}$ be the induced norm. In addition, suppose that $\mathrm{U} \subset \mathrm{V}$ is a finite dimensional linear subspace of this pre-Hilbert space. By the Fundamental Theorem 3.4, and in view of the strictness of the norm, we see that for any given element $f \in \mathrm{V}$, there exists a unique best approximation $\tilde{f} \in \mathrm{U}$.

Characterization Theorem. *$\tilde{f}$ is the best approximation of $f \in \mathrm{V}$ from* U *if and only if $\langle f - \tilde{f}, g\rangle = 0$ for all $g \in \mathrm{U}$.*

Proof. ($\Leftarrow$): Suppose $\langle f - \tilde{f}, g\rangle = 0$ for every $g \in \mathrm{U}$. We decompose g as $g = \tilde{f} + g'$, $g' \in \mathrm{U}$. Then $\|f - g\|^2 = \|(f - \tilde{f}) - g'\|^2 = \|f - \tilde{f}\|^2 + \|g'\|^2$, so that $\|f - \tilde{f}\|^2 \le \|f - g\|^2$.

($\Rightarrow$): Let $\tilde{f}$ be the best approximation, and suppose there exists a $g^* \in \mathrm{U}$ such that $\langle f - \tilde{f}, g^*\rangle = c \neq 0$.

Then taking $h := \tilde{f} + c\frac{g^*}{\|g^*\|^2} \in \mathrm{U}$, we see that

$$\|f - h\|^2 = \|f - \tilde{f}\|^2 - \frac{c}{\|g^*\|^2}\langle g^*, f - \tilde{f}\rangle - \frac{\overline{c}}{\|g^*\|^2}\langle f - \tilde{f}, g^*\rangle + |c|^2\frac{1}{\|g^*\|^2},$$

and thus $\|f - h\|^2 = \|f - \tilde{f}\|^2 - \frac{|c|^2}{\|g^*\|^2}$. This gives $\|f - h\| < \|f - \tilde{f}\|$, which contradicts the assumption that $\tilde{f}$ is a best approximation. $\square$

This characterization theorem immediately implies the following

Corollary. The error $\|f - \tilde{f}\|$ satisfies $\|f - \tilde{f}\|^2 = \|f\|^2 - \|\tilde{f}\|^2$. Indeed, $\|f\|^2 = \|(f - \tilde{f}) + \tilde{f}\|^2$, and the result follows since $\langle f - \tilde{f}, \tilde{f} \rangle = 0$.

5.2 The Normal Equations. Let U := span$(g_1, \ldots, g_n)$. We now show how the best approximation $\tilde{f} = \tilde{\alpha}_1 g_1 + \cdots + \tilde{\alpha}_n g_n$ can be computed immediately using the characterization theorem. By the theorem, $\langle f - \tilde{f}, g \rangle = 0$ for all $g \in$ U, and therefore in particular for $g := g_k$, $1 \leq k \leq n$. It follows that $\tilde{\alpha} = (\tilde{\alpha}_1, \ldots, \tilde{\alpha}_n)$ must be the solution of the set $\langle f - \sum_{j=1}^n \alpha_j g_j, g_k \rangle = 0$ of *normal equations*, which can be written out as

$$\sum_{j=1}^{n} \alpha_j \langle g_j, g_k \rangle = \langle f, g_k \rangle, \; 1 \leq k \leq n.$$

The solution of the normal equations is always unique since the linear independence of the elements $g_1, \ldots, g_n$ implies that the matrix $(\langle g_j, g_k \rangle)_{j,k=1}^n$, of the system is a positive definite Gram matrix with nonzero determinant; cf. e.g. W. H. Greub [1981].

We have seen that the best approximation can be easily computed from the normal equations. The deviation $\|f - \tilde{f}\|$ can also be easily computed since using Corollary 5.1 and the fact that

$$\|\tilde{f}\|^2 = \langle \tilde{f} - f + f, \tilde{f} \rangle = \langle \tilde{f} - f, \tilde{f} \rangle + \langle f, \tilde{f} \rangle = \langle f, \tilde{f} \rangle,$$

we get

$$\|f - \tilde{f}\|^2 = \|f\|^2 - \langle f, \tilde{f} \rangle,$$

so that

$$\|f - \tilde{f}\| = [\|f\|^2 - \sum_{1}^{n} \tilde{\alpha}_j \langle f, g_j \rangle]^{1/2}.$$

Mean Square Error. The linear space C$[a, b]$ endowed with the inner product $\langle f, g \rangle := \int_a^b f(x)g(x)dx$ and corresponding norm $\|f\| := \|f\|_2 =$ $= [\int_a^b [f(x)]^2]^{1/2}$ is a pre-Hilbert space. In this case it is common to average the error $\|f - \tilde{f}\|_2^2$ over the interval. The resulting quantity $\mu := \frac{\|f - \tilde{f}\|_2}{\sqrt{b-a}}$ is called the *mean square error*.

5.3 Orthonormal Systems. The solution of the normal equations is especially simple when the elements $g_1, \ldots, g_n$ are chosen to be orthonormal. In this case we have $\langle g_j, g_k \rangle = \delta_{jk}$, and the Gram matrix corresponding to the normal equations reduces to the identity matrix, and hence the solution of the normal equations 5.2 is simply

$$\tilde{\alpha}_k = \langle f, g_k \rangle, \; 1 \leq k \leq n.$$

In this case we have the further advantage that the dimension n of U need not be fixed in advance. The computation of $\tilde{\alpha}_\ell$ is independent of the values $\tilde{\alpha}_k$, $k < \ell$. In order to increase the accuracy of an approximation, we can simply increase the dimension of U as needed, without having to recompute the coefficients $\tilde{\alpha}_k$ which have already been computed.

We can always construct an orthonormal system (ONS) out of any given system $\{g_1, \ldots, g_n\}$ of linearly independent elements by the well-known orthonormalization method of E. Schmidt.

The Bessel Inequality. The formula for the deviation $\|f - \tilde{f}\|$ given in 5.2 implies the inequality $0 \le \|f\|^2 - \sum_1^n \tilde{\alpha}_j \langle f, g_j \rangle$. Now if the elements $g_1, \ldots, g_n$ form an ONS, then this inequality reduces to $\sum_1^n \tilde{\alpha}_j^2 \le \|f\|^2$. This inequality also remains correct even when the ONS $\{g_1, \ldots, g_n\}$ is extended to an ONS of infinite dimension, in which case we have the *Bessel Inequality*

$$\sum_{j=1}^{\infty} \tilde{\alpha}_j^2 \le \|f\|^2.$$

We now examine the question of whether the difference between the approximation $\tilde{f}_n := \sum_1^n \tilde{\alpha}_k g_k$ and the given f, measured in the norm, can be made arbitrary small by choosing n sufficiently large.

Convergence. Let V be a pre-Hilbert space, and suppose the elements g_1, $g_2, \ldots$ form a finite or infinite ONS in V. To answer the question of whether arbitrarily exact approximations to a given $f \in$ V can be constructed as linear combinations of the elements of the ONS, we introduce the following

Definition. An ONS $\{g_1, g_2, \ldots\}$ of elements of a pre-Hilbert space V is called *complete in* V provided that for every element $f \in$ V, there exists a sequence $(f_n)_{n=1,2,\ldots}$ with $f_n \in \operatorname{span}(g_1, \ldots, g_n)$ and $\lim_{n\to\infty} \|f - f_n\| = 0$.

If V is finite-dimensional, then of course every ONS is also finite, and every ONS which has dimension equal to that of V is complete.

The completeness of an ONS is the essential property needed in order to construct an arbitrarily accurate best approximation $\tilde{f}$ of any given element $f \in$ V. A complete ONS can also be characterized in the following way.

The Completeness Condition. Let $\{g_1, g_2, \ldots\}$ be a complete ONS. Consider a sequence (f_n), $f_n \in \operatorname{span}(g_1, \ldots, g_n)$, with $\lim_{n\to\infty} \|f - f_n\| = 0$, and let $\tilde{f}_n$ be the best approximants of f from the same linear subspaces. Then by 5.2 and 5.3, $\|f - \tilde{f}_n\| \le \|f - f_n\|$ for all n and $\|f - \tilde{f}_n\|^2 = \|f\|^2 - \sum_1^n \tilde{\alpha}_k^2$. Since $\lim_{n\to\infty} \|f - f_n\| = 0$, it follows that $\lim_{n\to\infty} \|f - \tilde{f}_n\| = 0$. This implies $\lim_{n\to\infty}(\|f\|^2 - \sum_1^n \tilde{\alpha}_k^2) = 0$, and hence $\sum_1^\infty \tilde{\alpha}_k^2 = \|f\|^2$.

On the other hand, if $\lim_{n\to\infty}(\|f\|^2 - \sum_1^n \tilde{\alpha}_k^2) = 0$, then it follows that $\lim_{n\to\infty} \|f - \tilde{f}_n\| = 0$, which implies the completeness of the ONS $\{g_1, g_2, \ldots\}$. We have established the following equivalence:

$$\{g_1, g_2, \ldots\} \text{ is a complete ONS } \Leftrightarrow \sum_{k=1}^{\infty} \tilde{\alpha}_k^2 = \|f\|^2.$$

We refer to this equivalence as the

Completeness Condition. *An ONS $\{g_1, g_2, \ldots\}$ is complete if and only if the following completeness condition holds:*

$$\sum_{k=1}^{\infty} \tilde{\alpha}_k^2 = \|f\|^2.$$

There is not universal agreement in the literature concerning the terminology used here. Some authors refer to an ONS which is *complete* in our sense as *closed*, while others call it *total*. For some authors the terminology *complete* means that any element which is orthogonal to all elements of an ONS must be the zero element $f = 0$. For pre-Hilbert spaces, this alternate form of completeness follows from our property of completeness, and hence as long as we remain in the pre-Hilbert space framework, this difference in terminology will cause no problems. The reader is, however, urged to examine the various terminology used in the literature carefully.

A similar situation persists regarding the completeness condition. It is frequently referred to as the *Parseval Equation*, and in the Russian literature as the *Parseval-Steklov-Equation*. The fact that there exist a variety of terminologies underscores the importance of these concepts.

5.4 The Legendre Polynomials. As an example of an ONS, in this section we study the system of polynomials which arise when we orthonormalize the monomials $g_j(t) := t^{j-1}$, $(j = 1, 2, \ldots)$, $t \in [-1, +1]$. We are looking for a system $\{L_k\}$ of polynomials $L_k \in \mathrm{P}_k$ which satisfy the orthogonality conditions $\langle L_k, L_\ell \rangle = \delta_{k\ell}$ for $k, \ell = 0, 1, \ldots$, with respect to the inner product $\langle L_k, L_\ell \rangle := \int_{-1}^{+1} L_k(t) L_\ell(t) dt$.

We want the polynomial L_n to satisfy the orthogonality condition $\langle L_n, L_k \rangle = 0$ for $k < n$. A sufficient condition for this is that $\langle L_n, g_j \rangle = 0$ for $j < n$, since this implies $\langle L_n, p_k \rangle = 0$ for all polynomials $p_k \in \mathrm{P}_k$, $k < n$, and thus also $\langle L_n, L_k \rangle = 0$. We shall use this observation to find the polynomials L_k.

Suppose $L_n = \frac{1}{\|\varphi_n\|} \varphi_n$, where we choose $\varphi_n(t) =: \dfrac{d^n \chi_n(t)}{dt^n}$, and χ_n is a function satisfying $\chi_n^{(n-k)}(t) := \int_{-1}^{t} \chi_n^{(n-k+1)}(\tau) d\tau$, $1 \le k \le n$. The function χ_n is called a generating function and satisfies $\chi_n^{(n-k)}(-1) = 0$. Now if $p \in \mathrm{P}_{n-1}$, we have

$$\int_{-1}^{+1} p(t) \chi_n^{(n)}(t) dt = p(t) \chi_n^{(n-1)}(t) \Big|_{-1}^{+1} - \cdots + (-1)^{n-1} p^{(n-1)}(t) \chi_n(t) \Big|_{-1}^{+1}.$$

In order for the orthogonality condition $\langle \varphi_n, g_j \rangle = \langle \chi_n^{(n)}, g_j \rangle = 0$ to be satisfied for $j = 1$ with $p(t) = g_j(t) := 1$, we must have $\chi_n^{(n-1)}(+1) = 0$. For $j = 2, \ldots, n$ we must have

$$\sum_{i=1}^{j} (-1)^{i-1} (j-1) \cdots (j-i+1) \chi_n^{(n-i)}(+1) = 0,$$

which means that we also need $\chi_n^{(n-k)}(+1) = 0$ for $k = 2, \ldots, n$. We conclude that χ_n must be of the form $\chi_n(t) = c_n(t^2-1)^n$ with some normalization constant c_n, and thus that $\varphi_n(t) = c_n \frac{d^n(t^2-1)^n}{dt^n}$.
The polynomials

$$\hat{L}_n(t) = \frac{n!}{(2n)!} \frac{d^n(t^2-1)^n}{dt^n} = t^n + \cdots, \; n \geq 0$$

form an orthogonal system of polynomials with leading coefficients one.
Setting $\hat{\chi}_n(t) := (t^2-1)^n$ leads to the normalization condition

$$\|L_n\|_2^2 = 1 \Rightarrow c_n^2 \int_{-1}^{+1} [\hat{\chi}_n^{(n)}(t)]^2 dt = 1.$$

$$\int_{-1}^{+1} [\hat{\chi}_n^{(n)}(t)]^2 dt = \hat{\chi}_n^{(n-1)}(t)\hat{\chi}_n^{(n)}(t) - \hat{\chi}_n^{(n-2)}(t)\hat{\chi}^{(n+1)}(t) + \cdots +$$

$$+ (-1)^{n-1}\hat{\chi}_n(t)\hat{\chi}_n^{(2n-1)}(t) \,|_{-1}^{+1} + (-1)^n \int_{-1}^{+1} \hat{\chi}_n(t)\hat{\chi}_n^{(2n)}(t)dt =$$

$$= (-1)^n (2n)! \int_{-1}^{+1} \hat{\chi}_n(t)dt.$$

Hence we require that $c_n = [(-1)^n(2n)!I_n]^{-1/2}$, where $I_n := \int_{-1}^{+1} \hat{\chi}_n(t)dt$. Now

$$I_n = \int_{-1}^{+1} (t^2-1)^n dt = \int_{-1}^{+1} t^2(t^2-1)^{n-1} dt - I_{n-1} =$$

$$= t\frac{1}{2n}(t^2-1)^n \,|_{-1}^{+1} - \frac{1}{2n}\int_{-1}^{+1} (t^2-1)^n dt - I_{n-1} = -\frac{1}{2n} I_n - I_{n-1};$$

$$I_n = -\frac{2n}{2n+1} I_{n-1} = (-1)^n \frac{2n}{2n+1}\frac{2n-2}{2n-1} \cdots \frac{2}{3} I_0,$$

and since $I_0 = 2$, it follows that $I_n = (-1)^n \frac{2^n n!}{(2n+1)(2n-1)\cdots 3} 2$, and so

$$c_n = \left[(2n)! \frac{2^n n!}{(2n+1)(2n-1)\cdots 3} 2\right]^{-1/2} =$$

$$= \left[\frac{(2^n n!)^2}{2n+1} 2\right]^{-1/2} = \left(\frac{2n+1}{2}\right)^{\frac{1}{2}} \frac{1}{2^n n!}.$$

This leads to *Rodrigues' formula* for the normalized *Legendre polynomials*

$$L_n(t) = \frac{1}{2^n n!} \sqrt{\frac{2n+1}{2}} \frac{d^n(t^2-1)^n}{dt^n}.$$

This gives

$$L_0(t) = \frac{1}{\sqrt{2}}, \qquad L_1(t) = \sqrt{\frac{3}{2}}t,$$

$$L_2(t) = \frac{1}{2}\sqrt{\frac{5}{2}}(3t^2 - 1), \qquad L_3(t) = \frac{1}{2}\sqrt{\frac{7}{2}}(5t^3 - 3t).$$

Minimal Property of the Legendre Polynomials. As we did in 4.6 for the uniform norm, we can now consider the problem of finding a polynomial in P_{n-1} which best approximates the monomial $f(t) := t^n$ on $[-1, +1]$ with respect to the norm $\|\cdot\|_2$.

Suppose we look for the best polynomial approximation of f, $f(t) := t^n$, in the form $\tilde{p} = \tilde{\alpha}_1 g_1 + \cdots + \tilde{\alpha}_n g_n$. Then the coefficients are the solution of the system of normal equations 5.2

$$\langle f - (\alpha_1 g_1 + \cdots + \alpha_n g_n), g_k \rangle = 0, \; 1 \leq k \leq n.$$

The unique solution of this system is given by the Legendre polynomial with highest coefficient one:

$$\hat{L}_n = f - (\tilde{\alpha}_1 g_1 + \cdots + \tilde{\alpha}_n g_n), \text{ and so } \tilde{p} = f - \hat{L}_n.$$

This result can be reformulated as follows: on the interval $[-1, +1]$, the Legendre polynomials $\hat{L}_n$ have minimal norm in the sense that $\|\hat{L}_n\|_2 \leq \|p\|_2$ for all polynomials $p \in \hat{P}_n$, where

$$\hat{P}_n := \{p \in P_n | p(t) = t^n + a_{n-1}t^{n-1} + \cdots + a_0\}.$$

The Legendre polynomials with highest coefficient one are the best approximations of the function $f = 0$ on $[-1, +1]$ with respect to the norm $\|\cdot\|_2$.

5.5 Properties of Orthonormal Polynomials. The Legendre polynomials are only *one* example of a system of orthonormal polynomials. They arise when we choose the interval $[a, b] := [-1, +1]$ and the weight function $w(x) = 1$ for $x \in [-1, +1]$ in the definition of the inner product $\langle f, g \rangle := \int_a^b f(x)g(x)w(x)dx$.

We now establish a theorem on the location of the zeros for general orthonormal systems of polynomials. First we need the following

Lemma. *Every polynomial $p \in P_n$ can be uniquely expanded as a linear combination of the elements $\psi_0, \ldots, \psi_n$ of an orthonormal system of polynomials.*

Proof. If $p \in P_n$, then $p \in \text{span}(\psi_0, \ldots, \psi_n)$, and thus using the normal equations, we get $p = \sum_0^n \beta_k \psi_k$ with $\beta_k = \langle p, \psi_k \rangle$. □

It is well known that a polynomial is determined up to a multiplicative constant by its zeros. The following remarkable theorem describes the zeros of polynomials in an ONS.

Zero Theorem. *If the set of polynomials $\{\psi_0, \psi_1, \ldots\}$, $\psi_n \in \mathrm{P}_n$ forms an ONS on $[a, b]$ with respect to the weight function w, then each of these polynomials has only simple real zeros, all of which lie in (a, b).*

Proof. Since ψ_n is a polynomial of exact degree n, it has n zeros, say $x_{n1}, x_{n2}, \ldots, x_{nn}$, some of which may occur in complex conjugate pairs. Then $\langle \psi_n, \psi_0 \rangle = 0$ for $n > 0$; i.e., $\int_a^b (x - x_{n1}) \cdots (x - x_{nn}) w(x) dx = 0$. It follows that there exists ***at least one*** real zero with sign change in (a, b), i.e., of odd multiplicity. Let $\{x_{n\nu} \mid \nu \in \mathrm{H} \subset \mathrm{N} := \{1, \ldots, n\}\}$ be the set of *all* real zeros of odd multiplicity of ψ_n in (a, b), where each multiple zero appears only once in the set. Then the polynomial $\pi(x) := \prod_{\nu \in \mathrm{H}} (x - x_{n\nu})$, $\pi \in \mathrm{P}_n$, satisfies either $\psi_n(x)\pi(x) \geq 0$ or $\psi_n(x)\pi(x) \leq 0$ for all $x \in (a, b)$, and so $\langle \psi_n, \pi \rangle \neq 0$. Now if any zero of ψ_n is not simple, then $\pi \in P_{n-1}$, which is a contradiction since in view of the Lemma, $\langle \psi_n, p \rangle = 0$ for all $p \in \mathrm{P}_{n-1}$. We conclude that $H = N$, and ψ_n has only simple real zeros. □

To illustrate this result, consider the Chebyshev polynomials of the first kind discussed in 4.8. As shown there, the polynomials $\frac{1}{\sqrt{\pi}} T_0$, $\sqrt{\frac{2}{\pi}} T_k$ for $k = 1, 2, \ldots$ form an ONS on $[-1, +1]$ with respect to the weight function $w(x) := \frac{1}{\sqrt{1-x^2}}$. In 4.7 we showed that T_n has n simple real zeros located at the points $x_{n\nu} = \cos(\frac{2\nu - 1}{n} \pi)$, $1 \leq \nu \leq n$ in $(-1, +1)$.

Minimal Property. The minimal property of the Legendre polynomials of 5.4 can be extended to general systems of orthogonal polynomials. In particular, suppose we take the polynomial of n-th degree in a system of orthogonal polynomials and normalize it so that its leading coefficient is one. Then this polynomial has minimal norm compared to any other polynomial of n-th degree with leading coefficient one.

5.6 Convergence in C[*a*, *b*]. In this section we discuss the question of convergence of a sequence of best approximations in $\mathrm{C}[a, b]$ with respect to the norm $\|\cdot\|_2$. It will be convenient to focus on the concrete case of approximating a continuous function by polynomials. In this case the corresponding ONS on the interval $[a, b]$ can be obtained from the Legendre polynomials L_0, $L_1, \ldots$ discussed in 5.4 by a simple linear tranformation. Convergence with respect to the norm $\|\cdot\|_2$ is commonly referred to as *convergence in mean.*

We begin with the following

Lemma. *If $(f_n)_{n \in \mathbb{N}}$ is a sequence of continuous functions which converges uniformly, then it also converges in mean.*

Proof. Uniform convergence means that if N is chosen sufficiently large, then $|f(x) - f_n(x)| < \frac{\varepsilon}{\sqrt{b-a}}$, for all $n > N$, where ε is independent of $x \in [a, b]$. This implies that $\|f - f_n\|_2 = [\int_a^b |f(x) - f_n(x)|^2 dx]^{1/2} < \varepsilon$, and it follows that $\lim_{n \to \infty} \|f - f_n\|_2 = 0$. □

We can now establish the desired

Convergence Theorem. *Let $f \in C[a,b]$, and let $(\tilde{p}_n)_{n\in\mathbb{N}}$ be the sequence of best approximations from P_n of f with respect to the norm $\|\cdot\|_2$. Then $\tilde{p}_n$ converges to f in mean.*

Proof. By the Weierstrass Approximation Theorem 2.2, there exists a sequence $(p_n)_{n\in\mathbb{N}}$ of polynomials $p_n \in P_n$ which converge uniformly to f. By the Lemma, the uniform convergence implies convergence of this sequence in mean, so that $\lim_{n\to\infty} \|f - p_n\|_2 = 0$. Since $\|f - \tilde{p}_n\|_2 \leq \|f - p_n\|_2$, it follows that $\lim_{n\to\infty} \|f - \tilde{p}_n\|_2 = 0$. □

Corollary. *The system $\{L_0^*, L_1^*, \ldots\}$ of Legendre polynomials on the interval $[a,b]$ is complete in $(C[a,b], \|\cdot\|_2)$.*

Proof. By Lemma 5.5, $\tilde{f}_n = \sum_{k=0}^{n} \langle \tilde{f}_n, L_k^* \rangle L_k^*$, which gives the completeness of the ONS $\{L_0^*, L_1^*, \ldots\}$ in the sense of Definition 5.3. □

5.7 Approximation of Piecewise Continuous Functions. The problem of approximating a function with a finite number of jumps occurs frequently in practice. In this section we show that, relative to the norm $\|\cdot\|_2$, this problem can be handled by the same methods used to solve the problem for continuous functions.

In this case the appropriate linear space to consider is the space $C_{-1}[a,b]$ of all functions which are piecewise continuous on $[a,b]$. As usual, we say a function is piecewise continuous whenever it is continuous except for a finite set of finite jumps. Let $f, g \in C_{-1}[a,b]$, and suppose that $\xi_1, \ldots, \xi_{m-1}$ are the jump points of the function $f \cdot g$. Setting $\xi_0 := a$ and $\xi_m := b$, we now define the inner product

$$\langle f, g \rangle := \int_a^b f(x)g(x)dx = \sum_{\mu=0}^{m-1} \int_{\xi_\mu}^{\xi_{\mu+1}} f(x)g(x)dx,$$

and the corresponding norm $\|f\| := \|f\|_2 = \langle f, f \rangle^{1/2}$.

Since this defines a pre-Hilbert space, we know that for any given f in $C_{-1}[a,b]$ and any finite dimensional linear subspace U, there exists a unique best approximation $\tilde{f}$ of f in U, and that it can be computed from the normal equations.

In this pre-Hilbert space we also have the following

Theorem. *Let $f \in C_{-1}[a,b]$. Then the sequence $(\tilde{p}_n)_{n\in\mathbb{N}}$ of best approximations $\tilde{p}_n$ in P_n converges to f in mean.*

Proof. The proof is based on the idea of approximating the discontinuous function f by a continuous function, and then considering its best approximations.

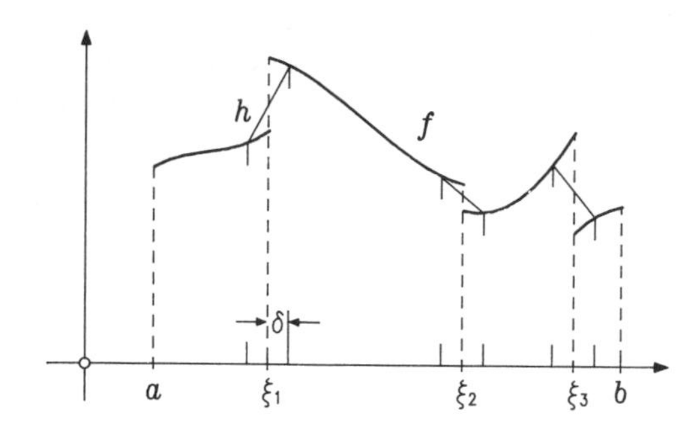

Given $f \in \mathrm{C}_{-1}[a,b]$ with jumps at the points $\xi_1, \ldots, \xi_{m-1}$, we construct the following associated continuous function h:

$$h(x) := \begin{cases} f(\xi_\mu - \delta) + \frac{f(\xi_\mu+\delta)-f(\xi_\mu-\delta)}{2\delta}[x - (\xi_\mu - \delta)] & \text{if } x \in [\xi_\mu - \delta, \xi_\mu + \delta], \\ & 1 \le \mu \le m-1, \\ f(x) & \text{otherwise,} \end{cases}$$

where $\delta \le \frac{1}{2}\min_{0\le\mu\le m-1}(\xi_{\mu+1} - \xi_\mu)$.

Let $\tilde{q}_n$ be the best approximation of h in P_n. Then for sufficiently large N, $\|h - \tilde{q}_n\|_2 < \frac{\varepsilon}{2}$ for all $n > N$. In addition,

$$\|f - \tilde{q}_n\|_2 = \|(f-h) + (h - \tilde{q}_n)\|_2 \le \|f - h\|_2 + \|h - \tilde{q}_n\|_2$$

and

$$\|f - h\|_2^2 = \sum_{\mu=0}^{m-1} \int_{\xi_\mu}^{\xi_{\mu+1}} [f(x) - h(x)]^2 dx = \sum_{\mu=1}^{m-1} \int_{\xi_\mu-\delta}^{\xi_\mu+\delta} [f(x) - h(x)]^2 dx.$$

Now with $M := \max_{x\in[a,b]} |f(x)|$, we have $|h(x) - f(x)| \le 2M$, for all x in $[a,b]$, independent of the choice of δ. It follows that $\|f-h\|_2^2 \le 4M^2(m-1)2\delta$, and so

$$\|f - h\|_2 < \frac{\varepsilon}{2} \quad \text{for} \quad \delta < \frac{\varepsilon^2}{32M^2(m-1)} \quad \text{and} \quad \|f - \tilde{q}_n\|_2 < \varepsilon.$$

Since the best approximation $\tilde{p}_n \in \mathrm{P}_n$ of f is closer to f than the best approximation $\tilde{q}_n \in \mathrm{P}_n$ of h, we conclude that $\|f - \tilde{p}_n\|_2 \le \|f - \tilde{q}_n\|_2 < \varepsilon$, and the theorem is proved. □

5.8 Trigonometric Approximation. There are many application areas where it is necessary to approximate some periodic process. For example, in modelling switching circuits, it is clear that piecewise continuous periodic functions are of particular interest.

Let $f \in C_{-1}[-\pi, +\pi]$ be 2π-periodic; i.e., $f(x) = f(x+2\pi)$ for all x. To approximate this kind of function, we construct a finite dimensional linear space spanned by a set of linearly independent 2π-periodic functions. The natural choice for the functions are the usual trigonometric functions. We already know that these functions form an orthogonal basis with respect to the norm $\|\cdot\|_2$. It remains only to normalize them appropriately to get a convenient

ONS of Trigonometric Functions. We defined the ONS $\{g_1, \ldots, g_{2m+1}\}$, $g_k : [-\pi, +\pi] \to \mathbb{R}$, $1 \leq k \leq 2m+1$, by

$$g_1(x) := \frac{1}{\sqrt{2\pi}}$$

$$g_{2j}(x) := \frac{1}{\sqrt{\pi}} \cos(jx), \; g_{2j+1}(x) := \frac{1}{\sqrt{\pi}} \sin(jx) \;\text{ for }\; 1 \leq j \leq m.$$

Then for any $f \in C_{-1}[-\pi, +\pi]$, its best approximation $\tilde{f}$ from the linear subspace $U_{2m+1} = \operatorname{span}(g_1, \ldots, g_{2m+1})$ can be found by solving the normal equations, and we get

$$\tilde{f}(x) = \sum_{k=1}^{2m+1} \tilde{\alpha}_k g_k(x) =: \frac{a_0}{2} + \sum_{j=1}^{m} [a_j \cos(jx) + b_j \sin(jx)]$$

with

$$a_j = \frac{1}{\pi} \int_{-\pi}^{+\pi} f(x) \cos(jx) dx, \quad 0 \leq j \leq m,$$

$$b_j = \frac{1}{\pi} \int_{-\pi}^{+\pi} f(x) \sin(jx) dx, \quad 1 \leq j \leq m.$$

The coefficients $a_0, a_1, \ldots, a_m, b_1, \ldots, b_m$ are the classical Fourier coefficients of the periodic function f, and thus the best approximation of f from U_{2m+1} is nothing more than the m-th partial sum of the Fourier series expansion of f. These partial sums are best approximations from certain subspaces, a fact which corresponds to the well-known minimal property of the partial sums which is discussed in the analysis literature.

The deviation in this case is

$$\|f - \tilde{f}\|_2 = [\|f\|_2^2 - \sum_{k=1}^{2m+1} \tilde{\alpha}_k^2]^{1/2} = [\|f\|_2^2 - \pi(\frac{a_0^2}{2} + \sum_{j=1}^{m}(a_j^2 + b_j^2))]^{1/2},$$

and the Bessel inequality is

$$\frac{a_0^2}{2} + \sum_{j=1}^{m}(a_j^2 + b_j^2) \leq \frac{1}{\pi}\|f\|_2^2.$$

Example. Consider the periodic function f defined by

$$f(x) := \begin{cases} -1 & \text{for } -\pi \le x < 0 \\ 0 & \text{for } x = 0 \\ +1 & \text{for } 0 < x < \pi \end{cases} , \quad f(x + 2\pi) = f(x).$$

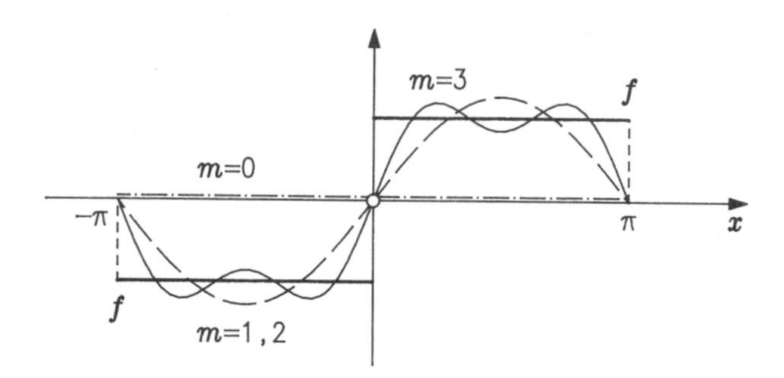

Since f is odd, $a_j = 0$ for $0 \le j \le m$, and we find that

$$b_j = \frac{2}{\pi} \int_0^\pi \sin(jx)dx = \begin{cases} \frac{4}{\pi j} & \text{for } j \text{ odd} \\ 0 & \text{for } j \text{ even.} \end{cases}$$

The corresponding best approximations for $m = 0, 1, 2, 3$ are shown in the figure above.

Convergence. If the periodic function f is continuous everywhere, then convergence in mean of the best approximations follows from the Weierstrass Approximation Theorem for Periodic Functions 2.4. The proof is analogous to the proof of the Convergence Theorem 5.6. This second approximation theorem of Weierstrass implies the existence of a sequence of trigonometric polynomials from U_{2m+1} that converge uniformly to f. This implies the convergence in mean, which in turn implies the convergence of the best approximations from U_{2m+1} with respect to the norm $\|\cdot\|_2$, and thus in mean. The extension to piecewise continuous functions can be carried out along the same lines as in 5.7, and we obtain the

Theorem. *Let $f \in C_{-1}[-\pi, +\pi]$ be periodic with period 2π. Then the sequence of best approximations with respect to $\|\cdot\|_2$ from the linear subspaces U_{2m+1} of trigonometric polynomials converges to f in mean.*

Corollary. The system of trigonometric functions is complete in the space of piecewise continuous periodic functions $(C_{-1}[-\pi, +\pi], \|\cdot\|_2)$ in the sense of Definition 5.3. It is also possible to consider approximating a non-periodic continuous function on $[a, b]$ by trigonometric polynomials. If we transform $[a, b]$ to $[-\pi, +\pi]$, then the situation is the same as for the periodic case; the periodic continuations outside of $[-\pi, +\pi]$ are of no concern. The trigonometric functions on $[a, b]$ obtained by transformation and renorming provide another complete ONS in $(\mathrm{C}_{-1}[a, b], \|\cdot\|_2)$.

Remarks. The sequence of best approximations with respect to $\|\cdot\|_2$ from U_{2m+1} of a continuous periodic function is in general different from the uniformly convergent sequence of trigonometric polynomials from U_{2m+1} which arise in the second approximation theorem of Weierstrass. These latter polynomials converge in $(\mathrm{C}[-\pi, +\pi], \|\cdot\|_\infty)$, while the former converge in mean, even for functions which are only piecewise continuous; that is, functions in $(\mathrm{C}_{-1}[-\pi, +\pi], \|\cdot\|_2)$. However, in this case the convergence is generally not uniform.

The seemingly inadequate convergence properties of the Fourier series expansion – overshooting of the approximation at jump points (Gibbs phenomenon), uniform convergence only under additional conditions, even for the continuous case, etc. – can be explained by the fact that the Chebyshev norm is not appropriate for orthogonal series. As we have seen, these kinds of problems do not arise if we work with norms induced by the associated inner product.

5.9 Problems. 1) a) Explain the geometric significance of the Characterization Theorem 5.1 in the case where a vector in $\mathbb{R}^3$ is to be approximated with respect to the Euclidean norm by a vector from $\mathbb{R}^2$.

b) Show that in a real pre-Hilbert space V, two elements $f, g \in \mathrm{V}$ satisfy $\langle f, g\rangle = 0$ if and only if $\|\alpha f + g\| \geq \|g\|$ for all $\alpha \in \mathbb{R}$.

2) Let $f \in \mathrm{C}[-1, +1]$, $f(x) := e^x$. Find the best approximations of f from P_k, $0 < k < 2$, with respect to the norm $\|\cdot\|_2$

a) using the normal equations;

b) by expansion of f in Legendre polynomials.

Compare the best approximations from P_0 and from P_1 with the results of Problems 3b) and 7b) in 4.12.

3) a) Let $f \in \mathrm{C}[-\pi, +\pi]$. Show that $\lim_{j\to\infty} \int_{-\pi}^{+\pi} f(x)\sin(jx)dx = 0$ and $\lim_{j\to\infty} \int_{-\pi}^{+\pi} f(x)\cos(jx)dx = 0$, $j \in \mathbb{N}$.

b) Let $f \in \mathrm{C}[-1, +1]$. Show that

$$\lim_{k\to\infty} \int_{-1}^{+1} f(x)L_k(x)dx = 0, \quad k \in \mathbb{N}.$$

4) Consider the pre-Hilbert space $(\mathrm{C}[-1, +1], \|\cdot\|)$ with norm induced by the inner product $\langle f, g\rangle := \int_{-1}^{+1} \sqrt{1-x^2} f(x)g(x)dx$. Show:

a) The functions

$$U_n(x) := \sqrt{\frac{2}{\pi}} \frac{\sin((n+1)\arccos(x))}{\sqrt{1-x^2}}$$

form an orthonormal system in this pre–Hilbert space.

b) The functions U_n are polynomials of degree n in x. (These functions are called *Chebyshev polynomials of the second kind*).

c) $T_n'(x) = n\, U_{n-1}(x)$.

5) Prove that the ONS of Legendre polynomials is also complete in the space $(C[-1,+1], \|\cdot\|_\infty)$, where completeness in this normed linear space is defined as in 5.3. Show the same holds for $(C[-1,+1], \|\cdot\|_1)$.

6) Consider the sequence $f_n(x) := [\frac{n}{1+n^4x^2}]^{\frac{1}{2}}$ in $(C[-1,+1], \|\cdot\|_2)$. Show that the sequence converges in mean to the element $f = 0$, but that pointwise convergence does not hold.

7) Let $f \in C(-\infty, +\infty)$ be periodic with $f(x) := x^2$ for $x \in [-\pi, +\pi]$.

a) Find the Fourier series expansion of f in terms of trigonometric functions, and sketch the best approximations of f from $\text{span}(g_1, g_2, g_3)$ and from $\text{span}(g_1, \ldots, g_5)$.

b) How can one use this expansion to compute the value of π, and how many terms are needed to get π to an accuracy of $5 \cdot 10^{-k}$?

6. The Method of Least Squares

In 1820 C. F. Gauss was assigned the task of measuring the kingdom of Hannover by King George IV. Fortunately, he already had earlier experience with the fitting of measurements, and in 1794 had developed some methods, including the method of least squares, for smoothing data in connection with geodetic and astronomical problems. Using this method, he succeeded in 1801 in computing the path of the planetoid Ceres to sufficient accuracy that it could be relocated after having been lost for over a year after its discovery by the astronomer G. Piazzi from Palermo, The first published results on the method, however, were due to A.-M. Legendre (1806). The problem had been well known for quite some time. In its simplest form, it amounts to the following: given a series of individual measurements, find an average value so that the deviation from the measurements is as small as possible. Laplace had suggested already in 1799 that one should minimize the sum of the absolute values of the errors. This method, which corresponds to approximation with respect to the norm $\|\cdot\|_1$ in the discrete case, has the advantage that the influence of a single large error in the measurements is suppressed; we have observed the same phenomenon in 4.11 for approximation of continuous functions. The calculation of the solution of this problem, however, is difficult. In contrast, Gauss proposed to minimize the sum of the squares of the errors. It is shown in Statistics that this proposal is appropriate for normally distributed measurement errors, and hence is a natural choice for our fitting problem. It is easy to see that for n measurements $y_1, \ldots, y_n$, the method produces the average value. Indeed, if we are looking for a number $\overline{y}$, which minimizes the sum of squares of the errors $(y-y_1)^2 + \cdots + (y-y_n)^2$, then a necessary condition for a minimum is that $(y-y_1)+\cdots+(y-y_n)=0$. This gives the solution $\overline{y}=\frac{1}{n}\sum_1^n y_\nu$.

The *method of least squares* of Gauss later developed into a more general *fitting method* which we now discuss in an appropriate pre-Hilbert space framework.

6.1 Discrete Approximation. Suppose we are given N pairs of data points $(x_1,y_1),\ldots,(x_N,y_N)$. The problem of discrete approximation is to find a linear combination of prescribed functions $g_1,\ldots,g_n$ whose values at the points $x_\nu \in [a,b]$, $1 \le \nu \le N$ approximate the values $y_1,\ldots,y_N$ as well as possible. As mentioned in the introduction, this kind of problem arises both in the fitting of experimental data, and in the approximation of a given function.

We restrict our attention here to approximation by a set of continuous functions $g_k \in \mathrm{C}[a,b]$, $1 \le k \le n$. The problem is to find a function $\tilde{f} \in \mathrm{U} = \mathrm{span}(g_1,\ldots,g_n)$ solving the following

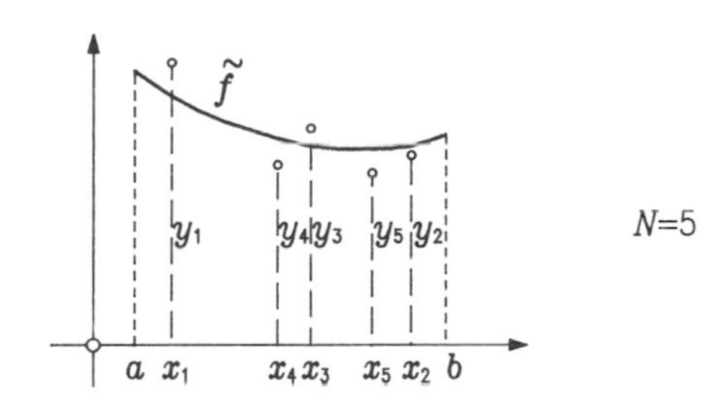

Fitting Problem. Find $\tilde{f} \in \mathrm{U}$ so that for all $g \in \mathrm{U}$,

$$\sum_{\nu=1}^{N}[y_\nu - \tilde{f}(x_\nu)]^2 \le \sum_{\nu=1}^{N}[y_\nu - g(x_\nu)]^2.$$

In order to apply our earlier results on approximation, we need to formulate this fitting problem in an appropriate pre-Hilbert space. We choose the Euclidean space $\mathrm{V} := \mathbb{R}^N$, where the inner product of of $\underline{u},\underline{v} \in \mathbb{R}^N$ is defined by $\langle \underline{u},\underline{v}\rangle := \sum_1^N u_\nu v_\nu$. This inner product induces the norm $\|\underline{u}\| := \|\underline{u}\|_2 = [\sum_1^N u_\nu^2]^{1/2}$. In this section we shall be working in both $\mathrm{C}[a,b]$ and $\mathbb{R}^N$. In order to avoid any confusion, we mark all vectors in $\mathbb{R}^N$ with a lower bar. Thus, for example, $g_k \in \mathrm{C}[a,b]$, but $\underline{g}_k \in \mathbb{R}^N$.

Setting $\underline{y} := (y_1,\ldots,y_N)^T$, $\underline{g}_k := (g_k(x_1),\ldots,g_k(x_N))^T$ and defining $\underline{g} := \sum_1^n \alpha_k \underline{g}_k$, we can now formulate the following approximation problem in $\mathbb{R}^N$:

Approximation Problem. Find $\underline{\tilde{f}} \in \mathrm{span}(\underline{g}_1,\ldots,\underline{g}_n)$ such that $\|\underline{y}-\underline{\tilde{f}}\|_2 \le \|\underline{y}-\underline{g}\|_2$ for all $\underline{g} \in \mathrm{span}(\underline{g}_1,\ldots,\underline{g}_n)$.

For $n > N$ the vectors $\underline{g}_1, \dots, \underline{g}_n$ are always linear dependent. Thus it only makes sense to consider $n \leq N$. In addition, we shall also assume for the present that the abscissae are distinct; i.e., $x_\nu \neq x_\mu$ for $\nu \neq \mu$.

By 5.1, this approximation problem has the unique solution

$$\tilde{\underline{f}} = \sum_{k=1}^{n} \tilde{\alpha}_k \underline{g}_k = \left(\sum_{k=1}^{n} \tilde{\alpha}_k g(x_1), \dots, \sum_{k=1}^{n} \tilde{\alpha}_k g(x_N) \right)^T .$$

Using the solution $\tilde{\alpha} = (\tilde{\alpha}_1, \dots, \tilde{\alpha}_n)$ of the approximation problem, we construct the solution of the fitting problem as $\tilde{f} = \sum_1^n \tilde{\alpha}_k g_k$. This function is uniquely defined whenever the

Normal Equations

$$\sum_{k=1}^{n} \alpha_k \langle \underline{g}_k, \underline{g}_\ell \rangle = \langle \underline{y}, \underline{g}_\ell \rangle, \quad 1 \leq \ell \leq n$$

used to compute $\tilde{\alpha}$ have a unique solution.

6.2 Solution of the Normal Equations. The solution of the system of normal equations is unique if and only if the Gram determinant defined by $\det((\langle \underline{g}_k, \underline{g}_\ell \rangle)_{k,\ell=1}^n$ is not zero. This is the case if and only if the vectors $\underline{g}_1, \dots, \underline{g}_n$ are linearly independent. For this condition to hold requires more than simply the linear independence of the functions $g_k \in \mathrm{U}$, $1 \leq k \leq n$. We need to require that U be a Haar space as discussed in 4.2. In this connection, we have the

Theorem. *Suppose $n \leq N$. Then the vectors $\underline{g}_k \in \mathbb{R}^N$, $1 \leq k \leq n$, are linearly independent for all choices of n distinct points $x_1, \dots, x_n$ in $[a, b]$ if and only if the functions $g_k \in \mathrm{U}$, $1 \leq k \leq n$, form a Chebyshev system.*

Proof. The linear independence of the vectors $\underline{g}_1, \dots, \underline{g}_n$ means that

$$\sum_{k=1}^{n} \beta_k \underline{g}_k = \underline{0} \Rightarrow \beta_k = 0 \text{ for } 1 \leq k \leq n,$$

which means that the linear system of equations

$$\sum_{k=1}^{n} \beta_k g_k(x_\nu) = 0,\ 1 \leq \nu \leq N,\ x_\nu \neq x_\mu \text{ for } \nu \neq \mu,$$

has only the trivial solution. Thus, the implication

$$\sum_{k=1}^{n} \beta_k g_k(x_\nu) = 0 \Rightarrow \beta_k = 0 \text{ for } 1 \leq k \leq n$$

must hold for *all* choices of N pairwise distinct abscissae $x_1, \dots, x_N$. This happens if and only if $g_1, \dots, g_n$ form a Chebyshev system. □

Combining the above results, we have the

Corollary. *If the functions $g_k \in \mathrm{U}$, $1 \le k \le n$ form a Chebyshev system, then for every choice of pairwise distinct points x_ν, $1 \le \nu \le N$ with $n \le N$, there is a unique solution $\tilde{\alpha} = (\tilde{\alpha}_1, \ldots, \tilde{\alpha}_n)$ of the normal equations 6.1, and the corresponding function $\tilde{f} = \sum_1^n \tilde{\alpha}_k g_k$ is the unique solution of both the fitting problem and the discrete approximation problem.*

The following two cases can occur:

(i) $n < N$: This is the usual approximation case where the solution $\tilde{f}$ of the fitting problem minimizes the squared sum of errors. If the vector $\underline{y}$ is not in $\mathrm{span}(\underline{g}_1, \ldots, \underline{g}_n)$, then $\|\underline{y} - \underline{\tilde{f}}\|_2 > 0$.

If however, $\underline{y} \in \mathrm{span}(\underline{g}_1, \ldots, \underline{g}_n)$, then the approximation problem reduces to the problem of expanding $\underline{y}$ in terms of the basis vectors $\underline{g}_1, \ldots, \underline{g}_n$. Since $\underline{\tilde{f}} = \underline{y}$, in this case $\|\underline{y} - \underline{\tilde{f}}\|_2 = 0$, and $\tilde{f}$ satisfies $\tilde{f}(x_\nu) = y_\nu$ at all points x_ν, $1 \le \nu \le N$.

In this latter case, $\tilde{f}$ is said to possess the *interpolation property.* This situation arises, for example, if the points (x_ν, y_ν) lie on a straight line, and the basis $g_1, \ldots, g_n$ consists of the polynomials $g_k(x) := x^{k-1}$. In this case, the unique solution of the fitting problem is the linear polynomial $\tilde{f}(x) = \tilde{\alpha}_1 + \tilde{\alpha}_2 x$, describing the straight line on which all of the points $(x_1, y_1), \ldots, (x_N, y_N)$ lie.

(ii) $n = N$: In this case we always have $\underline{y} \in \mathrm{span}(\underline{g}_1, \ldots, \underline{g}_n)$. Now the approximation problem reduces to an *interpolation problem.* The unique solution $\tilde{f}$ interpolates in the sense that $\tilde{f}(x_\nu) = y_\nu$ for all $1 \le \nu \le N$. We discuss the interpolation problem in detail in Chapter 5.

6.3 Fitting by Polynomials. The monomials are the standard example of a Chebyshev system, and thus are a natural choice for solving the fitting problem. We begin by calculating the solution of the problem of fitting a straight line to a given set of N points $(x_1, y_1), \ldots, (x_N, y_N)$. This corresponds to approximating with linear polynomials.

In this case we choose $g_1(x) := 1$, $g_2(x) := x$, and setting $\underline{g}_1 := (1, \ldots, 1)$ and $\underline{g}_1 := (x_1, \ldots, x_N)$, the Normal Equations 6.1 become

$$\alpha_1 N + \alpha_2 \sum_{\nu=1}^{N} x_\nu = \sum_{\nu=1}^{N} y_\nu$$

$$\alpha_1 \sum_{\nu=1}^{N} x_\nu + \alpha_2 \sum_{\nu=1}^{N} x_\nu^2 = \sum_{\nu=1}^{N} y_\nu x_\nu .$$

The solution of this system is

$$\tilde{\alpha}_1 = \frac{(\sum_1^N y_\nu)(\sum_1^N x_\nu^2) - (\sum_1^N x_\nu)(\sum_N^1 y_\nu x_\nu)}{N \sum_1^N x_\nu^2 - (\sum_1^N x_\nu)^2},$$

$$\tilde{\alpha}_2 = \frac{N \sum_1^N x_\nu y_\nu - (\sum_1^N y_\nu)(\sum_1^N x_\nu)}{N \sum_1^N x_\nu^2 - (\sum_1^N x_\nu)^2},$$

and the corresponding fitting polynomial is $\tilde{f}(x) = \tilde{\alpha}_1 + \tilde{\alpha}_2 x$.

In statistics the problem arises of describing the dependence of a random variable on given values of the variable. In this setting, the determination of a best approximation using the method of least squares is referred to as *regression*. It is called *linear regression* provided that the best approximation is to be computed as a linear combination of given functions. If the fitting function is required to be continuous, then this is precisely the fitting problem we discussed above. The least squares polynomial fit of degree 1 computed above is called a *regression line*.

It is easy to see that the *center of gravity* $(\xi, \eta) := (\frac{1}{N}\sum_1^N x_\nu, \frac{1}{N}\sum_1^N y_\nu)$ of the N points $(x_1, y_1), \ldots, (x_N, y_N)$ lies on the regression line. Now if we consider y as the independent variable and x as the dependent variable, then in the same way we can compute the corresponding regression line $\tilde{\varphi}(y) = \tilde{\beta}_1 + \tilde{\beta}_2 y$. Clearly, the center of gravity lies on this regression line as well, and hence is the intersection of the two regression lines. The size of the angle between the two lines is a measure of whether the quantities x_ν and y_ν, $1 \leq \nu \leq N$, are approximately linearly related. If the angle is small, then we say that there is a *linear correlation*. More on this concept can be found in standard books on statistics.

In computing $\tilde{\varphi}$, it may happen that $y_\nu = y_\mu$ for $\nu \neq \mu$. Up to now we have explicitly excluded this case. We now analyze it, and characterize those situations where it plays a role.

6.4 Coalescent Data Points. We now allow the possibility that $x_\nu = x_\mu$ for some $\nu \neq \mu$.

First we note that this generalization does not affect the existence of a solution to the approximation problem 6.1 in $\mathbb{R}^N$. Since this problem involves finding the best approximation $\underline{\tilde{f}}$ of $\underline{f}$ from the subspace spanned by the vectors $\underline{g}_1, \ldots, \underline{g}_n$ in the pre-Hilbert space $(\mathbb{R}^N, \|\cdot\|_2)$, it always has a unique solution. It can happen, of course, that the vectors $\underline{g}_1, \ldots, \underline{g}_n$ become linearly dependent which reduces the dimension of $\operatorname{span}(\underline{g}_1, \ldots, \underline{g}_n)$, but this does not affect the existence of a unique solution to the approximation problem in $\mathbb{R}^N$.

On the other hand, it might very well happen that the normal equations no longer have a unique solution, which would imply that the fitting problem doesn't either.

In order to analyze this case, let $\mathrm{H} := \{1, \ldots, N\}$, and consider the set $\mathrm{H}' := \mathrm{H} \setminus \{\mu \in \mathrm{H} \mid x_\nu = x_\mu \text{ for some } \nu \in \mathrm{H} \text{ with } \mu > \nu\}$. If we think of H as the indices of the data points, then H′ is a set of indices corresponding to distinct data points, and the number $N' \leq N$ of elements in H′ is the number of *distinct* points x_ν, $\nu \in \mathrm{H}$.

When $x_\nu = x_\mu$, then the ν-th and μ-th components of all vectors $\underline{g}_1, \ldots, \underline{g}_n$ have the same values: $g_k(x_\nu) = g_k(x_\mu)$ for $k = 1, \ldots, n$. This means that we have linear independence of $\underline{g}_1, \ldots, \underline{g}_n$, i.e.,

$$\sum_{k=1}^{n} \beta_k \underline{g}_k = \underline{0} \quad \Rightarrow \quad \beta_k = 0 \text{ for } 1 \leq k \leq n,$$

provided that

$$\sum_{k=1}^{n} \beta_k g_k(x_\nu) = 0 \text{ for all } \nu \in \mathrm{H}' \quad \Rightarrow \quad \beta_k = 0 \text{ for } 1 \leq k \leq n.$$

Now if $n \leq N'$, then as in 6.2, this condition will be satisfied if the functions $g_1, \ldots, g_n$ form a Chebyshev system. In this case there is a unique solution of the normal equations, and we have the following

Generalization of Corollary 6.2. *Suppose the functions $g_1, \ldots, g_n \in \mathrm{U}$ form a Chebyshev system. Then even if the points x_ν are no longer pairwise distinct, the fitting problem has a unique solution $\tilde{f} \in \mathrm{U}$ as long as $n \leq N'$.*

The solutions of the normal equations and the fitting problem are, however, no longer unique if $n > N'$, since in this case the vectors $\underline{g}_1, \ldots, \underline{g}_n$ are always linearly dependent. The matrix corresponding to the normal equations in this case has rank N', and so $(n - N')$ is the dimension of its solution space. We still have $\underline{\tilde{f}}$ uniquely defined, but $\tilde{f} = \sum_1^n \tilde{\alpha}_k g_k$ is no longer the unique best approximation in U. The fitting problem possesses a $(n - N')$-dimensional manifold of solutions.

Example.
$$(x_1, y_1) := (1,1) \quad (x_3, y_3) := (2,1)$$
$$(x_2, y_2) := (1,2) \quad (x_4, y_4) := (2,3)$$

In this example there are two double points $x_1 = x_2$ and $x_3 = x_4$. We have $N = 4$ and $N' = 2$. Let $n = 3$ and choose $g_1(x) := 1$, $g_2(x) := x$, $g_3(x) := x^2$. This leads to

$$\underline{g}_1 = (1,1,1,1),\ \underline{g}_2 = (1,1,2,2),\ \underline{g}_3 = (1,1,4,4),\ \underline{y} = (1,2,1,3),$$

and the normal equations

$$\alpha_1 \langle \underline{g}_1, \underline{g}_1 \rangle + \alpha_2 \langle \underline{g}_2, \underline{g}_1 \rangle + \alpha_3 \langle \underline{g}_3, \underline{g}_1 \rangle = \langle \underline{y}, \underline{g}_1 \rangle$$
$$\alpha_1 \langle \underline{g}_1, \underline{g}_2 \rangle + \alpha_2 \langle \underline{g}_2, \underline{g}_2 \rangle + \alpha_3 \langle \underline{g}_3, \underline{g}_2 \rangle = \langle \underline{y}, \underline{g}_2 \rangle$$

become

$$4\alpha_1 + 6\alpha_2 + 10\alpha_3 = 7$$
$$6\alpha_1 + 10\alpha_2 + 18\alpha_3 = 11$$

with the solution $(\tilde{\alpha}_1, \tilde{\alpha}_2, \tilde{\alpha}_3) = (1 + 2\alpha_3, \frac{1}{2} - 3\alpha_3, \alpha_3)$.

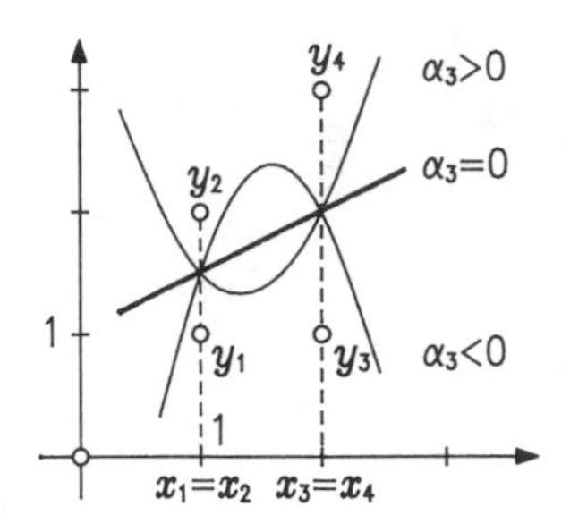

It follows that $\underline{\tilde{f}} = \tilde{\alpha}_1\underline{g}_1 + \tilde{\alpha}_2\underline{g}_2 + \tilde{\alpha}_3\underline{g}_3 = (\frac{3}{2}, \frac{3}{2}, 2, 2)$ is the unique solution of the approximation problem in $\mathbb{R}^4$. All functions of the form

$$\tilde{f} = (1 + 2\alpha_3)g_1 + (\frac{1}{2} - 3\alpha_3)g_2 + \alpha_3 g_3,$$

respectively,

$$\tilde{f}(x) = (1 + 2\alpha_3) + (\frac{1}{2} - 3\alpha_3)x + \alpha_3 x^2,$$

with $\alpha_3 \in \mathbb{R}$ are solutions of the fitting problem; i.e., are best approximations $\tilde{f} \in \mathrm{U}$.

We note that $\tilde{f}(1) = \frac{3}{2}$ and $\tilde{f}(2) = 2$ for all values $\alpha_3 \in \mathbb{R}$. The set of best approximations $\tilde{f}$ corresponds to the family of parabolas which pass through the points $(1, \frac{3}{2})$ and $(2, 2)$.

6.5 Discrete Approximation by Trigonometric Functions. For least squares fitting of periodic functions, it is natural to use trigonometric functions. In Section 5.8 we have introduced the associated orthogonal system $\{g_1, \ldots, g_{2m+1}\}$, $g_1(x) := 1$, $g_{2j}(x) := \cos(jx)$, $g_{2j+1}(x) := \sin(jx)$ for $1 \leq j \leq m$, and the corresponding ONS obtained by an appropriate normalization. By 4.2, this system of functions forms a Chebyshev system on $[-\pi, +\pi)$, and thus the results of 6.2 can be applied. Hence, if $n \leq N'$, $n = 2m + 1$, then the unique best approximation $\tilde{f} \in \mathrm{U}$ can be computed from the normal equations.

When the data points x_ν, $1 \leq \nu \leq N$, are equidistant, we get a remarkable special result. In this case, the system $\{\underline{g}_1, \ldots, \underline{g}_{2m+1}\}$ of vectors $\underline{g}_\ell \in \mathbb{R}^N$ for $1 \leq \ell \leq 2m + 1$ is then also an orthogonal system for $n \leq N$, and so the normal equations

$$\sum_{k=1}^{2m+1} \alpha_k \langle \underline{g}_k, \underline{g}_\ell \rangle = \langle \underline{y}, \underline{g}_\ell \rangle, \quad 1 \leq \ell \leq 2m + 1$$

have the solution $\tilde{\alpha}_k = \frac{1}{\|\underline{g}_k\|_2^2} \langle \underline{y}, \underline{g}_k \rangle$. To see this, we first prove the following

Orthogonality Relation in $\mathbb{R}^N$. *Let $x_\nu := (\nu - 1)\frac{2\pi}{N}$, $1 \le \nu \le N$, be N equidistant abscissae in the interval $[0, 2\pi)$. Then the associated vectors*

$$\begin{aligned} \underline{g}_1 &:= (1, \ldots, 1), \\ \underline{g}_{2\mu} &:= (\cos(\mu x_1), \ldots, \cos(\mu x_N)), \ 1 \le \mu \le m, \\ \underline{g}_{2\mu+1} &:= (\sin(\mu x_1), \ldots, \sin(\mu x_N)), \ 1 \le \mu \le m, \end{aligned}$$

$n = 2m + 1 \le N$, form an orthogonal system; i.e., $\langle \underline{g}_j, \underline{g}_\ell \rangle = 0$ for $j \ne \ell$, $1 \le j, \ell \le n$.

Proof. We note that

$$\sum_{\nu=1}^{N} [\cos(\mu x_\nu) + i \sin(\mu x_\nu)] = \sum_{\nu=1}^{N} e^{i\mu x_\nu} = \sum_{\nu=1}^{N} e^{i\mu(\nu-1)2\pi/N} = \frac{1 - e^{i\mu 2\pi}}{1 - e^{i\mu 2\pi/N}} = 0$$

for $\mu = 1, \ldots, N-1$. Also, $\langle \underline{g}_1, \underline{g}_\ell \rangle = 0$ for $\ell = 2, \ldots, n$, and $\langle \underline{g}_1, \underline{g}_1 \rangle = N$.

Using trigonometric identities, we get that for μ, $\kappa = 1, \ldots, m = \frac{n-1}{2}$,

$$\langle \underline{g}_{2\mu}, \underline{g}_{2\kappa+1} \rangle = \sum_{\nu=1}^{N} \cos(\mu x_\nu) \sin(\kappa x_\nu) =$$

$$= \frac{1}{2} \sum_{\nu=1}^{N} [\sin((\mu - \kappa) x_\nu) + \sin((\mu + \kappa) x_\nu)] = 0;$$

$$\langle \underline{g}_{2\mu}, \underline{g}_{2\kappa} \rangle = \sum_{\nu-1}^{N} \cos(\mu x_\nu) \cos(\kappa x_\nu) =$$

$$= \frac{1}{2} \sum_{\nu=1}^{N} [\cos((\mu - \kappa) x_\nu) + \cos((\mu + \kappa) x_\nu)] = \begin{cases} \frac{N}{2} & \text{for } \mu = \kappa \\ 0 & \text{for } \mu \ne \kappa; \end{cases}$$

$$\langle \underline{g}_{2\mu+1}, \underline{g}_{2\kappa+1} \rangle = \sum_{\nu=1}^{N} \sin(\mu x_\nu) \sin(\kappa x_\nu) =$$

$$= \frac{1}{2} \sum_{\nu=1}^{N} [\cos((\mu - \kappa) x_\nu) - \cos((\mu + \kappa) x_\nu)] = \begin{cases} \frac{N}{2} & \text{for } \mu = \kappa \\ 0 & \text{for } \mu \ne \kappa. \end{cases}$$

It follows that

$$\|\underline{g}_k\|_2^2 = \begin{cases} N & \text{for } k = 1 \\ \frac{N}{2} & \text{for } k = 2, \cdots, n, \end{cases}$$

and hence with the usual notation, we find that the solution of the normal equations is given by

$$\frac{\tilde{a}_0}{2} := \tilde{\alpha}_1, \quad \tilde{a}_\mu := \tilde{\alpha}_{2\mu} \quad \text{and} \quad \tilde{b}_\mu := \tilde{\alpha}_{2\mu+1} \quad \text{for } \mu = 1, \ldots, m$$

$$\tilde{a}_\mu = \frac{2}{N}\sum_{\nu=1}^{N} y_\nu \cos(\mu x_\nu), \quad 0 \le \mu \le m,$$

$$\tilde{b}_\mu = \frac{2}{N}\sum_{\nu=1}^{N} y_\nu \sin(\mu x_\nu), \quad 1 \le \mu \le m.$$

□

The solution of the least squares fitting problem is then given by

$$\tilde{f}(x) = \frac{\tilde{a}_0}{2} + \sum_{\mu=1}^{m} \tilde{a}_\mu \cos(\mu x) + \sum_{\mu=1}^{m} \tilde{b}_\mu \sin(\mu x).$$

If $n = 2m+1 < N$, we have the best approximation $\tilde{f} \in \text{span}(g_1, \ldots, g_{2m+1})$. If $2m+1 = N$, then $\tilde{f}$ solves the interpolation problem, since in this case, $\tilde{f}(x_\nu) = y_\nu$ for $\nu = 1, \ldots, 2m+1$.

The amount of computation needed to determine the coefficients $\tilde{a}_\mu$ and $\tilde{b}_\mu$ can be reduced by taking advantage of the symmetry properties of the trigonometric functions. This possibility was already observed in 1903 by C. Runge. In particular, he examined the interpolation problem with $n = N$ for an even number of data points; we treat this problem in 5.5.4. Runge gives computational algorithms for $n = 12$ and for $n = 24$ which were frequently used in the time of the mechanical calculator. Algorithms for other choices of data points were developed later. The question of efficiently computing these least squares fits enjoyed a renewed interest in the 1960's, since the same problem arises in the numerical implementation of the Fourier transform. The resulting algorithms are referred to as the *Fast Fourier Transform* (FFT).

CARL DAVID TOLMÉ RUNGE (1856 - 1927) was the first holder of the chair in applied mathematics at the University of Göttingen, after having been a professor at the Technical University in Hannover from 1886 to 1904. The creation of this position was the result of the efforts of Felix Klein, who thereby helped establish applied mathematics as a recognized part of mathematics. Runge studied in Munich and in Berlin, and was especially influenced by Weierstrass. After working on differential geometry, algebra, and function theory, his extensive interests led him to problems in physics, geodesy and astronomy, and hence to the problem of applying numerical methods to practical situations. Runge significantly influenced the development of applied mathematics. One of the three theses which he defended as part of his dissertation in 1880 in Berlin was entitled "The value of a mathematical discipline depends on its applicability to the natural sciences." He later explained "It was not the aim of my thesis to assert that every theorem must have a practical application. I only meant to observe that mathematics for its own sake is on the same level as chess or other games. They gain in value only through relationships to natural science. In my opinion, before a mathematician choses to work in a given area, he should ask himself if it has applications before he devotes his time and effort to it. Men such as Gauss, Lagrange, Jacobi etc. have undoubtedly done so." A hundred years later, this Credo for an applied mathematician remains incisive. (Translated citation from I. Runge [1949]). In honoring Runge on his 70-th birthday, E. Trefftz characterized him as follows: "If Runge has succeeded in building bridges between mathematics and technical science, then it rests on two characteristics of a true applied mathematician. First, his deep mathematical knowledge, which is manifest in his early papers in pure mathematics, and which always reappears in his applications of pure mathematics to problems in applied mathematics. And secondly, the untiring energy with which he developed his methods to the point where they had real practical applicabililty, and not just to the point where the mathematician considers it "simple", but to the point where the practitioner loses his distaste for the mathematical mechanics" (Z. angew. Math. Mech. 6, 423 - 424 (1926)).

6.6 Problems. 1) Using the method of least squares, find all best approximations from P_2 and from P_3 for the following data:
$(x_1, y_1) = (-1, 0)$; $(x_2, y_2) = (-1, 1)$; $(x_3, y_3) = (0, 1)$; $(x_4, y_4) = (1, 2)$; $(x_5, y_5) = (1, 3)$.

2) Using the method of least squares, find the best approximations of the data $(x_1, y_1) = (1, 2)$, $(x_2, y_2) = (2, 1)$, $(x_3, y_3) = (3, 3)$ from P_1 and from P_2, and sketch the solutions.

3) Find the best least squares approximations to the data

x_ν	1 2 1 3 1 2 3 2 3
y_ν	0 2 2 2 1 1 0 0 1

from span$(1, e^x)$, from P_2, and from P_3.

4) Consider the set $\{\frac{1}{\sqrt{2}}T_0, T_1, \ldots, T_{n-1}, \frac{1}{\sqrt{2}}T_n\}$ of Chebyshev polynomials of the first kind. Show that they form an ONS with respect to the discrete inner product

$$\langle f, g\rangle := \frac{1}{n}[f(x_0)g(x_0) + 2\sum_{\nu=1}^{n-1} f(x_\nu)g(x_\nu) + f(x_n)g(x_n)]$$

with $x_\nu := \cos(\frac{\nu\pi}{n})$, $0 \leq \nu \leq n$.

5) Let $n \in \mathbb{N}$, $n \geq 1$, and for $f, g : [-n, n] \to \mathbb{R}$, let their discrete inner product be defined by $\langle f, g\rangle := \sum_{-n}^{+n} f(\nu)g(\nu)$. Find the system $\{q_0, q_1, q_2\}$ of orthonormal polynomials $q_0 \in P_0$, $q_1 \in P_1$ and $q_2 \in P_2$ with respect to $\langle\cdot,\cdot\rangle$.

6) Let $f \in C[-\pi, +\pi]$, $f(x) := x^2$, be extended periodically. Find the best approximation from span$(1, \cos x, \sin x, \cos(2x), \sin(2x))$ with respect to the norm on $\mathbb{R}^6$ induced by the inner product $\langle f, g\rangle := \sum_1^6 f(x_\nu)g(x_\nu)$ with $x_\nu := (\nu - 1)\frac{2\pi}{6}$, $1 \leq \nu \leq 6$.

Compare the result with that of Problem 7a) in 5.9.

7) Let $a_{\mu_1}x_1 + a_{\mu_2}x_2 = b_\mu$, $1 \leq \mu \leq n$ and $n > 2$ be an over-determined linear system of equations for (x_1, x_2). Find an approximate solution which minimizes $\sum_1^n (a_{\mu_1}x_1 + a_{\mu_2}x_2 - b_\mu)^2$. Is the solution unique?

8) Find the plane which best approximates the data (x_ν, y_ν, z_ν) in $\mathbb{R}^3$, $1 \leq \nu \leq N$ in the least squares sense. Discuss existence and uniqueness of the solution.

5
Interpolation

The process of constructing a function which takes on given *data values* at given *data points* is called *interpolation*. In a certain sense, interpolation is a special case of discrete approximation, but the subject deserves a separate and detailed treatment. The results of the theory of interpolation are a basic part of the constructive theory of functions, and moreover, provide the basis for a wide variety of methods for numerical integration, numerical treatment of differential equations, and the discretization of general operator equations.

1. The Interpolation Problem

In Chapter 4 we have seen that approximation by a linear combination of prescribed functions is well understood from both the theoretical and practical standpoints. In discussing interpolation, we shall again concentrate almost exclusively on linear combinations.

1.1 Interpolation in Haar Spaces. In order to formulate the problem of interpolation by a linear combination of prescribed functions, we assume that $\{g_0, \ldots, g_n\}$ is a Chebyshev system of functions, and that we are given $(n+1)$ pairs of numbers (x_ν, y_ν), $0 \le \nu \le n$, with $x_\nu \ne x_\mu$ for all $\nu \ne \mu$. We are looking for a function $\tilde{f} \in \operatorname{span}(g_0, \ldots, g_n)$ satisfying the interpolation conditions $\tilde{f}(x_\nu) = y_\nu$ for $\nu = 0, \ldots, n$. Using Corollary 4.6.2 case (ii), we have the

Theorem. *Suppose $\{g_0, \ldots, g_n\}$ is a Chebyshev system in a function space, and that we are given $(n+1)$ pairs of numbers $(x_0, y_0), \ldots, (x_n, y_n)$, $x_\nu \ne x_\mu$, for $\nu \ne \mu$. Then there exists a unique function $\tilde{f} \in \operatorname{span}(g_0, \ldots, g_n)$ satisfying the interpolation conditions $\tilde{f}(x_\nu) = y_\nu$ for $\nu = 0, \ldots, n$.*

Solution of the Interpolation Problem. As in 4.6.2, $\tilde{f}$ can be computed by solving the normal equations, but this is a somewhat complicated approach, since the problem can also be solved more directly. In particular, if the element $f = \alpha_0 g_0 + \cdots + \alpha_n g_n$ is to satisfy the interpolation conditions

$f(x_\nu) = y_\nu$ for $\nu = 0, \ldots, n$, then the coefficients must satisfy the following system of equations:

$$\alpha_0 g_0(x_\nu) + \cdots + \alpha_n g_n(x_\nu) = y_\nu$$

for $\nu = 0, \ldots, n$. By Theorem 4.6.2, the vectors $\underline{g}_j = (g_j(x_0), \ldots, g_j(x_n))^T$ are linearly independent in $\mathbb{R}^{n+1}$, and so $\det(\underline{g}_0, \ldots, \underline{g}_n) \neq 0$. This assures that the system has a unique solution $\tilde{\alpha} = (\tilde{\alpha}_0, \ldots, \tilde{\alpha}_n)$, and the solution of the interpolation problem is then

$$\tilde{f}(x) = \tilde{\alpha}_0 g_0(x) + \cdots + \tilde{\alpha}_n g_n(x).$$

1.2 Interpolation by Polynomials. Because of its simplicity, the Chebyshev system of monomials is especially attractive for solving the interpolation problem. This is the classical interpolation method, and we now discuss it in more detail.

In terms of polynomials, we can now state Theorem 1.1 as follows:

Theorem. *Given $(n+1)$ pairwise distinct data points $x_0, \ldots, x_n$ and associated values $y_0, \ldots, y_n$, there is a unique polynomial of degree at most n which takes on these values.*

Proof. The functions $g_j(x) := x^j$, $0 \le j \le n$, span P_n. □

Direct Proof. This theorem can also be established directly. The determinant of the linear system of equations

$$a_0 + a_1 x_\nu + \cdots + a_n x_\nu^n = y_\nu, \ 0 \le \nu \le n$$

for determining the coefficients $\tilde{a} = (\tilde{a}_0, \ldots, \tilde{a}_n)$ of the interpolating polynomial $p(x) = a_0 + a_1 x + \cdots + a_n x^n$ is precisely the Vandermonde determinant

$$\det(x_\nu^\kappa)_{\nu,\kappa=0,\ldots,n} = \prod_{0 \le \nu < \mu \le n} (x_\mu - x_\nu),$$

which is nonzero since $x_\nu \neq x_\mu$ for $\nu \neq \mu$. □

Remark. The interpolating polynomial $\tilde{p} \in \mathrm{P}_n$ is of *exact* degree n, when $\tilde{a}_n \neq 0$. This need not always be the case. For example, consider the problem of interpolating the sine function on the interval $[-\frac{\pi}{2}, +\frac{\pi}{2}]$ at the points $(x_0, y_0) := (-\frac{\pi}{2}, -1)$, $(x_1, y_1) := (0, 0)$, $(x_2, y_2) := (\frac{\pi}{2}, 1)$. Then it is obvious that $\tilde{p}(x) := \frac{2}{\pi} x$ is the unique interpolating polynomial in P_2, but $\tilde{a}_2 = 0$.

We note that if we look at polynomials of higher degree, then there are always arbitrarily many of them which interpolate at a given set of data points $x_0, \ldots, x_n$. In particular, defining the ***data point polynomial*** $\Phi(x) := (x - x_0) \cdots (x - x_n)$, we have the following

Lemma. *A polynomial $p \in \mathrm{P}_m$, $m > n$, satisfies the interpolation conditions $p(x_\nu) = y_\nu$ for $\nu = 0, \ldots, n$ if and only if it has the form*

$$p(x) = \tilde{p}(x) + \Phi(x)q(x), \quad q \in \mathrm{P}_{m-n-1}.$$

Proof. ($\Rightarrow$): Since $\Phi(x_\nu) = 0$, $0 \le \nu \le n$, it follows that $p(x_\nu) = \tilde{p}(x_\nu)$, and thus if p interpolates, then so does $\tilde{p}$.
($\Leftarrow$): If $p(x_\nu) - \tilde{p}(x_\nu) = 0$ for $\nu = 0, \ldots, n$, then we can write $p(x) - \tilde{p}(x) = \Phi(x)q(x)$ for some suitable $q \in \mathrm{P}_{m-(n+1)}$. □

1.3 The Remainder Term. Up to now, it has been irrelevant whether the data values y_ν are related to each other in some way or not. This question does become important, however, if we want to say something about how the interpolant behaves between the data points.

In the case where the data values y_ν are the values at $x_\nu \in [a, b]$ for $\nu = 0, \ldots, n$ of some function f defined on an interval $[a, b]$, then it makes sense to investigate the deviation of the interpolating polynomial $\tilde{p}$ from f. To do this, we will certainly need to make some assumptions on the behavior of f.

We now assume that $y_\nu := f(x_\nu)$ for some $f \in \mathrm{C}_{n+1}[a, b]$. We want to find an expression for the *remainder term* $R_n := f - \tilde{p}$ which can be used to estimate its values

$$R_n(f; x) = f(x) - \tilde{p}(x)$$

for $x \neq x_\nu$.

Given $x \neq x_\nu$, let $\eta : [a, b] \to \mathbb{R}$ be defined by

$$\eta(t) := f(t) - \tilde{p}(t) - \frac{f(x) - \tilde{p}(x)}{\Phi(x)} \Phi(t), \; \eta \in \mathrm{C}_{n+1}[a, b].$$

We have

$$\eta(x_\nu) = f(x_\nu) - \tilde{p}(x_\nu) - \frac{f(x) - \tilde{p}(x)}{\Phi(x)} \Phi(x_\nu) = 0, \quad \text{for } \nu = 0, ..., n$$

and

$$\eta(x) = f(x) - \tilde{p}(x) - \frac{f(x) - \tilde{p}(x)}{\Phi(x)} \Phi(x) = 0.$$

This implies that the function η has at least $(n+2)$ zeros $x_0, \ldots, x_n, x$ in the interval $[\min_\nu(x_\nu, x), \max_\nu(x_\nu, x)] \subset [a, b]$. By Rolle's Theorem, it follows that $\eta^{(n+1)}$ has at least one zero $\xi \in (\min_\nu(x_\nu, x), \max_\nu(x_\nu, x))$, where the point ξ depends on x. Now

$$\eta^{(n+1)}(t) = f^{(n+1)}(t) - \frac{f(x) - \tilde{p}(x)}{\Phi(x)}(n+1)!\,;$$

$$\eta^{(n+1)}(\xi(x)) = 0 \quad \Rightarrow \quad f^{(n+1)}(\xi(x)) = \frac{f(x) - \tilde{p}(x)}{\Phi(x)}(n+1)!\,,$$

and thus $f^{(n+1)}(\xi(x))$ is continuous in x for $x \neq x_\nu$. Now setting

$$f^{(n+1)}(\xi(x)) := \frac{f'(x_\nu) - \tilde{p}'(x_\nu)}{\Phi'(x_\nu)}(n+1)! \text{ for } x := x_\nu,\ 0 \leq \nu \leq n,$$

we see that $f^{(n+1)}(\xi(x))$ becomes a continuous function for all $x \in [a,b]$, and we have the

Remainder Term Formula

$$\boxed{R_n(f;x) = \frac{f^{(n+1)}(\xi(x))}{(n+1)!}(x - x_0)\cdots(x - x_n).}$$

We note that

$$R_n(f;x_\nu) = 0 \text{ for } \nu = 0, \ldots, n \text{ and}$$

$$f(x) = \tilde{p}(x) + \frac{f^{(n+1)}(\xi(x))}{(n+1)!}\,\Phi(x).$$

1.4 Error Bounds. Assuming that f satisfies the bound

$$\sup_{x \in [a,b]} |f^{(n+1)}(x)| = \|f^{(n+1)}\|_\infty \leq M_{n+1},$$

the Remainder Term Formula 1.3 leads to the

Remainder Term Bound

$$|R_n(f;x)| \leq \frac{M_{n+1}}{(n+1)!}|(x - x_0)\cdots(x - x_n)|,$$

which in turn leads to the general

Interpolation Error Bound

$$\boxed{\|f - \tilde{p}\| \leq \frac{M_{n+1}}{(n+1)!}\|\Phi\|,}$$

valid for all of the norms $\|\cdot\|_p$, $1 \leq p \leq \infty$.

Remark. This error bound is typical in that it requires the assumption that f is $(n+1)$-times continuously differentiable. It is possible to get these error bounds under slightly weaker assumptions of f. In particular, the Remainder Term Formula 1.3 holds for all $f \in C_n[a,b]$ such that $f^{(n)}$ is differentiable in (a,b), and the Remainder Term Bound 1.4 holds under the same hypotheses as long as $f^{(n+1)}$ is bounded. In general, however, if we weaken the hypotheses on the differentiability of f, then we must be satisfied with weaker error bounds.

On the other hand, this error estimate cannot in general be improved by assuming more differentiability. In fact, it is optimal in the sense that we can explicitly find a function for which the bound is achieved. Indeed, we can take $f := \Phi$, since then $M_{n+1} := (n+1)!$, and we have the error bound $|R_n(f;x)| \le |(x-x_0)\cdots(x-x_n)| = |\Phi(x)|$. But in this case $\tilde{p} = 0$, since this is the only polynomial in P_n which agrees with Φ at the points $x_0, \ldots, x_n$, and we have $|R_n(f;x)| = |\Phi(x)|$.

In the case where $\|\cdot\| := \|\cdot\|_\infty$ on the interval $I := [\min_\nu x_\nu, \max_\nu x_\nu]$, we can now derive a particularly useful bound on the interpolation error $\|f - \tilde{p}\|$. We shall accomplish this by showing that the data point polynomial Φ constructed above satisfies

$$\|\Phi\|_\infty = \max_{x \in I} |(x-x_0)\cdots(x-x_n)| \le \frac{n!}{4} h^{n+1},$$

where h is *the maximal distance between neighboring data points.* Up to now, we have made no use of the distribution of data points.

Proof. Let $x_\nu < x_\mu$ be two neighboring data points, and consider a point $x \in [x_\nu, x_\mu]$. Then $|(x-x_\nu)(x-x_\mu)| \le \frac{h^2}{4}$, and a straightforward estimate of the size of the other terms in Φ leads to the stated bound on $\|\Phi\|_\infty$. □

Combining these facts, we find that the interpolation error satisfies the **Uniform Error Bound**

$$\boxed{\|f - \tilde{p}\|_\infty \le \frac{\|f^{(n+1)}\|_\infty}{4(n+1)} h^{n+1}.}$$

Comment. In order to interpret this bound correctly, we should think of the interpolating polynomial $\tilde{p} \in P_n$ as depending on h for fixed n. Then this error bound shows that the accuracy behaves like $O(h^{n+1})$ as we vary the interpolation interval $[x_0, x_n]$. This observation plays a role in the case where the interpolation interval is variable, or when piecewise combining interpolating polynomials of a fixed highest degree to form a continuous interpolating function $\tilde{f}$. The idea of working with piecewise polynomial approximations is the subject of Chapter 6.

The above error bound was obtained using a very rough estimate of the size of Φ, but nevertheless, it gives the correct order in terms of h.

Bounds for the Derivatives. The argument which led to the Remainder Formula 1.3 can be extended to the derivatives $(f - \tilde{p})^{(k)}$ for $k = 1, \ldots, n$.

For $k := 1$, the function $(f' - \tilde{p}')$ has at least n zeros $\xi_1, \ldots, \xi_n$, one between each successive pair of data points. Let $\psi(x) := (x-\xi_1)\cdots(x-\xi_n)$. Then the analysis in 1.3 leads to the inequality

$$\|f' - \tilde{p}'\| \le \frac{M_{n+1}}{n!} \|\psi\|.$$

This bound can also be written in terms of h, since clearly

$$\|\psi\|_\infty = \max_{x \in I} |(x-\xi_1)\cdots(x-\xi_n)| \le n!\, h^n,$$

and thus we obtain the error bound

$$\|f' - \tilde{p}'\|_\infty \le \|f^{n+1}\|_\infty \, h^n.$$

This process can be extended to higher derivatives, and we obtain the general

Error bound for Derivatives

$$\boxed{\|f^{(k)} - \tilde{p}^{(k)}\|_\infty \le \frac{\|f^{n+1}\|_\infty \, n!}{(k-1)!(n+1-k)!} h^{n+1-k}}$$

for $k = 1, \ldots, n$ (Problem 8).

1.5 Problems. 1) Let $g_0, \ldots, g_n \in C[a,b]$ be elements of a Chebyshev system, and let $x_0, \ldots, x_n \in [a,b]$ be pairwise distinct data points. For any two elements $f, g \in C[a,b]$, let $\langle f, g\rangle := \sum_0^n f(x_\nu)g(x_\nu)$ (cf. Problem 5 in 4.6.6). Show directly that if $\tilde{f} \in \operatorname{span}(g_0, \ldots, g_n)$ satisfies the normal equations for the best approximation of f with respect to $\langle\cdot,\cdot\rangle$, then $\tilde{f}$ interpolates f at $x_0, \cdots, x_n$.

2) Consider the interpolation problem using the space $\operatorname{span}(g_0, g_1)$ with $g_0(x) := 1$, $g_1(x) := x^2$ for the points

a) $(x_0, y_0) := (-\frac{1}{2}, 1)$; $(x_1, y_1) := (1, 2)$.

b) $(x_0, y_0) := (-1, 1)$; $(x_1, y_1) := (1, 2)$.

c) $(x_0, y_0) := (0, -1)$; $(x_1, y_1) := (1, 2)$.

Why is this interpolation problem not always uniquely solvable when $x_0 \ne x_1$ are arbitrary points in $[-1, +1]$, while it is if $x_0, x_1 \in [0, 1]$ holds?

3) Suppose pairwise distinct data points $x_0, \ldots, x_n$ are prescribed. Show that the coefficients $a_0, \ldots, a_n$ of the interpolation polynomial $\tilde{p} \in P_n$ depend continuously on the data values $y_0, \ldots, y_n$.

4) Let $f \in C_1[a,b]$, and suppose that $x_0, \ldots, x_n \subset [a,b]$ are pairwise distinct data points. Show that for every $\varepsilon > 0$, there exists a polynomial p such that $\|f - p\|_\infty < \varepsilon$ and, simultaneously, satisfies the interpolation conditions $p(x_\nu) = f(x_\nu)$, $0 \le \nu \le n$.

5) Consider the interpolation of the function $f \in C[a,b]$, $f(x) := |x|$, with $a < 0, b > 0$ at the pairwise distinct data points $x_0, \ldots, x_n \in [a,b]$ by a polynomial $\tilde{p} \in P_n$. Show that for all choices of n, $\sup_{x\in I} |f'(x) - \tilde{p}'(x)| \ge 1$, where $I := [a,b] \setminus \{0\}$.

6) a) In a table of base 10 logarithms, the entries are given to 5 places at equally spaced points at a distance of 10^{-3} apart. Is it reasonable to use linear interpolation on this table?

b) Let $\tilde{p} \in P_2$ be the polynomial which interpolates the sine function at the endpoints and at the midpoint of the interval $[0, \frac{\pi}{6}]$. Estimate the maximal error. Do the same for $[\frac{\pi}{6}, \frac{\pi}{2}]$.

7) How small must the maximal distance between two neighboring data points be to insure that the polynomial interpolant $\tilde{p} \in P_5$ of the exponential

function on $[-1,+1]$ satisfies $\|f-\tilde{p}\|_\infty \leq 5\cdot 10^{-8}$ and $\|f'-\tilde{p}'\|_\infty \leq 5\cdot 10^{-7}$ simultaneously?

8) Give the details of the proof of the Error Bound 1.4 for derivatives.

2. Interpolation Methods and Remainders

Section 1 was devoted to the basic theory of polynomial interpolation. In this and in the two following sections, we will present a detailed treatment of results which relate to practical aspects of interpolation. We begin with two classical methods for computing an interpolating polynomial which are exceptional for their remarkable simplicity.

2.1 The Method of Lagrange. Following Lagrange, we seek to represent the unique interpolating polynomial $\tilde{p} \in \mathrm{P}_n$ in the form

$$\tilde{p}(x) = \ell_{n0}(x)y_0 + \cdots + \ell_{nn}(x)y_n.$$

Now if we require that $\ell_{n\kappa} \in \mathrm{P}_n$ and $\ell_{n\kappa}(x_\nu) = \delta_{\kappa\nu}$ for $\kappa,\nu = 0,\ldots,n$, it follows that $\tilde{p}$ satisfies the interpolation conditions $\tilde{p}(x_\nu) = y_\nu$. Theorem 1.2 implies that these properties uniquely define the functions $\ell_{n\kappa}$. Since $\ell_{n\kappa}$ must have the zeros $x_0,\ldots,x_{\kappa-1},x_{\kappa+1},\ldots,x_n$ and since $\ell_{n\kappa}(x_\kappa)=1$ should hold, we immediately see that $\ell_{n\kappa}$ must be the following

Lagrange Polynomial

$$\ell_{n\kappa}(x) = \prod_{\substack{\nu=0\\ \nu\neq\kappa}}^{n} \frac{x-x_\nu}{x_\kappa - x_\nu}.$$

In terms of the data point polynomial $\Phi(x) = \Pi_0^n(x-x_\kappa)$ introduced in 1.2, we can write

$$\ell_{n\kappa}(x) = \begin{cases} \frac{\Phi(x)}{(x-x_\kappa)\,\Phi'(x_\kappa)} & \text{for } x \neq x_\kappa \\ 1 & \text{for } x = x_\kappa. \end{cases}$$

We also observe that $\sum_{\kappa=0}^{n} \ell_{n\kappa}(x) = 1$, since interpolation of $f(x)=1$ leads to $\tilde{p}=1$ for every n.

Using the Lagrange form, it is possible to write down the interpolating polynomial $\tilde{p}$ without solving a system of equations to compute the coefficients. We point out one disadvantage of the method, however. If the number of data points is increased, then there is no way to make use of the polynomial $\tilde{p}$ which has already been computed. The following older method avoids this disadvantage.

2.2 The Method of Newton. Suppose we write the interpolating polynomial $\tilde{p} \in \mathrm{P}_n$ in the form

$$\tilde{p}(x) = \gamma_0 + \gamma_1(x - x_0) + \gamma_2(x - x_0)(x - x_1) + \cdots$$
$$\cdots + \gamma_n(x - x_0) \cdots (x - x_{n-1}).$$

Then the coefficients $\gamma_0, \ldots, \gamma_n$ can be successively computed from the interpolation conditions $\tilde{p}(x_\nu) = y_\nu$ for $\nu = 0, \ldots, n$ as follows:

$$\tilde{p}(x_0) = y_0 \Rightarrow \gamma_0 = y_0$$

$$\tilde{p}(x_1) = y_1 \Rightarrow \gamma_1 = \frac{y_1 - y_0}{x_1 - x_0} \quad \text{etc.}$$

It is also possible to prove the existence and uniqueness of interpolating polynomials using the Lagrange and Newton forms. Indeed, clearly both formulae are polynomials of maximal degree n which satisfy the interpolation conditions. Now to establish the uniqueness, suppose that two polynomials $\tilde{p}, \tilde{q} \in \mathrm{P}_n$ both satisfy the interpolation conditions: $\tilde{p}(x_\nu) = \tilde{q}(x_\nu) = y_\nu$, $0 \leq \nu \leq n$. Then the polynomial $\tilde{p} - \tilde{q} \in \mathrm{P}_n$ has $(n+1)$ zeros at the points $x_0, \ldots, x_n$. By the Fundamental Theorem of Algebra, it follows that $\tilde{p} - \tilde{q} = 0$, which implies that $\tilde{p} = \tilde{q}$ and thereby the uniqueness.

The method of Newton has the advantage that if we add additional data points $x_{n+1}, \ldots, x_{n+m}$, then we have to compute the new coefficients $\gamma_{n+1}, \ldots, \gamma_{n+m}$, but the old coefficients $\gamma_0, \ldots, \gamma_n$ remain unchanged. Since the order of the data points is arbitrary, we can even extend the interval in which interpolation takes place by the addition of data points. We may also want to add data points to increase the density of points in the interval, and thereby the accuracy of the interpolation.

Sir ISAAC NEWTON (1642–1727) was interested in interpolation as a means to approximately compute integrals (cf. 7.1.5). JOSEF LOUIS DE LAGRANGE (1736–1813) was led to interpolation problems in his study of recurrent series.

2.3 Divided Differences. The coefficient γ_1 in the Newton form of the interpolating polynomial has the form of a difference quotient. We call this quotient a *divided difference of first order* and denote it by the symbol

$$[x_1 x_0] := \frac{y_1 - y_0}{x_1 - x_0}.$$

In order to find a unified way of expressing the other γ_ν, $2 \leq \nu \leq n$, we now introduce the *divided difference of m-th order*

$$[x_m x_{m-1} \ldots x_0] := \frac{[x_m \ldots x_1] - [x_{m-1} \ldots x_0]}{x_m - x_0}.$$

Let $y := f(x)$, and as above let $y_\nu := f(x_\nu)$. We form the divided differences of the function f using $x \neq x_\nu$ along with the data points $x_0, \dots, x_n$:

$$[x_0 x] = \frac{f(x_0) - f(x)}{x_0 - x}$$

$$[x_1 x_0 x] = \frac{[x_1 x_0] - [x_0 x]}{x_1 - x}$$

$$\vdots$$

$$[x_n x_{n-1} \dots x_0 x] = \frac{[x_n \dots x_0] - [x_{n-1} \dots x]}{x_n - x}.$$

Where necessary to avoid confusion, we will write $[x_m \dots x_0]f$ instead of $[x_m \dots x_0]$ for the divided difference of a given function f.

Starting with $[x_n \dots x]$, by repeated substitution of the above formulae, we obtain the

Newton Identity

$$\begin{aligned} f(x) =& f(x_0) + [x_1 x_0](x - x_0) + [x_2 x_1 x_0](x - x_0)(x - x_1) + \cdots \\ &+ [x_n \dots x_0](x - x_0) \cdots (x - x_{n-1}) + [x_n \dots x](x - x_0) \cdots (x - x_n). \end{aligned}$$

The Newton identity is an expansion of f using the symbolism of divided differences. It holds for every arbitrary function f, with no further assumptions on f. The identity decomposes f into the sum of a polynomial $p \in \mathrm{P}_n$ and a remainder term

$$r(x) = f(x) - p(x) = [x_n \dots x](x - x_0) \cdots (x - x_n).$$

The remainder term satisfies $r(x_\nu) = 0$ for $\nu = 0, \dots, n$, and hence $f(x_\nu) = = p(x_\nu)$ so that p is the interpolating polynomial $\tilde{p} \in \mathrm{P}_n$.

Comparing with the Newton formula above, we see that

$$\gamma_0 = f(x_0)$$

$$\gamma_1 = [x_1 x_0]$$

$$\gamma_2 = [x_2 x_1 x_0]$$

$$\vdots$$

$$\gamma_n = [x_n \dots x_0],$$

while the remainder r can be written in terms of the data point polynomial Φ as

$$r(x) = [x_n \dots x_0 x]\, \Phi(x).$$

In developing the Newton method and the divided differences, we have made no assumption about the order of the points $x_0, \dots, x_m$. The interpolating polynomial $\tilde{p} \in \mathrm{P}_n$ is uniquely defined. Its highest coefficient is $[x_n \dots x_0]f$, and this value is independent of the order of the data points. As a consequence, the divided differences have the

Symmetry Property. *The divided difference $[x_m ... x_0]$ is independent of the order of the points.*

The uniqueness of the interpolation polynomial implies the

Linearity of the Divided Differences. *If $f = \alpha u + \beta v$, then*

$$[x_m \dots x_0]f = \alpha[x_m \dots x_0]u + \beta[x_m \dots x_0]v.$$

Next we consider the divided difference of a function f which is the product $f = u \cdot v$ of two functions; i.e., $f(x) = u(x) \cdot v(x)$ for $x \in [a, b]$. In this case the divided difference satisfies the

Leibniz Rule.

$$[x_{j+k} \dots x_j]f = \sum_{\iota=j}^{j+k}([x_j \dots x_\iota]u)([x_\iota \dots x_{j+k}]v),$$

where $[x_\iota] := f(x_\iota)$.

Proof. Let

$$\varphi(x) := \left(\sum_{\iota=j}^{j+k}(x - x_j)\cdots(x - x_{\iota-1})[x_j \dots x_\iota]u\right) \cdot$$

$$\cdot \left(\sum_{\kappa=j}^{j+k}(x - x_{\kappa+1})\cdots(x - x_{j+k})[x_\kappa \dots x_{j+k}]v\right),$$

where $(x - x_i)\cdots(x - x_\ell) := 1$ for $\ell < i$. The expressions in the parentheses are the polynomials interpolating u with respect to the data points $x_j, \dots, x_{j+k}$, and v with respect to $x_{j+k}, \dots, x_j$. It follows that

$$\varphi(x_\iota) = u(x_\iota)v(x_\iota) = f(x_\iota) \text{ for } \iota = j, \dots, j + k.$$

Now abbreviating $\varphi(x) =: (\sum_{\iota=j}^{j+k} a_\iota(x))(\sum_{\kappa=j}^{j+k} b_\kappa(x))$, we can see that

$$(\sum_{\iota=j}^{j+k} a_\iota)(\sum_{\kappa=j}^{j+k} b_\kappa) = \sum_{\iota,\kappa=j}^{j+k} a_\iota b_\kappa = \sum_{\iota\leq\kappa} a_\iota b_\kappa + \sum_{\iota>\kappa} a_\iota b_\kappa.$$

Since $\sum_{\iota>\kappa} a_\iota(x_\lambda)b_\kappa(x_\lambda)$ vanishes for $\lambda = j, \dots, j + k$, we have

$$\sum_{\iota\leq\kappa} a_\iota(x_\lambda)b_\kappa(x_\lambda) = f(x_\lambda).$$

Comparing the highest coefficients of the polynomial $\sum_{\iota\leq\kappa} a_\iota b_\kappa \in \mathrm{P}_k$ and the polynomial which interpolates f at the points $x_j, ..., x_{j+k}$ leads immediately to the Leibniz rule. □

2.4 The General Peano Remainder Formula. In this section we study a remainder formula which is more general than the one discussed in 1.3, and which can also be applied in other cases besides interpolation. The essential component of the Interpolation Remainder Formula 1.3 is the expression $f^{(n+1)}(\xi)$. The presence of this term reflects the fact that every element $f \in \mathrm{P}_n$ is represented exactly by its interpolating polynomial $\tilde{p} \in \mathrm{P}_n$, since $f^{(n+1)} = 0$ for all $f \in \mathrm{P}_n$. G. PEANO observed that this fact can be exploited to construct a general remainder formula. To formulate his result, we shall make use of concepts from functional analysis.

We can think of the remainder term $r(x) = f(x) - \tilde{p}(x)$ as the result of applying a linear functional which operates on functions $f \in \mathrm{C}_{n+1}[a,b]$, and which annihilates polynomials $p \in \mathrm{P}_n \subset \mathrm{C}_{n+1}[a,b]$. Peano showed that a linear function which possesses this property can always be written in terms of an expression involving the $(n+1)$-st derivative of f.

We now work in the space $\mathrm{C}_{m+1}[a,b]$, $m \geq 0$, and consider linear functionals in a sufficiently general form to give error bounds for a variety of numerical processes including interpolation and numerical integration (cf. Chapter 7), for example.

Let L be a composite linear functional of the form

$$Lf = \int_a^b [w_0(x)f(x) + w_1(x)f'(x) + \cdots + w_k(x)f^{(k)}(x)]dx +$$

$$+ \sum_{\kappa=1}^{k_0} \beta_{\kappa 0} f(x_{\kappa 0}) + \sum_{\kappa=1}^{k_1} \beta_{\kappa 1} f'(x_{\kappa 1}) + \cdots + \sum_{\kappa=1}^{k_k} \beta_{\kappa k} f^{(k)}(x_{\kappa k}),$$

involving both integral and point functionals. We suppose that $k \leq m+1$ so that all of the derivatives of f appearing in Lf actually exist. Moreover, we also assume that L annihilates all $f \in \mathrm{P}_m$. In addition, suppose that the functions w_μ are elements of $\mathrm{C}_{-1}[a,b]$, $0 \leq \mu \leq k$, and that all of the data points $x_{\nu\mu}$ appearing in the point functionals lie in $[a,b]$. Define the function q_m by $q_m(x,t) := (x-t)_+^m$, where

$$(x-t)_+^m := \begin{cases} (x-t)^m & \text{for } x \geq t \\ 0 & \text{for } x < t \end{cases} \quad ; \text{ here } (x-t)_+^0 := 1 \text{ for } x \geq t.$$

We then have the

Peano Kernel Formula. *If $Lf = 0$ for all $f \in \mathrm{P}_m$, then*

$$Lf = \int_a^b f^{(m+1)}(t)K_m(t)dt,$$

$$K_m(t) := \frac{1}{m!} L\, q_m(\cdot, t),$$

for all $f \in C_{m+1}[a,b]$.

The m-th derivative of $q_m(\cdot,t)$ at $x = t$ is to be replaced by the right-hand derivative.

Proof. Let

$$f(x) = f(a) + f'(a)(x-a) + \cdots + \frac{f^{(m)}(a)(x-a)^m}{m!} + \frac{1}{m!}\int_a^x f^{(m+1)}(t)(x-t)^m\,dt$$

be the Taylor formula with integral remainder.
Then

$$\int_a^x f^{(m+1)}(t)(x-t)^m\,dt = \int_a^b f^{(m+1)}(t)(x-t)_+^m\,dt,$$

and thus

$$Lf = \frac{1}{m!}L\Big(\int_a^b f^{(m+1)}(t)\,q_m(\cdot,t)\,dt\Big)$$
$$= \frac{1}{m!}\int_a^b f^{(m+1)}(t)L\,q_m(\cdot,t)\,dt,$$

since the functional L and the integration can be interchanged. □

K_m is called the *Peano kernel* belonging to L.

Corollary. If $K_m(t)$ does not change sign in $[a,b]$, then

$$Lf = \frac{f^{(m+1)}(\xi)}{(m+1)!}L(x^{m+1}), \quad \xi \in (a,b).$$

Proof. By the mean-value theorem of calculus,

$$Lf = f^{(m+1)}(\xi)\int_a^b K_m(t)dt.$$

Applying this to $f(x) := x^{m+1}$, gives

$$Lf = (m+1)!\int_a^b K_m(t)dt,$$

and the result follows. □

The Peano Kernel Formula for the remainder term can be used for error bounds as well as for a study of the behavior of the error. In particular, it can be applied also for the case when f only has a low order of differentiability. Indeed, if a functional L annihilates all polynomials $f \in P_m$, then it also annihilates all $f \in P_\mu$ with $\mu < m$, and thus a remainder formula can be obtained involving the lower derivative $f^{(\mu+1)}$. This will allow us to obtain error bounds under weaker assumptions on f.

GIUSEPPE PEANO (1858–1932) worked at the University of Turin. He became known for his contributions to formal logic, especially for the system of Axioms named after him, and for his work in analysis. In several papers which appeared in the years 1913–1918, he studied the problem of representing the remainder term for numerical methods in integral form. Starting with remainder terms for various quadrature formula, he then developed the more general Peano Kernel Formula. G. Peano also devoted a series of other papers to problems in numerical analysis.

As a first application of the Peano Kernel Formula, we apply it to the interpolating polynomial in the Lagrange Form 2.1 to obtain the

Remainder Formula of Kowalewski. Suppose $x_0, \ldots, x_n \in [a, b]$ are pairwise disjoint data points, and for a fixed $x \in [a, b]$, let

$$Lf := R_n(f; x) = f(x) - \sum_{\nu=0}^{n} f(x_\nu)\ell_{n\nu}(x)$$

be the corresponding error functional. Using the fact that $\sum_{\nu=0}^{n} \ell_{n\nu}(x) = 1$, the Peano Kernel Formula gives

$$\begin{aligned} m!K_m(x,t) = (Lq_m(\cdot,t))(x) &= (x-t)_+^m - \sum_{\nu=0}^{n}(x_\nu - t)_+^m \ell_{n\nu}(x) \\ &= \sum_{\nu=0}^{n}[(x-t)_+^m - (x_\nu - t)_+^m]\ell_{n\nu}(x). \end{aligned}$$

Now since

$$\int_a^b [(x-t)_+^m - (x_\nu - t)_+^m] f^{(m+1)}(t)dt =$$

$$\int_a^x [(x-t)^m - (x_\nu - t)^m] f^{(m+1)}(t)dt + \int_{x_\nu}^x (x_\nu - t)^m f^{(m+1)}(t)dt,$$

it follows that

$$\begin{aligned} m! \int_a^b K_m(x,t) f^{(m+1)}(t)dt &= \\ &= \int_a^x f^{(m+1)}(t) \sum_{\nu=0}^{n}[(x-t)^m - (x_\nu - t)^m]\ell_{n\nu}(x)dt + \\ &+ \sum_{\nu=0}^{n} \ell_{n\nu}(x) \int_{x_\nu}^x (x_\nu - t)^m f^{(m+1)}(t)dt. \end{aligned}$$

Here

$$\sum_{\nu=0}^{n}[(x-t)^m - (x_\nu - t)^m]\ell_{n\nu}(x) = (x-t)^m - \sum_{\nu=0}^{n}(x_\nu - t)^m \ell_{n\nu}(x).$$

Now for $m \le n$, $\sum_{\nu=0}^{n}(x_\nu - t)^m \ell_{n\nu}(x)$ is the interpolating polynomial $\tilde{p} \in \mathrm{P}_n$ associated with $f(x) := (x-t)^m$, and so the term in the square brackets vanishes. This leaves

$$Lf = f(x) - \tilde{p}(x) = \int_a^b K_m(x,t) f^{(m+1)}(t)dt,$$

and thus the remainder is given by

$$R_n(f;x) = \frac{1}{m!}\sum_{\nu=0}^{n} \ell_{n\nu}(x)\int_{x_\nu}^{x}(x_\nu - t)^m f^{(m+1)}(t)dt,\ 0 \le m \le n.$$

Example. Consider interpolation by $\tilde{p} \in \mathrm{P}_2$, i.e., $n = 2$, and let $x_0 = -1$, $x_1 = 0$, $x_2 = 1$. Then for $f \in \mathrm{C}_3[-1,+1]$ we have $m = 2$, and the remainder term becomes

$$R_2(f;x) = \frac{1}{2}\sum_{\nu=0}^{2}\ell_{2\nu}(x)\int_{x_\nu}^{x}(x_\nu - t)^2 f'''(t)dt$$

with $\ell_{20}(x) = \frac{1}{2}x(x-1)$, $\ell_{21}(x) = 1 - x^2$, $\ell_{22}(x) = \frac{1}{2}x(x+1)$. Then

$$2R_2(f;x) = \ell_{20}(x)\int_{-1}^{x}(-1-t)^2 f'''(t)dt + \ell_{21}(x)\int_0^x t^2 f'''(t)dt + \\ + \ell_{22}(x)\int_1^x (1-t)^2 f'''(t)dt.$$

Now if $x \le 0$, then

$$2R_2(f;x) = \ell_{20}(x)\int_{-1}^{x}(1+t)^2 f'''(t)dt - \ell_{21}(x)\int_x^0 t^2 f'''(t)dt - \\ - \ell_{22}(x)\int_x^0 (1-t)^2 f'''(t)dt - \ell_{22}(x)\int_0^1 (1-t)^2 f'''(t)dt,$$

and we get

$$R_2(f;x) = \frac{1}{2}\int_{-1}^{+1} K_2(x,t) f'''(t)dt,$$

where

$$K_2(x,t) = \begin{cases} \ell_{20}(x)(1+t)^2 & \text{for } -1 \le t \le x \\ -\ell_{21}(x)t^2 - \ell_{22}(x)(1-t)^2 & \text{for } x \le t \le 0 \\ -\ell_{22}(x)(1-t)^2 & \text{for } 0 \le t \le 1. \end{cases}$$

Similarly, for $x \ge 0$, we get the Peano Kernel

$$K_2(x,t) = \begin{cases} \ell_{20}(x)(1+t)^2 & \text{for } -1 \le t \le 0 \\ \ell_{20}(x)(1+t)^2 + \ell_{21}(x)t^2 & \text{for } 0 \le t \le x \\ -\ell_{22}(x)(1-t)^2 & \text{for } x \le t \le 1. \end{cases}$$

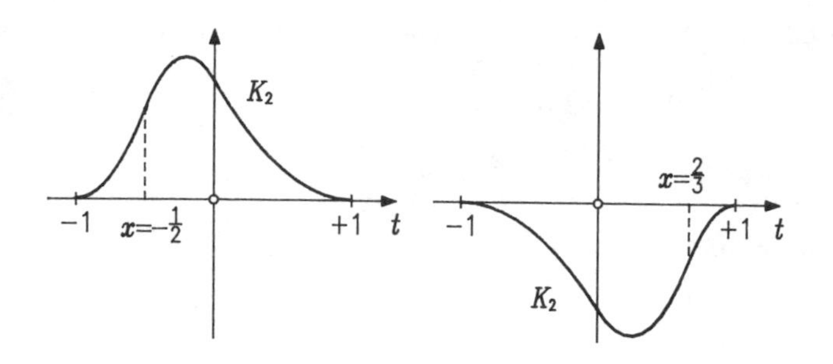

Now let $m = 1$. Then

$$\begin{aligned}R_2(f;x) &= \sum_{\nu=0}^{2} \ell_{2\nu}(x) \int_{x_\nu}^{x} (x_\nu - t) f''(t)dt = \\ &= -\ell_{20}(x) \int_{-1}^{x} (1+t) f''(t)dt - \ell_{21}(x) \int_{0}^{x} t f''(t)dt + \\ &\quad + \ell_{22}(x) \int_{1}^{x} (1-t) f''(t)dt.\end{aligned}$$

Hence

$$R_2(f;x) = \int_{-1}^{+1} K_1(x,t) f''(t)dt$$

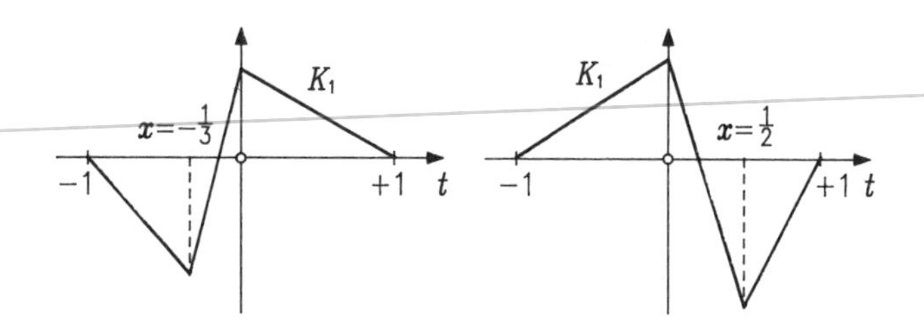

with the Peano Kernel

$$K_1(x,t) = \begin{cases} -\ell_{20}(x)(1+t) & \text{for } -1 \le t \le x \\ \ell_{21}(x)t - \ell_{22}(x)(1-t) & \text{for } x \le t \le 0 \\ -\ell_{22}(x)(1-t) & \text{for } 0 \le t \le 1 \end{cases} \quad \text{if } x \le 0,$$

$$K_1(x,t) = \begin{cases} -\ell_{20}(x)(1+t) & \text{for } -1 \le t \le 0 \\ -\ell_{20}(x)(1+t) - \ell_{21}(x)t & \text{for } 0 \le t \le x \\ -\ell_{22}(x)(1-t) & \text{for } x \le t \le 1 \end{cases} \quad \text{if } x \ge 0.$$

Finally, for $m = 0$ we obtain

$$
\begin{aligned}
R_2(f;x) &= \sum_{\nu=0}^{2} \ell_{2\nu}(x) \int_{x_\nu}^{x} f'(t)dt = \\
&= \ell_{20}(x) \int_{-1}^{x} f'(t)dt + \ell_{21}(x) \int_{0}^{x} f'(t)dt + \ell_{22}(x) \int_{1}^{x} f'(t)dt \\
&= \int_{-1}^{+1} K_0(x,t) f'(t)dt
\end{aligned}
$$

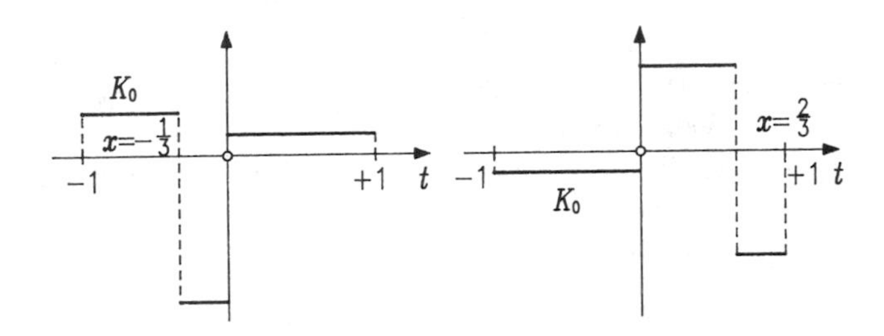

with the piecewise constant Peano Kernel

$$
K_0(x,t) = \begin{cases} \ell_{20}(x) & \text{for } -1 < t \leq x \\ -\ell_{21}(x) - \ell_{22}(x) & \text{for } \quad x < t \leq 0 \\ -\ell_{22}(x) & \text{for } \quad 0 < t \leq 1 \end{cases} \quad \text{if } x \leq 0,
$$

$$
K_0(x,t) = \begin{cases} \ell_{20}(x) & \text{for } -1 < t \leq 0 \\ \ell_{20}(x) + \ell_{21}(x) & \text{for } \quad 0 < t \leq x \\ -\ell_{22}(x) & \text{for } \quad x < t \leq 1 \end{cases} \quad \text{if } x \geq 0.
$$

Additional applications of the Peano Kernel Formula can be found in 3.3 where numerical differentiation is treated, and in Chapter 7 where numerical integration is discussed.

2.5 A Derivative-Free Error Bound. The Error Bound 1.4 requires knowing a bound for $|f^{(n+1)}(x)|$. The need to estimate a higher derivative makes the practical application of this bound difficult. We have already mentioned that the Peano Kernel Formula allows working with lower order derivatives. In this section we mention another approach based on the Cauchy integral formula (cf. e.g. R. Remmert [1990]), which is thus necessarily restricted to functions which are holomorphic in a subset G of the complex plane which completely contains the interval $[a, b]$.

Suppose $x \in [a,b]$, $[a,b] \subset G$, and that Γ is a closed, rectifiable curve with no multiple points which lies entirely inside of G, and which encloses $[a,b]$. Then

$$f^{(m)}(x) = \frac{m!}{2\pi i}\int_\Gamma \frac{f(\zeta)}{(\zeta - x)^{m+1}}d\zeta.$$

Now if we choose Γ to be the circle $|z - \frac{a+b}{2}| = \rho$, $\rho > \frac{b-a}{2}$, then we get the bound

$$|f^{(m)}(x)| \leq \frac{m!}{2\pi}\frac{1}{(\rho - \frac{b-a}{2})^{m+1}} \max_{|z-\frac{a+b}{2}|=\rho} |f(z)| 2\pi\rho,$$

which implies

$$\max_{x\in[a,b]} |f^{(m)}(x)| \leq m! \frac{\rho}{(\rho - \frac{b-a}{2})^{m+1}} \max_{z\in\Gamma} |f(z)|.$$

Despite the fact that this method does not in general give very good bounds, it is nevertheless remarkable for its simplicity. In addition to this Cauchy integral formula approach, there are other more refined ways of getting derivative-free error bounds by working in the complex domain, although they too are restricted to holomorphic functions.

2.6 Connection to Analysis. If f is an $(n+1)$-times continuously differentiable function, then comparing the remainder term in the Newton identity 2.3 with that in 1.3, we see that

$$[x_n \dots x_0 x] = \frac{1}{(n+1)!} f^{(n+1)}(\xi),$$

for all $f \in C_{n+1}[a,b]$.

Combining this result with the observation in Remark 1.4 concerning the weakening of the requirement that f be $(n+1)$-times continuously differentiable, we get the

Extended Mean-Value Theorem of Calculus. *Suppose $f \in C_{m-1}[a,b]$ is m-times differentiable in (a,b). Then for every choice of pairwise distinct numbers $x_0, \dots, x_m \in [a,b]$, there exists a number ξ located in the interval $(\min_\mu x_\mu, \max_\mu x_\mu) \subset (a,b)$ such that*

$$[x_m \dots x_0] = \frac{1}{m!} f^{(m)}(\xi).$$

The Newton approach to interpolation can also be used to provide a natural derivation of the Taylor formula. In particular, assuming that f satisfies the differentiability hypotheses of the extended mean-value theorem, if we let all of the interpolation points x_ν in the Newton identity converge to $x_0 \in (a,b)$, then we get the *Taylor Formula*

$$\begin{aligned} f(x) =& f(x_0) + f'(x_0)(x - x_0) + \cdots + \frac{f^{(n)}(x_0)}{n!}(x - x_0)^n + \\ &+ \frac{f^{(n+1)}(\xi)}{(n+1)!}(x - x_0)^{n+1}, \quad \xi \in (a,b). \end{aligned}$$

Thus, the Taylor formula with the Lagrange remainder term arises as the limiting case of an interpolation process. Instead of $(n+1)$ distinct interpolation points, we now have one $(n+1)$-fold interpolation point at x_0. In this case the interpolating polynomial becomes a polynomial which interpolates f and all of its derivatives up to the n-th at x_0. B. Taylor (1685–1731) himself followed this path in his "Methodus incrementorum".

The Taylor polynomial is an approximation of f which, in general, is very good in the neighborhood of the initial point x_0. It is well known that the Taylor polynomial is of fundamental importance in analysis. Its practical applicability is, however, restricted by the fact that derivatives must be computed, and even worse, often we don't even have an explicit formula for the function f.

In contrast to the Taylor polynomial, the interpolating polynomial matches only function values. But since this happens at $(n+1)$ distinct points, the interpolating polynomial can provide a usable approximation of f on larger intervals. A word of caution is in order, however. Interpolating polynomials of higher degree can strongly oscillate, so that the quality of the approximation does not always improve with increasing degrees. We will discuss this point in detail in Section 4.

As a rule, it is recommended to use interpolating polynomials only of low degree, say up to the third or fourth degree, and if necessary to use several of them pieced together. This idea is the basis of the theory of *splines*, which will be the objects of interest in Chapter 6.

We can also establish the following

Leibniz Rule for Derivatives of a Product $f(x) = u(x)v(x)$:

$$f^{(k)}(x) = \sum_{\kappa=0}^{k} \binom{k}{\kappa} u^{(\kappa)}(x) v^{(k-\kappa)}(x)$$

by passing to the limit in the Leibniz Rule 2.3 for divided differences. Indeed, letting $x_\imath \to x_j$ for $\imath = j+1, \cdots, j+k$, we get

$$\frac{1}{k!} f^{(k)}(x_j) = \sum_{\imath=j}^{j+k} \frac{u^{(\imath-j)}(x_j)}{(\imath-j)!} \frac{v^{(j+k-\imath)}(x_j)}{(j+k-\imath)!},$$

i.e.,

$$f^{(k)}(x) = \sum_{\kappa=0}^{k} k! \frac{u^{(\kappa)}(x)}{\kappa!} \frac{v^{(k-\kappa)}(x)}{(k-\kappa)!},$$

which leads to the the Leibniz Rule for derivatives, since

$$\frac{k!}{\kappa!(k-\kappa)!} = \binom{k}{\kappa}.$$

2.7 Problems. 1) Let $I_n f := \tilde{p} \in \mathrm{P}_n$ be the polynomial interpolating f at the interpolation points $x_0, \ldots, x_n \in [a,b]$, and consider the operator

$$I_n : (\mathrm{C}[a,b], \|\cdot\|_\infty) \to (\mathrm{P}_n, \|\cdot\|_\infty).$$

Show that

a) I_n is linear and bounded.

b) $\sup\{\|I_n f\|_\infty \mid \|f\|_\infty = 1\} = \|\sum_{\nu=0}^n |\ell_{n\nu}|\,\|_\infty$.

2) Determine the interpolating polynomial of degree 2 in both the Lagrange and Newton forms for the functions $f(x) := \frac{2}{1+x^2}$ and $f(x) := \cos(\pi x)$ using the interpolation points $x_0 = -1$, $x_1 = 0$, $x_2 = 1$. Repeat the problem using polynomials of degree 3 and the additional interpolation point $x_3 = \frac{1}{2}$.

3) Establish the symmetry property of the divided difference 2.3 by using induction to show that $[x_m \ldots x_0] = \sum_{\mu=0}^m \frac{y_\mu}{\Phi'(x_\mu)}$.

4) Show that for distinct points $x_0, \ldots, x_n$, the monomials $g_k(x) := x^k$ satisfy

$$[x_0 \ldots x_n] g_k = \sum_{\nu=0}^n \frac{x_\nu^k}{\Phi'(x_\nu)} = \begin{cases} 0 & \text{for } 0 \le k \le n-1 \\ 1 & \text{for } k = n \\ x_0 + \cdots + x_n & \text{for } k = n+1. \end{cases}$$

5) Using the Peano kernel K_2 in Example 2.4, show that the interpolation remainder term satisfies $|R_2(f;x)| \le \sigma(x) \max_{x\in[-1,+1]} |f'''(x)|$, where σ is an appropriate function. Compare this result with the Error Bound 1.4.

6) Compute the Peano representation of the remainders for linear interpolation at $x_0 = a$, $x_1 = b$ under the hypotheses that $f \in \mathrm{C}_2[a,b]$ and that $f \in \mathrm{C}_1[a,b]$, respectively. Show that $|R_1(f;x)| \le \max_{x\in[a,b]} |f'(x)|(b-a)$.

7) Interpolate the function $f \in \mathrm{C}[-1,+1]$ at the interpolation points $x_0 = -1$, $x_1 = 0$, $x_2 = 1$, where $f(x) := \begin{cases} 0 \text{ for } & -1 \le x \le 0 \\ x^2 \text{ for } & 0 \le x \le 1 \end{cases}$, and find a formula for the error using the Peano kernel K_0 of Example 2.4. How large is the maximal deviation $\|f - \tilde{p}\|_\infty$, and where is it assumed?

3. Equidistant Interpolation Points

The Newton interpolation formula can be further simplified if we take equidistant interpolation points. We choose the indexing of these points so that $x_\nu = x_0 + \nu h$, $0 \le \nu \le n$, for some fixed *step size* h, and introduce the corresponding **m -th Forward Difference**

$$\Delta^0 y_\nu := y_\nu$$
$$\Delta^m y_\nu := \Delta^{m-1} y_{\nu+1} - \Delta^{m-1} y_\nu \quad \text{for } m \ge 1.$$

The divided differences introduced in 2.3 then become

$$
\begin{aligned}
[x_1 x_0] &= \frac{y_1 - y_0}{x_1 - x_0} = \frac{\Delta y_0}{h} \quad (\Delta y_0 := \Delta^1 y_0), \\
[x_2 x_1 x_0] &= \frac{\frac{\Delta y_1}{h} - \frac{\Delta y_0}{h}}{2h} = \frac{1}{2}\frac{\Delta^2 y_0}{h^2}, \\
&\vdots \\
[x_m \ldots x_0] &= \ldots = \frac{1}{m!}\frac{\Delta^m y_0}{h^m}.
\end{aligned}
$$

In this case the Extended Mean-Value Theorem 2.6 becomes simply

$$\frac{\Delta^m f(x_0)}{h^m} = f^{(m)}(\xi), \quad \xi \in (x_0, x_m)$$

for all $f \in C_m[x_0, x_m]$.

Now the interpolating polynomial of Newton takes the form

$$\tilde{p}(x) = y_0 + \frac{\Delta y_0}{h}(x - x_0) + \cdots + \frac{\Delta^n y_0}{n!h^n}(x - x_0)\cdots(x - x_{n-1}),$$

and for $f \in C_{n+1}[a, b]$, the remainder term r in the Newton identity $f = \tilde{p} + r$ for $x, x_\nu \in [a, b]$ becomes

$$r(x) = \frac{f^{(n+1)}(\xi)}{(n+1)!}(x - x_0)\cdots(x - x_n), \ \xi \in (\min(x, x_0), \max(x, x_n)).$$

3.1 The Difference Table. The coefficients of the interpolating polynomial can now be computed by setting up the following simple *Difference Table:*

$$
\begin{array}{cccccc}
x_0 & y_0 & & & & \\
 & & \Delta y_0 & & & \\
x_1 & y_1 & & \Delta^2 y_0 & & \\
 & & \Delta y_1 & & \cdot & \\
x_2 & y_2 & & \Delta^2 y_1 & & \cdot \\
 & & \Delta y_2 & & \cdot & \\
x_3 & y_3 & & \Delta^2 y_2 & & \cdot \\
 & & \cdot & & \cdot & \\
\cdot & \cdot & & \cdot & & \cdot
\end{array}
$$

We have chosen the indices on the forward differences appearing in this table so that differences with the same indices are located on downward sloping diagonal lines, which accounts for the fact that these differences are sometimes referred to as *descending* differences. The scheme can be extended arbitrarily far by choosing additional interpolation points $x_{n+1}, x_{n+2}, \cdots$.

The numbering of the forward differences has been chosen so that the interpolating polynomial is built up using the leftmost points at each stage.

Depending on the particular interpolation problem at hand, this may or may not be the best way to proceed. Clearly, this approach is appropriate if we are constructing polynomials approximating the solution of an initial-value problem in ordinary differential equations, stepping forward in the x direction from the initial value $y(x_0)$. On the other hand, if we are trying to solve a boundary-value problem numerically using polynomial approximations built up stepwise starting at the right boundary, then clearly it would be more convenient to represent the polynomial in a form which starts on the right and works leftward. As a step towards this goal, we now consider some alternative forms of differences.

3.2 Representations of Interpolating Polynomials. We can derive an especially simple formula for the interpolating polynomial using descending differences if we introduce the

$$\text{Transformation } t: \ [x_0, x_n] \to [0, n], \quad t(x) := \frac{x - x_0}{h}.$$
$$\text{Notation } p^*(t) := \tilde{p}(x(t)), \ r^*(t) := r(x(t)), \ f^*(t) := f(x(t)).$$

Then we have

The Gregory-Newton Interpolation Formula I.

$$p^*(t) = y_0 + \Delta y_0 \binom{t}{1} + \Delta^2 y_0 \binom{t}{2} + \cdots + \Delta^n y_0 \binom{t}{n}$$

with remainder

$$r^*(t) = \binom{t}{n+1} \frac{d^{n+1} f^*(t)}{dt^{n+1}}\Big|_{t=\tau}, \quad \tau \in (\min(t, 0), \max(t, n)).$$

On the other hand, if we want the interpolating polynomial to be expanded using the rightmost points at each step, then we may take $x_{n-\nu} = x_n - \nu h$, $0 \le \nu \le n$, and introduce the

Backward Differences

$$\nabla^0 y_{n-\nu} := y_{n-\nu}$$
$$\nabla^m y_{n-\nu} := \nabla^{m-1} y_{n-\nu} - \nabla^{m-1} y_{n-\nu-1} \ \text{ for } \ m \ge 1,$$

which leads to the difference table

$$\begin{array}{ccccc}
\cdot & \cdot & & \cdot & \cdot \\
 & & \cdot & & \cdot \\
x_{n-2} & y_{n-2} & & \nabla^2 y_{n-1} & \\
 & & \nabla y_{n-1} & & \cdot \\
x_{n-1} & y_{n-1} & & \nabla^2 y_n & \\
 & & \nabla y_n & & \\
x_n & y_n & & &
\end{array}$$

which can be extended arbitrarily far upwards. In this case one speaks of *ascending* differences.

The transformation $t : [x_0, x_n] \to [-n, 0]$, $t(x) := \frac{x - x_n}{h}$ leads to the

Gregory-Newton Interpolation Formula II

$$p^*(t) = y_n + \nabla y_n \binom{t}{1} + \nabla^2 y_n \binom{t+1}{2} + \cdots + \nabla^n y_n \binom{t+n-1}{n}$$

with remainder

$$r^*(t) = \binom{t+n}{n+1} \frac{d^{n+1} f^*(t)}{dt^{n+1}}\Big|_{t=\tau}, \quad \tau \in (\min(t, -n), \max(t, 0)).$$

It can also make sense to develop the interpolating polynomial using the interpolation point in the center. The other interpolation points will then enter successively in a symmetrical way. In this case we write $x_\nu = x_0 + \nu h$, $(\nu = 0, \pm 1, \ldots, \pm k)$, and introduce the

Central Differences

$$\begin{aligned} \delta^0 y_\nu &:= y_\nu \\ \delta^m y_{\nu+\frac{1}{2}} &:= \delta^{m-1} y_{\nu+1} - \delta^{m-1} y_\nu \quad \text{for } m \geq 1 \text{ and odd}, \\ \delta^m y_\nu &:= \delta^{m-1} y_{\nu+\frac{1}{2}} - \delta^{m-1} y_{\nu-\frac{1}{2}} \quad \text{for } m \geq 2 \text{ and even}. \end{aligned}$$

The corresponding difference table has the form

$\cdot$	$\cdot$		$\cdot$	$\cdot$
		$\cdot$		$\cdot$
x_{-1}	y_{-1}		$\delta^2 y_{-1}$	$\cdot$
		$\delta y_{-1/2}$		$\cdot$
x_0	y_0		$\delta^2 y_0$	$\cdot$
		$\delta y_{1/2}$		$\cdot$
x_1	y_1		$\delta^2 y_1$	$\cdot$
		$\cdot$		$\cdot$
$\cdot$	$\cdot$		$\cdot$	$\cdot$

Using the mean differences $\overline{\delta}^m y_0 := \frac{1}{2}(\delta^m y_{1/2} + \delta^m y_{-1/2})$ for $m \geq 1$ and odd, and the transformation

$$t : [x_{-k}, x_{+k}] \to [-k, +k], \quad t(x) := \frac{x - x_0}{h},$$

lead to the

Stirling Interpolation Formula

$$\begin{aligned} p^*(t) = y_0 + \overline{\delta} y_0 t + \delta^2 y_0 \frac{t^2}{2!} + \overline{\delta}^3 y_0 \frac{t(t^2-1)}{3!} + \cdots \\ + \delta^{2k} y_0 \frac{t^2(t^2-1)\cdots(t^2-(k-1)^2)}{(2k)!} \end{aligned}$$

with remainder term

$$r^*(t) = \binom{t+k}{2k+1} \frac{d^{2k+1} f^*(t)}{dt^{n+1}}\Big|_{t=\tau}, \quad \tau \in (\min(t,-k), \max(t,k)).$$

We note that the same numerical values always appear in the various difference tables. The only difference is the notation, which in each case is adjusted to the way in which the interpolating polynomial is to be built up. Thus the various formulae for the interpolating polynomial differ only formally. In all cases we are led to one and the same polynomial which passes through the prescribed data values.

JAMES GREGORY (1638–1675), like Newton, was interested in the interpolation problem in connection with approximate integration. JAMES STIRLING (1692–1770) was interested in finding ways to streamline the computations needed to construct the Newton interpolating polynomial.

3.3 Numerical Differentiation. In order to compute approximations to the derivatives of a function $f \in C_j[a,b]$, $j \geq 1$, we start with an interpolating polynomial. We take the Stirling Interpolation Formula 3.2 which expands the polynomial around a data point x_ν in the middle of the interval. Setting

$$\tilde{p}' = \frac{dp^*}{dt}\frac{dt}{dx} = \frac{1}{h}\frac{dp^*}{dt}, \quad t(x) = \frac{x - x_\nu}{h},$$

we find that the first derivative of the interpolating polynomial is

$$h\tilde{p}'(x) = \overline{\delta} y_\nu + t\delta^2 y_\nu + \frac{3t^2 - 1}{3!}\overline{\delta}^3 y_\nu + \cdots.$$

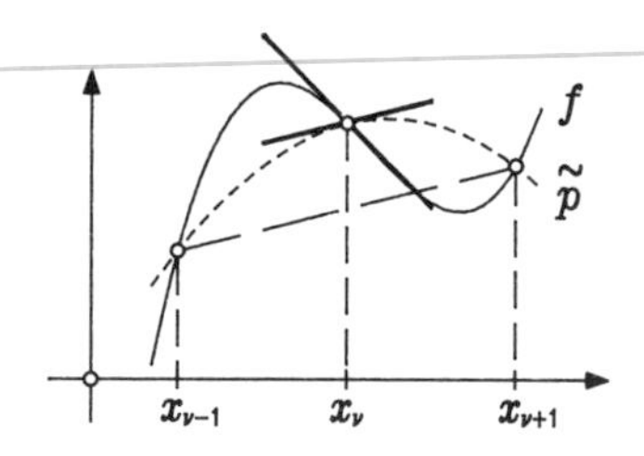

For example, using the interpolating polynomial $\tilde{p} \in P_2$, it follows that $h\tilde{p}'(x) = \overline{\delta} y_\nu + t\delta^2 y_\nu$ at x_ν, and we have the

1st Approximation of the First Derivative

$$(*) \qquad \tilde{p}'(x_\nu) = \frac{1}{2h}(y_{\nu+1} - y_{\nu-1}), \; 1 \leq \nu \leq n-1.$$

Using $\tilde{p} \in P_4$ gives $h\tilde{p}'(x_\nu) = \overline{\delta} y_\nu - \frac{1}{3!}\overline{\delta}^3 y_\nu$, and we have the

2nd Approximation of the First Derivative

$$\tilde{p}'(x_\nu) = \frac{1}{12h}(-y_{\nu+2} + 8y_{\nu+1} - 8y_{\nu-1} + y_{\nu-2}),\ 2 \le \nu \le n-2.$$

We proceed similarly for the second derivative. With $\tilde{p}''(x) = \frac{1}{h^2}\frac{d^2p^*}{dt^2}$ and $\tilde{p} \in \mathrm{P}_2$, we get the

1st Approximation of the Second Derivative

$$\tilde{p}''(x_\nu) = \frac{1}{h^2}(y_{\nu+1} - 2y_\nu + y_{\nu-1}),\ 1 \le \nu \le n-1,$$

while $\tilde{p} \in \mathrm{P}_4$ leads to the

2nd Approximation of the Second Derivative

$$\tilde{p}''(x_\nu) = \frac{1}{12h^2}(-y_{\nu+2} + 16y_{\nu+1} - 30y_\nu + 16y_{\nu-1} - y_{\nu-2}),\ 2 \le \nu \le n-2.$$

It is clear how to get formulae for the higher derivatives.

Error Bound for $(*)$**.** In order to estimate the error in this approximation, we use the Peano Kernel Theorem 2.4. Let

$$Lf := R_n(f; x_\nu) = f'(x_\nu) - \tilde{p}'(x_\nu).$$

Assuming that $f \in C_3[a,b]$, $(m = 2)$, then the 1st Approximation of the First Derivative with $n = 2$ satisfies

$$R_2(f; x_\nu) = \int_a^b K_2(t) f'''(t)\,dt,$$

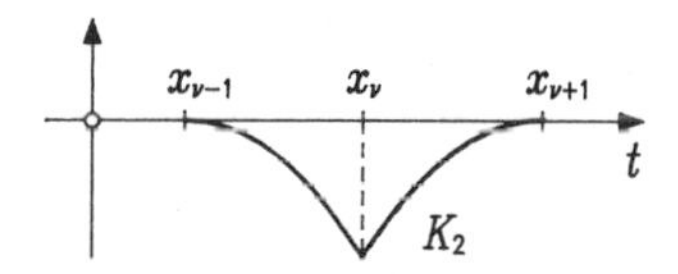

with

$$K_2(t) = \frac{1}{2}(R_2(q_2; x_\nu))(t) = (x_\nu - t)_+ - \frac{1}{4h}[(x_{\nu+1} - t)_+^2 - (x_{\nu-1} - t)_+^2];$$

i.e.,

$$K_2(t) = \begin{cases} (x_\nu - t) - \frac{1}{4h}(x_{\nu+1} - t)^2 & \text{for } x_{\nu-1} \le t \le x_\nu \\ -\frac{1}{4h}(x_{\nu+1} - t)^2 & \text{for } x_\nu \le t \le x_{\nu+1}. \end{cases}$$

Since K_2 is of one sign,

$$R_2(f;x_\nu) = f'''(\xi)\int_{x_{\nu-1}}^{x_{\nu+1}} K_2(t)dt, \; x_{\nu-1} < \xi < x_{\nu+1},$$

and we are led to the

Error Expression

$$\boxed{R_2(f;x_\nu) = -\frac{h^2}{6}f'''(\xi).}$$

On the other hand, if we assume only that $f \in C_2[a,b]$, $(m = 1)$, then the Peano kernel becomes

$$K_1(t) = (R_2(q_1;x_\nu))(t) = \begin{cases} 1 - \frac{1}{2h}(x_{\nu+1} - t) & \text{for} \quad x_{\nu-1} < t \leq x_\nu \\ -\frac{1}{2h}(x_{\nu+1} - t) & \text{for} \quad x_\nu < t \leq x_{\nu+1}, \end{cases}$$

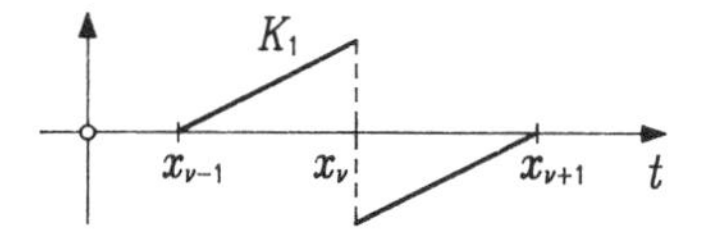

which gives us the weaker

Estimate

$$|R_2(f;x_\nu)| \leq \frac{h}{2}\max_x |f''(x)|, \quad x \in [x_{\nu-1}, x_{\nu+1}].$$

The general Error Bound for Derivatives presented in 1.4 can also be applied to estimate the accuracy of $(*)$. It gives the bound $|R_2(f;x)| \leq$ $\leq h^2 \max_{x\in[x_{\nu-1},x_{\nu+1}]} |f'''(x)|$, which is somewhat worse than the Error Expression given above for R_2, but has the advantage that it holds for all points x in the interval, rather than just for the interpolation points themselves. It also gives the correct order with respect to the step size h.

As one might expect, the order of approximation of the first derivative is one less than the order of approximation of the interpolation formula itself. This phenomenon is called the *roughening* effect of numerical differentiation.

The Peano Kernel Theorem can be used to obtain similar error formulae and error bounds for the other numerical differentiation formulae presented above.

One-Sided Derivatives. So far we have used the Stirling interpolation polynomial to derive our numerical differentiation formulae. These formulae make use of data points on both sides of x_ν, and hence cannot be used to compute derivatives at the end of the interval $[x_0, x_n]$. To this end, we can use either the interpolation formulae Gregory-Newton I or Gregory-Newton II given in 3.2. For example, consider

$$\tilde{p}'(x_0) = \frac{\Delta y_0}{h} + \frac{\Delta^2 y_0}{2h^2}(x_0 - x_1) + \cdots + \frac{\Delta^n y_0}{n!h^n}(x_0 - x_1)\cdots(x_0 - x_{n-1}),$$

with $x_\nu = x_0 + \nu h$ for $1 \le \nu \le n$. Then using $\tilde{p} \in \mathrm{P}_1$, $n = 1$, we get the

1st Right-Sided Approximation of the First Derivative

$$\tilde{p}'(x_0) = \frac{\Delta y_0}{h} = \frac{1}{h}(y_1 - y_0).$$

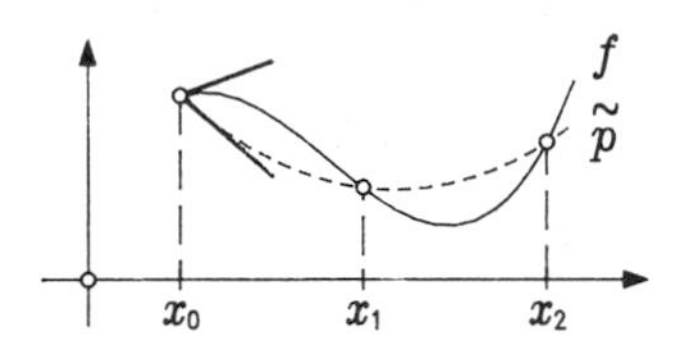

Using $\tilde{p} \in \mathrm{P}_2$, $n = 2$, leads to the

2nd Right-Sided Approximation of the First Derivative

$$\tilde{p}'(x_0) = \frac{1}{2h}(-y_2 + 4y_1 - 3y_0).$$

For this approximation the error is $R_2(f; x_0) = \int_{x_0}^{x_2} K_2(t) f'''(t)dt$ for $f \in C_3[a, b]$, $m = 2$, with the unsymmetric Peano Kernel

$$K_2(t) = \begin{cases} \frac{1}{4h}[(x_2 - t)^2 - 4(x_1 - t)^2] & \text{for } x_0 \le t \le x_1 \\ \frac{1}{4h}(x_2 - t)^2 & \text{for } x_1 < t \le x_2, \end{cases}$$

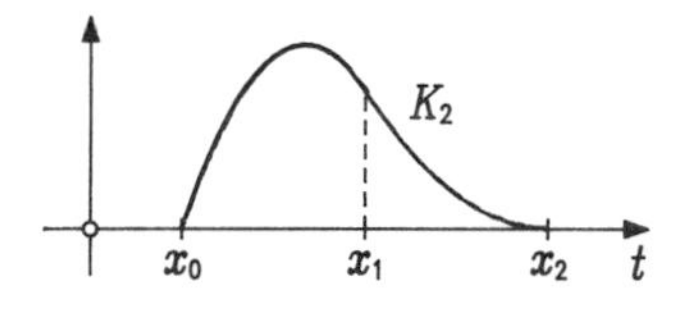

and leads to the error bound

$$R_2(f; x_0) = \frac{h^2}{3} f'''(\xi_0), \quad x_0 < \xi_0 < x_2.$$

Because it is one-sided, this formula is slightly less accurate then the approximation (*), but the order $O(h^2)$ remains the same.

The development of similar left-sided formulae, formulae for higher derivatives, and formulae based on other data points is left to the reader.

3.4 Problems. 1) Interpolate the function f of Problem 7 in 2.7 at the points $x_0 = -1$, $x_1 = -\frac{1}{3}$, $x_2 = \frac{1}{3}$, $x_3 = 1$. Find the remainder term $f - \tilde{p}$, and draw a sketch of it.

2) Show:

a) The operator Δ^n annihilates all functions $f \in \mathrm{P}_{n-1}$.

b) For given numbers $y_0, \ldots, y_n$,

$$\Delta^n y_0 = \nabla^n y_n = \sum_{\nu=0}^{n} (-1)^{n-\nu} \binom{n}{\nu} y_\nu .$$

3) Interpolation of a function $f \in \mathrm{C}_{n+1}[a,b]$ by $\tilde{p} \in \mathrm{P}_n$ leads to the remainder term $R_n(f;x) = \frac{f^{(n+1)}(\xi(x))}{(n+1)!} \Phi(x)$.

a) Use this to show that the remainder term for the derivative $f' - \tilde{p}'$ at a data point x_ν can be written as

$$R_n'(f;x_\nu) = \frac{f^{(n+1)}(\xi_\nu)}{(n+1)!} \Phi'(x_\nu), \quad \xi_\nu \in (\min_{0 \le \nu \le n} x_\nu, \max_{0 \le \nu \le n} x_\nu).$$

Hint: It suffices to use the fact that $f^{(n+1)}(\xi(x))$ can be extended to be a continuous function as was done in 1.3.

b) Apply this representation to obtain the errors of the formulae 3.3 for the first derivatives.

4) Find a formula for the error of the 1st Approximation of the Second Derivative 3.3, starting with the Taylor expansion of f.

5) The operator Δ^3 annihilates all elements $f \in \mathrm{P}_2$. Find the Peano Kernel K_2 for the formula $\Delta^3 y_0 = \int_{x_0}^{x_3} K_2(t) f'''(t)dt$ for $f \in \mathrm{C}_3[x_0, x_3]$, and use it to establish the Extended Mean-value Theorem 2.6.

6) Find an approximate value for $f'(\frac{1}{4})$, $f \in \mathrm{C}_3[-1,+1]$, by computing the derivative of the polynomial which interpolates at the points $x_0 = -1$, $x_1 = 0$, $x_2 = 1$. Find a formula for the error using the Peano Kernel.

4. Convergence of Interpolating Polynomials

Interpolation using polynomials would appear to be a natural way of approximating a function on the basis of a finite number of function values. Indeed, it seems reasonable to expect that every continuous function can be approximated arbitrarily well with respect to the Chebyshev norm by interpolating polynomials provided that the number of data points is sufficiently large. In this section we shall see that, although the Weierstrass approximation theorem asserts that there always exist arbitrarily exact polynomial approximations, not every interpolation process can be used to find them.

The question of when interpolating polynomials converge is much more difficult then one might at first glance expect. We shall see that there are a whole range of possibilities from uniform convergence to divergence at every point, and that in order to guarantee convergence, we will have to make a careful choice of the location of the interpolation points, as well as some rather strong assumptions on the analytic properties of the function to be approximated.

We begin by examining the question of how to choose the interpolation points to make the interpolation error as small as possible.

4.1 Best Interpolation. Let $f \in C_{n+1}[a, b]$. Consider the problem of making the Interpolation Error 1.3

$$r(x) = \frac{f^{(n+1)}(\xi(x))}{(n+1)!}\,\Phi(x)$$

as small as possible. To achieve this, we would have to have an explicit expression for the derivative $f^{(n+1)}$, and, moreover, would have to know how the point ξ depends on x. Since, in general, we do not have this information, we now consider the related simpler problem of making the Error Bound 1.4 for the interpolation error

$$\|r\| \le M_{n+1} \|\,\Phi\,\|$$

as small as possible; i.e., we seek to choose the interpolation points $x_0, \ldots, x_n$ so as to find the minimum of the norm $\|\,\Phi\,\|$ of the data point polynomial $\Phi(x) = (x - x_0) \cdots (x - x_n)$. The position of the optimal points depends, of course, on the norm being used.

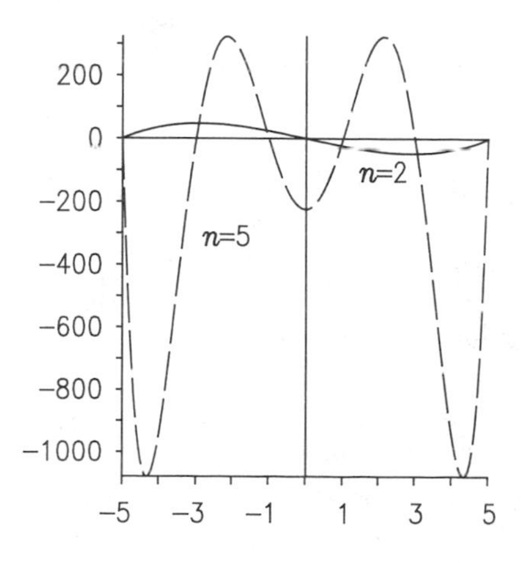

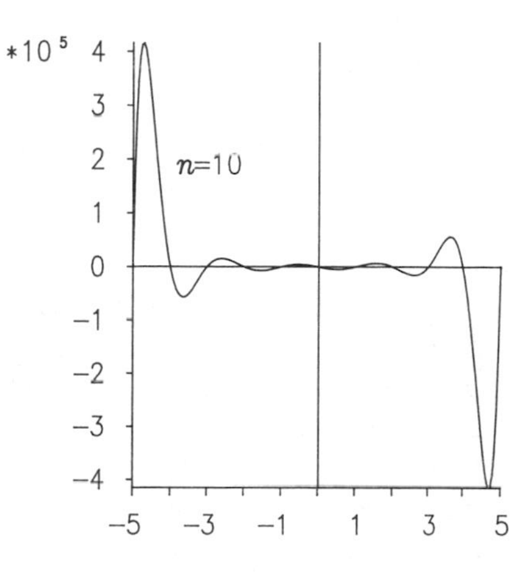

The sketch above shows the behavior of the polynomial Φ on the interval $[-5,+5]$ for equally spaced interpolation points $x_0, \ldots, x_n$ for $n = 2$, $n = 5$ and $n = 10$. The extremely rapid growth of $\| \Phi \|$ with increasing n (note the scale on the vertical axis in the case $n = 10$), shows that there is hope that $\| \Phi \|$ can be greatly reduced.

We assume now that $[a,b] := [-1,+1]$, and consider the uniform norm. It was shown in 4.4.7 that among all p in the space $\hat{\mathrm{P}}_{n+1}$ of polynomials with leading coefficient one, the Chebyshev polynomial of the first kind $\hat{T}_{n+1}$ is extremal in the sense that $\|\hat{T}_{n+1}\|_\infty \leq \|p\|_\infty$. It follows that in order to minimize $\| \Phi \|_\infty$, we need only choose the interpolation points $x_0, \ldots, x_n$ to be the zeros of the Chebyshev polynomial $\hat{T}_{n+1}$. In contrast with the equally spaced points shown in the sketch, the zeros of $\hat{T}_{n+1}$ tend to cluster near the ends of the interval, and thereby tend to reduce the large values of $\Phi(x)$ there.

Now suppose that $\| \cdot \| := \| \cdot \|_2$. The minimal property 4.5.4 of the Legendre polynomials asserts that $\|\hat{L}_{n+1}\|_2 \leq \|p\|_2$ for all $p \in \hat{\mathrm{P}}_{n+1}$. It follows that $\| \Phi \|_2$ is minimized if we choose the interpolation points to be the zeros of the Legendre polynomial. As in the uniform norm case above, these zeros also tend to be closer together near the ends of the interval than in the center. The zeros of the first few Legendre polynomials are tabulated in 7.3.6.

4.2 Convergence Problems. The study of the convergence properties of interpolating polynomials gave rise to an extensive series of research papers. In this section we discuss several examples to show the range of behavior which can occur.

Given a continuous function $f \in \mathrm{C}[a,b]$, it is natural to conjecture that the sequence of interpolating polynomials corresponding to equally spaced interpolation points converges to f as the number of points increases. S. N. Bernstein [1912] (cf. also I. P. Natanson [1965], Vol. III, p. 30) gave the following counterexample showing that this conjecture is false: The sequence of polynomials interpolating the function $f(x) = |x|$ at equally spaced points in $[-1,+1]$ diverges for all $0 < |x| < 1$. In this connection, we note that we always have convergence for $x = \pm 1$, since we are assuming that the end points of the interval are included in every equally spaced set of interpolation points. Moreover, it is obvious that there are always subsequences which also converge at isolated points. For example, $x = 0$ is an interpolation point whenever the number of interpolation points is odd, and thus the corresponding subsequence of interpolating polynomials converges there. The convergence of the complete sequence at $x = 0$ is, however, nontrivial. This function is by no means pathological; in fact it is everywhere differentiable except at $x = 0$.

We turn now to analytic functions. C. Runge [1901] investigated the function $f(x) = \frac{1}{1+x^2}$ on $[-5,+5]$. He showed that there exists a constant

$c \doteq 3.63$ such that the sequence of polynomials interpolating f at equidistant points converges only for $|x| \leq c$, and is divergent for all other x. This behavior can be explained by the fact that although $f(z)$ is an analytic function in a domain including $[-5, +5]$, it has singularities at $z_{1,2} = \pm i$.

The following example illustrates still another convergence behavior. Consider the continuous function $f : [0, 1] \to \mathbb{R}$ defined by $f(x) := x \sin(\frac{\pi}{x})$ for $x \in (0, 1]$ with $f(0) := 0$. Let $\tilde{p}_n \in \mathrm{P}_n$ be the polynomials interpolating f at $x_{n\nu} := \frac{1}{\nu+1}$ for $0 \leq \nu \leq n$. Since $f(x_{n\nu}) = 0$ for $\nu = 0, \ldots, n$ holds, it follows that all of the interpolating polynomials are the same constant function $\tilde{p}_n(x) \equiv 0$. Then trivially $\lim_{n\to\infty} \tilde{p}_n = 0$. In this case the sequence of interpolating polynomials converges uniformly, but if x is not one of the interpolation points, then the limit is not equal to $f(x)$.

4.3 Convergence Results. Suppose we are given a sequence of pairwise distinct interpolation points $x_{n0}, \ldots, x_{nn}$, and let $\tilde{p}_n \in \mathrm{P}_n$ be the corresponding interpolating polynomials satisfying $\tilde{p}_n(x_{n\nu}) = f(x_{n\nu})$ for $\nu = 0, \ldots, n$. We arrange the points in a triangular array S as follows:

$$\mathrm{S}: \quad \begin{array}{cccc} x_{00} & & & \\ x_{10} & x_{11} & & \\ \vdots & \vdots & \ddots & \\ x_{n0} & x_{n1} & \cdots & x_{nn} \\ \vdots & & & \vdots \end{array} .$$

In view of the examples given in 4.2, we expect to have to make a strong assumption on the properties of f in order to be able to establish convergence. The Runge Example in 4.2 suggests that the behavior of the extension of the real-valued function f to the complex plane $\mathbb{C}$ influences the convergence of the sequence of interpolating polynomials. Starting with $f : [a, b] \to \mathbb{R}$, we now assume that its holomorphic extension to the complex plane is an entire function. This means that the power series expansion of $f(z) = \sum_0^\infty a_j z^j$ converges in the entire complex plane. The assumption $f(x) \in \mathbb{R}$ for $x \in [a, b] \subset \mathbb{R}$ assures that all coefficients a_j are real.

We now have the following

Convergence Theorem. *Let f be an entire function which is real-valued for real variables, and let* S *be an arbitrary system of interpolating points $x_{n\nu} \in [a, b]$ for $n = 0, 1, \ldots$ and $0 \leq \nu \leq n$. Then the sequence $(\tilde{p}_n)_{n\in\mathbb{N}}$ of corresponding interpolating polynomials converges uniformly to f.*

Proof. To establish the convergence, we consider the Remainder Term 1.3

$$r_n(x) = \frac{f^{(n+1)}(\xi)}{(n+1)!} \Phi(x), \ \Phi(x) = (x - x_{n0}) \cdots (x - x_{nn}),$$

and apply the Cauchy integral formula.

Let $x \in [a,b]$, and let Γ_x be a circle around x of radius $\rho = 2(b-a)$. Let $M(x) := \max_{z\in\Gamma_x} |f(z)|$ and $M := \sup_{x\in[a,b]} M(x) < \infty$. Then from the Cauchy integral formula

$$\frac{f^{(k)}(x)}{k!} = \frac{1}{2\pi i}\int_{\Gamma_x} \frac{f(\zeta)}{(\zeta - x)^{k+1}} d\zeta,$$

we get the Cauchy estimate

$$|\frac{f^{(k)}(x)}{k!}| \leq \frac{1}{2\pi} M(x)\, 2\pi\rho \frac{1}{\rho^{k+1}} = \frac{M(x)}{\rho^k}.$$

This implies that uniformly for all $x \in [a,b]$,

$$|\frac{f^{(n+1)}(x)}{(n+1)!}| \leq \frac{M}{2^{n+1}(b-a)^{n+1}}.$$

Using $\| \Phi \|_\infty \leq (b-a)^{n+1}$, we see that

$$\|r_n\|_\infty \leq \frac{M}{2^{n+1}},$$

and hence that $\lim_{n\to\infty} \|r_n\|_\infty = 0$. □

Example. Let $f(x) := e^x$ in $x \in [0,1]$. Here $f^{(n+1)}(x) = e^x$ which leads to the estimate

$$\left|\frac{f^{(n+1)}(\xi)}{(n+1)!}\right| < \frac{e}{(n+1)!},$$

and hence

$$\|r_n\|_\infty < \frac{e}{(n+1)!} \| \Phi \|_\infty \leq \frac{e}{(n+1)!} \to 0 \quad \text{for } n \to \infty.$$

The following theorem gives a different kind of result on the convergence of interpolating polynomials under the weaker hypothesis that f is only a continuous function.

Theorem of Marcinkiewicz. *Given any function $f \in \mathrm{C}[a,b]$, then there always exists a triangular array* S *of interpolation points $x_{n\nu} \in [a,b]$ for $n = 0,1,\ldots$ and $0 \leq \nu \leq n$ such that the corresponding sequence $(\tilde{p}_n)_{n\in\mathbb{N}}$ of interpolating polynomials converges uniformly to f.*

Proof. By the Alternation Theorem 4.4.3, it follows that the uniquely defined best approximations $\tilde{p}_n \in \mathrm{P}_n$ with respect to the norm $\|\cdot\|_\infty$ always interpolate f on at least $(n+1)$ points $a < \xi_{n0} < \cdots < \xi_{nn} < b$; i.e., $\tilde{p}_n(\xi_{n\nu}) = f(\xi_{n\nu})$ for $0 \leq \nu \leq n$. Now the Convergence Theorem 4.4.9 asserts that this sequence of polynomials converges uniformly to f. We have shown that the triangular array of interpolation points $\xi_{n\nu}$ for $n = 0, 1, \ldots$ and $0 \le \nu \le n$ satisfy the assertion of the theorem. □

In contrast to the Theorem of Marcinkiewicz, we have the

Theorem of Faber. *Suppose* S *is a given triangular array of interpolation points* $x_{n\nu} \in [a, b]$ *for* $n = 0, 1, \ldots$ *and* $0 \le \nu \le n$. *Then there exists a function* $f \in C[a, b]$ *such that the sequence* $(\tilde{p}_n)_{n\in\mathbb{N}}$ *of interpolating polynomials does not converge uniformly to* f.

On the Proof. The proof involves explicitly constructing an appropriate continuous function f. The details, which we do not give here, can be found in G. Faber [1914]; cf. also I. P. Natanson ([1965], Vol. III, p. 27). □

Remark. The theorem of Faber shows that no triangular array of interpolation points can work for every continuous function. The theorem of Marcinkiewicz assures that for any given such function, there always exists an array which does work in the sense that the corresponding sequence of interpolating polynomials converges uniformly to f, although our proof does not provide a usable method for actually constructing the array.

Convergence in Mean. So far in this section we have concentrated on uniform convergence, i.e. convergence with respect to the Chebyshev norm $\|\cdot\|_\infty$. In Lemma 4.5.6 we showed that convergence in mean, i.e., convergence with respect to $\|\cdot\|_2$, always follows from uniform convergence. On the other hand, since convergence in mean is weaker, it is to be expected that more can be said about convergence in this norm. Indeed, in this case we can show the following, for example: Let $\{\psi_1, \psi_2, \ldots\}$ be a system of polynomials which form an orthonormal system with respect to a weight function w on $[a, b]$. By the Zero Theorem 4.5.5, the zeros of this polynomial are always simple, real, and lie in (a, b). Now if we arrange these points into a triangular array of interpolation points, then the corresponding sequence of interpolating polynomials associated with a given continuous function f always converges to f with respect to the norm $\|f\| := (\int_a^b w(x) f^2(x) dx)^{\frac{1}{2}}$. In the case where the system is the set of Legendre polynomials $\{\psi_1, \psi_2, \ldots\}$, we get convergence in mean on the interval $[-1, +1]$.

In contrast to the situation for the Chebyshev norm, for convergence in the mean, it turns out that it is possible to construct a triangular array of interpolating points which works for all continuous functions; for the detailed proof, see Problems 5 and 6.

4.4 Problems. 1) Let $f \in C[a, b]$ and let $x_\nu := a + \nu\frac{b-a}{n}$ for $0 \le \nu \le n$ be an equally spaced set of interpolation points. Let s_n be the piecewise linear polygon which interpolates f at these points. Show $\lim_{n\to\infty} \|s_n - f\|_\infty = 0$.

2) Show: a) If we interpolate the function $f(x) := \frac{1}{1+x}$ at the points $x_\nu = \frac{\nu}{n}$, $0 \le \nu \le n$, then the sequence $(\tilde{p}_n)_{n\in\mathbb{N}}$ of interpolating polynomials $\tilde{p}_n \in P_n$ converges uniformly to f on the interval $[0, 1]$.

b) The same holds for the interpolation points $x_\nu = \alpha^\nu$ with $\alpha < 1$.

c) Sketch the remainder $f - \tilde{p}_n$ in the cases a) and b) for $n = 1, \ldots, 5$.

3) Let $f \in C_\infty[0, \infty)$, and suppose $|f^{(k)}(x)| \leq 1$ for $x \geq 0$ and for $k \in \mathbb{N}$. Let $\tilde{p}_n \in P_n$ be the polynomials which interpolate f at the points $x_\nu = \nu h$, $0 \leq \nu \leq n$, for fixed step size h. Find h_0 such that for all $h \leq h_0$, the uniform convergence assertion $\lim_{n\to\infty} \tilde{p}_n(x) = f(x)$ for $0 \leq x \leq 1$ holds.

4) Let $f \in C_{n+1}[-1, +1]$ and suppose $\tilde{p} \in P_n$ are the associated interpolating polynomials with respect to the points $x_0 \ldots, x_n$. Show:

a) If $x_0, \ldots, x_n$ are the zeros of the Chebyshev polynomials T_{n+1}, then

$$\|f - \tilde{p}\|_\infty \leq \frac{1}{2^n(n+1)} \|f^{(n+1)}\|_\infty.$$

b) If $x_0, \ldots, x_n$ are the zeros of the Legendre polynomials L_{n+1}, then

$$\|f - \tilde{p}\|_2 \leq \sqrt{\frac{2}{2n+3}} \cdot \frac{1}{(2n+1)(2n-1)\cdots 1} \|f^{(n+1)}\|_\infty.$$

5) Let $x_0, \ldots, x_n$ be the zeros of the Legendre polynomials L_{n+1}. Show:

a) For every polynomial $p \in P_n$,

$$\|p\|_2 \leq \sqrt{2} \max_{0 \leq \nu \leq n} |p(x_\nu)|.$$

Hint: Start with the Lagrange interpolation formula, and use the orthogonality of the Legendre polynomials (cf. also 7.3.1–7.3.2).

b) Let $f \in C[-1, +1]$ and suppose $\tilde{p}_n \in P_n$ are the polynomials which interpolate at $x_0, \ldots, x_n$. Then

$$\lim_{n\to\infty} \|f - \tilde{p}_n\|_2 = 0.$$

Hint: Compare $\tilde{p}_n$ with the best approximations $\tilde{q}_n$ with respect to $\|\cdot\|_\infty$, and apply the Approximation Theorem of Weierstrass.

6) Extend the result of Problem 5 b) to orthogonal systems of polynomials with respect to a general weight function. Use this to establish that if the points $x_0, \ldots, x_n$ are the zeros of the Chebyshev polynomials T_{n+1}, then $\lim_{n\to\infty} \|f - \tilde{p}_n\| = 0$ with respect to the norm $\|\cdot\|$ constructed from the corresponding weight function.

5. More on Interpolation

In this section we present several additional results on interpolation. We begin by discussing some practical aspects of computing with interpolating polynomials. The first question which we treat is the following: How can the value $p(\xi)$ of a polynomial p at a given point ξ be most efficiently computed? In 1.4.4 we have already noted that the "naive algorithm" can be replaced by a significantly more efficient one:

5.1 Horner's Scheme. The value $p(\xi)$ of a polynomial $p(x) = a_0 + a_1 x + \cdots + a_n x^n$ can be computed as

$$p(\xi) = a_0 + \xi(a_1 + \xi(a_2 + \cdots + \xi a_n)\cdots).$$

This leads to the algorithm

a_n	a_{n-1}	a_{n-2}	$\cdots$	a_1	a_0
	$+a'_n\xi$	$+a'_{n-1}\xi$	$\cdots$	$+a'_2\xi$	$+a'_1\xi$
$a_n =: a'_n$	a'_{n-1}	a'_{n-2}	$\cdots$	a'_1	$a'_0 = p(\xi)$

Multiplying out, we see that this algorithm leads to the expansion

$$p(x) = a'_0 + (x - \xi)(a'_1 + a'_2 x + \cdots + a'_n x^{n-1}).$$

From this it follows that

$$p'(\xi) = a'_1 + a'_2\xi + \cdots + a'_n\xi^{n-1}.$$

Now the value $p'(\xi)$ can be easily computed by another application of the algorithm. But then with

$$a''_j := a'_j + a''_{j+1}\xi \text{ for } j = 1, \cdots, n-1 \text{ and } a''_n := a'_n,$$

we have the representation

$$p(x) = a'_0 + (x - \xi)a''_1 + (x - \xi)^2(a''_2 + a''_3 x + \cdots + a_n x^{n-2}) \Rightarrow$$

$$\frac{1}{2}p''(\xi) = a''_2 + a''_3\xi + \cdots + a_n\xi^{n-2} \text{ etc.}$$

The complete Horner Algorithm can be described as in the table on the following page. It leads to the following expansion of p around the point ξ:

$$p(x) = a'_0 + a''_1(x - \xi) + a'''_2(x - \xi)^2 + \cdots + a_n(x - \xi)^n.$$

5.2 The Aitken-Neville Algorithm. The complete Horner Algorithm can be used to compute the value of a polynomial and the values of all of its derivatives at a given point, starting with the coefficients of the polynomial. In certain applications of interpolation, we may be interested in finding the values of interpolating polynomials at a fixed point ξ, and at no others.

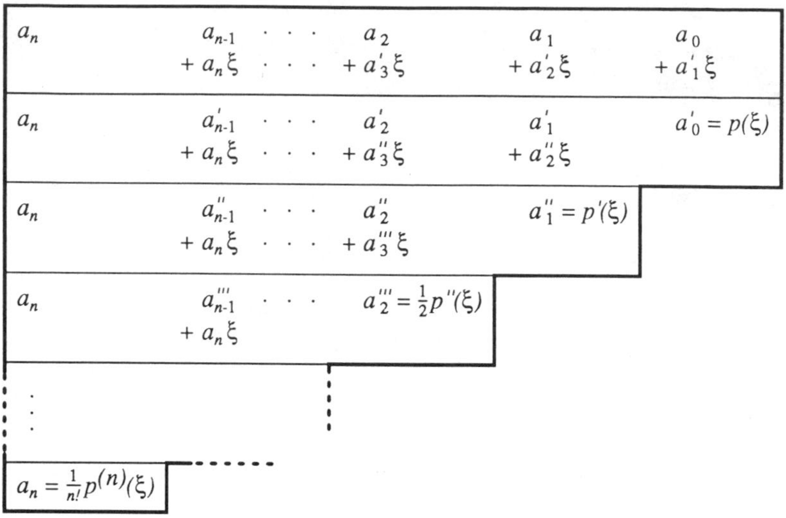

a_n	a_{n-1} $+ a_n\xi$	$\cdots$	a_2 $+ a'_3\xi$	a_1 $+ a'_2\xi$	a_0 $+ a'_1\xi$
a_n	a'_{n-1} $+ a_n\xi$	$\cdots$	a'_2 $+ a''_3\xi$	a'_1 $+ a''_2\xi$	$a'_0 = p(\xi)$
a_n	a''_{n-1} $+ a_n\xi$	$\cdots$	a''_2 $+ a'''_3\xi$	$a''_1 = p'(\xi)$	
a_n	a'''_{n-1} $+ a_n\xi$	$\cdots$	$a'''_2 = \frac{1}{2}p''(\xi)$		
$\vdots$	$\vdots$				
$a_n = \frac{1}{n!}p^{(n)}(\xi)$					

In this section we examine a scheme for computing the values at ξ of a sequence of polynomials of increasing degrees, without actually computing all of the coefficients of each polynomial. This allows us to successively add interpolation points until we have computed $p(\xi)$ to some desired accuracy.

Suppose $p_1 \in \mathrm{P}_n$ is the polynomial which interpolates a given function f at the points $x_m, \ldots, x_{m+n}$, and that $p_2 \in \mathrm{P}_n$ is the polynomial which interpolates f at the points $x_{m+1}, \ldots, x_{m+n+1}$. We now show how to combine these two polynomials in a simple way to obtain a polynomial q of degree $(n+1)$ which interpolates f at the combined set of points $x_m, \ldots, x_{m+n+1}$. It is easy to see that the polynomial

$$q(x) := \frac{1}{x_{m+n+1} - x_m} \begin{bmatrix} p_1(x) & x_m - x \\ p_2(x) & x_{m+n+1} - x \end{bmatrix}$$

satisfies $q(x_\nu) = y_\nu := f(x_\nu)$ for $\nu = m, \ldots, m+n+1$. Thus,

$$p(x_m, \ldots, x_{m+n}; \xi) := p_1(\xi), \quad p(x_{m+1}, \ldots, x_{m+n+1}; \xi) := p_2(\xi),$$
$$p(x_m, \ldots, x_{m+n+1}; \xi) := q(\xi),$$

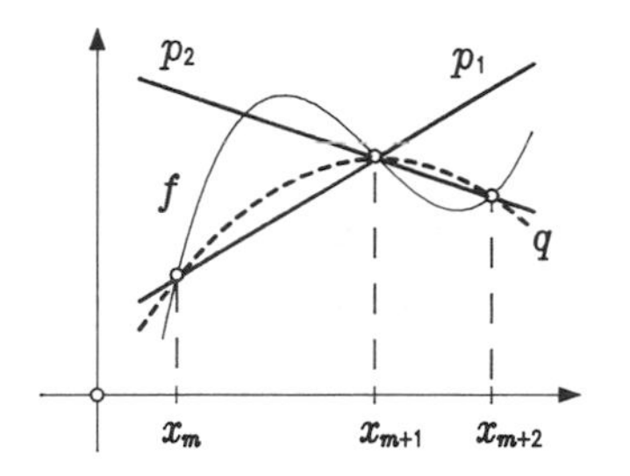

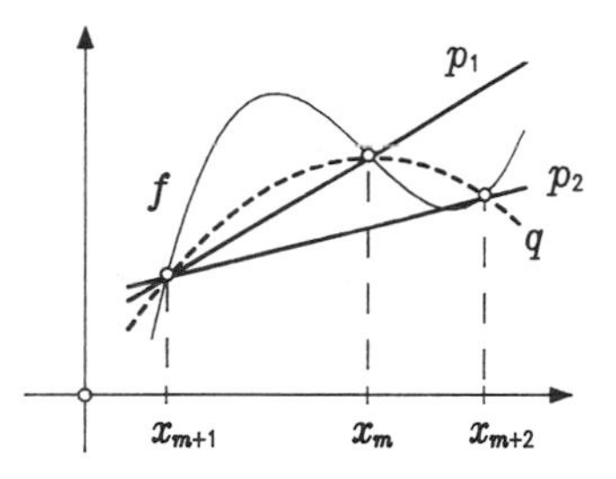

which leads to the following scheme for computing the values of polynomials p of increasing degrees at ξ:

x_ν	y_ν	$p \in \mathrm{P}_1$	$p \in \mathrm{P}_2$	$p \in \mathrm{P}_3$
x_0	y_0			
		$p(x_0, x_1; \xi)$		
x_1	y_1		$p(x_0, x_1, x_2; \xi)$	
		$p(x_1, x_2; \xi)$		$p(x_0, \cdots, x_3; \xi)$
x_2	y_2		$p(x_1, x_2, x_3; \xi)$	
		$p(x_2, x_3; \xi)$		$p(x_1, \cdots, x_4; \xi)$
x_3	y_3		$p(x_2, x_3, x_4; \xi)$	
$\vdots$	$\vdots$	$\vdots$	$\vdots$	$\vdots$

5.3 Hermite Interpolation. It is reasonable to expect that one way to improve the accuracy of an interpolant is to make it interpolate both values and derivatives f. We discuss this type of interpolation using Chebyshev systems of differentiable functions. As a natural extension of Definition 4.4.2, we introduce the

Definition. Suppose $\{g_0, \ldots, g_k\}$ is a set of $(k+1)$ linearly independent functions $g_\kappa \in \mathrm{C}_k[a,b]$, $0 \le \kappa \le k$, such that every nontrivial function g in $\mathrm{span}(g_0, \ldots, g_k)$ possesses at most k zeros in $[a,b]$, counting multiple zeros according to their multiplicity. Then we call $\{g_0, \ldots, g_k\}$ an *Extended Chebyshev System.*

Here the multiplicity of a zero is defined in the usual way: We say ξ is a zero of g of multiplicity $m \le k$ provided that $g(\xi) = g'(\xi) = \cdots = g^{(m-1)}(\xi) = 0$, but $g^{(m)}(\xi) \neq 0$.

The Hermite Interpolation Problem. Let $\{g_0, \ldots, g_k\}$ with $g_\kappa \in \mathrm{C}_k[a,b]$ for $\kappa = 0, \ldots, k$, be an Extended Chebyshev system, and let $f \in \mathrm{C}_k[a,b]$. Given pairwise distinct points $x_\nu \in [a,b]$, $0 \le \nu \le n$, find a function $\tilde{f}$ in $\mathrm{span}(g_0, \ldots, g_k)$ which satisfies the Hermite interpolation conditions

$$\tilde{f}^{(j)}(x_\nu) = f^{(j)}(x_\nu) \quad \text{for } j = 0, \ldots, m_\nu - 1.$$

Here it is assumed that the numbers m_ν describing the *multiplicity of the interpolation points* x_ν satisfy $\sum_{\nu=0}^{n} m_\nu = k+1$.

The questions of existence and uniqueness of Hermite interpolants can be answered in the same way as for simple interpolation:

Theorem. *The Hermite interpolation problem using an Extended Chebyshev system has a unique solution.*

Proof. We can write each function $g \in \operatorname{span}(g_0, \ldots, g_k)$ in the form $g(x) = \sum_{\kappa=0}^{k} \alpha_\kappa g_\kappa(x)$. Now writing out the interpolation conditions leads to the linear system of $k+1 = \sum_{\nu=0}^{n} m_\nu$ equations

$$\sum_{\kappa=0}^{k} \alpha_\kappa g_\kappa^{(j)}(x_\nu) = f^{(j)}(x_\nu),$$
$$0 \le j \le m_\nu - 1 \quad \text{and} \quad 0 \le \nu \le n,$$

in the unknowns $\alpha_0, \ldots, \alpha_k$. We now show that this system of equations always has a unique solution by following the argument in the proof of Theorem 4.6.2, but taking account of the multiple zeros. Indeed, if $\det(g_\kappa^{(j)}(x_\nu)) = 0$, then the homogeneous system of equations $\sum_{\kappa=0}^{k} \alpha_\kappa g_\kappa^{(j)}(x_\nu) = 0$ would have a nontrivial solution, which is impossible since any linear combination of the functions in the Chebyshev system $\{g_0, \ldots, g_k\}$ can have at most k zeros, counting multiplicities. □

As before, the most important case is interpolation by polynomials. In this case we can give a simple formula for the error of interpolation.

Remainder Term for Polynomials. Let $\tilde{p} \in \mathrm{P}_k$ be the solution of the Hermite interpolation problem. If $f \in C_{k+1}[a,b]$, then the remainder term $r = f - \tilde{p}$ can be written in the form

$$r(x) = \frac{f^{(k+1)}(\xi)}{(k+1)!} \Phi_H(x)$$

with

$$\Phi_H(x) := \prod_{\nu=0}^{n} (x - x_\nu)^{m_\nu}.$$

To prove this result, we replace the function Φ which appears in the derivation 1.3 of the remainder term for simple polynomial interpolation by Φ_H, and then argue as before, but taking account of the multiple zeros of Φ_H.

Simple Hermite Interpolation. The simplest case of Hermite interpolation involves finding a polynomial $\tilde{p} \in \mathrm{P}_k$ such that $\tilde{p}^{(j)}(x_\nu) = f^{(j)}(x_\nu)$ for $j = 0, 1$ and for $\nu = 0, \ldots, n$; i.e., we are interpolating both f and its derivative f' at each interpolation point. In this case we have $k = 2n+1$ and $\tilde{p} \in \mathrm{P}_{2n+1}$, and we look for p in the form

$$\tilde{p}(x) = \sum_{\nu=0}^{n} [\psi_{2n+1,\nu}(x) f(x_\nu) + \chi_{2n+1,\nu}(x) f'(x_\nu)],$$

where $\psi_{2n+1,\nu}$, $\chi_{2n+1,\nu} \in \mathrm{P}_{2n+1}$ are chosen so that

$$\psi_{2n+1,\mu}(x_\nu) = \delta_{\mu\nu} \quad \text{and} \quad \psi'_{2n+1,\mu}(x_\nu) = 0$$

and

$$\chi_{2n+1,\mu}(x_\nu) = 0 \quad \text{and} \quad \chi'_{2n+1,\mu}(x_\nu) = \delta_{\mu\nu},$$

for $0 \leq \mu, \nu \leq n$.

These conditions imply that $\chi_{2n+1,\mu}(x) = \ell^2_{n\mu}(x)(x - x_\mu)$, and $\psi_{2n+1,\mu}(x)$ must be of the form $\psi_{2n+1,\mu}(x) = \ell^2_{n\mu}(x)(c_{2n+1,\mu} x + d_{2n+1,\mu})$, where the coefficients $c_{2n+1,\mu}$ and $d_{2n+1,\mu}$ can be determined from $\psi_{2n+1,\mu}(x_\mu) = 1$ and $\psi'_{2n+1,\mu}(x_\mu) = 0$. We find that

$$c_{2n+1,\mu} = -2 \sum_{\substack{\nu=0 \\ \nu \neq \mu}}^{n} \frac{1}{x_\mu - x_\nu} \quad \text{and} \quad d_{2n+1,\mu} = 1 - c_{2n+1,\mu} x_\mu.$$

The remainder term becomes

$$r(x) = \frac{f^{(2n+2)}(\xi)}{(2n+2)!}(x - x_0)^2 \cdots (x - x_n)^2,$$

$\xi \in (\min(x, x_\nu), \max(x, x_\nu))$.

Hermite interpolation requires that *all* of the derivatives $f^{(j)}(x_\nu)$ with $0 \leq j \leq m_\nu - 1$ be interpolated at each interpolation point. This problem can be generalized by requiring only that a subset of these derivatives be interpolated; i.e., gaps are allowed. This kind of generalized interpolation was treated already in 1906 in a paper of G. D. Birkhoff. We do not have space here to discuss Birkhoff interpolation; details can be found in the book of Lorentz-Jetter-Riemenschneider [1983].

5.4 Trigonometric Interpolation. In 4.6.5 we have already discussed a method for computing the coefficients of a trigonometric polynomial interpolating data on equally spaced points. Our discussion there was in connection with least squares under the assumption that $n = 2m + 1$, which means that the number N of interpolation points must be odd. Although on the basis of symmetry this is the most natural case, it is also interesting to consider $N = n = 2m$.

In this case, the Orthogonality Conditions 4.6.5 remain unaltered for $\mu, \kappa = 1, \ldots, \frac{n}{2} - 1$, but we now have

$$\langle \underline{g}_{2m}, \underline{g}_{2m} \rangle = \sum_{\nu=1}^{2m} \cos^2(m x_\nu) = n,$$

since now $x_\nu = (\nu-1)\frac{2\pi}{2m}$ for $\nu = 1, \dots, n$, and thus $mx_\nu = (\nu-1)\pi$ so $\cos^2(mx_\nu) = 1$.

It follows that the trigonometric interpolating polynomial can be written as

$$\tilde{f}(x) = \frac{\tilde{a}_0}{2} + \sum_{\mu=1}^{m} \tilde{a}_\mu \cos(\mu x) + \sum_{\mu=1}^{m-1} \tilde{b}_\mu \sin(\mu x),$$

with coefficients

$$\tilde{a}_\mu = \frac{2}{n}\sum_{\nu=1}^{n} y_\nu \cos(\mu x_\nu) \quad \text{for } \mu = 0, 1, \dots, m-1,$$

$$\tilde{b}_\mu = \frac{2}{n}\sum_{\nu=1}^{n} y_\nu \sin(\mu x_\nu) \quad \text{for } \mu = 1, \dots, m-1,$$

$$\tilde{a}_m = \frac{1}{n}\sum_{\nu=1}^{n} y_\nu (-1)^{\nu-1}.$$

Here we have adopted the numbering of the interpolation points (x_ν, y_ν) for $1 \le \nu \le n$ used in 4.6.5. The coefficient $\tilde{b}_m$ does not appear since $\sin(mx_\nu) = 0$ for $1 \le \nu \le n$.

5.5 Complex Interpolation. So far we have restricted our discussion of interpolation to real functions. It frequently happens in analysis, however, that the behavior of a function for real values only becomes clear once we study its properties in the complex plane. This was the case, for example, with the convergence results in 4.3, as well as for the example of Runge presented in 4.2. Thus it will be useful to take a quick look at interpolation in the complex plane, even though it is of lesser practical importance.

The simple polynomial interpolation problem in the complex case is as follows: Suppose we are given $(n+1)$ pairs of complex numbers (z_ν, w_ν), $0 \le \nu \le n$, where the data points z_ν are pairwise distinct. Find a complex polynomial $\tilde{p}$ of degree at most n, which satisfies the interpolation conditions $\tilde{p}(z_\nu) = w_\nu$ for $\nu = 0, \dots, n$.

As in 1.2, the existence and uniqueness of $\tilde{p}$ follow from the interpolation equations. The Lagrange Formula 2.1 and the Newton Formula 2.2 can also be directly carried over.

If we are dealing with interpolation of a holomorphic function $f(z)$, then we can also represent the interpolating polynomial in terms of a complex integral which leads to a convenient remainder formula. We have the

Integral Representation. *Let f be holomorphic in a simply connected domain G, and let Γ be a closed rectifiable curve with no multiple points which lies entirely in G. Suppose that the pairwise distinct interpolation points z_ν all lie inside of Γ, $0 \le \nu \le n$. Let $\tilde{p} \in \mathrm{P}_n$ be the interpolating polynomial; i.e., $\tilde{p}(z_\nu) = f(z_\nu)$ for $\nu = 0, \dots, n$. Then*

$$\tilde{p}(z) = \frac{1}{2\pi i}\int_\Gamma \frac{\Phi(\zeta) - \Phi(z)}{\zeta - z}\frac{f(\zeta)}{\Phi(\zeta)}d\zeta, \quad z \in \mathrm{G},$$

where $\Phi(z) = (z - z_1) \cdots (z - z_n)$.

Proof. Using the residue theorem (cf. e.g. R. Remmert [1990]) and taking account of the form of Φ, it follows that $\tilde{p} \in \mathbf{P}_n$. Now since $\Phi(z_\nu) = 0$, we also have

$$\tilde{p}(z_\nu) = \frac{1}{2\pi i} \int_\Gamma \frac{f(\zeta)}{\zeta - z_\nu} d\zeta = f(z_\nu), \quad 0 \leq \nu \leq n,$$

which shows that the interpolation conditions are satisfied. □

Remainder Term. The integral representation also leads to a closed form expression for the remainder term $r = f - \tilde{p}$. Since

$$f(z) - \tilde{p}(z) = \frac{1}{2\pi i} \int_\Gamma \left[\frac{f(\zeta)}{\zeta - z} - \frac{\Phi(\zeta) - \Phi(z)}{\zeta - z} \frac{f(\zeta)}{\Phi(\zeta)} \right] d\zeta,$$

it follows that

$$r(z) = \frac{1}{2\pi i} \int_\Gamma \frac{\Phi(z)}{\Phi(\zeta)} \frac{f(\zeta)}{\zeta - z} d\zeta.$$

The existence and uniqueness of Hermite interpolation in the complex plane can also be carried over from the real case. The integral representation also holds in this case, where the polynomial $\Phi(z)$ has zeros at the interpolation points of appropriate multiplicities.

5.6 Problems. 1) Program the complete Horner Algorithm 5.1, and for the polynomial $p(x) = 3x^5 - 7x^4 + 2x^2 + 4x + 12$, compute the coefficients of its expansions around the points $\xi := 2$, $\xi := -1$, $\xi = -3$. See also Example 2 in 1.4.4.

2) Inverse Interpolation: To solve the equation $\sin(x) = 0.75$, exchange the roles of the data points and data values. Find an approximation to the solution of this equation lying in the interval $(\frac{\pi}{4}, \frac{\pi}{3})$ by interpolation at the points $\sin(0) = 0$, $\sin(\frac{\pi}{6}) = \frac{1}{2}$, $\sin(\frac{\pi}{4}) = \frac{1}{2}\sqrt{2}$, $\sin(\frac{\pi}{3}) = \frac{1}{2}\sqrt{3}$, $\sin(\frac{\pi}{2}) = 1$.

3) Let $x_0, \ldots, x_{n+k}$ be pairwise distinct points, and let $y_0, \ldots, y_{n+k}$ be associated data values. Find the associated interpolating polynomial $p \in \mathbf{P}_{n+k}$ using the interpolating polynomials $p_\kappa \in \mathbf{P}_n$ which interpolate the values $y_0, \ldots, y_{n-1}, y_{n+\kappa}$ at the points $x_0, \ldots, x_{n-1}, x_{n+\kappa}$ for all $0 \leq \kappa \leq k$.

4) Using the algorithm of Aitken-Neville, find an approximation to

a) $\exp(0.53)$, using the points $x_\nu = 0.3 + \nu h$, $h = 0.1$ for $0 \leq \nu \leq 5$;

b) $f(1.4)$ for $f(x) := \frac{1}{x^2}$, using the points $x_0 = 0.2$, $x_1 = 0.5$, $x_2 = 1.0$, $x_3 = 1.5$, $x_4 = 2.0$, $x_5 = 3.0$.

Check the accuracy of the approximations, and explain the result b).

5) a) Approximate the function $f(x) = \sin(\frac{\pi}{2}x)$ for $x \in [0, 1]$ by simple cubic Hermite interpolation using the points $x_0 = 0$ and $x_1 = 1$. What is the maximum relative interpolation error in the intervals $[0, \frac{1}{4}]$, $[\frac{1}{4}, \frac{3}{4}]$, and $[\frac{3}{4}, 1]$?

b) Interpolate the same function in the intervals $[0, \frac{1}{2}]$ and $[\frac{1}{2}, 1]$ using simple cubic Hermite interpolation at the endpoints of the intervals.

6) Discuss existence and uniqueness for the following interpolation problems, and when they exist, find the interpolants:

a) Find $p \in \mathrm{P}_3$ such that $p(0) = p(1) = 1$, $p''(0) = 0$ and $p'(1) = 1$.

b) Find $p \in \mathrm{P}_2$ such that $p(-1) = p(1) = 1$ and $p'(0) = 0$.

c) Find $p \in \mathrm{P}_2$ such that $p(-1) = p(1) = 1$ and $p'(0) = 1$.

d) Find $p \in \mathrm{P}_2$ such that $p(0) = 1$, $p'(0) = 1$ and $\int_{-1}^{+1} p(x)dx = 1$.

7) Find the Lagrange polynomials $\lambda_{n\nu}$ for trigonometric interpolation using an odd number $n = 2m + 1$ of points, so that the interpolant can be written as $\tilde{f}(x) = \sum_{\nu=1}^{n} y_\nu \lambda_{n\nu}(x)$.

8) Let f be holomorphic inside the simply connected domain $\mathrm{G} \subset \mathbb{C}$, and let Γ be a closed rectifiable curve with no multiple points which lies in G. Let $z_0, \ldots, z_n \in \mathrm{G}$ be pairwise distinct points lying inside of Γ. Show:

$$[z_0 \cdots z_n]f = \frac{1}{2\pi i} \int_\Gamma \frac{f(z)}{(z - z_0) \cdots (z - z_n)} dz.$$

6. Multidimensional Interpolation

The interpolation problem can be generalized to several dimensions in a natural way. To show how this works, in this section we discuss the two-dimensional case. Here the interval $[a, b]$ is replaced by a closed domain $\overline{\mathrm{G}}$ in the (x, y)-plane, and we desire to interpolate a function $f : \overline{\mathrm{G}} \to \mathbb{R}$. Geometrically, we can interpret this as finding an approximation to a surface in $\mathbb{R}^3$ which lies over $\overline{\mathrm{G}}$.

Two-dimensional problems are far more complicated than those in one dimension, even if we restrict ourselves to polynomial interpolation. To get full analogs of the one-dimensional results, we either have to work on special domains, or restrict the location of the interpolation points.

6.1 Various Interpolation Problems. In view of the well-known Taylor expansion of a function of two variables, is is natural to ask the following question: Is it possible to develop a reasonable interpolation process using the linear space

$$\mathrm{P}_{(n)} := \{p \mid p(x, y) = \sum_{0 \le \mu + \kappa \le n} a_{\mu\kappa} x^\mu y^\kappa, a_{\mu\kappa} \in \mathbb{R}\}$$

of all polynomials of degree at most n?

It is easy to see that the $1+2+\cdots+(n+1) = \binom{n+2}{2}$ linearly independent functions $g_{\mu\kappa}$, $g_{\mu\kappa}(x, y) := x^\mu y^\kappa$, $0 \le \mu + \kappa \le n$, form a basis for $\mathrm{P}_{(n)}$, and thus $\dim(\mathrm{P}_{(n)}) = \binom{n+2}{2}$. This suggests that for a given set of data points (x_λ, y_λ) and associated data values $f(x_\lambda, y_\lambda)$, $1 \le \lambda \le \binom{n+2}{2}$, we should look for a polynomial $p \in \mathrm{P}_{(n)}$ with $p(x_\lambda, y_\lambda) = f(x_\lambda, y_\lambda)$ for $1 \le \lambda \le \binom{n+2}{2}$.

Without going into the question of where the data points can be located, in general, we now prove the following

Theorem. *Suppose that $x_0, \ldots, x_n$ are pairwise distinct, and that the same holds for $y_0, \ldots, y_n$. Then there exists a unique polynomial $p \in \mathrm{P}_{(n)}$ which takes on prescribed values $f(x_\rho, y_\sigma)$ at the data points (x_ρ, y_σ) for $0 \leq \rho + \sigma \leq n$.*

Proof. For the proof, we show that the problem of finding $p \in \mathrm{P}_{(n)}$ such that $p(x_\rho, y_\sigma) = 0$ at the $\binom{n+2}{2}$ interpolation points has the unique solution $\tilde{p} = 0$.

Writing $p(x,y) = \sum_{0 \leq \mu+\kappa \leq n} a_{\mu\kappa} x^\mu y^\kappa$ in the equivalent form $p(x,y) = \sum_{\lambda=0}^{n} q_\lambda(x) y^{n-\lambda}$ with $q_\lambda \in \mathrm{P}_\lambda$, then the conditions $p(x_\rho, y_\sigma) = 0$ for $0 \leq \rho + \sigma \leq n$ can be used as follows:

a) $p(x_0, y_\sigma) = 0$ for $0 \leq \sigma \leq n$ and the fact that $p(x_0, \cdot) \in \mathrm{P}_n$ implies that $p(x_0, y) = 0$ for all y. Thus $q_\lambda(x_0) = 0$ for $0 \leq \lambda \leq n$, and it follows that $q_0 = 0$ and thus that $p(x, \cdot) \in \mathrm{P}_{n-1}$.

b) $p(x_1, y_\sigma) = 0$ for $0 \leq \sigma \leq n-1$ and the fact that $p(x_1, \cdot) \in \mathrm{P}_{n-1}$ implies $p(x_1, y) = 0$ for all y. Thus $q_\lambda(x_1) = 0$ for $1 \leq \lambda \leq n$, which together with $q_\lambda(x_0) = 0$ implies $q_1 = 0$, and thus that $p(x, \cdot) \in \mathrm{P}_{n-2}$.

c) The theorem now follows by continuing this process until we get $q_n = 0$. □

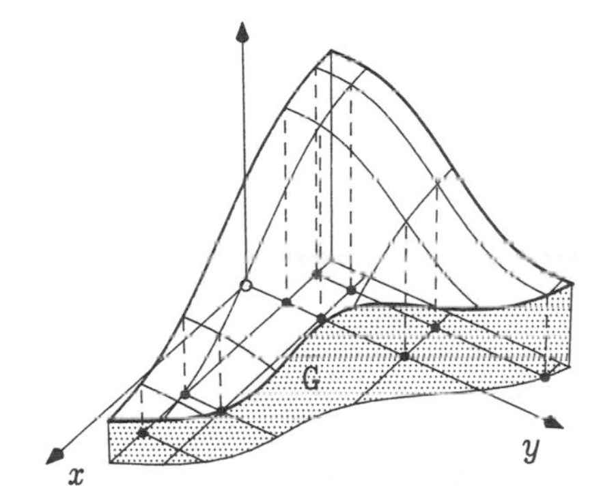

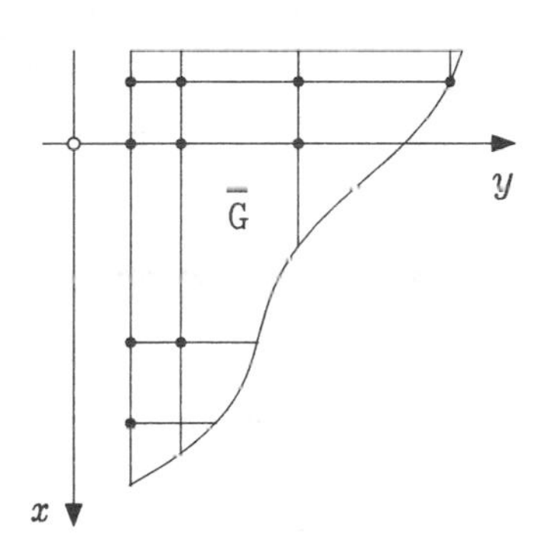

This interpolation problem corresponds to interpolating values at points which can be arranged on a rectangular grid as shown in the above sketch.

We now consider an interpolation problem of particular practical importance where the $(n+1)(k+1)$ interpolation points (x_ν, y_κ), $0 \leq \nu \leq n$ and $0 \leq \kappa \leq k$ lie on a rectangular grid in a rectangular domain. To solve this problem, we now define the linear space of all polynomials of degree at most n in x and degree at most k in y as follows:

$$\mathrm{P}_{nk} := \{p \mid p(x,y) = \sum_{\substack{0 \leq \nu \leq n \\ 0 \leq \kappa \leq k}} a_{\nu\kappa} x^\nu y^\kappa, \ a_{\nu\kappa} \in \mathbb{R}\}.$$

This space will be used throughout the remainder of this section.

6.2 Interpolation on Rectangular Grids. Given a rectangular domain $\overline{G} := \{(x,y) \in \mathbb{R}^2 | a \le x \le b, c \le y \le d\}$, let

$$a = x_0 < x_1 < \cdots < x_n = b, \quad c = y_0 < y_1 < \cdots < y_k = d.$$

This defines $(n+1)(k+1)$ interpolation points (x_ν, y_κ) lying at the corners of a rectangular grid. Then the interpolation problem is as follows: Given values $f(x_\nu, y_\kappa)$, find a polynomial $p \in \mathrm{P}_{nk}$ such that

$$p(x_\nu, y_\kappa) = f(x_\nu, y_\kappa)$$

for $\nu = 0, \ldots, n$ and $\kappa = 0, \ldots, k$.

Existence of an Interpolating Polynomial. Starting with the univariate Lagrange polynomials $\ell_{n\nu}(x)$ of degree n and $\ell_{k\kappa}(y)$ of degree k, we define

$$\ell_{\nu\kappa}^{nk}(x,y) := \ell_{n\nu}(x)\ell_{k\kappa}(y).$$

Then clearly

$$p(x,y) = \sum_{\substack{0 \le \nu \le n \\ 0 \le \kappa \le k}} f(x_\nu, y_\kappa)\ell_{\nu\kappa}^{nk}(x,y)$$

is a polynomial of degree at most n in x and of degree at most k in y which satisfies the interpolation conditions.

Uniqueness of the Interpolation Polynomial. The uniqueness follows from the uniqueness of the interpolation polynomial in one dimension. In particular, if

$$q(x,y) = \sum_{\nu=0}^{n} \sum_{\kappa=0}^{k} a_{\nu\kappa} x^\nu y^\kappa$$

is an interpolating polynomial, then for each $\imath = 0, \ldots, n$,

$$q_\imath(y) := q(x_\imath, y) = \sum_{\kappa=0}^{k} \Big(\sum_{\nu=0}^{n} a_{\nu\kappa} x_\imath^\nu\Big) y^\kappa = \sum_{\kappa=0}^{k} b_{\imath\kappa} y^\kappa$$

is a one dimensional polynomial which as a function of y interpolates the values $f(x_\imath, y_\kappa)$, $0 \le \kappa \le k$. The coefficients $b_{\imath\kappa}$ are therefore uniquely defined. But now for each $0 \le \kappa \le k$ we can determine the coefficients $a_{\nu\kappa}$ from the system of equations

$$\sum_{\nu=0}^{n} a_{\nu\kappa} x_\imath^\nu = b_{\imath\kappa}, \quad 0 \le \imath \le n.$$

Each of these systems of equations is uniquely solvable since their corresponding determinants are Vandermonde determinants. We have established the

Theorem. *There exists a unique polynomial $p \in \mathrm{P}_{nk}$ of degree at most n in x and degree at most k in y which satisfies the $(n+1)(k+1)$ interpolation conditions $p(x_\nu, y_\kappa) = f(x_\nu, y_\kappa)$ for $\nu = 0, \ldots, n$ and $\kappa = 0, \ldots, k$.*

Bivariate Lagrange Polynomials. We have already shown how to write down an explicit formula for the bivariate interpolating polynomial with the help of the Lagrange polynomials $\ell^{nk}_{\nu\kappa}$. Since $\ell^{nk}_{\nu\kappa} = \ell_{n\nu} \cdot \ell_{k\kappa}$, we can infer their behavior from the one-dimensional case. The following sketches show two typical Lagrange polynomials. The one on the left is linear in x and cubic in y. The one on the right is quadratic in both x and y.

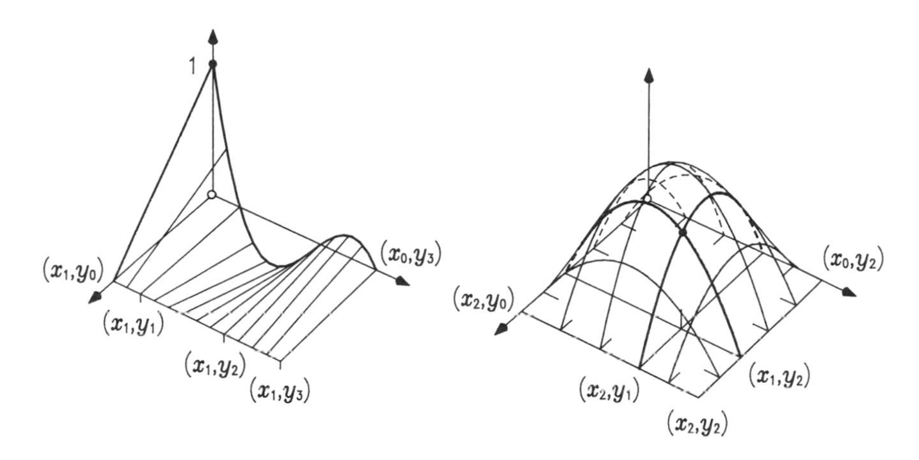

6.3 Bounding the Interpolation Error. Error bounds for interpolation on a rectangular grid can be obtained by using results from the one-dimensional case. For a fixed value $x \in [a, b]$, let $P_y f$ be the polynomial interpolating $f(x, y)$ at the data points $y_0, \ldots, y_k$; i.e.,

$$(P_y f)(x, y) = \sum_{\kappa=0}^{k} f(x, y_\kappa)\ell_{k\kappa}(y).$$

Similarly, for fixed $y \in [c, d]$, let $P_x f$ interpolate $f(x, y)$ at the points $x_0, \ldots, x_n$; i.e.,

$$(P_x f)(x, y) = \sum_{\nu=0}^{n} f(x_\nu, y)\ell_{n\nu}(x).$$

Then

$$(Pf)(x, y) := (P_x P_y f)(x, y) = \sum_{\substack{0 \leq \nu \leq n \\ 0 \leq \kappa \leq k}} f(x_\nu, y_\kappa)\ell_{n\nu}(x)\ell_{k\kappa}(y) = \tilde{p}(x, y)$$

with $P_x P_y f = P_y P_x f = \tilde{p}$.

Moreover, writing $D_x^j g := \frac{\partial^j g}{\partial x^j}$ and $D_y^j g := \frac{\partial^j g}{\partial y^j}$, we also have

$$(D_x^j P_y f)(x,y) = (P_y D_x^j f)(x,y)$$
$$\text{and} \quad (D_y^j P_x f)(x,y) = (P_x D_y^j f)(x,y).$$

Now let $f \in C_{n+k+2}(\overline{G})$. Then

$$\|f - Pf\| \leq \|f - P_y f\| + \|P_y f - Pf\|.$$

Using 1.4, for the first term we get

$$(*) \qquad \|f - P_y f\|_\infty \leq \frac{\|D_y^{k+1} f\|_\infty}{4(k+1)} h_y^{k+1}$$

for $k \geq 1$, where $h_y := \max_{0 \leq \kappa \leq k-1} |y_{\kappa+1} - y_\kappa|$. Similarly, for the second term,

$$\|P_y f - Pf\|_\infty = \|P_y f - P_x(P_y f)\|_\infty \leq \frac{\|D_x^{n+1}(P_y f)\|_\infty}{4(n+1)} h_x^{n+1}$$

for $n \geq 1$, where $h_x := \max\limits_{0 \leq \nu \leq n-1} |x_{\nu+1} - x_\nu|$.
Now since $D_x^{n+1}(P_y f) = P_y(D_x^{n+1} f)$, this expression is nothing more than the polynomial of degree at most k which interpolates $D_x^{n+1} f$ as a function of y, and hence satisfies

$$\|D_x^{n+1} f - P_y(D_x^{n+1} f)\|_\infty \leq \frac{\|D_y^{k+1}\, D_x^{n+1} f\|_\infty}{4(k+1)} h_y^{k+1}.$$

From this we get

$$\|D_x^{n+1}(P_y f)\|_\infty \leq \|D_x^{n+1} f\|_\infty + \frac{\|D_y^{k+1}\, D_x^{n+1} f\|_\infty}{4(k+1)} h_y^{k+1},$$

and combining the above results leads to the
Error Estimate

$$\boxed{\|f - Pf\|_\infty \leq \frac{\|D_x^{n+1} f\|_\infty}{4(n+1)} h_x^{n+1} + \frac{\|D_y^{k+1} f\|_\infty}{4(k+1)} h_y^{k+1} + \frac{\|D_x^{n+1}\, D_y^{k+1} f\|_\infty}{16(n+1)(k+1)} h_x^{n+1}\, h_y^{k+1}.}$$

In the case where $n = k$ and $h_x = h_y =: h$, this error bound simplifies to

$$\|f - Pf\|_\infty \leq$$
$$\leq \frac{h^{n+1}}{4(n+1)} \left[\|D_x^{n+1} f\|_\infty + \|D_y^{n+1} f\|_\infty + \frac{\|D_x^{n+1}\, D_y^{n+1} f\|_\infty}{4(n+1)} h^{n+1} \right].$$

This bound is especially useful when we want to construct an interpolating function by piecing together two-dimensional interpolating polynomials. In this connection, see Comment 1.4, which has a direct analog here.

Still a further simplification is possible in the bilinear case where $n = 1$ and $k = 1$. The linear interpolant of a function $g : [y_0, y_1] \to \mathbb{R}$ is given by

$$(Pg)(y) = g(y_0)\frac{y_1 - y}{y_1 - y_0} + g(y_1)\frac{y - y_0}{y_1 - y_0},$$

and hence

$$|(Pg)(y)| \leq |g(y_0)|\frac{y_1 - y}{y_1 - y_0} + |g(y_1)|\frac{y - y_0}{y_1 - y_0} \leq \|g\|_\infty$$

for $y_0 \leq y \leq y_1$. This gives

$$\|P_y f - Pf\|_\infty = \|P_y(f - P_x f)\|_\infty \leq \|f - P_x f\|_\infty,$$

and so

$$\|f - Pf\|_\infty \leq \|f - P_y f\|_\infty + \|f - P_x f\|_\infty.$$

Using the error bound (*), we see that for all $f \in C_2([x_0, x_1] \times [y_0, y_1])$, we have the

Error Bound for Bilinear Interpolation

$$\boxed{\|f - Pf\|_\infty \leq \frac{1}{8}(\|D_x^2 f\|_\infty\, h_x^2 + \|D_y^2 f\|_\infty\, h_y^2).}$$

This estimate requires weaker hypotheses on the differentiability of f than the general error bound given above.

6.4 Problems. 1) Suppose we choose to interpolate the function f in $C_1([x_0, x_1] \times [y_0, y_1])$ by the constant $p(x, y) = f(\frac{x_0+x_1}{2}, \frac{y_0+y_1}{2})$. Establish the error bound $\|f - p\|_\infty \leq \frac{1}{2}(\|D_x f\|_\infty h_x + \|D_y f\|_\infty h_y)$.

2) Find the bilinear interpolating polynomial satisfying the conditions $p(0,0) = 1$, $p(1,0) = p(0,1) = p(1,1) = 0$. Sketch the resulting surface $p(x, y)$ by sketching families of lines which lie on the surface, and find the intersection curve of the surface with the plane $y = x$. Explain why this surface is called a "hyperbolic paraboloid".

3) Suppose we are given pairwise distinct points (x_λ, y_λ), $1 \leq \lambda \leq \binom{n+2}{2}$. For every $f : \mathbb{R}^2 \to \mathbb{R}$, there exists a $p \in P_{(n)}$, satisfying the interpolation conditions $p(x_\lambda, y_\lambda) = f(x_\lambda, y_\lambda)$ for $1 \leq \lambda \leq \binom{n+2}{2}$. Show:

a) p is unique.

b) There exist uniquely defined functions $\ell_\lambda \in P_{(n)}$, $1 \leq \lambda \leq \binom{n+2}{2}$, such that $p(x, y) = \sum_{1 \leq \lambda \leq \binom{n+2}{2}} f(x_\lambda, y_\lambda)\ell_\lambda(x, y)$.

c) Each ℓ_λ has exact degree n.

Hint: To solve c), prove that assuming $\ell_\lambda \in P_{(n-1)}$ contradicts the uniqueness of ℓ_λ.

4) a) Interpolate $f(x,y) = \sin(\pi x)\sin(\pi y)$ on $(x,y) \in [0,1] \times [0,1]$ on a rectangular grid with the step size $h_x = h_y = \frac{1}{2}$ using a polynomial which is quadratic in x and in y.

b) Assuming we interpolate with a polynomial with the same degree n in both x and y, what degree n is needed to guarantee an accuracy of $\pm 1 \cdot 10^{-2}$?

5) As a followup to 5.3, find a formula for a bicubic Hermite interpolating polynomial which interpolates f at the four corners (x_ν, y_κ) of a rectangle in the sense that

$$\begin{aligned} &p(x_\nu, y_\kappa) = f(x_\nu, y_\kappa),\ (D_x p)(x_\nu, y_\kappa) = (D_x f)(x_\nu, y_\kappa), \\ &(D_y p)(x_\nu, y_\kappa) = (D_y f)(x_\nu, y_\kappa),\ (D_x D_y p)(x_\nu, y_\kappa) = (D_x D_y f)(x_\nu, y_\kappa), \\ &0 \le \nu, \kappa \le 1. \end{aligned}$$

Is this interpolating polynomial unique?

6) The *method of finite elements* involves approximating functions over triangulations of a domain. For example, we can construct a surface approximation by piecing together the linear polynomials $p_\mu : \mathrm{T}_\mu \to \mathbb{R}$, $p_\mu(x,y) = a_\mu x + b_\mu y + c_\mu$, which interpolate f at the vertices of the μ-th triangle T_μ. This results in a continuous surface $\tilde{f}$. We can define a basis for the space of all such surfaces as follows. Suppose that the vertices of the triangles lie at the points π_ν, $0 \le \nu \le n$. For each point π_ν, let q_ν be the pyramid function defined by the condition $q_\lambda(\pi_\nu) = \delta_{\lambda\nu}$. Then the set of pyramid functions q_ν, $0 \le \nu \le n$, form a basis.

Find a formula for the surface $\tilde{f}$ which interpolates a given surface f at the points π_ν, $0 \le \nu \le n$. Explicitly construct a typical pyramid q_ν.

7) Let $\overline{\mathrm{G}} \subset \mathbb{R}^2$ be a set with polygon boundary, and given $f \in \mathrm{C}_2(\overline{\mathrm{G}})$, let $M_2 := \max\{\|f_{xx}\|_\infty, \|f_{xy}\|_\infty, \|f_{yy}\|_\infty\}$. Suppose for some triangulation of $\overline{\mathrm{G}}$ that h is the maximal side length of any triangle. Show that the error bound $\|f - \tilde{f}\|_\infty \le (\frac{3}{2} + \sqrt{3}) M_2 h^2$ holds on $\overline{\mathrm{G}}$. Use the Taylor expansion of $(f - \tilde{f})$ and the error bound for derivatives 1.4.

6

Splines

A *spline* is a function which is piecewise defined on intervals such that the pieces are joined together smoothly. The terminology was introduced by I. J. Schoenberg [1946], although these kinds of functions had been used earlier by several other authors. For example, the Euler method for constructing a piecewise polynomial approximation to the solution of an initial-value problem for ordinary differential equations (and which is often used to establish the Peano Theorem on the existence of solutions of such problems) can be regarded as a simple application of splines. In this regard we should also mention the papers of C. Runge [1901], W. Quade and L. Collatz [1938], J. Favard [1940] and R. Courant [1943], among others. The theory of splines is a good example of an area in mathematics which was developed in response to practical needs. One of the early problems which gave impetus to the development of splines was the need for usuable methods for constructing smooth approximations on the basis of tabulated data arising in ballistics. The subject has steadily developed over the past thirty years, and at present there are several thousand research papers on splines and their applications. In view of this large literature, it is clear that within the framework of this book, we will only be able to give an introduction to a part of the theory. Our discussion will focus on splines constructed from polynomial pieces.

1. Polynomial Splines

By working with polynomial splines, we can retain the convenient properties of low degree polynomials, while at the same time achieving the advantages of a smooth, flexible approximation class. We begin with the definition.

1.1 Spline Spaces. A set of points $\Omega_n := \{x_\nu\}_{\nu=0}^n$, where $a = x_0 < x_1 < \cdots < x_n = b$ which *partitions* a given interval $[a, b] \subset \mathbb{R}$ into subintervals is called a *knot set*. We call the points $x_1, \ldots, x_{n-1}$ *interior knots*, and the points x_0 and x_n *boundary knots*.

Definition of Polynomial Splines. Let ℓ be a nonnegative integer. A function $s : [a,b] \to \mathbb{R}$ is called a *polynomial spline of degree* ℓ provided that it possesses the following properties:

a) $s \in C_{\ell-1}[a,b]$;
b) $s \in P_\ell$ for $x \in [x_\nu, x_{\nu+1})$, $0 \le \nu \le n-1$.

Here, as before, the space $C_{-1}[a,b]$ is to be understood as the space of piecewise continuous functions on $[a,b]$.

We denote the set of all polynomial splines of degree ℓ associated with the partition Ω_n by $S_\ell(\Omega_n)$. In this book we will deal exclusively with polynomial splines, which we will often refer to simply as *splines*.

Remark. Every polynomial of degree ℓ is automatically a spline in the set $S_\ell(\Omega)$ for any partition Ω. The converse is not true, of course.

Example 1. Suppose we are given $(n+1)$ data points $(x_0, y_0), \ldots, (x_n, y_n)$. Then the polygon consisting of straight lines joining successive data points is a spline $s \in S_1(\Omega_n)$; see the example on the left in the figure below.

Example 2. The functions

$$q_{\ell\nu} : [a,b] \to \mathbb{R}, \quad 0 \le \nu \le n-1$$

defined by

$$q_{\ell\nu}(x) = (x - x_\nu)_+^\ell = \begin{cases} (x-x_\nu)^\ell & \text{for } x \ge x_\nu \\ 0 & \text{for } x < x_\nu, \end{cases}$$

and introduced above in 5.2.4 in connection with the Peano Kernel Theorem, are splines of degree ℓ associated with the partition Ω_n. Here, we have written $q_{\ell\nu}(x) := q_\ell(x, x_\nu)$ for convenience. A typical set of such splines is depicted on the right in the figure below. Clearly, these functions are not polynomials on all of $[a,b]$.

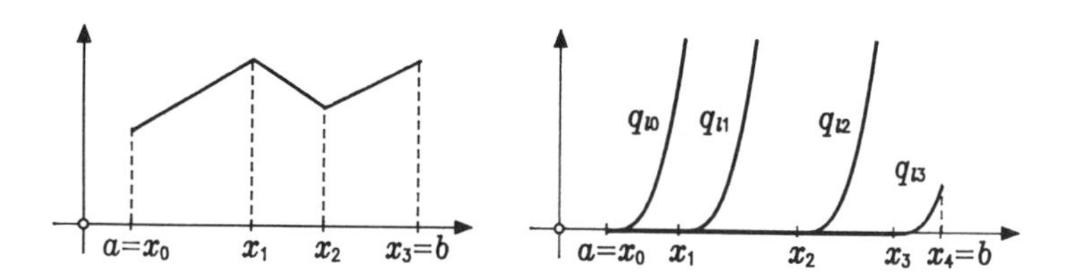

We next investigate the structure of the set $S_\ell(\Omega_n)$ of splines. It follows immediately from the definition that this set is a linear subspace of $C_{\ell-1}[a,b]$. We now look for a basis.

1.2 A Basis for the Spline Space. Using the functions $q_{\ell 1}, \ldots, q_{\ell,n-1}$, given in Example 2 of 1.1, we can now identify the dimension of $S_\ell(\Omega_n)$ and a basis for it.

Theorem. *The set $S_\ell(\Omega_n)$ is a linear space of dimension $(n+\ell)$, and a basis is given by the functions $\{p_0, \dots, p_\ell, q_{\ell 1}, \dots, q_{\ell,n-1}\}$ where $p_\lambda(x) := x^\lambda$, $0 \le \lambda \le \ell$.*

Proof. We show that every $s \in S_\ell(\Omega_n)$ has a unique expansion of the form

$$s(x) = \sum_{\lambda=0}^{\ell} a_\lambda x^\lambda + \sum_{\nu=1}^{n-1} b_\nu (x - x_\nu)_+^\ell, \qquad x \in [a,b].$$

We accomplish this by proceeding interval by interval, starting on the left. Suppose $s \in S_\ell(\Omega_n)$. Then clearly s is a polynomial of degree ℓ for x in the first interval $I_1 := [x_0, x_1]$; i.e., $s(x) = a_0 + a_1 x + \cdots + a_\ell x^\ell$. It follows that the expansion

$$s(x) = \sum_{\lambda=0}^{\ell} a_\lambda x^\lambda + \sum_{\nu=1}^{k-1} b_\nu (x - x_\nu)_+^\ell$$

for $I_k := [x_0, x_k]$ holds for $k = 1$, where we define $\sum_{\nu=1}^{0} b_\nu (x - x_\nu)_+^\ell := 0$.

Now consider

$$\rho(x) := s(x) - \sum_{\lambda=0}^{\ell} a_\lambda x^\lambda - \sum_{\nu=1}^{k-1} b_\nu (x - x_\nu)_+^\ell.$$

Then $\rho \in C_{\ell-1}(I_{k+1})$ and $\rho = 0$ for $x \in I_k$. Moreover, for $x \in [x_k, x_{k+1}]$, $\rho \in P_\ell$, and so ρ can be considered as a solution of the differential equation $y^{(\ell+1)}(x) = 0$ with initial conditions $y(x_k) = y'(x_k) = \cdots = y^{(\ell-1)}(x_k) = 0$. The solution of this initial-value problem is unique up to a multiplicative constant, and can be written in the form $\rho(x) = b_k (x - x_k)_+^\ell$ for $x \ge x_k$. We have shown the expansion holds for all $k \le n$. For $k = n$ it gives us the desired basis representation for the entire interval $I_n = [a,b]$. Counting the number of linearly independent elements $p_0, \dots, q_{\ell,n-1}$, we immediately see that $\dim(S_\ell) = n + \ell$. □

The basis given in this theorem for the spline space $S_\ell(\Omega_n)$ involves what are called *one-sided splines*.

1.3 Best Approximations in Spline Spaces. Since, as we have seen in the previous section, $S_\ell(\Omega_n)$ is a finite dimensional linear space, it follows immediately from the Fundamental Theorem 4.3.4 that for any function $v \in V$, where V is a normed linear space containing $S_\ell(\Omega_n)$, there always exists a best approximation of v from the spline space. We shall be primarily interested in choosing V as one of the spaces $(C[a,b], \|\cdot\|_\infty)$ or $(C[a,b], \|\cdot\|_2)$. It should be emphasized that we are fixing the spline space by choosing the degree of the polynomial pieces and the locations of the knots.

If we are working in the strictly normed space $(C[a,b], \|\cdot\|_2)$, then we know that the best approximation is unique. The space $(C[a,b], \|\cdot\|_\infty)$ is not strictly normed, and so we must approach the uniqueness in another way. In view of the Uniqueness Theorem 4.4.4, it is natural to ask whether $S_\ell(\Omega_n)$ is a Haar space. We can answer this question immediately in the negative, since as shown in Example 2 of 1.1, there exist splines which vanish on an entire interval without vanishing on all of $[a,b]$. This means that the m-dimensional space $S_\ell(\Omega_n)$ with $m = n+\ell$ cannot be a Haar space, since in view of Definition 4.4.2, such spaces are precisely characterized by the requirement that any function in them can have at most $(m-1)$ isolated zeros. Uniqueness has to be established in another way.

Zeros of Splines. While spline spaces are not Haar spaces, it is nevertheless interesting and useful to investigate their zero properties. To do this, we have to differentiate between subintervals $[x_\nu, x_{\nu+1}]$ where s vanishes identically, and subintervals where this is not the case. To this end we introduce the

Definition. A point $\xi \in [x_\nu, x_{\nu+1}) \subset [a,b]$, $0 \le \nu \le n-1$, is called an *essential zero* of a spline $s \in S_\ell(\Omega_n)$ provided that $s(\xi) = 0$, but s does not vanish for all $x \in [x_\nu, x_{\nu+1})$. If $s(b) = 0$, we define b to be an essential zero.

By this definition, if $[x_\nu, x_{\nu+\mu}]$ is a subinterval of maximal length where s vanishes, then $x_{\nu+\mu}$ is an essential zero of s of multiplicity ℓ. Indeed, since $s \in C_{\ell-1}[a,b]$, it follows that $s(x_{\nu+\mu}) = s'(x_{\nu+\mu}) = \cdots = s^{(\ell-1)}(x_{\nu+\mu}) = 0$.

For essential zeros of a splines we now have the

Zero Theorem. *A spline $s \in S_\ell(\Omega_n)$ can have at most $(n+\ell-1)$ essential zeros in $[a,b]$, where each zero is counted according to its multiplicity.*

Proof. Let r be the number of essential zeros in $[a,b]$. By Rolle's Theorem, $s^{(\ell-1)} \in S_1(\Omega_n)$ has at least $r-(\ell-1) = r-\ell+1$ essential zeros. Now the piecewise linear spline $s^{(\ell-1)}$ is continuous, and can have at most n essential zeros in $[a,b]$. It follows that $r-\ell+1 \le n$, and so $r \le n+\ell-1$. □

Supplement. *The bound $r \le n+\ell-1$ is optimal.*

Proof. We claim the bound $r = n+\ell-1$ is assumed. To show this, consider the spline

$$s(x) = \left(\frac{x-a}{x_1-a}\right)^\ell + \sum_{\nu=1}^{n-1} b_\nu (x-x_\nu)_+^\ell$$

whose coefficients are defined recursively by

$$b_\nu := \frac{1}{(x_{\nu+1}-x_\nu)^\ell}\left[(-1)^\nu - \left(\frac{x_{\nu+1}-a}{x_1-a}\right)^\ell - \sum_{\mu=1}^{\nu-1} b_\mu (x_{\nu+1}-x_\mu)^\ell\right]$$

for $\nu = 1,\ldots,n-1$.

It follows that $s(x_\mu) = (-1)^{\mu-1}$ for $\mu = 1,\ldots,n$, and thus s possesses at least one zero in every interval $(x_\nu, x_{\nu+1})$, $1 \le \nu \le n-1$. In addition, $x := a$ is an ℓ-fold zero. All together, these are $(n+\ell-1)$ essential zeros in $[a,b]$. □

The Zero Theorem shows that, with respect to its essential zeros, a spline $s \in S_\ell(\Omega_n)$ behaves like a polynomial from the space $P_{n+\ell-1}$ of the same dimension $(n+\ell)$ as $S_\ell(\Omega_n)$.

To conclude this section we now present a sharpening of this theorem which holds for certain splines in $S_\ell(\Omega_n)$, and which will be useful later on in 4.3.

Corollary. *If the spline $s \in S_\ell(\Omega_n)$ is such that $s(x) = 0$ for $x \in [x_0, x_\sigma]$ and for $x \in [x_\tau, x_n]$, $0 < \sigma < \tau < n$ and $\tau - \sigma \geq \ell + 1$, but does not vanish identically on any other interval, then the number r of essential zeros of s in (x_σ, x_τ) satisfies the sharper bound*

$$r \leq \tau - (\sigma + \ell + 1).$$

Proof. Let $\Omega_{(\tau-\sigma)} := \{x_\sigma, \ldots, x_\tau\}$. Applying the Zero Theorem to the spline space $S_\ell(\Omega_{(\tau-\sigma)})$, we find that the number r of essential zeros of a spline s in this space satisfies $r \leq \tau - \sigma + \ell - 1$. Since $s(x_\sigma) = s'(x_\sigma) = \cdots = s^{(\ell-1)}(x_\sigma) = 0$ and $s(x_\tau) = s'(x_\tau) = \cdots = s^{(\ell-1)}(x_\tau) = 0$, it follows that the knots x_σ and x_τ are both ℓ-fold zeros. It follows that s can have at most $r \leq \tau - \sigma + \ell - 1 - 2\ell = \tau - (\sigma + \ell + 1)$ zeros in (x_σ, x_τ). Now since $\Omega_{(\tau-\sigma)} \subset \Omega_n$, the assertion also holds for a spline in $S_\ell(\Omega_n)$. □

Extension. The Corollary can be extended still further to cover the case $\tau - \sigma < \ell + 1$. It is precisely the contents of Remark 3.1 below to show that in this case, $s(x) = 0$ for all $x \in (x_\sigma, x_\tau)$.

1.4 Problems. 1) Suppose we are given $p, q \in P_\ell$ and $\hat{x} \in \mathbb{R}$, and suppose that $p^{(\kappa)}(\hat{x}) = q^{(\kappa)}(\hat{x})$ for $0 \leq \kappa \leq k$. Show that the difference can be written as $p(x) - q(x) = \sum_{k+1}^{\ell} \alpha_\lambda (x - \hat{x})^\lambda$.

2) Given $-1 \leq \mu < \ell$, define the linear space

$$S_\ell^\mu(\Omega_n) := \{s \in C_\mu[a,b] \mid s \in P_\ell \text{ for } x \in [x_\nu, x_{\nu+1}], 0 \leq \nu \leq n-1\}.$$

Show that the elements $\{p_0, \ldots, p_\ell\}$ together with $\{q_{\lambda 1}, \ldots, q_{\lambda, n-1}\}$ for $\lambda = \mu + 1, \ldots, \ell$ form a basis for $S_\ell^\mu(\Omega_n)$.

3) Let $\Omega_2 := \{0, \frac{1}{2}, 1\}$ and $\ell = 1$. By working directly with the one-sided basis respresentation, find the best approximation of the function $f(x) := x^2$ on $[0,1]$ from $S_1(\Omega_2)$ with respect to the norm $\|\cdot\|_2$, and sketch the result.

4) Suppose the cubic spline $s \in S_3(\Omega_3)$ is such that $s(x) = 0$ for x in $[x_0, x_1]$ and for x in $[x_2, x_3]$. Show that then we also have $s(x) = 0$ for $x \in [x_1, x_2]$

a) by direct calculation, and

b) by application of the Zero Theorem.

5) Let $\Omega_2 := \{0, 1, 2\}$, $\ell \in \mathbb{N}$ and $f_\ell : [0,2] \to \mathbb{R}$,

$$f_\ell(x) := \begin{cases} \sin((\ell+2)\pi x) & \text{for } x \in [0,1] \\ 0 & \text{otherwise.} \end{cases}$$

Show that the splines $g_\alpha \in S_\ell(\Omega_2)$ with $g_\alpha(x) := \alpha(x-1)_+^\ell$ are best approximations of f_ℓ with respect to $\|\cdot\|_\infty$ for every $\alpha \in [-1,+1]$.

2. Interpolating Splines

The discussion in the previous section indicates that some care is needed in formulating interpolation problems which can be uniquely solved using splines. We shall focus mainly on interpolation using splines of odd degree, but later in this section, we also include a discussion of quadratic interpolating splines. Linear, quadratic, and cubic splines are the most widely used in applications. Spline interpolation methods are remarkable in that they use low order polynomials to produce globally smooth interpolants, while at the same time avoiding the disadvantages of high degree polynomials.

2.1 Splines of Odd Degree. The simplest example of a spline of odd degree is the linear spline. Given $(n+1)$ data points $(x_0, y_0), \ldots, (x_n, y_n)$ with $x_0 < x_1 < \cdots < x_n$, we have already seen in Example 1 of 1.1 that the linear spline interpolant is uniquely defined in each subinterval as the straight line interpolating at the two endpoints, and hence globally is the unique polygon obtained by joining the data points together.

We now turn to the more interesting case of splines of odd degree, say $\ell = 2m-1$ for $m \geq 2$.

Since $\dim(S_{2m-1}) = n + 2m - 1$, if we require interpolation at each of the $(n+1)$ knots $x_0, \ldots, x_n$, then there remain $(2m-2)$ free parameters which can be used in various ways. We shall show below that the following three ways of using these extra parameters lead to well-defined interpolation problems:

(i) **Interpolation with Hermite End Conditions.**
Given $f \in C_m[a,b]$, find $s \in S_{2m-1}(\Omega_n)$ such that

a) $s(x_\nu) = f(x_\nu) \qquad$ for $\nu = 0, \ldots, n$

and

b) $s^{(\mu)}(a) = f^{(\mu)}(a)$ and $s^{(\mu)}(b) = f^{(\mu)}(b)$ for $\mu = 1, \ldots, m-1$.

(ii) **Interpolation with Natural End Conditions.**
Given $f \in C_m[a,b]$ with $2 \leq m \leq n+1$, find $s \in S_{2m-1}(\Omega_n)$ such that

a) $s(x_\nu) = f(x_\nu) \qquad$ for $\nu = 0, \ldots, n$

and

b) $s^{(\mu)}(a) = s^{(\mu)}(b) = 0 \quad$ for $\mu = m, \ldots, 2m-2$.

(iii) **Interpolation with Periodic End Conditions.**
Let $f \in C_m[a,b]$ be such that $f^{(\kappa)}(a) = f^{(\kappa)}(b)$ for $\kappa = 0, \ldots, m-1$. Find $s \in S_{2m-1}(\Omega_n)$ such that

a) $s(x_\nu) = f(x_\nu) \qquad$ for $\nu = 0, \ldots, n$

and

b) $s^{(\mu)}(a) = s^{(\mu)}(b)$ for $\mu = 1, \ldots, 2m-2$.

In order to show that problems (i) – (iii) are uniquely solvable, we derive the following

Integral Relation. *Let $f \in C_m[a,b]$, $m \geq 2$, and let $s \in S_{2m-1}(\Omega_n)$ be an interpolating spline such that the difference $f(x) - s(x) =: d(x)$ satisfies the boundary condition*

$$\sum_{\mu=0}^{m-2}(-1)^\mu s^{(m+\mu)}(a) d^{(m-\mu-1)}(a) = \sum_{\mu=0}^{m-2}(-1)^\mu s^{(m+\mu)}(b) d^{(m-\mu-1)}(b).$$

Then the following integral relation holds:

$$\int_a^b [f^{(m)}(x)]^2 dx = \int_a^b [f^{(m)}(x) - s^{(m)}(x)]^2 dx + \int_a^b [s^{(m)}(x)]^2 dx.$$

Remark on the boundary condition. We have formulated the boundary condition in a sufficiently general way to include all three cases (i) – (iii). For example, in the case $m = 2$, which corresponds to cubic splines, this boundary condition becomes $s''(a)d'(a) = s''(b)d'(b)$. This equation is satisfied, for example, if

$$d'(a) = d'(b) = 0, \qquad \text{corresponding to splines of type (i);}$$

or if

$$s''(a) = s''(b) = 0, \qquad \text{corresponding to splines of type (ii);}$$

or if

$$s''(a) = s''(b) \text{ and } d'(a) = d'(b), \text{corresponding to splines of type (iii).}$$

Proof of the integral relation. We have to show that

$$\hat{J} := \int_a^b [f^{(m)}(x)s^{(m)}(x) - (s^{(m)}(x))^2] dx = 0.$$

Now

$$\hat{J} = \int_a^b s^{(m)}(x) d^{(m)}(x) dx = s^{(m)}(x) d^{(m-1)}(x)|_a^b - \int_a^b s^{(m+1)}(x) d^{(m-1)}(x) dx,$$

and by repeated integration by parts,

$$\hat{J} = \sum_{\mu=0}^{m-3}(-1)^\mu s^{(m+\mu)}(x) d^{(m-\mu-1)}(x)|_a^b + (-1)^{m-2}\int_a^b s^{(2m-2)}(x) d''(x) dx.$$

Since s only lies in $C_{2m-2}[a,b]$, for the next integration by parts we have to split the interval into pieces, and we get

$$\int_a^b s^{(2m-2)}(x)d''(x)dx = \sum_{\nu=0}^{n-1}[(s^{(2m-2)}(x)d'(x) - s^{(2m-1)}(x)d(x))|_{x_\nu}^{x_{\nu+1}} + \\ + \int_{x_\nu}^{x_{\nu+1}} s^{(2m)}(x)d(x)dx] = \\ = s^{(2m-2)}(x)d'(x)|_a^b,$$

since $s^{(2m)} = 0$ and $d(x_\nu) = 0$ for $\nu = 0, \ldots, n$. We have shown that

$$\hat{J} = \sum_{\mu=0}^{m-2}(-1)^\mu s^{(m+\mu)}(x)d^{(m-\mu-1)}(x)|_a^b,$$

and using the boundary condition, we get $\hat{J} = 0$ as was to be shown. □

We can now prove that the interpolation problems (i) – (iii) have unique solutions.

Theorem. *The interpolation problems (i) – (iii) are always uniquely solvable.*

Proof. If we look for the interpolating spline s as a linear combination of the one-sided splines as in 1.2, then each of the interpolation problems (i) – (iii) reduces to solving a system of $(n+\ell)$ linear equations for the $(n+\ell)$ unknown coefficients $(a_0, \ldots, a_\ell, b_1, \ldots, b_{n-1})$. We now show that in each case the determinant of the system is nonzero by proving that the associated homogeneous system of equations has only the trivial solution.

In all three cases (i) – (iii), the homogeneous system corresponds to interpolating the function $f = 0$. But then clearly $s = 0$ is an interpolating spline. It remains to show that it is unique. The integral relation implies that if $f^{(m)} = 0$, then $s^{(m)} = 0$ for any interpolating spline s. Now if s is written in the form

$$s(x) = \sum_{\lambda=0}^{2m-1} a'_\lambda \frac{x^\lambda}{\lambda!} + \sum_{\nu=1}^{n-1} b'_\nu \frac{(x - x_\nu)_+^{2m-1}}{(2m-1)!},$$

this means that the derivative $s^{(m)} \in S_{m-1}(\Omega_n)$ satisfies

$$s^{(m)}(x) = \sum_{\lambda=m}^{2m-1} a'_\lambda \frac{x^{\lambda-m}}{(\lambda-m)!} + \sum_{\nu=1}^{n-1} b'_\nu \frac{(x - x_\nu)_+^{m-1}}{(m-1)!} = 0$$

for all $x \in [a,b]$. Since the one-sided splines appearing in this equation are linearly independent, it follows that

$$a'_m = \cdots = a'_{2m-1} = b'_1 = \cdots = b'_{n-1} = 0,$$

and thus that $s(x) = a_0 + a_1 x + \cdots + a_{m-1}x^{m-1}$. We now consider the following three cases:

Interpolation condition (i): $s(a) = s'(a) = \cdots = s^{(m-1)}(a) = 0$ implies $a_0 = a_1 = \cdots = a_{m-1} = 0$.

Interpolation condition (ii): $s(x_0) = s(x_1) = \cdots = s(x_n) = 0$ implies $a_0 = a_1 = \cdots = a_{m-1} = 0$ for $m \leq n+1$.

Interpolation condition (iii): $s(a) = s(b), \cdots, s^{(m-2)}(a) = s^{(m-2)}(b)$ implies $a_1 = a_2 = \cdots = a_{m-1} = 0$, and from $s(a) = 0$ it also follows that $a_0 = 0$.

We have shown that in all three cases $s = 0$ is the unique spline interpolating $f = 0$; i.e., the homogeneous system of equations possesses only the trivial solution. □

2.2 An Extremal Property of Splines. The integral relation for splines of degree $(2m-1)$ implies the following

Extremal Property. Let $f \in C_m[a,b]$, $m \geq 2$, and let $\tilde{s} \in S_{2m-1}(\Omega_n)$ be the interpolating spline with respect to one of the interpolation conditions (i) – (iii). Let g be an arbitrary function in $C_m[a,b]$, which satisfies the same interpolation conditions as $\tilde{s}$, and which in case (iii) is also periodic. Then

$$\|\tilde{s}^{(m)}\|_2 \leq \|g^{(m)}\|_2.$$

Proof. It is easy to see that the integral relation holds not only for f, but for any function $g \in C_m[a,b]$ satisfying the conditions above. Now dropping the first term on the right-hand side of the relation gives the desired result. □

Cubic Splines. Cubic splines ($m = 2$) are by far the most heavily used of the spline spaces, and hence deserve a more complete discussion.

The extremal property of cubic splines says that

$$\int_a^b [\tilde{s}''(x)]^2 dx \leq \int_a^b [g''(x)]^2 dx.$$

We now give geometric and mechanical interpretations of this property.

Geometric Interpretation. The curvature κ of a curve lying in the (x,y)-plane defined by a function $y = g(x)$ is useful for describing the geometric properties of the curve. It is known from differential geometry that the local curvature in this case is given by the formula

$$\kappa(x) = \frac{g''(x)}{(1 + [g'(x)]^2)^{\frac{3}{2}}}.$$

If we now assume that $|g'(x)| \ll 1$ for $x \in [a,b]$, then the value $\|\kappa\|_2^2$ is approximately equal to $\int_a^b [g''(x)]^2 dx$. The extremal property of cubic splines now asserts that the interpolating cubic spline $\tilde{s}$ minimizes the norm of the curvature $\|\kappa\|_2$ over the class of all functions $g \in C_2[a,b]$ satisfying the interpolation conditions.

Mechanical Interpretation. In mechanics it is shown that the local bending moment of a homogeneous, isotropic beam whose center line is given by a function $y = g(x)$ has the value

$$M(x) = c_1 \frac{g''(x)}{(1+[g'(x)]^2)^{\frac{3}{2}}},$$

where c_1 is an appropriate constant. Now assuming that $|g'(x)| \ll 1$ for all $x \in [a,b]$, this moment expression is linearized, and we obtain the approximation $c_3 \int_a^b [g''(x)]^2 dx$ for the bending energy $E(g) = c_2 \int_a^b M^2(x)dx$ of the beam. A beam which is forced to go through fixed "interpolation points" in a way which exerts only forces perpendicular to the beam will assume a position of minimal energy. The extremal property asserts that the cubic interpolating spline approximates the centerline position of such a beam.

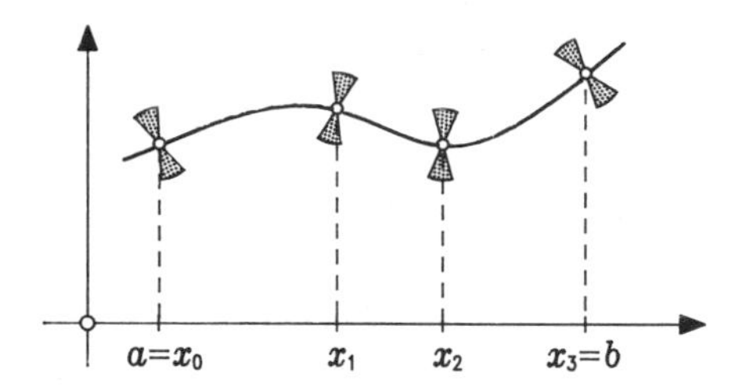

Natural Splines. Clearly, outside of the interval $[a,b]$ the mechanical spline is not constrained, and hence for $x \leq a$ and $b \leq x$ it assumes the "natural" shape of a straight line, which corresponds to the no-energy case where $g''(x) = 0$. In this sense, the end conditions $s''(a) = 0$ and $s''(b) = 0$ are "natural" end condition for the problem. Hence, splines which satisfy the boundary conditions (ii) are called ***natural splines***.

Comment. We can now explain Schoenberg's choice of the word "spline function". A spline is a mechanical instrument consisting of a flexible rod which can be used to draw smooth curves through prescribed points. These kinds of instruments have been used both for technical drawing and for navigation. We have seen above that spline functions model these mechanical splines.

2.3 Quadratic Splines. The space $S_2(\Omega_n)$ of quadratic splines corresponding to the partition Ω_n with $(n+1)$ knots $x_0, \cdots, x_n$ has dimension $(n+2)$. Now if we require that a spline in this space interpolate at each knot, then there remains just *one* free parameter, and therefore it is impossible to enforce a set of symmetric end conditions as was the case for splines of odd degree.

We now give two interpolation problems which lead to uniquely defined quadratic splines, and which do have symmetric end conditions. To achieve this, we have to give up the requirement that the spline interpolate *at the knots*.

Let

$$\Omega_{n-1} : a = \xi_0 < \xi_1 < \cdots < \xi_{n-1} = b$$

and

$$\Omega_n : a = x_0 < x_1 < \cdots < x_n = b$$

be partitions of $[a, b]$ such that

$$x_0 = \xi_0 < x_1 < \xi_1 < \cdots < x_{n-1} < \xi_{n-1} = x_n.$$

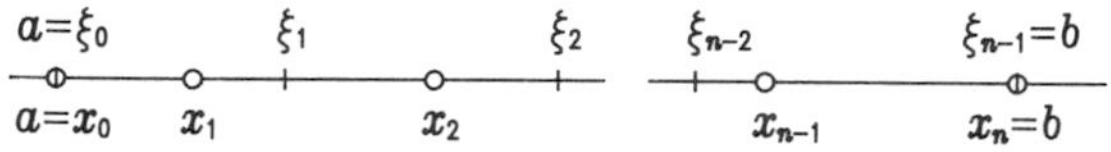

$a=\xi_0$ ξ_1 ξ_2 ξ_{n-2} $\xi_{n-1}=b$

$a=x_0$ x_1 x_2 x_{n-1} $x_n=b$

The spline space $S_2(\Omega_{n-1})$ has dimension $(n+1)$, while $S_2(\Omega_n)$ has dimension $(n+2)$.

We now consider the following two interpolation problems:

Interpolation Problem (i) for Quadratic Splines. Find $s \in S_2(\Omega_{n-1})$ such that

$$s(x_\nu) = f(x_\nu) \quad \text{for } \nu = 0, \cdots, n,$$

where f is a prescribed function.

Interpolation Problem (ii) for Quadratic Splines. Let $f \in C_1[a, b]$. Find $s \in S_2(\Omega_n)$ such that

$$s(\xi_\nu) = f(\xi_\nu) \quad \text{for } \nu = 0, \ldots, n-1$$

and

$$s'(\xi_0) = f'(\xi_0), s'(\xi_{n-1}) = f'(\xi_{n-1}).$$

We now have the

Theorem. *Both interpolation problems (i) and (ii) have unique solutions.*

Proof for Problem (i). The assertion will be established if we can show that the homogeneous interpolation problem has only the trivial solution.

Since $s \in C_1[a, b]$, we can apply Rolle's Theorem in each of the intervals $(x_\nu, x_{\nu+1})$, $0 \le \nu \le n-1$, to deduce that in each $(x_\nu, x_{\nu+1})$ there is at least one point x_ν^* where $s'(x_\nu^*) = 0$. This means that s satisfies the $(2n+1)$ equations

$$s(x_\nu) = 0 \quad \text{for } \nu = 0, \ldots, n \text{ and } s'(x_\nu^*) = 0 \quad \text{for } \nu = 0, \ldots, n-1.$$

Since the partition Ω_{n-1} contains $(n-1)$ subintervals, at least *one* of them, say $[\xi_\mu, \xi_{\mu+1}]$, must contain two of these points. Thus we have an interval containing x_μ^*, $x_{\mu+1}$ and $x_{\mu+1}^*$ such that $s'(x_\mu^*) = 0$, $s(x_{\mu+1}) = 0$ and $s'(x_{\mu+1}^*) = 0$. These three conditions imply that the quadratic polynomial s must be identically zero on the interval $[\xi_\mu, \xi_{\mu+1}]$, and so $s(\xi_\mu) = s'(\xi_\mu) = 0$ and $s(\xi_{\mu+1}) = s'(\xi_{\mu+1}) = 0$. Now the same argument can be applied to the neighboring subintervals $[\xi_{\mu-1}, \xi_\mu]$ and $[\xi_{\mu+1}, \xi_{\mu+2}]$, using the interpolation conditions $s(x_\mu) = 0$ and $s(x_{\mu+2}) = 0$, respectively. We conclude that $s = 0$ on these intervals, and repeating the argument a finite number of times, we get $s = 0$ on the entire interval $[a, b]$.

Proof for Problem (ii). The argument here is similar to that used above for Problem (i). By Rolle's Theorem, s' vanishes at least once in each of the $(n-1)$ intervals $(\xi_\nu, \xi_{\nu+1})$, $0 \leq \nu \leq n-2$, as well as at ξ_0 and in ξ_{n-1}, for a total of at least $(n+1)$ zeros. It follows that in at least *one* of the n subintervals of the partition Ω_n, s' vanishes twice. The only quadratic polynomial which can do this is $s = 0$. But then $s(x_\mu) = s'(x_\mu) = 0$ and $s(x_{\mu+1}) = s'(x_{\mu+1}) = 0$, and since $s(\xi_{\mu-1}) = 0$ and $s(\xi_{\mu+1}) = 0$, we can argue that $s = 0$ also holds in the neighboring subintervals $[x_{\mu-1}, x_\mu]$ and $[x_{\mu+1}, x_{\mu+2}]$, etc. □

2.4 Convergence. One of the disadvantages of the simple interpolating polynomial is the fact that its convergence behavior is unsatisfactory. Indeed, convergence of the sequence of interpolating polynomials of higher and higher degrees does not hold in all cases, even under the strong assumption of analyticity.

For splines, we consider a different kind of convergence question. Here we do not consider a sequence of splines of higher and higher degrees, but rather a sequence of splines of fixed degree with an increasing number of knots. We shall see that this leads to much better convergence properties. In general, under relatively weak hypotheses on f and the distribution of the knots, the interpolating spline converges uniformly to f.

We consider first the convergence properties of linear splines .

Convergence of Linear Splines. Since the linear interpolating spline can be given explicitly, it is possible to establish several of its properties directly. It suffices to consider the problem of interpolating by a straight line on a subinterval $[x_\nu, x_{\nu+1}]$. In this case the associated Lagrange polynomials $\ell_{1\nu}$ and $\ell_{1,\nu+1}$ of 5.2.1 are

$$\ell_{1\nu}(x) = \frac{x - x_{\nu+1}}{x_\nu - x_{\nu+1}} \quad \text{and} \quad \ell_{1,\nu+1}(x) = \frac{x - x_\nu}{x_{\nu+1} - x_\nu}.$$

It follows that for each $0 \leq \nu \leq n-1$, the linear spline which interpolates f can be written as

$$\tilde{s}(x) = f(x_\nu)\ell_{1\nu}(x) + f(x_{\nu+1})\ell_{1,\nu+1}(x)$$

on $x \in [x_\nu, x_{\nu+1}]$. Now let $f \in \mathrm{C}[a,b]$, and consider the deviation for x in $[x_\nu, x_{\nu+1}]$. In view of the identity $\ell_{1\nu}(x) + \ell_{1,\nu+1}(x) = 1$, by 5.2.1 we have

$$f(x) - \tilde{s}(x) = f(x)[\ell_{1\nu}(x) + \ell_{1,\nu+1}(x)] - f(x_\nu)\ell_{1\nu}(x) - f(x_{\nu+1})\ell_{1,\nu+1}(x) = \\ = \ell_{1\nu}(x)[f(x) - f(x_\nu)] + \ell_{1,\nu+1}(x)[f(x) - f(x_{\nu+1})],$$

and using $\ell_{1\nu}(x) \geq 0$, $\ell_{1,\nu+1}(x) \geq 0$, we get

$$|f(x) - \tilde{s}(x)| \leq \max\{|f(x) - f(x_\nu)|, |f(x) - f(x_{\nu+1})|\}.$$

It follows that

$$\max_{x\in[x_\nu,x_{\nu+1}]} |f(x) - \tilde{s}(x)| \leq \max_{x\in[x_\nu,x_{\nu+1}]} \max\{|f(x) - f(x_\nu)|, |f(x) - f(x_{\nu+1})|\},$$

and inserting the modulus of continuity ω_f of f, we get

$$\max_{x\in[x_\nu,x_{\nu+1}]} |f(x) - \tilde{s}(x)| \leq \omega_f(|x_{\nu+1} - x_\nu|).$$

Setting $h := \max_{\nu=0,\dots,n-1} |x_{\nu+1} - x_\nu|$, we are led to the uniform

Error Estimate.

$$\|f - \tilde{s}\|_\infty \leq \omega_f(h),$$

which implies the uniform convergence of the linear spline interpolants to $f \in \mathrm{C}[a,b]$ as $h \to 0$.

Here we have restricted our attention to linear splines; we return to the question of convergence of interpolating splines in 5.4.

2.5 Problems. 1) Find the cubic polynomial which takes on the values $p(x_\nu)$, $p''(x_\nu)$ and $p(x_{\nu+1})$, $p''(x_{\nu+1})$ at the end points of the interval $[x_\nu, x_{\nu+1}]$.

2) Find the interpolating cubic spline on the partition Ω_n for the cases (i), (ii) and (iii) in 2.1. Start with the formula from Problem 1), and use the continuity of s' in order to find a system of equations for the quantities $s''(x_\nu)$, $1 \leq \nu \leq n-1$.

3) Starting with the system of equations obtained in Problem 2), show that the interpolation problems (i), (ii) and (iii) can be uniquely solved using cubic splines.

4) Repeat Problems 1 and 2 for quadratic splines in case (i).

5) Suppose we interpolate the function $f \in \mathrm{C}_1[a,b]$ on an equally-spaced partition Ω_n using a spline $\tilde{s} \in \mathrm{S}_0(\Omega_n)$. How must the interpolation points $\xi_\nu \in (x_\nu, x_{\nu+1})$, $0 \leq \nu \leq n-1$, be chosen to assure that the factor α in the estimate $\|f - \tilde{s}\|_\infty \leq \alpha h \|f'\|_\infty$ is as small as possible?

6) Let Ω_n be a partition of $[0,1]$, and let $\tilde{s} \in \mathrm{S}_1(\Omega_n)$ be the spline interpolating the function $f(x) := \sqrt{x}$. Find the error $\|f - \tilde{s}\|_\infty$ for

a) an equally-spaced partition,

b) the partition $x_\nu := (\frac{\nu}{n})^4$, $0 \le \nu \le n$.

3. B-splines

In Section 1 we have given a basis for the $(n+\ell)$- dimensional space $S_\ell(\Omega_n)$ of splines of degree ℓ associated with a knot set Ω_n. It consists of polynomials together with certain "one-sided power functions" $q_{\ell\nu}$. In this section we discuss an alternative basis for the spline space which is much better suited for computations with splines. Schoenberg studied these "Basic Spline Curves", which later came to be called simply ***B-splines***. To introduce these splines, we work in an infinite-dimensional spline space, and show the existence of certain elements with compact support which can be used as basis functions.

3.1 Existence of B-splines. In order to avoid having to take account of end conditions, we consider the infinite knot set $\Omega_\infty := \{x_\nu\}_{\nu\in\mathbb{Z}}$, $x_\nu < x_{\nu+1}$ with $x_\nu \to -\infty$ for $\nu \to -\infty$ and $x_\nu \to \infty$ for $\nu \to \infty$. We shall show that for every $\nu \in \mathbb{Z}$, there is exactly one spline $s \in S_\ell(\Omega_\infty)$ such that

$$s(x) = 0 \quad \text{for} \quad x < x_\nu \quad \text{and} \quad x_{\nu+\ell+1} \le x$$

and for which the

Normalization Condition

$$\int_{-\infty}^{+\infty} s(x)dx = \int_{x_\nu}^{x_{\nu+\ell+1}} s(x)dx = 1$$

holds.

Proof. For x in the interval $[x_{\nu-1}, x_{\nu+\ell+2}]$, it is clear that the expansion of s in terms of the basis in 1.2 does not include any polynomial part since $s(x) = 0$ for all $x \le x_\nu$. It follows that we can write

$$s(x) = \sum_{\kappa=0}^{k} b_\kappa (x - x_{\nu+\kappa})_+^\ell$$

for some integer k still to be determined. Now taking $k := \ell+1$ and using the requirement that $s(x) = 0$ for $x \ge x_{\nu+\ell+1}$ while $(x - x_{\nu+\kappa})_+^\ell = (x - x_{\nu+\kappa})^\ell$ there, it follows that the coefficients $b_0, \dots, b_{\ell+1}$ are the unique solution of the nonsingular system of equations

$$\sum_{\kappa=0}^{\ell+1} b_\kappa (x - x_{\nu+\kappa})^\ell = 0 \quad \text{for} \quad x \ge x_{\nu+\ell+1}.$$

Indeed, setting the coefficients of the various powers of x to zero, we get the system of equations

$$(*)\qquad \begin{array}{lllll} b_0 & +b_1 & +\cdots & +b_{\ell+1} & =0 \\ b_0x_\nu & +b_1x_{\nu+1} & +\cdots & +b_{\ell+1}x_{\nu+\ell+1} & =0 \\ \vdots & & & \vdots & \\ b_0x_\nu^\ell & +b_1x_{\nu+1}^\ell & +\cdots & +b_{\ell+1}x_{\nu+\ell+1}^\ell & =0 \end{array}$$

for the coefficients $b_0, \ldots, b_{\ell+1}$.

The normalization condition leads to the equation

$$\sum_{\kappa=0}^{\ell+1} \frac{b_\kappa}{\ell+1}(x_{\nu+\ell+1} - x_{\nu+\kappa})^{\ell+1} = 1.$$

Expanding this out and using the $(\ell+1)$ equations $(*)$, this latter equation can be rewritten as an $(\ell+2)$-nd equation for $b_0, \ldots, b_{\ell+1}$:

$$b_0x_\nu^{\ell+1} + b_1x_{\nu+1}^{\ell+1} + \cdots + b_{\ell+1}x_{\nu+\ell+1}^{\ell+1} = (-1)^{\ell+1}(\ell+1).$$

Now the determinant of this system is a Vandermonde, and we conclude that the system has a unique solution. □

Remark. If we choose $k \le \ell$ above, then the conditions $s(x) = 0$ for $x < x_\nu$ and for $x_{\nu+k} \le x$ imply that $s = 0$. Indeed, in this case the system of equations $(*)$ has at least one less column, so that $(\ell+1)$ homogeneous equations for the $k+1 \le \ell+1$ unknowns $b_0, \ldots, b_k$ remain. The Vandermonde matrix of this system has full rank, and it follows that $b_0 = \cdots = b_k = 0$ is the only solution. This means that there are no nontrivial splines of degree ℓ whose support is a proper subset of the support of the B-spline of the same degree; i.e., the B-spline has *minimal support*.

This remark fills in the gap in Corollary 1.3 for $\tau - \sigma < \ell + 1$ discussed in Extension 1.3.

3.2 Local Bases. We have shown in 3.1 that B-splines exist, and if normalized are also unique. We now define them formally, and show that they have the properties described above.

We again start with the functions q_ℓ defined in 5.2.4, but now with the notation $q_\ell(t, x) := (t - x)_+^\ell$. For the time being, we write $t_\nu := x_\nu$ for the knots in order to make it clear that the divided difference appearing in the definition below is taken over these knots.

With this notation, we now have the

Definition. The B-spline of degree ℓ corresponding to knots t_ν in the knot set Ω_∞ is defined to be

$$B_{\ell\nu}(x) := (t_{\nu+\ell+1} - t_\nu)[t_\nu \cdots t_{\nu+\ell+1}]q_\ell(\cdot, x).$$

Comment. By the result of Problem 3 in 5.2.7, $B_{\ell\nu}$ can be written in the form

$$B_{\ell\nu}(x) = \sum_{k=\nu}^{\nu+\ell+1} \left(\prod_{\substack{r=\nu \\ r\neq k}}^{\nu+\ell+1} (t_k - t_r) \right)^{-1} (t_k - x)_+^\ell,$$

and thus is a spline.

In this definition of $B_{\ell\nu}$ we have used a different normalization than that used in 3.1. The usefulness of this alternative normalization will become clear later (see the Partition of Unity 3.3).

First we verify that, except for a normalization constant, the function $B_{\ell\nu}$ is the same B-spline whose existence was established in 3.1. For this, it suffices to show that $B_{\ell\nu}$ has support on $[t_\nu, t_{\nu+\ell+1}]$. To see this, consider $x < t_\nu$. Then $q_\ell(\cdot, x) \in \mathrm{P}_\ell$ with respect to t, and the divided difference $[t_\nu \dots t_{\nu+\ell+1}]q_\ell(\cdot, x)$ of $(\ell+1)$-st order is zero, as can be seen from the Extended Mean-Value Theorem 5.2.6. On the other hand, for $t_{\nu+\ell+1} \leq x$, we have $q_\ell(t, x) = 0$. We have shown that $B_{\ell\nu}(x) = 0$ for $x < t_\nu$ as well as for $t_{\nu+\ell+1} \leq x$. For $x \in (t_\nu, t_{\nu+\ell+1})$, $B_{\ell\nu}$ does not vanish, and in fact we have the

Positivity of the B−splines. By the Zero Theorem 1.3, the spline $B_{\ell\nu}$ has at most 2ℓ essential zeros in $[t_\nu, t_{\nu+\ell+1}]$. Each of the points t_ν and $t_{\nu+\ell+1}$ is an ℓ-fold zero, since $B_{\ell\nu} \in \mathrm{C}_{\ell-1}(-\infty, +\infty)$. It follows that there cannot be any other zeros in $(t_\nu, t_{\nu+\ell+1})$. Indeed, on $(t_{\nu+\ell}, t_{\nu+\ell+1})$ we have $B_{\ell\nu}(x) = (\prod_{r=\nu}^{\nu+\ell}(t_{\nu+\ell+1} - t_r))^{-1}(t_{\nu+\ell+1} - x)^\ell > 0$.

We now consider the space $\mathrm{S}_\ell(\Omega_n)$ of splines of degree ℓ defined on $[x_0, x_n]$. In the set of B-splines just defined, only $B_{\ell,-\ell}, \cdots, B_{\ell,n-1}$ have nonzero values in $[x_0, x_n]$. To show that these $(n+\ell)$ functions form a basis for $\mathrm{S}_\ell(\Omega_n)$, we must establish that they are linearly independent.

Linear Independence. The condition for the B-splines $B_{\ell,-\ell}, \cdots, B_{\ell,n-1}$ to be linearly independent is that the equation

$$s(x) = \beta_{-\ell} B_{\ell,-\ell}(x) + \cdots + \beta_{n-1} B_{\ell,n-1}(x) = 0 \quad \text{for all} \quad x \in [x_0, x_n]$$

can hold only if the coefficients satisfy $\beta_{-\ell} = \cdots = \beta_{n-1} = 0$. To see that this is the case, we introduce an additional knot $x_{-\ell-1} < x_{-\ell}$. By the definition of the B-splines, $s(x) = 0$ for $x \in [x_{-\ell-1}, x_{-\ell}]$. Now if $s(x) = 0$ on $[x_0, x_1]$, then in the notation of Corollary 1.3, applying it to $[x_{-\ell-1}, x_1]$, we have $\tau := 0$ and $\sigma := -\ell$, and thus $\tau - \sigma = \ell < \ell + 1$. By Extension 1.3, we conclude that $s(x) = 0$ for all $x \in [x_{-\ell-1}, x_1]$.

We have shown that $s(x) = 0$ for $x \in [x_0, x_n]$ is synonymous with the vanishing of $s(x)$ for all $x \in [x_{-\ell}, x_n]$, and thus the linear independence of $B_{\ell,-\ell}, \dots, B_{\ell,n-1}$ on $[x_{-\ell}, x_n]$ is sufficient for the linear independence on $[x_0, x_n]$. We now show how the linear independence on $[x_{-\ell}, x_n]$ can easily be established.

To see why the equation $s(x) = 0$ for $x \in [x_{-\ell}, x_n]$ can only hold for coefficients $\beta_{-\ell} = \cdots = \beta_{n-1} = 0$, we proceed stepwise. In the subinterval $[x_{-\ell}, x_{-\ell+1}]$, $s(x) = 0$ reduces to $\beta_{-\ell} B_{\ell,-\ell}(x) = 0$, which in turn implies that $\beta_{-\ell} = 0$. Now we proceed to the subinterval $[x_{-\ell+1}, x_{-\ell+2}]$, and show that we also have $\beta_{-\ell+1} = 0$. Repeating this process leads to the result.

We have now proved that the B-splines $B_{\ell,-\ell}, \ldots, B_{\ell,n-1}$ form a basis for the spline space $S_\ell(\Omega_n)$ on the interval $[x_0, x_n]$. We can restate this as the

Representation Theorem. *Every spline $s \in S_\ell(\Omega_n)$ on the interval $[x_0, x_n]$ has a unique expansion in terms of B-splines*

$$s = \sum_{\nu=-\ell}^{n-1} \alpha_\nu B_{\ell\nu}, \quad \alpha_\nu \in \mathbb{R}.$$

Warning. The argument that the linear independence of a system of functions on an interval $[a, b]$ follows from its linear independence on a larger interval $[a', b'] \supset [a, b]$ holds in general. The proof of the linear independence of the B-splines above, however, involves an argument in the reverse direction which is not correct in general.

3.3 Additional Properties of B-splines. We now present some formulae for B-splines which simplify dealing with them, and which are especially important for their computation. We begin with a formula which follows directly from Definition 3.2. The B-splines $B_{\ell\nu}$ form a

Partition of Unity

$$\sum_{\nu \in \mathbb{Z}} B_{\ell\nu}(x) = 1 \quad \text{for all} \quad x \in (-\infty, +\infty).$$

Proof. This assertion holds for $\ell = 0$ since by definition, $B_{0\nu}(x) = 1$ for x in $[x_\nu, x_{\nu+1})$ and $B_{0\nu}(x) = 0$ otherwise. Now for $\ell \geq 1$ we can write $B_{\ell\nu}(x)$ in the form

$$B_{\ell\nu}(x) = [t_{\nu+1} \ldots t_{\nu+\ell+1}] q_\ell(\cdot, x) - [t_\nu \ldots t_{\nu+\ell}] q_\ell(\cdot, x),$$

(cf. 5.2.3), it follows that for $x \in [t_\mu, t_{\mu+1}]$ we have

$$\begin{aligned}
\sum_{\nu \in \mathbb{Z}} B_{\ell\nu}(x) &= \sum_{\nu=\mu-\ell}^{\mu} B_{\ell\nu}(x) = \\
&= \sum_{\nu=\mu-\ell}^{\mu} [t_{\nu+1} \ldots t_{\nu+\ell+1}] q_\ell(\cdot, x) - \sum_{\nu=\mu-\ell}^{\mu} [t_\nu \ldots t_{\nu+\ell}] q_\ell(\cdot, x) \\
&= [t_{\mu+1} \ldots t_{\mu+\ell+1}] q_\ell(\cdot, x) - [t_{\mu-\ell} \ldots t_\mu] q_\ell(\cdot, x).
\end{aligned}$$

Now $[t_{\mu+1} \ldots t_{\mu+\ell+1}]q_\ell(\cdot, x) = 1$, since this is the divided difference of order ℓ with respect to t of a polynomial $(t-x)^\ell$. Moreover, since $q_\ell(t,x) = 0$ for $x \in [t_\mu, t_{\mu+1}]$, we have $[t_{\mu-\ell} \ldots t_\mu]q_\ell(\cdot, x) = 0$, and the assertion is proved. □

Another important property of the B-splines $B_{\ell\nu}$ for $\ell \geq 1$ is the following

Recursion Formula

$$\boxed{B_{\ell\nu}(x) = \frac{x - x_\nu}{x_{\nu+\ell} - x_\nu} B_{\ell-1,\nu}(x) + \frac{x_{\nu+\ell+1} - x}{x_{\nu+\ell+1} - x_{\nu+1}} B_{\ell-1,\nu+1}(x).}$$

This recursion formula follows by applying the Leibniz rule 5.2.3 to

$$q_\ell(t,x) = (t-x)_+^\ell = (t-x)(t-x)_+^{\ell-1} = (t-x)q_{\ell-1}(t,x).$$

This gives

$$\begin{aligned}[t_\nu \ldots t_{\nu+\ell+1}]q_\ell(\cdot, x) = (t_\nu - x)[t_\nu \ldots t_{\nu+\ell+1}]q_{\ell-1}(\cdot, x) + \\ + [t_{\nu+1} \ldots t_{\nu+\ell+1}]q_{\ell-1}(\cdot, x),\end{aligned}$$

and using

$$\begin{aligned}&[t_\nu \ldots t_{\nu+\ell+1}]q_{\ell-1}(\cdot, x) = \\ &= \frac{1}{t_{\nu+\ell+1} - t_\nu}([t_{\nu+1} \ldots t_{\nu+\ell+1}]q_{\ell-1}(\cdot, x) - [t_\nu \ldots t_{\nu+\ell}]q_{\ell-1}(\cdot, x))\end{aligned}$$

we get

$$\begin{aligned}[t_\nu \ldots t_{\nu+\ell+1}]q_\ell(\cdot, x) = \frac{t_{\nu+\ell+1} - x}{t_{\nu+\ell+1} - t_\nu}[t_{\nu+1} \ldots t_{\nu+\ell+1}]q_{\ell-1}(\cdot, x) - \\ - \frac{t_\nu - x}{t_{\nu+\ell+1} - t_\nu}[t_\nu \ldots t_{\nu+\ell}]q_{\ell-1}(\cdot, x)\end{aligned}$$

so that

$$\begin{aligned}&(t_{\nu+\ell+1} - t_\nu)[t_\nu \ldots t_{\nu+\ell+1}]q_\ell(\cdot, x) = \\ &= \frac{t_{\nu+\ell+1} - x}{t_{\nu+\ell+1} - t_{\nu+1}}(t_{\nu+\ell+1} - t_{\nu+1})[t_{\nu+1} \ldots t_{\nu+\ell+1}]q_{\ell-1}(\cdot, x) - \\ &- \frac{t_\nu - x}{t_{\nu+\ell} - t_\nu}(t_{\nu+\ell} - t_\nu)[t_\nu \ldots t_{\nu+\ell}]q_{\ell-1}(\cdot, x).\end{aligned}$$

Identifying $x := t$ gives the recursion formula.

Differentiation of splines can easily be accomplished with the use of the

Recursion Formula for Derivatives

$$\boxed{B'_{\ell\nu}(x) = \ell\left(\frac{B_{\ell-1,\nu}}{x_{\nu+\ell} - x_\nu} - \frac{B_{\ell-1,\nu+1}}{x_{\nu+\ell+1} - x_{\nu+1}}\right).}$$

We can derive this formula as follows. Inserting

$$q_\ell'(t,x) := \frac{dq_\ell(t,x)}{dx} = -\ell(t-x)_+^{\ell-1} = -\ell q_{\ell-1}(t,x)$$

into the formula for the derivative

$$\begin{aligned}B_{\ell\nu}'(x) &= [t_{\nu+1}\dots t_{\nu+\ell+1}]q_\ell'(\cdot,x) - [t_\nu \dots t_{\nu+\ell}]q_\ell'(\cdot,x)\\ &= -\ell([t_{\nu+1}\dots t_{\nu+\ell+1}]q_{\ell-1}(\cdot,x) - [t_\nu\dots t_{\nu+\ell}]q_{\ell-1}(\cdot,x))\\ &= -\ell\frac{B_{\ell-1,\nu+1}(x)}{t_{\nu+\ell+1}-t_{\nu+1}} + \ell\frac{B_{\ell-1,\nu}(x)}{t_{\nu+\ell}-t_\nu},\end{aligned}$$

which follows from the expansion of $B_{\ell\nu}$ used above in the proof of the partition of unity, we immediately get the recursion formula for $B_{\ell\nu}'$.

The recursion formulae lead to very fast and effective algorithms for computing with B-splines. The recursive computation starts with the trivial case of constant B-splines, which as noted in the proof of the partition of unity, satisfy $B_{0\nu}(x) = 1$ for $x \in [x_\nu, x_{\nu+1})$.

For later purposes it will be useful to give explicit formulae for B-splines for some of the most important spline spaces.

3.4 Linear B-splines. Linear B-splines can be easily described explicitly. They are piecewise linear in $[x_\nu, x_{\nu+2}]$, are continuous, and vanish outside of the interval $[x_\nu, x_{\nu+2}]$. In addition, they satisfy $B_{1,\nu-1}(x) + B_{1\nu}(x) = 1$ for all $x \in [x_\nu, x_{\nu+1}]$. These conditions imply that

$$B_{1\nu}(x) = \begin{cases} \dfrac{x-x_\nu}{x_{\nu+1}-x_\nu} & \text{for} \quad x_\nu \le x < x_{\nu+1} \\[2ex] \dfrac{x_{\nu+2}-x}{x_{\nu+2}-x_{\nu+1}} & \text{for} \quad x_{\nu+1} \le x < x_{\nu+2} \end{cases}$$

and $B_{1\nu}(x) = 0$ for $x < x_\nu$ and for $x_{\nu+2} \le x$.

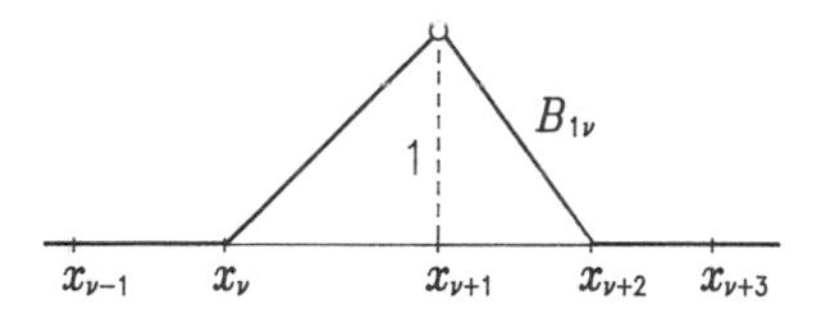

The linear spline which interpolates the values $y_0, \dots, y_n$ at the knots $x_0, \dots, x_n$ can now be written in the form

$$\tilde{s}(x) = \sum_{\nu=-1}^{n-1} y_{\nu+1}B_{1\nu}(x).$$

We have already used this representation in essence in 2.4, since $B_{1\nu}$ is composed of linear Lagrange polynomials 5.2.1. In the case of equally-spaced knots, the B-splines take the form

$$B_{1\nu}(x) = \frac{1}{h}\begin{cases} (x - x_\nu) & \text{for} \quad x_\nu \le x < x_{\nu+1} \\ (x_{\nu+2} - x) & \text{for} \quad x_{\nu+1} \le x < x_{\nu+2} \end{cases}$$

with $B_{1\nu}(x) = 0$ for $x < x_\nu$ and for $x_{\nu+2} \le x$. Here $x_\nu = x_0 + \nu h$, $1 \le \nu \le n$, with $h := \frac{b-a}{n}$.

3.5 Quadratic B-splines. We now restrict our attention to equally-spaced knots. The recurrence formula 3.3 can be used to find the quadratic B-spline $B_{2\nu}$ from two linear B-splines. It can also be constructed directly since it consists of three polynomial pieces, each of degree two, which means that there are nine free parameters to determine. The continuity conditions on $B_{2\nu}$ and $B'_{2\nu}$ at the interior knot give four linear equations, while the requirements $B_{2\nu}(x_\nu) = B'_{2\nu}(x_\nu) = 0$ and $B_{2\nu}(x_{\nu+3}) = B'_{2\nu}(x_{\nu+3}) = 0$ give an additional four. These conditions together with the Partition of Unity Formula 3.3 uniquely determine $B_{2\nu}$ as

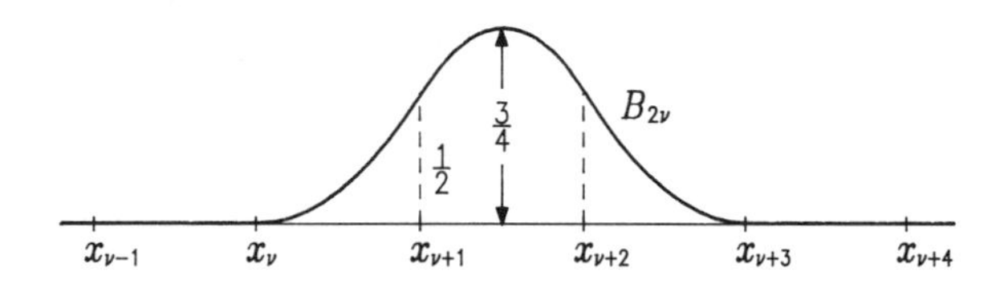

$$B_{2\nu}(x) = \frac{1}{2h^2}\begin{cases} (x - x_\nu)^2, & x_\nu \le x < x_{\nu+1} \\ h^2 + 2h(x - x_{\nu+1}) - 2(x - x_{\nu+1})^2, & x_{\nu+1} \le x < x_{\nu+2} \\ (x_{\nu+3} - x)^2, & x_{\nu+2} \le x < x_{\nu+3} \end{cases}$$

and $B_{2\nu}(x) = 0$ for $x < x_\nu$ and for $x_{\nu+3} \le x$.

3.6 Cubic B-splines. The cubic B-spline for equally-spaced knots is given by

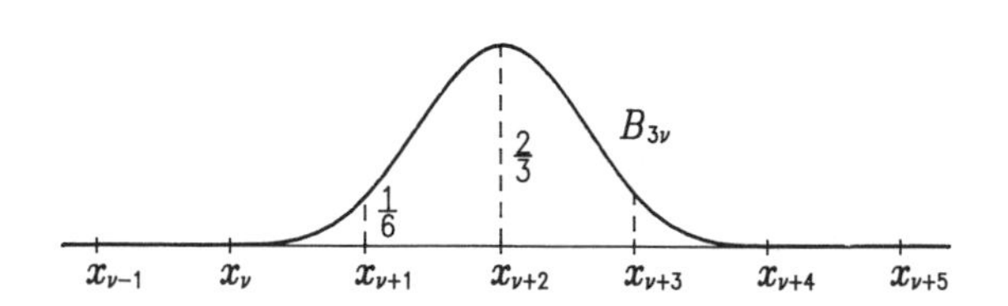

$$B_{3\nu}(x) = \frac{1}{6h^3}\begin{cases} (x-x_\nu)^3, & x_\nu \le x < x_{\nu+1} \\ h^3 + 3h^2(x-x_{\nu+1}) + \\ \quad +3h(x-x_{\nu+1})^2 - 3(x-x_{\nu+1})^3, & x_{\nu+1} \le x < x_{\nu+2} \\ h^3 + 3h^2(x_{\nu+3}-x) + \\ \quad +3h(x_{\nu+3}-x)^2 - 3(x_{\nu+3}-x)^3, & x_{\nu+2} \le x < x_{\nu+3} \\ (x_{\nu+4}-x)^3, & x_{\nu+3} \le x < x_{\nu+4} \end{cases}$$

and $B_{3\nu}(x) = 0$ for $x < x_\nu$ as well as for $x_{\nu+4} \le x$.

3.7 Problems. 1) Show that for the partition $x_\nu := \nu$, $\nu \in \mathbb{Z}$,

$$B_{\ell\nu}(x) = \sum_{j=0}^{\ell+1} (-1)^{\ell+1-j} \binom{\ell+1}{j} \frac{(\nu+j-x)_+^\ell}{\ell!}.$$

2) Show:

$$[x_\nu \dots x_{\nu+\ell+1}]f = \frac{1}{x_{\nu+\ell+1} - x_\nu} \frac{1}{\ell!} \int_{\mathbb{R}} B_{\ell\nu}(x) f^{(\ell+1)}(x)dx.$$

3) a) The B-splines $B_{\ell,-\ell}, \dots, B_{\ell,n-1}$ are linearly independent in the interval $[a,b]$. Are they linearly independent on all of $\mathbb{R}$?

b) Verify that the partition of unity formula for cubic B-splines holds at $x := x_\nu + \frac{h}{2}$.

4) Find a basis of quadratic B-splines for general non-equidistant partitions.

5) Write a program which uses the recurrence formula to compute the values $B_{\ell,\nu-\ell}(x), \dots, B_{\ell,\nu}(x)$ for a given $x \in [x_\nu, x_{\nu+1})$ and use it to compute values of quadratic and quintic B-splines at several points.

4. Computing Interpolating Splines

One approach to the numerical construction of an interpolating spline would be to write the spline as a linear combination of one-sided splines as in 1.2, and then determine the coefficients from the linear system of equations which arises when we write down the interpolation and end conditions. This method, however has the disadvantage that it leads to a poorly-conditioned system of equations.

A second possibility for constructing an interpolating spline is to make use of the fact that if $s \in S_\ell(\Omega_n)$, then $s^{(\ell-1)} \in S_1(\Omega_n)$. Using repeated integration and taking account of the interpolation conditions as well as the continuity of all derivatives of s up to the $(\ell - 1)$-st at each interior knot, we are again led to a linear system of equations for finding the parameters of the spline. This results, however, in a rather complicated algorithm; cf. Problem 2 in 2.5. We now discuss a much simpler and more efficient method.

We shall work with the local basis of B-splines. Every spline $s \in S_\ell(\Omega_n)$ can be written in the form $s(x) = \sum_{\nu=-\ell}^{n-1} \alpha_\nu B_{\ell\nu}(x)$ using the B-spline basis functions $B_{\ell\nu}$. We restrict our attention to the case of equally-spaced knots. To work with the B-splines, we introduce additional knots $x_{-\ell}, \ldots, x_{-1}$. Now the coefficients $\alpha_{-\ell}, \ldots, \alpha_{n-1}$ can be determined by solving the linear system of equations which arises when we write down the interpolation and end conditions. For the interpolation problems (i)–(iii) in 2.1, where the interpolation takes place at the knots, we can give the entries of the corresponding matrices explicitly.

We have already solved the interpolation problem using linear splines in 2.4. We now discuss the cubic case.

4.1 Cubic Splines. To set up the system of equations in the cubic case, we need the values of the B-splines $B_{3\nu}$, $\nu = -3, \ldots, n-1$, at the knots $x_0, \ldots, x_n$ as well as the values of the derivatives $B'_{3\nu}$ (or $B''_{3\nu}$, respectively) at x_0 for $\nu = -3, -2, -1$ and at x_n for $\nu = n-3, n-2, n-1$. Using 3.6 leads to the following table:

	x_ν	$x_{\nu+1}$	$x_{\nu+2}$	$x_{\nu+3}$	$x_{\nu+4}$
$B_{3\nu}(x)$	0	$\frac{1}{6}$	$\frac{2}{3}$	$\frac{1}{6}$	0
$B'_{3\nu}(x)$	0	$\frac{1}{2h}$	0	$-\frac{1}{2h}$	0
$B''_{3\nu}(x)$	0	$\frac{1}{h^2}$	$-\frac{2}{h^2}$	$\frac{1}{h^2}$	0

Now assume that s is given by the formula,

$$s(x) = \sum_{\nu=-3}^{n-1} \alpha_\nu B_{3\nu}(x).$$

Then in all three cases (i) – (iii), the interpolation conditions become

$$\sum_{\nu=\kappa-3}^{\kappa-1} \alpha_\nu B_{3\nu}(x_\kappa) = f(x_\kappa),\ 0 \le \kappa \le n.$$

The corresponding end conditions are

$$\text{Case (i):} \quad \sum_{\nu=\kappa-3}^{\kappa-1} \alpha_\nu B'_{3\nu}(x_\kappa) = f'(x_\kappa), \; \kappa = 0, n;$$

$$\text{Case (ii):} \quad \sum_{\nu=\kappa-3}^{\kappa-1} \alpha_\nu B''_{3\nu}(x_\kappa) = 0, \quad \kappa = 0, n;$$

$$\text{Case (iii):} \quad \sum_{\nu=-3}^{-1} \alpha_\nu B'_{3\nu}(x_0) = \sum_{\nu=n-3}^{n-1} \alpha_\nu B'_{3\nu}(x_n)$$

$$\sum_{\nu=-3}^{-1} \alpha_\nu B''_{3\nu}(x_0) = \sum_{\nu=n-3}^{n-1} \alpha_\nu B''_{3\nu}(x_n).$$

Now the system of equations for the computation of the coefficient vector $\tilde{\alpha} := (\tilde{\alpha}_{-3}, \ldots, \tilde{\alpha}_{n-1})^T$ of the interpolating spline $\tilde{s} \in S_3(\Omega_n)$ can be written in the form $B\alpha = b$, where the matrix B and the right-hand side b are as follows:

Case (i) (Hermite End Conditions)

$$B = \frac{1}{6}\begin{pmatrix} -\frac{3}{h} & 0 & \frac{3}{h} & & & & & & \\ 1 & 4 & 1 & 0 & & & & & \\ 0 & 1 & 4 & 1 & 0 & & & 0 & \\ & \ddots & \ddots & \ddots & \ddots & \ddots & & & \\ & & \ddots & \ddots & \ddots & \ddots & \ddots & & \\ & & & \ddots & \ddots & \ddots & \ddots & \ddots & \\ & & & & \ddots & \ddots & \ddots & \ddots & 0 \\ & 0 & & & & 0 & 1 & 4 & 1 \\ & & & & & & -\frac{3}{h} & 0 & \frac{3}{h} \end{pmatrix},$$

$$b = (f'(x_0), f(x_0), \ldots, f(x_n), f'(x_n))^T.$$

Case (ii) (Natural Spline)

$$B = \frac{1}{6}\begin{pmatrix} \frac{6}{h^2} & -\frac{12}{h^2} & \frac{6}{h^2} & & & & \\ 1 & 4 & 1 & 0 & & 0 & \\ 0 & \ddots & \ddots & \ddots & \ddots & & \\ & \ddots & \ddots & \ddots & \ddots & \ddots & \\ & & \ddots & \ddots & \ddots & \ddots & 0 \\ & 0 & & 0 & 1 & 4 & 1 \\ & & & & \frac{6}{h^2} & -\frac{12}{h^2} & \frac{6}{h^2} \end{pmatrix},$$

$$b = (0, f(x_0), \ldots, f(x_n), 0)^T.$$

Case (iii) (Periodic Spline)

$$B = \frac{1}{6}\begin{pmatrix} -\frac{3}{h} & 0 & \frac{3}{h} & 0 & \cdots & & 0 & \frac{3}{h} & 0 & -\frac{3}{h} \\ \frac{6}{h^2} & -\frac{12}{h^2} & \frac{6}{h^2} & 0 & \cdots & & 0 & -\frac{6}{h^2} & \frac{12}{h^2} & -\frac{6}{h^2} \\ 1 & 4 & 1 & 0 & \cdots & \cdot & & \cdot & \cdot & \\ & \ddots & \ddots & \ddots & \ddots & & & 0 & & \\ & & \ddots & \ddots & \ddots & \ddots & & & & \\ & & & \ddots & \ddots & \ddots & \ddots & & & \\ & & 0 & & \ddots & \ddots & \ddots & \ddots & & \\ & & & & & \ddots & \ddots & \ddots & \ddots & \\ & & & & & & 1 & 4 & 1 & 0 \\ & & & & & & & 1 & 4 & 1 \end{pmatrix},$$

$b = (0, 0, f(x_0), \ldots, f(x_{n-1}), f(x_0))^T$.

4.2 Quadratic Splines. We consider first Problem (i) of 2.3 in the equally-spaced case. Let $h := \frac{b-a}{n-1}$, and let the corresponding knots be denoted by $\xi_0, \ldots, \xi_{n-1}$ with $\xi_\nu = \xi_0 + \nu h$, $1 \leq \nu \leq n-1$. We are looking for a quadratic spline of the form

$$s(x) = \sum_{\nu=-2}^{n-2} \alpha_\nu B_{2\nu}(x)$$

which interpolates a given function f at the points $x_\kappa \in (\xi_{\kappa-1}, \xi_\kappa)$ for $\kappa = 1, \ldots, n-1$ and at $x_0 = \xi_0$ and in $x_n = \xi_{n-1}$. Then the coefficients $\tilde{a}_{-2}, \ldots, \tilde{a}_{n-2}$ of the interpolating spline are the solution of the system

$$\sum_{\nu=-2}^{n-2} \alpha_\nu B_{2\nu}(x_\kappa) = f(x_\kappa), \quad 0 \leq \kappa \leq n.$$

We now focus on the important example where the interior interpolation points are chosen at $x_\kappa := \xi_{\kappa-1} + \frac{h}{2}$. In this case the values of the splines $B_{2\nu}$ at the points $\xi_0 = x_0$, $\xi_{n-1} = x_n$, and $x_\kappa = \xi_{\kappa-1} + \frac{h}{2}$ for $1 \leq \kappa \leq n-1$ are given by

	ξ_ν	$\xi_{\nu+1}$	$\xi_{\nu+2}$	$\xi_{\nu+3}$
$B_{2\nu}(x)$	0	$\frac{1}{2}$	$\frac{1}{2}$	0

	$x_{\nu+1} = \xi_\nu + \frac{h}{2}$	$x_{\nu+2} = \xi_{\nu+1} + \frac{h}{2}$	$x_{\nu+3} = \xi_{\nu+2} + \frac{h}{2}$
$B_{2\nu}(x)$	$\frac{1}{8}$	$\frac{3}{4}$	$\frac{1}{8}$

The coefficient vector $\tilde{\alpha} = (\tilde{a}_{-2}, \ldots, \tilde{a}_{n-2})^T$ is now the solution of the system of equations $B\alpha = b$ with

$$B = \frac{1}{2} \begin{pmatrix} 1 & 1 & & & & \\ \frac{1}{4} & \frac{3}{2} & \frac{1}{4} & & 0 & \\ & \frac{1}{4} & \frac{3}{2} & \frac{1}{4} & & \\ & \ddots & \ddots & \ddots & \ddots & \\ & 0 & \ddots & \frac{1}{4} & \frac{3}{2} & \frac{1}{4} \\ & & & & 1 & 1 \end{pmatrix},$$

and $b = (f(x_0), \ldots, f(x_n))^T$.

We now turn to Problem (ii) of 2.3. In this case we choose the knots to be equally-spaced with $h := \frac{b-a}{n}$, and require that $s(\xi_\nu) = f(\xi_\nu)$ for $0 \le \nu \le n-1$ with $\xi_\nu \in (x_\nu, x_{\nu+1})$, $1 \le \nu \le n-2$, and $s'(\xi_0) = f'(\xi_0)$, $s'(\xi_{n-1}) = f'(\xi_{n-1})$.

Consider now the special choice $\xi_\nu = x_\nu + \frac{h}{2}$, $1 \le \nu \le n-2$. Then the required B-spline values are given in the following table:

	x_ν	$x_{\nu+1}$	$x_{\nu+2}$	$x_{\nu+3}$
$B'_{2\nu}(x)$	0	$\frac{1}{h}$	$-\frac{1}{h}$	0

This leads to the system $B\alpha = b$, where

$$B = \frac{1}{2} \begin{pmatrix} -2/h & 2/h & & & & & & \\ 1 & 1 & 0 & & & & 0 & \\ & 1/4 & 3/2 & 1/4 & & & & \\ & & \ddots & \ddots & \ddots & & & \\ & & & \ddots & \ddots & \ddots & & \\ & & & & \ddots & \ddots & \ddots & \\ & 0 & & & & 1/4 & 3/2 & 1/4 \\ & & & & & & 0 & 1 & 1 \\ & & & & & & & -2/h & 2/h \end{pmatrix},$$

and $b = (f'(\xi_0), f(\xi_0), \ldots, f(\xi_{n-1}), f'(\xi_{n-1}))^T$.

4.3 A General Interpolation Problem. Since the spline space $S_\ell(\Omega_n)$ has dimension $(n + \ell)$, it is natural to ask whether it is always possible to find a spline $s \in S_\ell(\Omega_n)$ which takes on given values at $(n+\ell)$ arbitrary pairwise distinct interpolation points $\xi_j \in [a,b]$, $1 \le j \le n+\ell$. This is, in fact, the most basic question to ask in connection with interpolation.

In as much as the solvability of this interpolation problem is equivalent to the solvability of a corresponding linear system of $(n+\ell)$ equations in $(n+\ell)$ unknowns, it might be expected that a solution always exists. We shall see, however, that in contrast to polynomial interpolation or interpolation by Haar systems, this problem is only solvable under certain restrictions on the location of the interpolation points. It was for this reason, along with the fact that the conditions on the location of the interpolation points are best understood in terms of B-splines, that in the earlier sections of this chapter we restricted our attention to special choices of the interpolation points.

Given $(n+1)$ knots $x_0, \ldots, x_n$, we extend them by choosing additional knots $x_{-\ell}, \ldots, x_{-1}$ and $x_{n+1}, \ldots, x_{n+\ell}$. Suppose that $B_{\ell,-\ell}, \ldots, B_{\ell,n-1}$ is the corresponding basis of B-splines for $S_\ell(\Omega_n)$. We now prove that the interpolation problem is uniquely solvable for arbitrary interpolation points $y_1, \ldots, y_{n+\ell}$ if and only if for each j, the interpolation point ξ_j lies in the interior of the support of the associated B-spline $B_{\ell,-\ell+j-1}$. This result was first established by I. J. Schoenberg and A. Whitney [1953].

Interpolation Theorem. *Fix the knots*

$$x_{-\ell} < \cdots < x_0 < \cdots < x_n < \cdots < x_{n+\ell}.$$

Then there exists a uniquely defined spline $s \in S_\ell(\Omega_n)$ interpolating given values $y_1, \ldots, y_{n+\ell}$ at $(n+\ell)$ interpolation points $\xi_1 < \xi_2 < \cdots < \xi_{n+\ell}$ if and only if $B_{\ell,-\ell+j-1}(\xi_j) \neq 0$ for $1 \le j \le n+\ell$.

Proof. For $\ell = 0$ the assertion of the theorem is obvious, and so from now on we assume that $\ell \ge 1$. We begin by showing that the conditions $B_{\ell,-\ell+j-1}(\xi_j) \neq 0$ are necessary.

Let $s(x) = \sum_{\nu=-\ell}^{n-1} \alpha_\nu B_{\ell\nu}(x)$. Then the interpolation conditions require that

$$\sum_{\nu=-\ell}^{n-1} \alpha_\nu B_{\ell\nu}(\xi_j) = y_j$$

for $j = 1, \ldots, n+\ell$. Now suppose that $B_{\ell,-\ell+j-1}(\xi_j) = 0$ for some j. Then either $\xi_j \le x_{-\ell+j-1}$ or $x_j \le \xi_j$. In the first case it follows that $B_{\ell\nu}(x) = 0$ for all $x \le \xi_j$ with $\nu \ge -\ell+j-1$. In view of this, the first j interpolation conditions now read

$$\sum_{\nu=-\ell}^{-\ell+j-2} \alpha_\nu B_{\ell\nu}(\xi_i) = y_i, \quad i = 1, \ldots, j.$$

These are j equations for the $(j-1)$ unknowns $\alpha_{-\ell}, \ldots, \alpha_{-\ell+j-2}$, and hence do not have solutions for every right-hand side. In the other case where

$x_j \le \xi_j$, then the last $n+\ell-(j-1)$ interpolation conditions have the form

$$\sum_{\nu=-\ell+j}^{n-1} \alpha_\nu B_{\ell\nu}(\xi_i) = y_i, \quad j \le i \le n+\ell.$$

As before, the number of equations is one higher than the number of unknowns, and hence there may be no solution.

We have shown that the conditions $B_{\ell,-\ell+j-1}(\xi_j) \neq 0$, $1 \le j \le n+\ell$, are *necessary* for a unique solution to exist. We now show that they are also *sufficient*. In particular, we show that if the conditions are satisfied, then the homogeneous interpolation problem with $y_i = 0$ for $i = 1, \dots, n+\ell$ has only the trivial solution $s = 0$, i.e., the corresponding coefficients in the B-spline expansion must be $\alpha_\nu = 0$, $\nu = -\ell, \dots, n-1$.

Suppose that $B_{\ell,-\ell+j-1}(\xi_j) \neq 0$ for $j = 1, \dots, n+\ell$, but $s \neq 0$. Then there is at least one interval $[x_\sigma, x_\tau]$ in which s has at most isolated zeros, while $s(x) = 0$ in $[x_{\sigma-1}, x_\sigma]$ and in $[x_\tau, x_{\tau+1}]$. It follows from Extension 1.3 that $\tau - \sigma \ge \ell + 1$. Now the supports of the B-splines $B_{\ell\sigma}, \dots, B_{\ell,\tau-(\ell+1)}$ lie entirely in $[x_\sigma, x_\tau]$. This implies that *at least* $\tau - (\sigma + \ell)$ of the interpolation points $\xi_\sigma, \dots, \xi_{\tau-(\ell+1)}$ lie in (x_σ, x_τ). Since $s(\xi_j) = 0$ at these points, it follows that s has isolated zeros there. But Corollary 1.3 applied to the partition $x_{\sigma-1} < \cdots < x_{\tau+1}$ gives the bound $r \le \tau - (\sigma + \ell + 1)$ on the number of zeros s can have in this interval. We conclude that $s(x) = 0$ for all $x \in [x_\sigma, x_\tau]$, and consequently that $s = 0$ is the only solution of the homogeneous interpolation problem. □

Practical Computation. The most convenient way to compute general interpolating splines is via the B-spline expansion. In this case the matrix of the linear system of equations for computing the coefficients $\tilde{\alpha}_{-\ell}, \dots, \tilde{\alpha}_{n-1}$ of the interpolating spline $\tilde{s} \in S_\ell(\Omega_n)$ is always a band matrix as in 4.1 and 4.2, and hence efficient methods can be applied to solve the system. For more details and an explicit program, see the book of C. de Boor ([1978], Chap. XIII).

4.4 Problems. 1) Given the function $f(x) := \ln(x)$, find the cubic splines which interpolate f with Hermite end conditions on the interval $[0.01, 1.01]$ with n equally-spaced knots for $n = 5$ and $n = 10$, and sketch the function and its approximations.

2) For the Runge example $f(x) := \frac{1}{1+x^2}$, $x \in [-5, +5]$, find the interpolating cubic spline with natural end conditions for $n = 10$ and $n = 20$. How can one reduce the size of the system of equations in this example by half?

3) Compare the amount of computational effort required to solve Problems 1 and 2 with that needed using the method described in Problem 2 in 2.5.

4) For the function $f(x) := \sin(2\pi x)$ find

a) the interpolating cubic spline on $x \in [0,1]$ with periodic end conditions for $n = 4$, 8 and 16;

b) the interpolating cubic spline on $x \in [0, \frac{1}{4}]$ with Hermite end conditions for $n = 2$ and for $n = 4$.

c) Compare the cubic splines from a) for $n = 8$ and from b) for $n = 2$ on the interval $[0, \frac{1}{4}]$ with the Hermite interpolating polynomial found in Problem 5b) in 5.5.6. Which of the three approximations gives the best approximation to $\sin(\frac{\pi}{3})$?

5) Given knots $x_\nu = x_0 + \nu h$, $0 \le \nu \le 4$, with $x_0 = 0$ and $h = \frac{1}{4}$, find the quadratic spline interpolating the function $f(x) := \exp(-\frac{1}{x^2})$, $f(0) = 0$, at the data points $\xi_\mu = x_0 + (\mu - 1)h'$, $1 \le \mu \le 6$ and $h' = \frac{1}{5}$.

5. Error Bounds and Spline Approximation

In 2.4 we derived an error bound for linear splines interpolating a given continuous function f in terms of the modulus of continuity of f. It is to be expected that stronger assumptions on the function f would lead to faster convergence of the sequence of interpolating splines $(\tilde{s}_n)_{n \in \mathbb{Z}_+}$, $\tilde{s}_n \in S_1(\Omega_n)$, and to better error bounds. In this section we explore the connection between the smoothness of a function and how well it can be approximated by splines. Since interpolation and approximation by splines are closely related, we treat the two simultaneously. We shall work mostly with the norm $\|\cdot\|_2$; error bounds for the Chebyshev norm lead to some very interesting results, but require methods which are beyond the scope of this book.

5.1 Error Bounds for Linear Splines. In 2.4 we showed that if f is a continuous function on $[a, b]$, then the interpolating spline $\tilde{s} \in S_1(\Omega_n)$ satisfies the Error Bound 2.4

$$\|f - \tilde{s}\|_\infty \le \omega_f(h).$$

This bound can be applied in practice whenever we can explicitly find the modulus of continuity. For example, if f is Hölder continuous on $[a, b]$, i.e., $|f(x) - f(z)| \le K|x - z|^\alpha$ for some $0 < \alpha < 1$, or Lipschitz bounded ($\alpha = 1$), then we get the error bound

$$\|f - \tilde{s}\|_\infty \le K h^\alpha.$$

This shows that whenever f is Lipschitz bounded, then the sequence of linear interpolating splines converges uniformly to f at a *linear* rate with respect to the maximal distance h between knots.

Now suppose $f \in C_1[a, b]$. By the Newton identity 5.2.3, we see that the error in the interpolating spline $\tilde{s} \in S_1(\Omega_n)$ can be written as

$$f(x) - \tilde{s}(x) = (x - x_\nu)(x - x_{\nu+1})[x_{\nu+1}\, x_\nu x]f,$$

for $[x_\nu, x_{\nu+1}]$, where

$$[x_{\nu+1}\, x_\nu\, x]f = [x_{\nu+1}\, x\, x_\nu]f = \frac{1}{x_{\nu+1} - x_\nu}([x_{\nu+1}x]f - [xx_\nu]f),$$

which leads to

$$[x_{\nu+1}\, x_\nu\, x]f = \frac{1}{x_{\nu+1} - x_\nu}\left(\frac{f(x) - f(x_{\nu+1})}{x - x_{\nu+1}} - \frac{f(x) - f(x_\nu)}{x - x_\nu}\right).$$

By the mean-value theorem, there exist points $\eta_\nu, \zeta_\nu \in (x_\nu, x_{\nu+1})$ such that

$$f'(\eta_\nu) = \frac{f(x) - f(x_{\nu+1})}{x - x_{\nu+1}} \quad \text{and} \quad f'(\zeta_\nu) = \frac{f(x) - f(x_\nu)}{x - x_\nu}.$$

It follows that

$$|f(x) - \tilde{s}(x)| \le \frac{x_{\nu+1} - x_\nu}{4} \max_{\eta,\zeta \in [x_\nu, x_{\nu+1}]} |f'(\eta) - f'(\zeta)|$$

on $[x_\nu, x_{\nu+1}]$, and hence

$$(*) \qquad \|f - \tilde{s}\|_\infty \le \frac{h}{4}\omega_{f'}(h)$$

uniformly on $[a, b]$.

When the derivative of f is Lipschitz bounded, we get

$$\|f - \tilde{s}\|_\infty \le \frac{K}{4}h^2,$$

i.e., we have *quadratic* convergence with respect to h.

If we assume a little more about f, namely that $f \in C_2[a, b]$, then it follows from 5.2.6 that

$$[x_{\nu+1}\, x_\nu\, x] = \frac{1}{2}f''(\xi), \quad \xi \in (x_\nu, x_{\nu+1}),$$

and substituting this in the error representation above, we find that

$$|f(x) - \tilde{s}(x)| \le \frac{(x_{\nu+1} - x_\nu)^2}{8} \max_{t \in [x_\nu, x_{\nu+1}]} |f''(t)|$$

for $x \in [x_\nu, x_{\nu+1}]$, which implies that

$$(**) \qquad \boxed{\|f - \tilde{s}\|_\infty \le \frac{h^2}{8}\|f''\|_\infty}$$

holds on $[a, b]$. This result also follows from the Error Bound 5.1.4. The form of the error expression used to derive this error bound shows that it is generally not possible to get better than quadratic convergence for arbitrary two-times continuously differentiable functions f.

5.2 On Uniform Approximation by Linear Splines. In 1.3 we have already seen that in a given spline space $S_\ell(\Omega_n)$, every continuous function f has a best approximation with respect to the Chebyshev norm $\|\cdot\|_\infty$. In general, the computation of such uniform best approximations by splines is an unsolved problem. In this section we consider the case of linear splines where we can show that the interpolating spline is close to being a best approximation.

If $\tilde{s} \in S_1(\Omega_n)$ is the spline which interpolates f at the knots, then

$$\|\tilde{s}\|_\infty = \max_\nu |\tilde{s}(x_\nu)| = \max_\nu |f(x_\nu)| \leq \|f\|_\infty, \; 0 \leq \nu \leq n.$$

Now if $s \in S_1(\Omega_n)$ is an arbitrary spline, then since $(\tilde{s} - s)$ interpolates the function $(f - s)$, it follows that

$$\|\tilde{s} - s\|_\infty \leq \|f - s\|_\infty.$$

This implies

$$\|f - \tilde{s}\|_\infty = \|(f - s) - (\tilde{s} - s)\|_\infty \leq \|f - s\|_\infty + \|\tilde{s} - s\|_\infty \leq 2\|f - s\|_\infty,$$

and so the distance $E_{S_1(\Omega_n)}(f) = \min_{s \in S_1(\Omega_n)} \|f - s\|_\infty$ satisfies the

Bound

$$E_{S_1(\Omega_n)}(f) \leq \|f - \tilde{s}\|_\infty \leq 2E_{S_1(\Omega_n)}(f).$$

This shows that the interpolating linear spline is a usable substitute for a best approximation of f from $S_1(\Omega_n)$.

5.3 Least Squares Approximation by Linear Splines. Given a continuous function f, let $\hat{s} \in S_\ell(\Omega_n)$ be its best approximation with respect to the norm $\|\cdot\|_2$. By 4.5.2, if we write $\hat{s}$ as a linear combination of the B-splines $B_{\ell,-\ell}, \ldots, B_{\ell,n-1}$ spanning $S_\ell(\Omega_n)$, then the coefficients can be computed from the normal equations

$$\sum_{\nu=-\ell}^{n-1} \alpha_\nu \langle B_{\ell\nu}, B_{\ell\mu} \rangle = \langle f, B_{\ell\mu} \rangle,$$

$\mu = -\ell, \ldots, n-1$, which in the linear case become

$$\sum_{\nu=-1}^{n-1} \alpha_\nu \langle B_{1\nu}, B_{1\mu} \rangle = \langle f, B_{1\mu} \rangle,$$

$\mu = -1, \ldots, n-1$. Here the inner product of any two functions $u, v \in C[a, b]$ is given by $\langle u, v \rangle := \int_a^b u(x)v(x)dx$. By the local support properties of the

B-splines, the matrix $B := (\langle B_{1\nu}, B_{1\mu}\rangle)_{\nu,\mu=-1}^{n-1}$ is banded. For *equally-spaced knots* at a spacing of h, this matrix is

$$B = \frac{h}{6}\begin{pmatrix} 2 & 1 & & & & 0 \\ 1 & 4 & 1 & & & \\ & \ddots & \ddots & \ddots & & \\ & & \ddots & \ddots & \ddots & \\ & 0 & & 1 & 4 & 1 \\ & & & & 1 & 2 \end{pmatrix}.$$

The Gram matrix B is nonsingular, and in fact is even diagonally dominant, and it follows that the coefficients $\hat{\alpha}_\nu$ of the best approximation $\hat{s} = \sum_{\nu=-1}^{n-1} \hat{\alpha}_\nu B_{1\nu}$ are uniquely determined by the normal equations.

Comparing the error $\|f - \hat{s}\|_\infty$ with the minimal deviation $E_{S_1(\Omega_n)}(f)$ with respect to the Chebyshev norm, we get the following

Estimate. Let $f \in C[a,b]$, and let $\hat{s} \in S_1(\Omega_n)$ be the best approximation of f with respect to the norm $\|\cdot\|_2$. Then

$$\|f - \hat{s}\|_\infty \leq 4E_{S_1(\Omega_n)}(f).$$

Proof. For each $\mu = 0, \ldots, n-2$, the corresponding row in the normal equations is

$$\frac{h}{6}\alpha_{\mu-1} + \frac{2h}{3}\alpha_\mu + \frac{h}{6}\alpha_{\mu+1} = \langle f, B_{1\mu}\rangle.$$

Now suppose that $\hat{\alpha}_\rho$ is a coefficient of $\hat{s}$ with largest absolute value; i.e., $|\hat{\alpha}_\rho| = \max_{\nu=-1,\ldots,n-1} |\hat{\alpha}_\nu|$. Then if $\rho \in \{0, \ldots, n-2\}$, we have

$$|2\hat{\alpha}_\rho| = |\frac{3}{h}\langle f, B_{1\rho}\rangle - (\frac{1}{2}\hat{\alpha}_{\rho-1} + \frac{1}{2}\hat{\alpha}_{\rho+1})| \leq \frac{3}{h}|\langle f, B_{1\rho}\rangle| + |\hat{\alpha}_\rho|,$$

and thus

$$|\hat{\alpha}_\rho| \leq \frac{3}{h}|\langle f, B_{1\rho}\rangle|.$$

Since

$$\frac{1}{h}|\langle f, B_{1\mu}\rangle| \leq \|f\|_\infty \frac{1}{h}\int_{x_\mu}^{x_{\mu+2}} B_{1\mu}(x)dx = \|f\|_\infty$$

for $\mu = 0, \ldots, n-2$, it follows that $|\hat{\alpha}_\rho| \leq 3\|f\|_\infty$.

If ρ is one of the numbers -1 or $(n-1)$, then the same bound can be obtained using the first and last equations in the system, respectively. We have shown that

$$\|\hat{s}\|_\infty \leq 3\|f\|_\infty$$

since

$$\|\hat{s}\|_\infty = \max_\nu |\hat{s}(x_\nu)| = \max_\nu |\hat{\alpha}_\nu|.$$

Now let s be any spline in $S_1(\Omega_n)$. Then

$$\|f-\hat{s}\|_\infty = \|(f-s)-(\hat{s}-s)\|_\infty \leq \|f-s\|_\infty + \|\hat{s}-s\|_\infty \leq$$
$$\leq \|f-s\|_\infty + 3\|f-s\|_\infty = 4\|f-s\|_\infty,$$

since $(\hat{s}-s)$ is a best approximation of the function $(f-s)$. □

5.4 Error Bounds for Splines of Higher Degree. In this section we give some error bounds for interpolating splines of higher degree. To illustrate the kind of result we are looking for, we begin with the case of the cubic spline interpolating a function $f \in C_4[a,b]$ using the Hermite end conditions (also called type (i)). In preparation, we first present a lemma which establishes a connection between this spline and the uniquely defined best approximation of f'' from $S_1(\Omega_n)$ with respect to the norm $\|\cdot\|_2$.

Lemma. *Let $f \in C_2[a,b]$ and let $\tilde{s} \in S_3(\Omega_n)$ be the interpolating cubic spline with Hermite end conditions. Then $\tilde{s}''$ is the best approximation of f'' from $S_1(\Omega_n)$ with respect to the norm $\|\cdot\|_2$; that is,*

$$\|f''-\tilde{s}''\|_2 \leq \|f''-s\|_2$$

for all $s \in S_1(\Omega_n)$.

Proof. In this proof we will denote the interpolating cubic spline of type (i) of a function $u \in C_2[a,b]$ by $\tilde{s}_u$.

Given an arbitrary spline $s \in S_1(\Omega_n)$, we define the function $\sigma(x) := = \int_a^b (x-t)_+\, s(t)dt$. Then $\sigma'' = s$, and so $\sigma \in S_3(\Omega_n)$, which implies that $\tilde{s}_\sigma$ is identical with σ, i.e., $\tilde{s}_\sigma = \sigma$.

Now suppose we write the Integral Relation 2.1 in the form

$$\|g''\|_2^2 = \|g''-\tilde{s}_g''\|_2^2 + \|\tilde{s}_g''\|_2^2$$

and express the function f as $f = g+\sigma$, where $g = f-\sigma$. Then since $\tilde{s}_{(f-\sigma)} = \tilde{s}_f - \tilde{s}_\sigma$, we get

$$\|f''-\sigma''\|_2^2 = \|f''-\sigma''-(\tilde{s}_f-\tilde{s}_\sigma)''\|_2^2 + \|\tilde{s}_f-\tilde{s}_\sigma\|_2^2$$

and

$$\|f''-s\|_2^2 = \|f''-\tilde{s}_f''\|_2^2 + \|\tilde{s}_f-\tilde{s}_\sigma\|_2^2.$$

This implies

$$\|f''-\tilde{s}_f''\|_2^2 \leq \|f''-s\|_2^2$$

for all $s \in S_1(\Omega_n)$. Equality holds here precisely when $\tilde{s}_f = \tilde{s}_\sigma$, i.e. when $s := \tilde{s}_f''$ and $\tilde{s}_f''$ is the best approximation of f'' from $S_1(\Omega_n)$. □

We now turn to the problem of estimating the error of the interpolating cubic spline $\tilde{s} \in S_3(\Omega_n)$ of type (i) for a function $f \in C_4[a,b]$ in the case of *equidistant knots.* For a given subinterval $[x_\nu, x_{\nu+1}]$ with $0 \le \nu \le n-1$, we may apply the Newton identity to $d := f - \tilde{s}$ to get

$$d(x) = d(x_\nu) + (x - x_\nu)[x_\nu x_{\nu+1}]d + (x - x_\nu)(x - x_{\nu+1})\frac{d''(\xi)}{2} =$$
$$= (x - x_\nu)(x - x_{\nu+1})\frac{d''(\xi)}{2}, \quad \xi \in (x_\nu, x_{\nu+1}),$$

where we have used the facts that $d(x_\nu) = 0$ and $[x_\nu, x_{\nu+1}]d = 0$.

But then,

$$|d(x)| \le \frac{h^2}{8} \max_{t \in [x_\nu, x_{\nu+1}]} |d''(t)|$$

for every point x in the subinterval $[x_\nu, x_{\nu+1}]$, and since this holds for every subinterval, using the Lemma and the Estimate 5.3, we get

$$\|f - \tilde{s}\|_\infty \le \frac{h^2}{8}\|f'' - \tilde{s}''\|_\infty \le \frac{h^2}{2} E_{S_1(\Omega_n)}(f'').$$

Now applying the error bound (**) in 5.1 to the expression on the right, we get

$$E_{S_1(\Omega_n)}(f'') \le \frac{h^2}{8}\|f^{(4)}\|_\infty,$$

which shows that the cubic interpolating spline of type (i) with *equidistant knots* satisfies the

Error Bound

$$\boxed{\|f - \tilde{s}\|_\infty \le \frac{h^4}{16}\|f^{(4)}\|_\infty.}$$

This error bound is optimal with respect to the order of the mesh size h, but not with respect to the constant $1/16$. Using a more refined argument, C. A. Hall [1968] showed that the constant can be improved to $5/384$, and that this value is best possible.

In order to get optimal error bounds for interpolating splines of various types and degrees, we would have to treat each case individually. Instead of doing that, we content ourselves here with establishing an error bound for splines of types (i) – (iii) of odd degree $(2m-1)$, $m \ge 2$, which is not optimal, but which does give information on the convergence of the derivatives of the spline. Given an arbitrary knot distribution, let $h := \max_\nu |x_{\nu+1} - x_\nu|$ be the mesh size. Then we have the

Error Estimate. *Let $f \in C_m[a,b]$, and let $\tilde{s} \in S_{2m-1}(\Omega_n)$ be an interpolating spline of type (i) – (iii), where $m \ge 2$. Then for $0 \le j \le m-1$,*

$$\|f^{(j)} - \tilde{s}^{(j)}\|_\infty \le \frac{m!}{\sqrt{m}}\,\frac{1}{j!}h^{m-j-\frac{1}{2}}\|f^{(m)}\|_2.$$

Proof. The difference $d := f - \tilde{s}$ lies in $C_m[a, b]$ and satisfies $d(x_\nu) = 0$ for $\nu = 0, \ldots, n$. We now investigate the location of the zeros of $d^{(j)}$.

By Rolle's Theorem, there is at least one zero of d' in each subinterval defined by the partition. Repeating the argument, we see that for each $i \leq m-1$, $d^{(i)}$ must have at least $(j - i + 1)$ zeros in each of the intervals $[x_\nu, x_{\nu+j}]$, and it follows that $d^{(j)}$ must have at least one zero in each of the intervals $[x_\nu, x_{\nu+j}]$, $\nu + j \leq n$.

Now suppose ζ_j is such that $|d^{(j)}(\zeta_j)| = \|d^{(j)}\|_\infty$. Then by the above, it follows that the zero ξ_j of $d^{(j)}$ which is closest to ζ_j must satisfy the bound $|\zeta_j - \xi_j| < < (j+1)h$.

It follows that for $j \leq m-2$,

$$\|d^{(j)}\|_\infty = \left| \int_{\xi_j}^{\zeta_j} d^{(j+1)}(x)dx \right| \leq (j+1)h\|d^{(j+1)}\|_\infty \leq$$
$$\leq (j+1)(j+2)h^2\|d^{(j+2)}\|_\infty \leq \cdots \leq$$
$$\leq (j+1)\cdots(m-1)h^{m-j-1}\|d^{(m-1)}\|_\infty = \frac{(m-1)!}{j!}h^{m-j-1}\|d^{(m-1)}\|_\infty.$$

The same inequality holds trivially for $j = m-1$.

Now using the Schwarz inequality, we get

$$\|d^{(m-1)}\|_\infty = \left| \int_{\xi_{m-1}}^{\zeta_{m-1}} d^{(m)}(x)dx \right| \leq (m\,h)^{\frac{1}{2}} \left| \int_{\xi_{m-1}}^{\zeta_{m-1}} (d^{(m)}(x))^2 dx \right|^{\frac{1}{2}} \leq$$
$$\leq (m\,h)^{\frac{1}{2}}\|d^{(m)}\|_2,$$

and in virtue of the Integral Relation 2.1, it follows that

$$\|d^{(m)}\|_2 \leq \|f^{(m)}\|_2.$$

Combining these results, we get the estimate

$$\|d^{(j)}\|_\infty \leq \frac{m!}{\sqrt{m}}\frac{1}{j!}h^{m-j-\frac{1}{2}}\|f^{(m)}\|_2. \qquad \square$$

Comment. The case $m := 1$ of linear splines is covered by the bound (*) in 5.1 which is directly applicable to give convergence results.

For $m := 2$, the above result shows that the cubic spline satisfies the error bounds

$$\|f - \tilde{s}\|_\infty \leq \sqrt{2}h^{\frac{3}{2}}\|f''\|_2$$

and

$$\|f' - \tilde{s}'\|_\infty \leq \sqrt{2}h^{\frac{1}{2}}\|f''\|_2.$$

Neither of these bounds is optimal, or of any practical use. Indeed, it can be shown that $\|f - \tilde{s}\|_\infty = O(h^2)$ and $\|f' - \tilde{s}'\|_\infty = O(h)$ for $f \in C_2[a, b]$. Our error estimate is still of interest in two respects, however. It provides a convergence result for $\|f - \tilde{s}\|_\infty$, and, moreover, shows that the derivative of the interpolating spline converges uniformly to f' as h goes to zero.

More generally, the above error bound leads to the

Convergence Theorem. *Let Ω_n be a sequence of partitions of $[a,b]$, and let $\tilde{s}_n \in S_{2m-1}(\Omega_n)$ be the splines of one of the types (i) – (iii) interpolating a function $f \in C_m[a,b]$. Then the sequence $(\tilde{s}_n)$ converges uniformly to f provided the maximal distance between knots in the partition goes to zero as $n \to \infty$. Moreover, if $m \geq 2$, then for each $j = 1, \ldots, m-1$, the sequences $(\tilde{s}_n^{(j)})$ of derivatives converge uniformly to the derivatives of $f^{(j)}$.*

Remark. The assertions in this convergence theorem are by no means as simple as we might have expected from our study of linear splines in 2.4. There we saw that uniform convergence of the sequence of linear interpolating splines holds for every continuous function f, as long as $h \to 0$ as $n \to \infty$. There are analogous results for higher degree splines interpolating functions $f \in C[a,b]$, but they require restrictions on the partitions. For example, in the case of cubic splines, we need to assume that the ratio of the maximal distance h between knots to the minimal distance between knots in the sequence (Ω_n) of partitions is uniformly bounded.

5.5 Least Square Splines of Higher Degree. Given a continuous function f, then as we saw in 5.3, the spline $\hat{s} \in S_\ell(\Omega_n)$ which fits f best in the least squares sense can be found by solving the Normal Equations 4.5.2:

$$\sum_{\nu=-\ell}^{n-1} \alpha_\nu \langle B_{\ell\nu}, B_{\ell\mu}\rangle = \langle f, B_{\ell\mu}\rangle,$$

$\mu = -\ell, \ldots, n-1$. For a given partition Ω_n, the matrix B of this system can be computed once and for all, while the vector on the right-hand side has to be computed for each choice of f.

For equidistant knots, B is symmetric and banded with nonzero elements only on a total of $2\ell + 1$ diagonals. In particular, it always has the form

$$B = \begin{pmatrix} b_{11} & \cdots & b_{1,\ell+1} & & & & \\ \vdots & \ddots & \vdots & \ddots & & 0 & \\ b_{1,\ell+1} & & b_{\ell+1,\ell+1} & & \ddots & & \\ & \ddots & & \ddots & & \ddots & \\ & & \ddots & & b_{\ell+1,\ell+1} & \cdots & b_{1,\ell+1} \\ & 0 & & \ddots & \vdots & \ddots & \vdots \\ & & & & b_{1,\ell+1} & & b_{11} \end{pmatrix}$$

and is completely determined already by the entries in the upper triangular part of the $(\ell+1)$ by $(\ell+1)$ principal matrix in the left-hand upper corner.

When $\ell = 0$, we have $B = h\,I$, where I is the identity matrix. In 5.3 we gave B for the case $\ell = 1$. For $\ell = 2$ the upper triangular matrix determining B is given by

$$\frac{h}{120}\begin{pmatrix} 6 & 13 & 1 \\ 0 & 60 & 26 \\ 0 & 0 & 66 \end{pmatrix},$$

and for $\ell = 3$ by

$$\frac{h}{5040}\begin{pmatrix} 20 & 129 & 60 & 1 \\ 0 & 1208 & 1062 & 120 \\ 0 & 0 & 2396 & 1191 \\ 0 & 0 & 0 & 2416 \end{pmatrix}.$$

5.6 Problems. 1) Let Ω_n be a partition of the interval $[a, b]$, and let $\tilde{s} \in \mathrm{S}_1(\Omega_n)$ be the spline interpolating a function $f \in \mathrm{C}_2[a, b]$. Using the Peano Kernel Formula, show that

$$\|f - \tilde{s}\|_\infty \leq \frac{h^2}{8}\|f''\|_\infty \quad \text{and} \quad \|f' - \tilde{s}'\|_\infty \leq \frac{h}{2}\|f''\|_\infty,$$

where h is the maximal distance between two neighboring knots.

2) Prove the following *Inequality of Wirtinger*: If $u \in \mathrm{C}_1[0, 2\pi]$ with $u(0) = u(2\pi)$ and $\int_0^{2\pi} u(t)dt = 0$, then

$$\int_0^{2\pi} [u(t)]^2 dt \leq \int_0^{2\pi} [\frac{du}{dt}]^2 dt.$$

Hint: Use the Fourier expansion of f.

In addition, show that for $f \in \mathrm{C}_1[a, b]$ with $f(a) = f(b) = 0$, this inequality implies

$$\pi^2 \int_a^b [f(x)]^2 dx \leq (b-a)^2 \int_a^b [f'(x)]^2 dx.$$

3) a) Show that the Integral Relation 2.1 also holds for linear splines $(m = 1)$.

b) Using this and the inequality established in Problem 2, show that the interpolating spline $\tilde{s}$ satisfies $\|f - \tilde{s}\|_2 \leq \frac{h}{\pi}\|f'\|_2$ for any $f \in \mathrm{C}_1[a, b]$. *Hint:* Establish the inequality $\|f - \tilde{s}\|_2 \leq \frac{h}{\pi}\|f' - \tilde{s}'\|_2$.

4) a) Prove the "Second Integral Relation":

$$\|f' - \tilde{s}'\|_2^2 = \langle f - \tilde{s}, f''\rangle$$

for any $f \in C_2[a,b]$, where $\tilde{s} \in S_1(\Omega_n)$ is the interpolating linear spline.

b)Using this and Problem 3b), show that

$$\|f - \tilde{s}\|_2 \leq \frac{h^2}{\pi^2}\|f''\|_2.$$

5) Let Ω_n be a partition of the interval $[a,b]$, and let $\tilde{s} \in S_3(\Omega_n)$ be the interpolating cubic spline with Hermite end conditions for a given function $f \in C_4[a,b]$. Prove that if $h := \max_{\nu=0,\dots,n-1} |x_{\nu+1} - x_\nu|$, then

$$\|f - \tilde{s}\|_2 \leq 4\frac{h^4}{\pi^4}\|f^{(4)}\|_2$$

by carrying out the following steps:

a) Using integration by parts and applying the inquality in Problem 2, show that

$$\|f - \tilde{s}\|_2 \leq \frac{h}{\pi}\|f' - \tilde{s}'\|_2;$$

b) applying Rolle's Theorem to $f - \tilde{s}$, show that

$$\|f' - \tilde{s}'\|_2^2 \leq \frac{(2h)^2}{\pi^2}\|f - \tilde{s}\|_2\|f^{(4)}\|_2,$$

which implies the desired bound.

6) Using the results of 5.3, solve Problem 3 in 1.4 of finding the best approximation on $f(x) := x^2$ on $[0,1]$ from $S_1(\Omega_n)$ with respect to $\|\cdot\|_2$ for an equidistant partition with $n = 5$ and $n = 10$. Work out the bounds in 5.2 and 5.3 and in the formula (**) in 5.1 for $\|f - \hat{s}\|_\infty$, and check their tightness and the order of convergence with respect to h numerically.

6. Multidimensional Splines

As we have already seen in our general discussion of multidimensional interpolation in 5.6, the generalization of spline methods to several dimensions leads to a series of new questions. In this section we treat rectangular domains in two dimensions. As in 5.6.2, we suppose we are given a set of $(n+1)(k+1)$ grid points associated with

$$a = x_0 < \cdots < x_n = b,$$
$$c = y_0 < \cdots < y_k = d.$$

We shall investigate approximations which can be represented as splines in the x- and y-directions.

6.1 Bilinear Splines. As analogs to the one-dimensional linear B-splines, we introduce the basis functions

$$B_{1\nu\kappa}(x,y) := \begin{cases} \frac{(x_{\nu+2}-x)(y_{\kappa+2}-y)}{(x_{\nu+2}-x_{\nu+1})(y_{\kappa+2}-y_{\kappa+1})} & \text{for } (x,y) \in \text{I} \\ \frac{(x-x_{\nu})(y_{\kappa+2}-y)}{(x_{\nu+1}-x_{\nu})(y_{\kappa+2}-y_{\kappa+1})} & \text{for } (x,y) \in \text{II} \\ \frac{(x-x_{\nu})(y-y_{\kappa})}{(x_{\nu+1}-x_{\nu})(y_{\kappa+1}-y_{\kappa})} & \text{for } (x,y) \in \text{III} \\ \frac{(x_{\nu+2}-x)(y-y_{\kappa})}{(x_{\nu+2}-x_{\nu+1})(y_{\kappa+1}-y_{\kappa})} & \text{for } (x,y) \in \text{IV} \end{cases}$$

for $\nu = -1, \ldots, n-1$ and $\kappa = -1, \ldots, k-1$. Here

$$\text{I} := [x_{\nu+1}, x_{\nu+2}] \times [y_{\kappa+1}, y_{\kappa+2}], \quad \text{II} := [x_{\nu}, x_{\nu+1}] \times [y_{\kappa+1}, y_{\kappa+2}],$$
$$\text{III} := [x_{\nu}, x_{\nu+1}] \times [y_{\kappa}, y_{\kappa+1}], \quad \text{IV} := [x_{\nu+1}, x_{\nu+2}] \times [y_{\kappa}, y_{\kappa+1}].$$

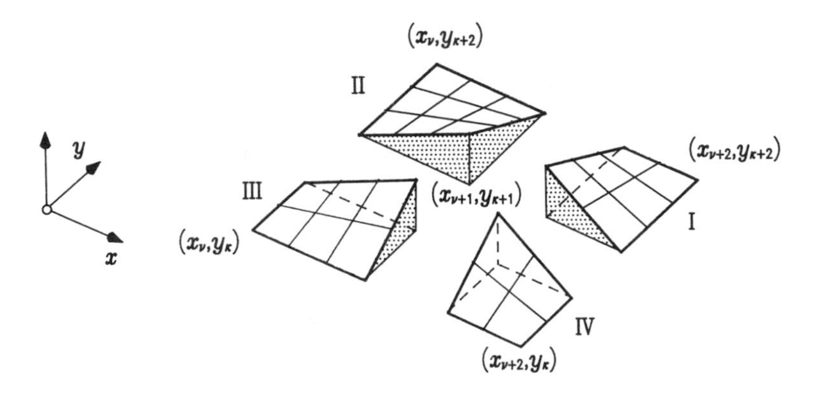

Using these basis splines, we consider expansions of the form

$$s = \sum_{\substack{-1 \le \nu \le n-1 \\ -1 \le \kappa \le k-1}} \alpha_{\nu\kappa} B_{1\nu\kappa}.$$

Since each of the basis functions $B_{1\nu\kappa}$ is a product $B_{1\nu\kappa}(x,y) = B^x_{1\nu}(x) B^y_{1\kappa}(y)$ of two linear B-splines, $B^x_{1\nu}$ in the x-direction and $B^y_{1\kappa}$ in the y-direction, we call s a *bilinear* spline. The unique spline which interpolates a given function $f : [a,b] \times [c,d] \to \mathbb{R}$ at all of the grid points is given by

$$s(x,y) = \sum_{\substack{-1 \le \nu \le n-1 \\ -1 \le \kappa \le k-1}} f(x_{\nu+1}, y_{\kappa+1}) B_{1\nu\kappa}(x,y).$$

In general, if $\text{X}_r := \text{span}(\varphi_1, \ldots, \varphi_r)$ and $\text{Y}_m := \text{span}(\psi_1, \ldots, \psi_m)$ are linear function spaces of dimension r and m, respectively, then the

Tensor-Product

$$X_r \otimes Y_m := \text{span}\{\varphi_\rho\psi_\mu \mid (\varphi_\rho\psi_\mu)(x,y) := \varphi_\rho(x)\psi_\mu(y), \quad \text{for } 1 \le \rho \le r, 1 \le \mu \le m\}$$

is defined to be the linear space of dimension $r \cdot m$ spanned by the products $\varphi_\rho\psi_\mu$. Thus the space of bilinear splines is the tensor-product space of dimension $(n+1)(k+1)$ obtained from the spaces $\text{span}(B^x_{1,-1},\dots,B^x_{1,n-1})$ and $\text{span}(B^y_{1,-1},\dots,B^y_{1,k-1})$.

To bound the error of a bilinear interpolating spline $s(x,y)$, we use the results of 5.6.3. In particular, if $f \in C_2([a,b]\times[c,d])$ and

$$h_x := \max_{\nu=0,\dots,n-1} |x_{\nu+1} - x_\nu|, \qquad h_y := \max_{\kappa=0,\dots,k-1} |y_{\kappa+1} - y_\kappa|,$$

then we have the

Error Bound

$$\|f - \tilde{s}\|_\infty \le \frac{1}{8}(h_x^2\|D_x^2 f\|_\infty + h_y^2\|D_y^2 f\|_\infty).$$

6.2 Bicubic Splines. Using the tensor-products of the cubic B-splines $\{B^x_{3,-3},\dots,B^x_{3,n-1}\}$ and $\{B^y_{3,-3},\dots,B^y_{3,k-1}\}$, each bicubic spline can be written as

$$s(x,y) = \sum_{\substack{-3\le\nu\le n-1\\ -3\le\kappa\le k-1}} \alpha_{\nu\kappa} B^x_{3\nu}(x) B^y_{3\kappa}(y).$$

We can make s satisfy the interpolation conditions

$$s(x_\nu, y_\kappa) = f(x_\nu, y_\kappa)$$

for $\nu = 0,\dots,n$ and $\kappa = 0,\dots,k$ by appropriately choosing the $(n+3)(k+3)$ coefficients $\alpha_{\nu\kappa}$, $-3 \le \nu \le n-1$, $-3 \le \kappa \le k-1$.

To determine all of the coefficients uniquely, we need to add boundary conditions. For example, adding the Hermite boundary conditions

$$\left.\begin{aligned} D_x s(x_0, y_\kappa) &= D_x f(x_0, y_\kappa) \\ D_x s(x_n, y_\kappa) &= D_x f(x_n, y_\kappa) \end{aligned}\right\}, 0 \le \kappa \le k,$$

$$\left.\begin{aligned} D_y s(x_\nu, y_0) &= D_y f(x_\nu, y_0) \\ D_y s(x_\nu, y_k) &= D_y f(x_\nu, y_k) \end{aligned}\right\}, 0 \le \nu \le n,$$

and the conditions

$$D_x D_y s(x_\nu, y_\kappa) = D_x D_y f(x_\nu, y_\kappa),$$

$\nu = 0, n$ and $\kappa = 0, k$ at the four corners, we get a total of

$$(n+1)(k+1) + 2(n+1) + 2(k+1) + 4 = (n+3)(k+3)$$

conditions which uniquely determine the coefficients $\alpha_{\nu\kappa}$. Natural and periodic splines can be treated similarly.

If $f \in C_4([a,b] \times [c,d])$, then we have the

Error Bound

$$\boxed{\|f - \tilde{s}\|_\infty \leq \frac{5}{384} h_x^4 \|D_x^4 f\|_\infty + \frac{4}{9} h_x^2 h_y^2 \|D_x^2 D_y^2 f\|_\infty + \frac{5}{384} h_y^4 \|D_y^4 f\|_\infty .}$$

This bound, whose proof can be found in the paper of C. A. Hall [1968], is optimal.

6.3 Spline-Blended Functions. Another way of generalizing one-dimensional splines to two dimensions, different from tensor-products, is to form certain *spline-blended functions*. They were originally developed for use in designing surfaces for technical products; see W. J. Gordon [1969].

Given a function $f : [a,b] \times [c,d] \to \mathbb{R}$ of two variables, let $\mathcal{P}_x f$ denote the spline which interpolates f with respect to the variable x, and let $\mathcal{P}_y f$ be the corresponding spline with respect to y. Then (cf. 6.2), the tensor-product spline s can be written as $s(x,y) = ((\mathcal{P}_x \mathcal{P}_y) f)(x,y)$. We can define a different approximation of f using the *Boolean Sum*

$$(\mathcal{P}_x \oplus \mathcal{P}_y) f := \mathcal{P}_x f + \mathcal{P}_y f - \mathcal{P}_x \mathcal{P}_y f.$$

This function includes the tensor-product spline as one term, but is itself not a spline since it contains other non-spline terms formed from the function f.

The symmetry $(\mathcal{P}_x \mathcal{P}_y) f = (\mathcal{P}_y \mathcal{P}_x) f$ of the tensor-product mapping shows that $(\mathcal{P}_x \oplus \mathcal{P}_y) = (\mathcal{P}_y \oplus \mathcal{P}_x)$. Moreover, it also implies that the Boolean sum satisfies

$$\begin{aligned}
((\mathcal{P}_x \oplus \mathcal{P}_y) f)(x_\nu, y) &= (\mathcal{P}_x f)(x_\nu, y) + (\mathcal{P}_y f)(x_\nu, y) - ((\mathcal{P}_x \mathcal{P}_y) f)(x_\nu, y) \\
&= f(x_\nu, y) + (\mathcal{P}_y f)(x_\nu, y) - (\mathcal{P}_y (\mathcal{P}_x f))(x_\nu, y) \\
&= f(x_\nu, y) + (\mathcal{P}_y f)(x_\nu, y) - (\mathcal{P}_y f)(x_\nu, y) \\
&= f(x_\nu, y)
\end{aligned}$$

on the grid line (x_ν, y) and

$$((\mathcal{P}_x \oplus \mathcal{P}_y) f)(x, y_\kappa) = f(x, y_\kappa)$$

on the grid line (x, y_κ). We have established the

Interpolation Property. *Spline-blended functions interpolate not only at the grid points* (x_ν, y_κ), *but also all along each of the grid lines* (x_ν, y), $0 \le \nu \le n$, *and* (x, y_κ), $0 \le \kappa \le k$.

Spline-blended functions $(\mathcal{P}_x \oplus \mathcal{P}_y)f =: \sigma$ can be computed with the help of B-splines. Indeed, we can write

$$\sigma(x,y) = \sum_{\nu=-\ell}^{n-1} \beta_\nu(y) B_{\ell\nu}(x) + \sum_{\kappa=-\ell}^{k-1} \gamma_\kappa(x) B_{\ell\kappa}(y) - \sum_{\substack{-\ell \lesssim \nu \lesssim n-1 \\ -\ell \lesssim \kappa \lesssim k-1}} \alpha_{\nu\kappa} B_{\ell\nu}(x) B_{\ell\kappa}(y),$$

where the numbers $\alpha_{\nu\kappa}$ are the coefficients of the tensor-product spline interpolating at the corners of the grid, and where the functions $\beta_\nu(y)$ and $\gamma_\kappa(x)$ can be found by computing $(\mathcal{P}_x f)(x,y)$ and $(\mathcal{P}_y f)(x,y)$.

In the linear case this expansion becomes especially simple:

$$\sigma(x,y) = \sum_{\nu=-1}^{n-1} f(x_{\nu+1}, y) B_{1\nu}(x) + \sum_{\kappa=-1}^{k-1} f(x, y_{\kappa+1}) B_{1\kappa}(y) - \\ - \sum_{\substack{-1 \lesssim \nu \lesssim n-1 \\ -1 \lesssim \kappa \lesssim k-1}} f(x_{\nu+1}, y_{\kappa+1}) B_{1\nu}(x) B_{1\kappa}(y).$$

Accuracy. The interpolation property of spline-blended functions suggests that we can expect high order of accuracy. We now show that the order of approximation by spline-blended functions can be estimated by using our earlier results on the accuracy of the one-dimensional interpolating splines from which the spline-blended function is constructed.

In particular, let $\mathcal{P}\varphi$ denote a one-dimensional spline which interpolates the function φ on the interval $\mathrm{I} \subset \mathbb{R}$ and satisfies the error bound

$$(*) \qquad \|\varphi - \mathcal{P}\varphi\|_\infty \le C h^r \|\varphi^{(r)}\|_\infty$$

for all r-times continuously differentiable functions φ, where C is a fixed constant. Then we get the

Error Bound for Spline-Blended Functions. *Let* $f \in \mathrm{C}_{2r}([a,b] \times [c,d])$, *and let* $\sigma := (\mathcal{P}_x \oplus \mathcal{P}_y)f$ *be a spline-blended function such that both* $\mathcal{P}_x f$ *and* $\mathcal{P}_y f$ *satisfy an error bound of the form* $(*)$. *Then the error is bounded by*

$$\|f - \sigma\|_\infty \le C_1 C_2 h_x^r h_y^r \|D_x^r D_y^r f\|_\infty .$$

Proof. We again make use of the commutativity properties

$$D_x^r(\mathcal{P}_y f) = \mathcal{P}_y(D_x^r f) \quad \text{and} \quad D_y^r(\mathcal{P}_x f) = \mathcal{P}_x(D_y^r f)$$

used already above in 5.6.3. This gives

$$\|f-\sigma\|_\infty = \|(f-\mathcal{P}_x f) - \mathcal{P}_y(f-\mathcal{P}_x f)\|_\infty \le C_2\, h_y^r \|D_y^r(f-\mathcal{P}_x f)\|_\infty =$$
$$= C_2\, h_y^r \|D_y^r f - \mathcal{P}_x(D_y^r f)\|_\infty \le C_1 C_2\, h_y^r h_x^r \|D_x^r D_y^r f\|_\infty. \qquad \square$$

Application. From (**) in 5.1, we know that the linear interpolating splines satisfy the error bound

$$\|\varphi - \tilde{s}\|_\infty \le \frac{h^2}{8}\|\varphi''\|_\infty.$$

It follows that the *linearly* spline-blended function satisfies

$$\boxed{\|f-\sigma\|_\infty \le \frac{h_x^2 h_y^2}{64}\|D_x^2 D_y^2 f\|_\infty}$$

where $r = 2$. If we use cubic interpolating splines with Hermite end conditions, we get a blended function σ which, as the reader may verify, satisfies the same Hermite boundary conditions given in 6.2 for bicubic splines. Using the Error Bound 5.4

$$\|\varphi - \tilde{s}\|_\infty \le \frac{h^4}{16}\|\varphi^{(4)}\|_\infty,$$

which holds for equidistant knots, we get

$$\boxed{\|f-\sigma\|_\infty \le \frac{h_x^4 h_y^4}{256}\|D_x^4 D_y^4 f\|_\infty.}$$

Here $r := 4$.

If the domain of the spline-blended function is a square, so that $h_x = h_y =: h$, then in the linear case we have

$$\|f-\sigma\|_\infty \le \frac{h^4}{64}\|D_x^2 D_y^2 f\|_\infty,$$

and in the cubic case we have

$$\|f-\sigma\|_\infty \le \frac{h^8}{256}\|D_x^4 D_y^4 f\|_\infty.$$

Comment. Compared with the tensor-product spline, the spline-blended function has an order of approximation which is twice as high. Error bounds of the order $O(h^8)$, which we have just shown to hold for functions blended with cubic splines, are exceptionally high, and lead to outstanding approximations. We must pay for this high order of approximation with the complicated structure of the spline-blended function. The choice of what kind of two-dimensional approximation to use in a given setting has to be made on a case-by-case basis.

The problem of approximating a given surface is only one of the problem areas which require two-dimensional approximations. As in one dimension, they are also needed in the numerical treatment of operator equations, especially those arising from partial differential equations and integral equations.

6.4 Problems. 1) Starting from Problem 4 in 5.6, show that for f in $C_2([a,b] \times [c,d])$, the interpolating bilinear spline satisfies:

$$\|f - \tilde{s}\|_2 \leq \frac{1}{\pi^2}(h_x^2\|D_x^2 f\|_2 + h_x h_y \|D_x D_y f\|_2 + h_y^2 \|D_y^2 f\|_2);$$
$$\|D_x(f - \tilde{s})\|_2 \leq \frac{1}{\pi}(h_x\|D_x^2 f\|_2 + 2h_y\|D_x D_y f\|_2);$$
$$\|D_y(f - \tilde{s})\|_2 \leq \frac{1}{\pi}(h_y\|D_y^2 f\|_2 + 2h_x\|D_x D_y f\|_2).$$

2) As in Problem 1, show that the bicubic spline interpolating a function $f \in C_4([a,b] \times [c,d])$ with Hermite boundary conditions satisfies

$$\|f - \tilde{s}\|_2 \leq \frac{4}{\pi^4}(h_x^4\|D_x^4 f\|_2 + h_x^2 h_y^2 \|D_x^2 D_y^2 f\|_2 + h_y^4 \|D_y^4 f\|_2).$$

3) Establish error bounds for $\|f - \sigma\|_2$ for functions blended with linear and cubic splines.

4) Suppose we approximate the function $f(x,y) = \sin(\pi x)\sin(\pi y)$ for $x \in [0,1]$, $y \in [0,1]$

a) by bilinear tensor-product splines with the following step sizes: $h_x = h_y = \frac{1}{2}$, $h_x = h_y = \frac{1}{3}$ and $h_x = h_y = \frac{1}{4}$;

b) by blending with linear splines with the same step size.

c) For both cases a) and b), calculate the error at $x = y = \frac{5}{12}$, and check numerically that the errors are of order $O(h^2)$ and $O(h^4)$, respectively.

7
Integration

The numerical computation of definite integrals is one of the oldest problems in mathematics. The problem, which in its earliest form involved finding the area of regions bounded by curved lines, has been around for thousands of years, long before the concept of integrals in the framework of analysis was developed in the 17th and 18th century. Certainly the best-known example of this problem was that of computing the area contained in a circle, which in turn led to a study of the number π and its computation. Using a numerical method involving the approximation of a circle by inscribed and circumscribed polygons, Archimedes (287–212 B.C.) was able to give the astonishingly good bounds $3\,\frac{10}{71} < \pi < 3\,\frac{1}{7}$. For more on this, see Chap. 5 of the book *Numbers* by H.-D. Ebbinghaus, et.al. [1990].

Numerical integration is often referred to as *numerical quadrature.* This nomenclature comes from the problem of computing the area of a circle, which can be thought of as finding a square with the same area. The numerical computation of two-dimensional integrals is often referred to as *numerical cubature*, while the calculation of n-dimensional integrals is called n-dimensional numerical integration. Both of these topics will also be treated in this chapter.

We now describe three situations where it is necessary to calculate approximations to definite integrals. The first is the case where the antiderivative of the function to be integrated cannot be expressed in terms of elementary functions. Typical examples of this include finding the value of $\int_0^\infty e^{-x^2}\,dx$, or finding the arclength of an ellipse. The second situation arises when the antiderivative can be written down, but is so complicated that the application of a quadrature method for its numerical evaluation is desirable. The third situation arises when the integrand is given only pointwise, for example as the result of measuring it.

The last case also arises when quadrature methods are applied in the numerical treatment of differential- or integral-equations. Numerous methods for discretizing such equations rest on numerical integration methods. We have seen that numerical quadrature plays a central role in solving problems from a wide variety of mathematical applications.

1. Interpolatory Quadrature

We begin by discussing the idea of approximately computing the value of a definite integral $\int_a^b f(x)dx$ by replacing the integrand f by an approximation $\tilde{f}$ which is simple to integrate. At first, we require only that the function f be Riemann integrable. Then it is reasonable to approximate the integrand using interpolating polynomials.

To this end, let $a = x_0 < x_1 < \cdots < x_n = b$ be a partition of the interval. As a first method, we use a piecewise constant function to interpolate the function f. This leads to the

1.1 Rectangle Rules. Suppose we interpolate f in each of the subintervals $(x_\nu, x_{\nu+1})$, $0 \le \nu \le n-1$, by the constant $f(x_\nu^*)$, $x_\nu^* \in [x_\nu, x_{\nu+1}]$. Then the sum

$$Qf := \sum_{\nu=0}^{n-1} f(x_\nu^*)(x_{\nu+1} - x_\nu)$$

provides an approximate value for the definite integral $\int_a^b f(x)dx$. Since sums of this type can be interpreted geometrically as the sum of rectangles, we refer to them as *rectangle rules.* The following two specific rules are particularly natural:

(a) $x_\nu^* := x_\nu$ which gives $Q_a f := \sum_0^{n-1} f(x_\nu)(x_{\nu+1} - x_\nu)$.
(b) $x_\nu^* := x_{\nu+1}$ which gives $Q_b f := \sum_0^{n-1} f(x_{\nu+1})(x_{\nu+1} - x_\nu)$.

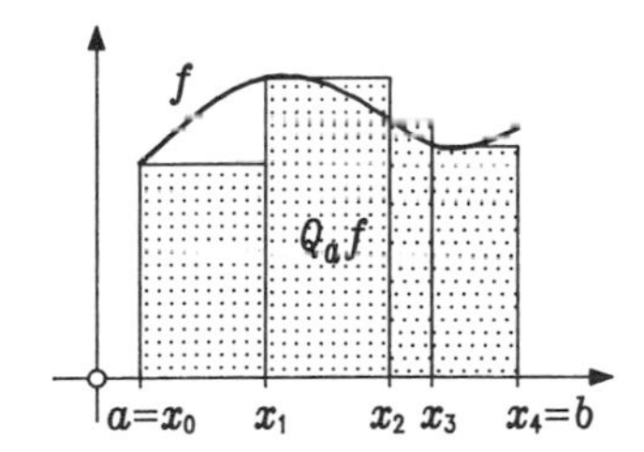

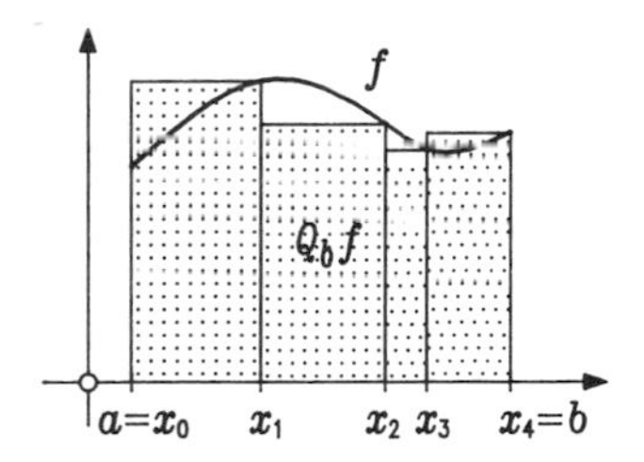

Another natural choice is to take $x_\nu^* := \frac{x_{\nu+1}+x_\nu}{2}$, which leads to the

Midpoint Rule $$Q_M f := \sum_{\nu=0}^{n-1} f\left(\frac{x_{\nu+1}+x_\nu}{2}\right)(x_{\nu+1} - x_\nu).$$

If the partition of the interval is equally spaced, then we get

$$h := x_{\nu+1} - x_\nu = \frac{b-a}{n}, \ 0 \le \nu \le n-1,$$

$$Q_a f = h \sum_{\nu=0}^{n-1} f(x_\nu), \quad Q_b f = h \sum_{\nu=1}^{n} f(x_\nu) \text{ and}$$

$$Q_M f = h \sum_{\nu=0}^{n-1} f(x_\nu + \frac{h}{2}).$$

The possibility of finding a useful error bound for this method depends on what assumptions we are willing to make on f. For example, if f is only assumed to be continuous, i.e., $f \in C[a,b]$, then in the equally-spaced case we have

$$\left| \int_{x_\nu}^{x_{\nu+1}} f(x)dx - hf(x_\nu^*) \right| = \left| \int_{x_\nu}^{x_{\nu+1}} [f(x) - f(x_\nu^*)]dx \right| \le$$
$$\le \max_{x \in [x_\nu, x_{\nu+1}]} |f(x) - f(x_\nu^*)| h,$$

and using the modulus of continuity ω_f of f, we get the

Error Bounds

$$\left| \int_a^b f(x)dx - Qf \right| \le \omega_f(h) \cdot (b-a) \ \text{ for } Q := Q_a \text{ and } Q := Q_b$$

and

$$\left| \int_a^b f(x)dx - Qf \right| \le \omega_f(\frac{h}{2}) \cdot (b-a) \ \text{ for the midpoint rule.}$$

If we assume that $f \in C_1[a,b]$, then using the mean-value theorem, we get

$$f(x) = f(x_\nu^*) + f'(\xi_\nu)(x - x_\nu^*), \quad \xi_\nu \in (\min(x, x_\nu^*), \max(x, x_\nu^*)),$$

and thus

$$\int_{x_\nu}^{x_{\nu+1}} f(x)dx - hf(x_\nu^*) = \int_{x_\nu}^{x_{\nu+1}} f'(\xi_\nu)(x - x_\nu^*)dx.$$

Now for $x_\nu^* := x_\nu$, this implies

$$\int_{x_\nu}^{x_{\nu+1}} f(x)dx - hf(x_\nu) = f'(\xi_\nu^*)\frac{h^2}{2}, \quad \xi_\nu^* \in (x_\nu, x_{\nu+1}),$$

since ξ_ν is a continuous function of x. It follows that the corresponding total quadrature error $R_n f$ defined by

$$\int_a^b f(x)dx = Q_a f + R_n f$$

is

$$R_n f = \frac{h^2}{2}\sum_{\nu=0}^{n-1} f'(\xi_\nu^*) = \frac{h}{2}(b-a)\frac{1}{n}\sum_{\nu=0}^{n-1} f'(\xi_\nu^*) = \frac{h}{2} f'(\xi)(b-a),$$

where $\xi \in (a,b)$ by the intermediate value theorem. For the quadrature formula Q_b with $x_\nu^* := x_{\nu+1}$, we get $R_n f = -\frac{h}{2} f'(\hat{\xi})(b-a)$, $\hat{\xi} \in (a,b)$, and hence both of the quadrature rules Q_a and Q_b satisfy the

Error Bound for the Rectangle Rules

$$\boxed{|R_n f| \le \frac{h}{2} \max_{x\in[a,b]} |f'(x)|(b-a).}$$

For the midpoint rule where $x_\nu^* := \frac{x_\nu + x_{\nu+1}}{2}$, we have

$$\left| \int_{x_\nu}^{x_{\nu+1}} f(x)dx - hf\left(\frac{x_\nu + x_{\nu+1}}{2}\right) \right| \le \int_{x_\nu}^{x_{\nu+1}} |f'(\xi_\nu)| \left| x - \frac{x_\nu + x_{\nu+1}}{2} \right| dx,$$

and thus,

$$|R_n f| \le \frac{h}{4} \max_{x\in[a,b]} |f'(x)|(b-a).$$

We can get a better estimate for the midpoint rule, however, if we assume that $f \in C_2[a,b]$. In this case,

$$f(x) = f(x_\nu + \frac{h}{2}) + f'(x_\nu + \frac{h}{2})[x - (x_\nu + \frac{h}{2})] + \frac{1}{2} f''(\xi_\nu)[x - (x_\nu + \frac{h}{2})]^2,$$

and so

$$\int_{x_\nu}^{x_{\nu+1}} f(x)dx - hf(x_\nu + \frac{h}{2}) = \frac{h^3}{24} f''(\xi_\nu^*), \quad \xi_\nu^* \subset (x_\nu, x_{\nu+1}),$$

and it follows as above that the quadrature error of the midpoint rule is given by

$$R_n f = \frac{h^3}{24}\sum_{\nu=0}^{n-1} f''(\xi_\nu^*) = \frac{h^2}{24} f''(\xi)(b-a), \quad \xi \in (a,b).$$

This leads to the

Error Bound for the Midpoint Rule

$$\boxed{|R_n f| \le \frac{h^2}{24} \max_{x\in[a,b]} |f''(x)|(b-a).}$$

It should be noted that this bound is of the order $O(h^2)$. Even though the midpoint rule, along with the rectangle rules Q_a and Q_b, are all based on approximating f by polynomials $p \in \mathrm{P}_0$, it nevertheless gives *quadratic* convergence in h.

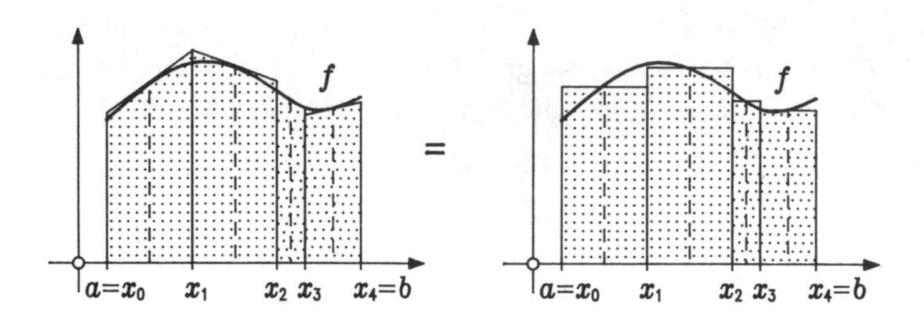

It is easy to see, however, that the midpoint rule can also be interpreted as follows: It arises from the approximation of the function f on the interval $[x_\nu, x_{\nu+1}]$ by the piecewise function composed of the tangent lines at the points $(x_\nu + \frac{h}{2})$. The fact that polynomials $p \in \mathrm{P}_1$ are involved explains the quadratic convergence. This interpretation also explains why the midpoint rule is sometimes referred to as the *Tangent Trapezoidal Rule.*

1.2 The Trapezoidal Rule. If we interpolate the function f by a piecewise function which is linear on each of the n subintervals $[x_\nu, x_{\nu+1}]$, $0 \le \nu \le n-1$, then the sum of the areas of the corresponding trapezoids is

$$T_n f := \sum_{\nu=0}^{n-1} \frac{y_\nu + y_{\nu+1}}{2}(x_{\nu+1} - x_\nu),$$

where $y_\nu := f(x_\nu)$. This approximation to the definite integral is called the *Trapezoidal Rule*, and we denote the remainder by $R_n f$; i.e.,

$$\int_a^b f(x)dx = T_n f + R_n f.$$

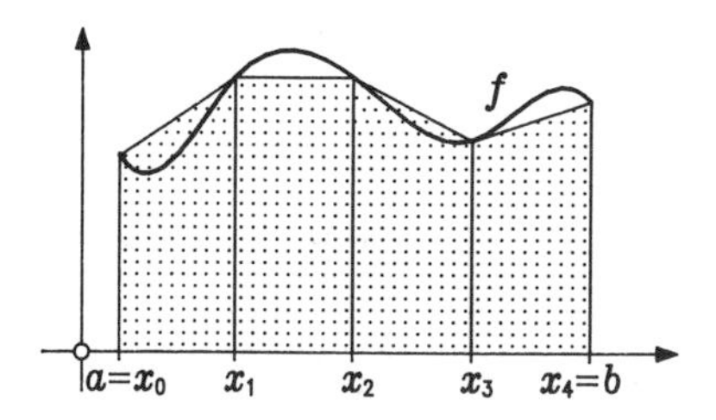

For equidistant nodes, the trapezoidal rule takes the form

$$\boxed{T_n f = h[\frac{1}{2}y_0 + \sum_{\nu=1}^{n-1} y_\nu + \frac{1}{2}y_n].}$$

To estimate $R_n f$, we make use of the Peano Kernel Formula 5.2.4. First, suppose that $f \in C_1[a,b]$. Then since the error functional R annihilates all elements $f \in P_1$, and thus also all $f \in P_0$, we have

$$Rf := \int_{x_\nu}^{x_{\nu+1}} f(x)dx - \frac{h}{2}[f(x_\nu) + f(x_{\nu+1})] = \int_{x_\nu}^{x_{\nu+1}} K_0(t) f'(t)dt$$

with $m := 0$, where $K_0(t) = Rq_0(\cdot,t)$ and $q_0(x,t) = \begin{cases} 1 & \text{for } t \le x \\ 0 & \text{for } x < t \end{cases}$. Now

$$\begin{aligned} Rq_0(\cdot,t) &= \int_{x_\nu}^{x_{\nu+1}} q_0(x,t)dx - \frac{h}{2}[q_0(x_\nu,t) + q_0(x_{\nu+1},t)] = \\ &= \int_{x_\nu}^{t} q_0(x,t)dx + \int_{t}^{x_{\nu+1}} q_0(x,t)dx - \frac{h}{2}[0+1] \text{ for } x_\nu < t \le x_{\nu+1}, \end{aligned}$$

and thus

$$Rq_0(\cdot,t) = \int_{t}^{x_{\nu+1}} q_0(x,t)dx - \frac{h}{2} = x_{\nu+1} - t - \frac{h}{2} = x_\nu + \frac{h}{2} - t.$$

It follows that

$$R_n f = \int_a^b f(x)dx - T_n f = \int_a^b K_0(t) f'(t)dt$$

with the Peano kernel

$$K_0(t) = (x_\nu + \frac{h}{2}) - t \text{ for } x_\nu < t \le x_{\nu+1},\ 0 \le \nu \le n-1.$$

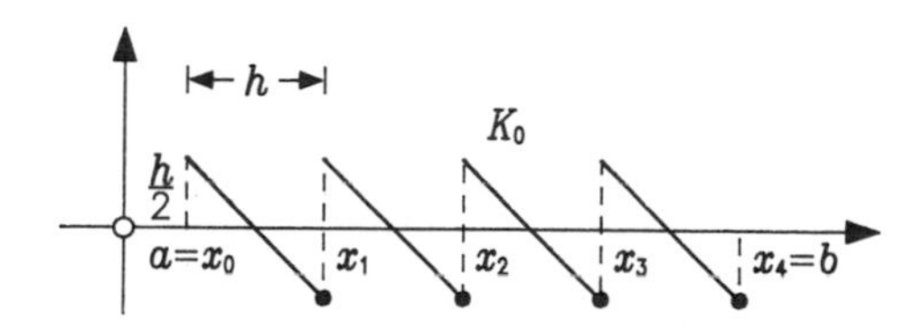

This leads immediately to the

Estimate

$$|R_nf| \le \frac{h}{4} \max_{x\in[a,b]} |f'(x)|(b-a)$$

for the quadrature error which arises in using the trapezoidal rule T_n on a function $f \in C_1[a,b]$.

Since the trapezoidal rule is also exact for all linear functions, we can take $m = 1$ in the Peano kernel formula provided $f \in C_2[a,b]$, and thus derive a better error bound. After integrating by parts, we get

$$(*) \qquad \int_a^b f(x)dx - T_nf = f'(t)\hat{K}(t)|_a^b - \int_a^b \hat{K}(t)f''(t)dt,$$

where $\hat{K}(t) := \int_t K_0(\tau)d\tau = -\frac{1}{2}[(x_\nu + \frac{h}{2}) - t]^2 + c_1$, $t \in [x_\nu, x_{\nu+1}]$, and $c_1 \in \mathbb{R}$ is arbitrary.

Now if we choose $c_1 := \frac{h^2}{8}$, then $\hat{K}(a) = \hat{K}(b) = 0$, and we have

$$R_nf = \int_a^b f(x)dx - T_nf = -\int_a^b \hat{K}(t)f''(t)dt.$$

This is the desired Peano kernel formula, where $K_1(t) = -\hat{K}(t)$ is given by

$$K_1(t) = \frac{1}{2}[(x_\nu + \frac{h}{2}) - t]^2 - \frac{h^2}{8} = \frac{1}{2}(x_\nu - t)(x_{\nu+1} - t)$$

for $t \in [x_\nu, x_{\nu+1}]$, $0 \le \nu \le n-1$.

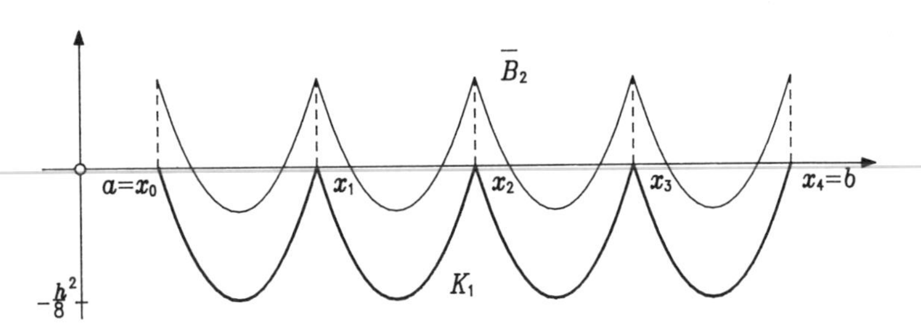

Since K_1 is of one sign, this implies that

$$R_nf = f''(\xi) \int_a^b K_1(t)dt = f''(\xi)n \cdot \frac{h^3}{12}, \quad \xi \in (a,b),$$

which leads immediately to the

Quadrature Error

$$R_nf = \frac{h^2}{12} f''(\xi)(b-a)$$

and the

Error Bound

$$\boxed{|R_n f| \leq \frac{h^2}{12} \max_{x\in[a,b]} |f''(x)|(b-a).}$$

1.3 The Euler-MacLaurin Expansion. In this section we show how variants of the trapezoidal rule can be obtained if we choose c_1 in the formula (*) in 1.2 so that $\int_a^b \hat{K}(t)f''(t)dt = 0$ for all $f \in \mathrm{P}_2$. The appropriate choice is $c_1 := \frac{h^2}{24}$, and we get

$$(*) \qquad \int_a^b f(x)dx = T_n f - \frac{h^2}{12}[f'(b) - f'(a)] + h^2 \int_a^b \overline{B}_2(t)f''(t)dt$$

with

$$\overline{B}_2(t) := \frac{1}{2}\Big[\frac{t-x_\nu}{h} - \frac{1}{2}\Big]^2 - \frac{1}{24} \quad \text{for } t \in [x_\nu, x_{\nu+1}],\ 0 \leq \nu \leq n-1.$$

If f is sufficiently differentiable, we can use integration by parts on the last term to develop this expansion still further. For this purpose, it is convenient to define the *Bernoulli polynomials* $B_j : [0,1] \to \mathrm{I\!R}$, $(j = 0, 1, \ldots)$, by the recurrence relation

$$B_0(\xi) := 1, \ \frac{d}{d\xi}B_j(\xi) := B_{j-1}(\xi) \ \text{ with}$$

$$\int_0^1 B_j(\xi)d\xi = 0 \ \text{ for } j = 1, 2, \ldots.$$

This gives $B_1(\xi) = \xi - \frac{1}{2}$, $B_2(\xi) = \frac{1}{2}(\xi - \frac{1}{2})^2 - \frac{1}{24}$. It now follows that the kernel $\overline{B}_2$ appearing in (*) above satisfies $\overline{B}_2(t) = B_2(\frac{t-x_\nu}{h})$ for t in $[x_\nu, x_{\nu+1}]$, for all $0 \leq \nu \leq n-1$. Thus $\overline{B}_2$ is the periodic function which is obtained by piecing together transformed Bernoulli polynomials as shown in the figure above.

Properties of the Bernoulli Polynomials.

(i) *Symmetry.*
Since $\int_0^1 B_j(\xi)d\xi = B_{j+1}(1) - B_{j+1}(0) = 0$ for $j \geq 1$, we see that $B_{j+1}(0) = B_{j+1}(1)$. We now show that, in general, we even have the stronger form of symmetry

$$B_j(\xi) = (-1)^j B_j(1-\xi)$$

for $j \geq 0$. We prove this by induction:
The assertion is valid for $j = 0$ and for $j = 1$. In addition, we have

$$B_{j+1}(\xi) - B_{j+1}(0) = \int_0^\xi B_j(\eta)d\eta = (-1)^j \int_0^\xi B_j(1-\eta)d\eta =$$

$$= (-1)^{j+1} \int_1^{1-\xi} B_j(\theta)d\theta = (-1)^{j+1}[B_{j+1}(1-\xi) - B_{j+1}(1)].$$

Now for $m \geq 1$, if

(a) $j = 2m + 1$: $B_{j+1}(0) = B_{j+1}(1)$, and so
$B_{j+1}(\xi) = (-1)^{j+1} B_{j+1}(1-\xi)$;

(b) $j = 2m$: $B_{j+1}(0) = B_{j+1}(1)$, and so
$B_{j+1}(\xi) + B_{j+1}(1-\xi) = 2B_{j+1}(0)$.

Since $\int_0^1 B_{j+1}(\xi)d\xi = \int_0^1 B_{j+1}(1-\xi)d\xi$,
it follows that $2B_{j+1}(0) = 2\int_0^1 B_{j+1}(0)d\xi = 2\int_0^1 B_{j+1}(\xi)d\xi = 0$,
and thus $B_{j+1}(0) = 0$ and $B_{j+1}(\xi) + B_{j+1}(1-\xi) = 0$, which again implies $B_{j+1}(\xi) = (-1)^{j+1} B_{j+1}(1-\xi)$.

(ii) *Zeros.*

(a) Since $B_{j+1}(0) = B_{j+1}(1)$ and $B_j(\xi) = (-1)^j B_j(1-\xi)$ for all $j \geq 1$, it follows that $B_j(0) = B_j(1) = 0$ for all $j = 2m+1$, $m \geq 1$.

(b) From $B_j(\xi) = (-1)^j B_j(1-\xi)$, it follows that
$B_j(\frac{1}{2}) = 0$ for all $j = 2m+1$, $m \geq 0$.

(c) For $j = 2m$, $m \geq 0$, we have $B_j(0) = B_j(1) \neq 0$.
Indeed, the assertion is correct for $m = 0$ and for $m = 1$. Now if B_{2m+1} has only the simple zeros $0, \frac{1}{2}, 1$ in $[0,1]$ – which holds for B_3 – then B_{2m+2} can only have extrema at these points, and only one simple zero in $(0, \frac{1}{2})$ and $(\frac{1}{2}, 1)$. Thus in view of (ii)(b), B_{2m+3} has precisely the zeros $0, \frac{1}{2}, 1$ in $[0,1]$, etc.

(d) For $m \geq 1$, the argument (c) shows that on $[0,1]$, B_{2m} has precisely one simple zero in $(0, \frac{1}{2})$ and one in $(\frac{1}{2}, 1)$, while B_{2m+1} has precisely the simple zeros $0, \frac{1}{2}, 1$.

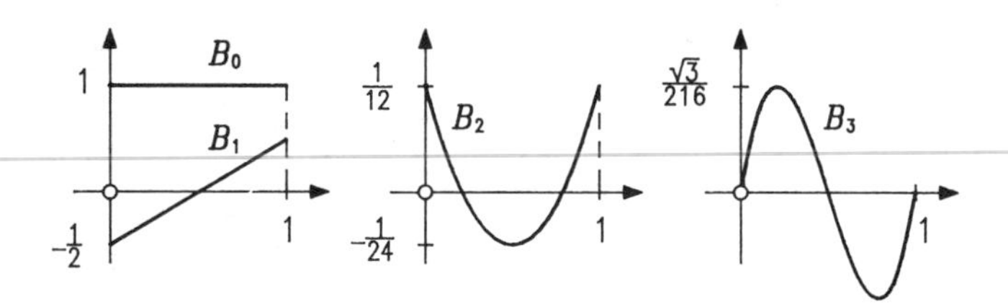

Using these properties of Bernoulli polynomials, and successively integrating by parts in $(*)$ leads to the

Euler-MacLaurin Expansion

$$\int_a^b f(x)dx = T_n f - \frac{h^2}{12}[f'(b) - f'(a)] + \frac{h^4}{720}[f'''(b) - f'''(a)] - \cdots$$
$$- h^{2m} B_{2m}(0)[f^{(2m-1)}(b) - f^{(2m-1)}(a)] +$$
$$+ h^{2m} \int_a^b \overline{B}_{2m}(t) f^{(2m)}(t)dt$$

with the periodic *Bernoulli functions*

$$\overline{B}_j(t) := B_j(\frac{t - x_\nu}{h}) \quad \text{for} \quad t \in [x_\nu, x_{\nu+1}].$$

The numbers $B_j := j!B_j(0)$ are called the Bernoulli numbers, and also appear in the power series $\frac{z}{e^z-1} = \sum_0^\infty \frac{B_j}{j!}z^j$. For details, see e.g. the book of R. Remmert [1990].

Combining the last two terms in the Euler-MacLaurin expansion, we get

$$\begin{aligned}
&- h^{2m}B_{2m}(0)[f^{(2m-1)}(b) - f^{(2m-1)}(a)] + h^{2m}\int_a^b \overline{B}_{2m}(t)f^{(2m)}(t)dt = \\
&= -h^{2m}\int_a^b [B_{2m}(0) - \overline{B}_{2m}(t)]f^{(2m)}(t)dt = \\
&= -h^{2m}f^{(2m)}(\xi)\int_a^b [B_{2m}(0) - \overline{B}_{2m}(t)]dt = \\
&= -h^{2m}f^{(2m)}(\xi)\frac{B_{2m}}{(2m)!}(b-a), \qquad a < \xi < b,
\end{aligned}$$

since either $B_{2m}(0) \geq \overline{B}_{2m}(t)$ or $B_{2m}(0) \leq \overline{B}_{2m}(t)$ for all $t \in [a,b]$, and $\int_a^b \overline{B}_{2m}(t)dt = 0$.

The Euler-MacLaurin expansion

$$\int_a^b f(x)dx - T_n f = \alpha_2 h^2 + \alpha_4 h^4 + \cdots + \alpha_{2m-2}h^{2m-2} + O(h^{2m})$$

provides an expansion of the quadrature error of the trapezoidal rule in powers of the step size h. The coefficients in the expansion depend only on f and on the limits of integration.

LEONHARD EULER (1707–1783) was the leading mathematician of the 18-th century. A discussion of his mathematical work can be found in W. Walter [1989]. Euler, like Gauss, contributed very significantly to the entire spectrum of mathematics, as well as to numerous application areas such as mechanics, hydrodynamics, optics, and astronomy. His teacher JOHANN BERNOULLI in Basel spoke of the twenty year old Euler "whose cleverness promises great things, given the ease and inventiveness with which he finds his way under our guidance to the heart of any mathematical argument." (Citation from E. A. Feldmann: Über einige mathematische Sujets im Briefwechsel Leonhard Eulers mit Johann Bernoulli, Zum Werk Leonhard Eulers, E. Knobloch, I. S. Louhivaara, and J. Winkler, eds., Birkhäuser Verlag 1984, 39–66.) During 1727–1741 Euler was at the Academy in St. Petersburg, and then moved to Berlin. A lack of appreciation by King Friedrich II of Prussia drove him back in 1766 to St. Petersburg, where he worked until his death. Indeed, King Friedrich paid this great mathematician a salary of only 1,600 Taler, while Voltaire got 20,000! His successor in Berlin was Lagrange. Euler's grave is located in the Newski cemetary in Leningrad. Concerning the depth and importance of his work, it was Gauss' opinion that "Studying Euler's papers remains the best way to learn about the various areas of mathematics, and it cannot be replaced by anything else."

The Bernoulli polynomials which arise in the Euler-MacLaurin expansion go back to JAKOB BERNOULLI (1654–1705), the brother of Johann Bernoulli. The Bernoulli family of Basel produced a number of important mathematicians of the 17th and 18th centuries, who made especially significant contributions to the development and application of the still new infinitesimal calculus.

COLIN MACLAURIN (1698–1746) published the expansion formula in about 1737, independently from Euler, who had presented it in 1730.

1.4 Simpson's Rule. Assuming $n = 2k$, we can get a better order of approximation to a definite integral than that given by the trapezoidal rule if we interpolate the integrand f on each of the intervals $[x_{2\kappa}, x_{2\kappa+2}]$ by a polynomial $\tilde{p} \in \mathrm{P}_2$. The required polynomial is

$$\tilde{p}(x) = y_{2\kappa} + \frac{\Delta y_{2\kappa}}{h}(x - x_{2\kappa}) + \frac{\Delta^2 y_{2\kappa}}{2h^2}(x - x_{2\kappa})(x - x_{2\kappa+1}) \quad \text{for}$$

$x \in [x_{2\kappa}, x_{2\kappa+2}]$, for which

$$\int_{x_{2\kappa}}^{x_{2\kappa+2}} \tilde{p}(x)dx = \frac{h}{3}(y_{2\kappa} + 4y_{2\kappa+1} + y_{2\kappa+2}).$$

This approximation formula for the value of a definite integral based on three equally-spaced nodes was developed by JOHANNES KEPLER (1571–1630) in 1612 in Linz. He developed the method to compute the capacity of some barrels of wine he was interested in buying. For this reason, the formula is also referred to as "Kepler's Barrel Rule". Since a volume was to be calculated, this problem may seem to be one involving cubature, but since the cross-sections of the barrel were circular, it is equivalent to a quadrature problem. Kepler treated the problem in his paper "Stereometria doliorum". The word "dolium" means barrel in Latin. We should keep in mind that the concept of integral was not developed until the end of the 17th century.

Extending Kepler's barrel rule to the interval $[x_0, x_{2k}]$ leads to a quadrature formula $\int_{x_0}^{x_{2k}} f(x)dx = S_{2k}f + R_{2k}f$ involving

Simpson's Rule

$$S_{2k}f = \frac{h}{3}(y_0 + 4y_1 + 2y_2 + \cdots + 4y_{2k-1} + y_{2k}).$$

The name of THOMAS SIMPSON (1710–1761) is known in mathematics for his quadrature formula, although his work in other areas such as geometry, trigonometry, probability theory, and astronomy are actually more important. The names sine, cosine, tangent and cotangent for the trigonometric functions were invented by Simpson.

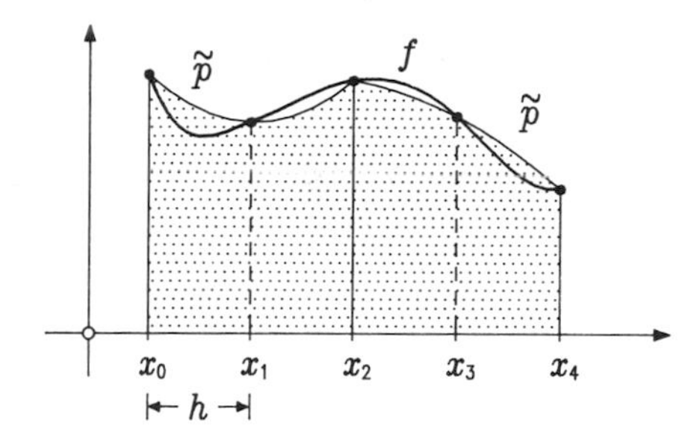

Suppose now that we raise the degree of the interpolating polynomial $\tilde{p} \in \mathrm{P}_2$ appearing in Kepler's barrel rule by one, and look for a polynomial $\pi \in \mathrm{P}_3$ which also interpolates at the additional data point $x^* \in (x_{2\kappa}, x_{2\kappa+2})$ with $x^* \neq x_{2\kappa+1}$. Then π is given by

$$\pi(x) = \tilde{p}(x) + ([x^* x_{2\kappa+2} x_{2\kappa+1} x_{2\kappa}]f)(x - x_{2\kappa})(x - x_{2\kappa+1})(x - x_{2\kappa+2}).$$

Now since

$$\int_{x_{2\kappa}}^{x_{2\kappa+2}} (x - x_{2\kappa})(x - x_{2\kappa+1})(x - x_{2\kappa+2})dx = 0,$$

it follows that

$$\int_{x_{2\kappa}}^{x_{2\kappa+2}} \pi(x)dx = \int_{x_{2\kappa}}^{x_{2\kappa+2}} \tilde{p}(x)dx,$$

and we observe that Kepler's barrel rule is actually exact for every polynomial $\pi \in \mathrm{P}_3$. We encountered a similar phenomenon above for the midpoint rule.

To bound the error, we now apply the Peano kernel formula, assuming that $f \in C_4[a, b]$. This leads to the formula

$$\begin{aligned} Rf &= \int_{x_{2\kappa}}^{x_{2\kappa+2}} f(x)dx - \frac{h}{3}[f(x_{2\kappa}) + 4f(x_{2\kappa+1}) + f(x_{2\kappa+2})] \\ &= \int_{x_{2\kappa}}^{x_{2\kappa+2}} K_3(t) f^{(4)}(t)dt \end{aligned}$$

with $K_3(t) = \frac{1}{3!} Rq_3(\cdot, t)$ and $q_3(x, t) = (x - t)_+^3$.

For the sake of simplicity, we now suppose that the interval of integration is $[x_{2\kappa}, x_{2\kappa+2}] := [-h, h]$. Then for $-h < t < 0$,

$$Rq_3(\cdot, t) = \int_t^h (x - t)^3 dx - \frac{h}{3}[-4t^3 + (h - t)^3],$$

and so $K_3(t) = \frac{1}{3!} Rq_3(\cdot, t) = \frac{1}{4!}(h - t)^4 - \frac{h}{3 \cdot 3!}[-4t^3 + (h - t)^3]$.

Computing $K_3(t)$ for $0 < t < h$, we see that it is an even function; i.e., $K_3(t) = K_3(-t)$. We conclude that the Peano kernel for Kepler's barrel rule is given by

$$K_3(t) = \begin{cases} -\frac{1}{72}(h+t)^3(h-3t) & \text{for} \quad -h \le t \le 0 \\ -\frac{1}{72}(h-t)^3(h+3t) & \text{for} \quad 0 \le t \le h. \end{cases}$$

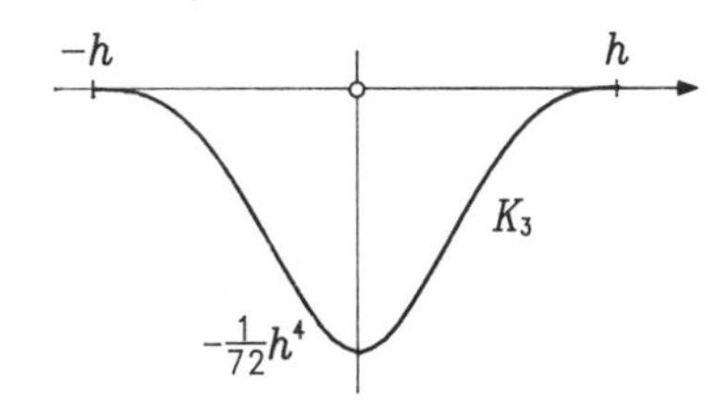

Since K_3 has one sign in $[-h, +h]$, it follows that

$$Rf = f^{(4)}(\tau) \int_{-h}^{+h} K_3(t)dt = -\frac{h^5}{90} f^{(4)}(\tau), \; -h < \tau < h.$$

Now combining the k intervals of length $2h$, we get the

Error Formula for Simpson's Rule

$$R_{2k}f = -\frac{h^4}{180} f^{(4)}(\xi)(b-a), \quad a < \xi < b,$$

which leads to the

Error Bound for Simpson's Rule

$$\boxed{|R_{2k}f| \le \frac{h^4}{180} \max_{x \in [a,b]} |f^{(4)}(x)|(b-a).}$$

In view of the symmetrical position of the nodes x_ν, $0 \le \nu \le n$, and the symmetry of the coefficients γ_ν, $0 \le \nu \le n$, in Simpson's rule

$$S_n f = \sum_{\nu=0}^{n} \gamma_\nu f(x_\nu),$$

we say that it belongs to the class of *symmetrical quadrature formulae*. In general, a quadrature formula is called symmetric provided that the coefficients γ_ν satisfy the symmetry condition $\gamma_\nu = \gamma_{n-\nu}$, $0 \le \nu \le n$, while the nodes satisfy the symmetry condition $x_\nu - a = b - x_{n-\nu}$, $0 \le \nu \le n$.

The quadrature formula treated in 1.2 and in 1.1 are also symmetric. Since the value of $\int_a^b f(x)dx$ is not changed if we make the change of variables $t = (a+b) - x$ centered about the middle of the interval of integration, it seems natural to use a quadrature formula with the same symmetry, and so symmetric quadrature formulae are the most important.

In general, a Peano kernel associated with a quadrature formula of the form $Q_n f = \sum_0^n \gamma_\nu f(x_\nu)$ can be written as the

Kernel Representation

$$K_m(t) = \frac{(b-t)^{m+1}}{(m+1)!} - \frac{1}{m!}\sum_{\nu=0}^{n} \gamma_\nu (x_\nu - t)_+^m$$

or as

$$K_m(t) = \frac{(a-t)^{m+1}}{(m+1)!} - \frac{(-1)^{m+1}}{m!}\sum_{\nu=0}^{n} \gamma_\nu (t - x_\nu)_+^m,$$

as can be seen from Definition 5.2.4 and from the fact that

$$\frac{(b-t)^{m+1}}{(m+1)!} - \frac{(a-t)^{m+1}}{(m+1)!} = \int_a^b \frac{(x-t)^m}{m!} dx = \frac{1}{m!}\sum_{\nu=0}^{n} \gamma_\nu (x_\nu - t)^m.$$

It follows that a symmetric quadrature formula satisfies the

Symmetry Condition

$$K_m(t) = (-1)^{m+1} K_m(a+b-t),$$

which simplifies the computation of Peano kernels. For example, if f is in $C_2[a,b]$, we can estimate the quadrature error of Simpson's rule using the Peano kernel $K_1(t)$. We find that

$$K_1(t) = \frac{1}{6}(h-t)(h-3t) \text{ for } 0 \le t \le h$$

and $K_1(-t) = K_1(t)$.

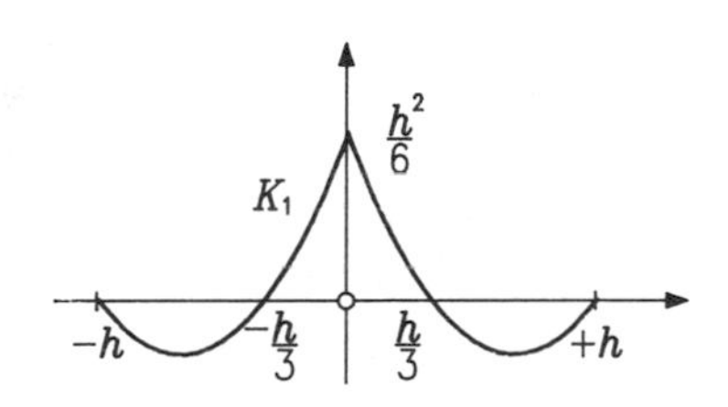

But then

$$\left| \int_{-h}^{+h} K_1(t) f''(t) dt \right| \leq \max_{t \in [-h,+h]} |f''(t)| \int_{-h}^{+h} |K_1(t)| dt$$

leads to the error bound

$$|R_{2k} f| \leq \frac{4h^2}{81} \max_{x \in [a,b]} |f''(x)| (b-a).$$

For $f \in C_2[a,b]$, this bound is comparable to those obtained for the trapezoidal rule and the tangent trapezoidal rule.

1.5 Newton-Cotes Formulae. In the previous sections we have constructed a variety of quadrature formula by first interpolating the integrand and then integrating it. This idea can be used to develop further formulae. Quadrature formulae obtained in this way are called *Newton-Cotes formulae.*

According to remarks of Newton, ROGER COTES (1682–1716) also worked on interpolation. How highly Newton valued Cotes' work can be judged from his remark on Cotes' early death: "Had Cotes lived, we might have known something".

We conclude this section by mentioning a quadrature formula which can be used when $n = 3k$, and is obtained by interpolating f by a piecewise cubic polynomial whose pieces are joined together at the data points $x_{3\kappa}$, $1 \leq \kappa \leq k$.

Newton's $\frac{3}{8}$ -Rule

$$\int_{x_0}^{x_{3k}} f(x) dx = \frac{3h}{8}(y_0 + 3y_1 + 3y_2 + 2y_3 + \cdots + 3y_{n-1} + y_n) + R_{3k} f$$

and its associated error formula (Problem 5)

$$R_{3k} f = -\frac{h^4}{80} f^{(4)}(\xi)(b-a), \ a < \xi < b, \ \text{ for } \ f \in C_4[a,b].$$

Newton himself called this quadrature formula "pulcherrima", the most beautiful, because of the fact that its coefficients are almost all the same. This property implies that the formula is relatively insensitive to random errors in the values of f at an individual node. This formula does not give a higher order error term with respect to the step size h as compared with Simpson's rule. This is to be expected, since we already know that for equally-spaced interpolation by polynomials of even degree, the order of the error of the associated quadrature formula is two higher than that of the interpolation process, while for odd degree interpolation, it is just one higher.

1.6 Unsymmetric Quadrature Formulae. It is not necessary that the interpolation interval and the integration interval always be the same. In particular, it may make sense to use an interpolating polynomial based on points which lie outside of the integration interval. This idea opens the door to a variety of additional symmetric and unsymmetric quadrature formulae.

Suppose we want to compute the value of $\int_{x_0}^{x_1} f(x)dx$. Using only the two end points as nodes, then the only possible formulae available are symmetric Newton-Cotes formulae, whose highest order accuracy is $O(h^2)$. On the other hand, the polynomial $\tilde{p} \in \mathrm{P}_2$ which interpolates in the sense that $\tilde{p}(x_k) = f(x_k)$ for $0 \leq k \leq 2$ gives

$$\int_{x_0}^{x_1} f(x)dx = \int_{x_0}^{x_1} \tilde{p}(x)dx + Rf,$$

where for equally-spaced nodes, we have

$$\int_{x_0}^{x_1} \tilde{p}(x)dx = \frac{h}{12}[5f(x_0) + 8f(x_1) - f(x_2)].$$

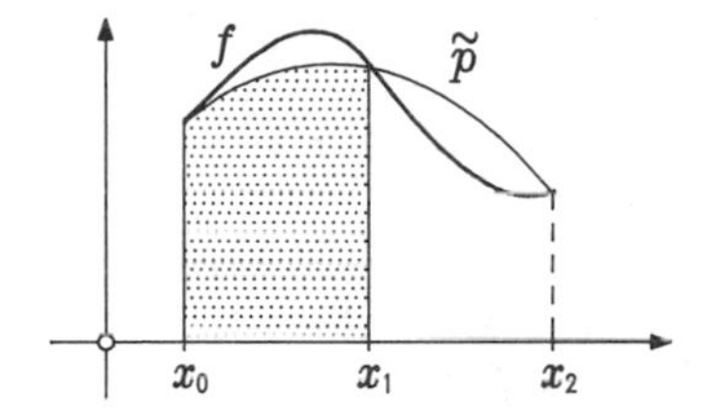

Using the expression

$$r(x) = \frac{f'''(\xi)}{3!}(x - x_0)(x - x_1)(x - x_2), \quad \xi \in (x_0, x_2)$$

for the interpolation error, we find that

$$Rf = \int_{x_0}^{x_1} r(x)dx = \frac{h^4}{4!}f'''(\hat{\xi}), \quad \hat{\xi} \in (x_0, x_2),$$

provided $f \in \mathrm{C}_3[x_0, x_2]$. If f has less differentiability, then the associated quadrature error can be estimated with the help of the corresponding Peano kernel. In comparing this error bound as a function of h with those given in 1.1–1.5 for symmetric Newton-Cotes formulae, one should keep in mind that here we are considering integration over just one subinterval. This kind of formula is appropriate, for example, for integration over boundary intervals.

1.7 Problems. 1) The Peano Remainder Formula 1.2 implies that for $f \in \mathrm{C}_1[0, n]$, $n \in \mathbb{Z}_+$,

$$\sum_{\nu=0}^{n} f(\nu) = \frac{1}{2}[f(0) + f(n)] + \int_0^n f(x)dx + \sum_{\nu=1}^{n} \int_{\nu-1}^{\nu} (x - \nu + \frac{1}{2})f'(x)dx.$$

Use this to establish the existence of the *Euler constant*

$$C := \lim_{n\to\infty}\left[\sum_{\nu=1}^{n}\frac{1}{\nu} - \log(n)\right],$$

and show that $C = \frac{1}{2} - \int_0^\infty \frac{x-[x]-\frac{1}{2}}{(1+x)^2}dx$.

2) The Euler-MacLaurin expansion implies the *quadrature formula of Chevilliet*

$$\int_a^b f(x)dx = \frac{1}{2}(b-a)[f(a)+f(b)] - \frac{1}{12}(b-a)^2[f'(b)-f'(a)] + Rf.$$

Show $Rp = 0$ for all $p \in \mathrm{P}_3$, and under the assumption that $f \in \mathrm{C}_4[a,b]$, give a formula for the error Rf using the Peano kernel K_3.

3) Find the Peano kernels K_0 and K_2 for Simpson's rule, and use them to estimate the quadrature error.

4) Carry out the details of the derivation of the Symmetry Condition 1.4.

5) Derive the error formula for the Newton $\frac{3}{8}$-rule 1.5 using the Peano kernel.

6) Under the assumption that $f \in \mathrm{C}_2[x_0, x_2]$, estimate the error of the unsymmetric quadrature formula 1.6.

7) Integrate the polynomial which interpolates f at the equally-spaced nodes x_0, x_1, x_2 to determine the coefficients of the associated quadrature formula $\int_{x_2}^{x_3} f(x)dx = \sum_{\nu=0}^{2}\gamma_\nu f(x_\nu) + Rf$.

Remark: Formulae of this type are used in the so-called "explicit multistep methods" for solving ordinary differential equations.

2. Extrapolation

In the introduction to this chapter we mentioned Archimedes' approach to estimating π by approximating the area of a circle of radius one by inscribed and circumscribed polygons. If F_n is the area of a regular circumscribed n-gon, then it is easy to see that the sequence F_6, F_{12}, $F_{24}, \cdots$ converges monotonely to π. It was already known to C. HUYGENS (1629–1695) that the rate of convergence of this sequence is of order $(\frac{1}{n})^2$; i.e., it is quadratic. By taking appropriate linear combinations of pairs of successive terms, he succeeded in constructing a sequence which converges with order $(\frac{1}{n})^4$. In this section we want to systematically apply this idea to quadrature formulae.

2.1 The Romberg Method. Suppose we have a quadrature formula $(Q_0 f)(h)$ with $h = \frac{b-a}{n}$, and an expansion of its quadrature error in the form

$$\int_a^b f(x)dx - (Q_0 f)(h) = \alpha_2 h^2 + \alpha_4 h^4 + \cdots + \alpha_{2m-2}h^{2m-2} + O(h^{2m}).$$

We have already seen this kind of expansion in 1.3. Then for any $0 < q < 1$,

$$\int_a^b f(x)dx - (Q_0 f)(qh) = \alpha_2(qh)^2 + \alpha_4(qh)^4 + \cdots + \alpha_{2m-2}(qh)^{2m-2} + O(h^{2m}),$$

and thus

$$\int_a^b f(x)dx - \frac{(Q_0 f)(qh) - q^2(Q_0 f)(h)}{1-q^2} = \hat{\alpha}_4\, h^4 + \cdots + \hat{\alpha}_{2m-2}\, h^{2m-2} + O(h^{2m})$$

with $\hat{\alpha}_{2\mu+2} := -q^2 \frac{1-q^{2\mu}}{1-q^2} \alpha_{2\mu+2}$ for $1 \le \mu \le m-2$.

This shows that the linear combination

$$(Q_1 f)(h) := \frac{(Q_0 f)(qh) - q^2(Q_0 f)(h)}{1-q^2}$$

provides a quadrature formula which exhibits a much better error behavior than the original quadrature formula $(Q_0 f)(h)$.

This method of reducing the step size by an appropriate factor q, and then forming a linear combination to find a formula with a higher order term can be continued. We now apply this idea with the trapezoidal rule as the starting quadrature formula $Q_0 f$. Assuming that f is sufficiently often differentiable, the Euler-MacLaurin Expansion 1.3 provides an expansion of the error in powers of h^2. Taking $q := \frac{1}{2}$, we get the

Romberg Method. Starting with $(Q_0 f)(h) := (T_0 f)(h)$,

$$(T_0 f)(h) := \frac{h}{2}(y_0 + 2y_1 + \cdots + 2y_{n-1} + y_n) \quad \text{and}$$
$$(T_0 f)(\frac{h}{2}) = \frac{h}{4}(y_0 + 2y_{1/2} + 2y_1 + \cdots + 2y_{n-1} + 2y_{n-1/2} + y_n)$$

we get

$$(T_1 f)(h) := \frac{4(T_0 f)(\frac{h}{2}) - (T_0 f)(h)}{3}.$$

After computing $T_0 f(\frac{h}{4})$, we can form

$$(T_1 f)(h/2) = \frac{4(T_0 f)(\frac{h}{4}) - (T_0 f)(\frac{h}{2})}{3}$$

and then

$$(T_2 f)(h) := \frac{16(T_1 f)(\frac{h}{2}) - (T_1 f)(h)}{15}.$$

It is easy to see that the expansion of the error of the quadrature formula $T_2 f$ starts with a term of order $O(h^6)$. Now writing

$$T_j^k f := (T_j f)(\frac{h}{2^k}),$$

we can construct further formulae by the rule

$$T_j^k = \frac{4^j T_{j-1}^{k+1} - T_{j-1}^k}{4^j - 1}.$$

The following table shows how $T_j^k f$ depends on previous values. The first column contains the values produced by the trapezoidal rule for the step sizes $\frac{h}{2^k}$, $(k = 0, 1, \ldots)$. Each successive column is computed from its predecessor using the rule above. We refer to this table as the

Romberg Table

$$\begin{array}{ccccc}
T_0^0 & & & & \\
\vdots & \ddots & & & \\
T_0^1 & \ldots & T_1^0 & & \\
\vdots & \ddots & \vdots & \ddots & \\
T_0^2 & \ldots & T_1^1 & \ldots & T_2^0 \\
\vdots & & \vdots & & \vdots \quad \ddots
\end{array}$$

This method was suggested by W. Romberg 1955, and led to a series of further results on these kinds of quadrature formulae.

As can be easily checked, the numbers appearing in the second column of the Romberg Table are precisely those produced by Simpson's Rule 1.4 using the step sizes $\frac{h}{2^k}$, $(k = 1, 2, \ldots)$. The numbers in the third column are those produced by the Newton-Cotes formula

$$\int_a^b f(x)dx = \frac{2\hat{h}}{45}[7y_0 + 32y_{1/4} + 12y_{1/2} + 32y_{3/4} + 14y_1 + \cdots + 7y_n] + Rf,$$

starting with $\hat{h} := \frac{h}{4}$. This formula is the so-called *Boolean* or *Milne Rule*. It arises by interpolating f by a polynomial $p \in \mathrm{P}_4$, and is exact for all $p \in \mathrm{P}_5$. The additional columns of the table cannot, however, be identified with Newton-Cotes formulae. In this sense, the Romberg method is fundamentally different from the other numerical integration methods discussed earlier. This will become even clearer from the following

2.2 Error Analysis. Each column of the Romberg table arises as a linear combination of the values in the previous column, and thus in fact as a linear combination of the values $T_0^k f$, $(k = 0, 1, \ldots)$, in the first column. The Romberg method is designed in such a way that the error of the approximation $T_j f$ is of the order $O(h^{2j+2})$. Now using the Euler-MacLaurin Expansion 1.3, we see that the errors of the quadrature formulae in the j-th column have the form

$$\int_a^b f(x)dx - T_j^k f = h^{2j+2} \int_a^b b_{2j+2,k}(x) f^{(2j+2)}(x)dx$$

for $f \in C_{2j+2}[a,b]$. Here $b_{2j+2,k}$ is a function which is pieced together as linear combinations of the periodic Bernoulli functions $\overline{B}_{2j+2}$ with respect to the step sizes $\frac{h}{2^k}$ for $k = 0, \dots, j$. It follows that the quadrature formulae $T_j^{\cdot} f$ produce the exact value of $\int_a^b f(x)dx$ whenever $f^{(2j+2)} = 0$. In other words, the quadrature operators $T_j^{\cdot}$ integrate all polynomials $p \in P_{2j+1}$ exactly.

We now compare with the Newton-Cotes formulae which also exactly integrate polynomials up to a given degree. The number of nodes involved in the method $T_j^0 f$ is $(1 + 2^j)$. For a Newton-Cotes formula, the number of nodes determines the order of exactness, while for the Romberg method, the j-th column of the table has an order of exactness $(2j+1)$. For $j = 1$ and $j = 2$ the orders of Newton-Cotes and Romberg are the same; that is, the columns $T_1^{\cdot}$ and $T_2^{\cdot}$ together with the trapezoidal rule $T_0^{\cdot}$ are Newton-Cotes formulae, as we have already noted in 2.1. But for $j \geq 3$, the Newton-Cotes formulae have a higher order of exactness than the column $T_j^{\cdot}$.

In this connection, we should note that the Newton-Cotes formula for $j = 3$ is obtained by integration of an interpolating polynomial over 9 nodes, and is the first in the series of such formulae which involves negative coefficients. The appearance of negative coefficients adversely affects the stability of a quadrature formula. It can be shown, that with increasing interpolation degrees, negative coefficients appear in the Newton-Cotes formulae again and again. This fact is of key importance for the convergence behavior of Newton-Cotes formulae. We return to the question of convergence in Section 4 of this chapter.

A quadrature formula in which only positive coefficients appear is called a *positive quadrature formula*. All Newton-Cotes formulae which are obtained by using interpolating polynomials with at most eight data points are positive formulae. In addition, it can be shown that all columns of the Romberg table represent positive quadrature formulae. This is not necessarily the case for the more general extrapolation methods introduced in the following section; see e.g. the book of H. Braß ([1977], p. 199 ff.).

Examples of the Romberg method ($h = \frac{1}{2^k}$).

1. Romberg table for $J_1 = \int_0^1 \cos(\frac{\pi}{2}x)dx$. This integrand is infinitely differentiable on $[0,1]$.

$k=0$	0.50000000			
1	0.60355339	0.63807119		
2	0.62841744	0.63670546	0.63661441	
3	0.63457315	0.63662505	0.63661969	0.63661977

The true value is $J_1 = \frac{2}{\pi} \doteq 0.63661977$.

2. Romberg method for $J_2 = \int_0^1 x^{3/2}dx$. Here the integrand is only one-times continuously differentiable on $[0,1]$.

0	0.500000					
1	0.426777	0.402369				
2	0.407018	0.400432	0.400303			
3	0.401812	0.400077	0.400054	0.400050		
4	0.400463	0.400014	0.400009	0.400009	0.400009	
5	0.400118	0.400002	0.400002	0.400002	0.400002	0.400002

The true value is $J_2 = 0.4$, and for $k = 5$, the error Rf satisfies $|Rf| = 2 \cdot 10^{-6}$.

2.3 Extrapolation. The Romberg method can be generalized in various ways. We need not start with the trapezoidal rule, and it is not necessary that the step sizes be reduced in a geometric progression. Moreover, it suffices to assume that the integrand f is only Riemann integrable, and the same convergence results remain valid. Of course, in this case we can no longer use the Euler-MacLaurin Expansion as the key to the method. We thus take another approach to extrapolation with respect to the step size due to L. F. Richardson and J. A. Gaunt [1927].

Let $[a,b] := [0,1]$, and suppose $f : [0,1] \to \mathbb{R}$ is a Riemann integrable function. Let $(h_k)_{k\in\mathbb{N}}$ with $h_0 \le 1$ be a monotone decreasing sequence of step sizes converging to zero. In addition, let Q_0 be a quadrature operator such that $Q_0^k f := (Q_0 f)(h_k)$ satisfies $\lim_{k\to\infty} Q_0^k f = \int_a^b f(x)dx$.

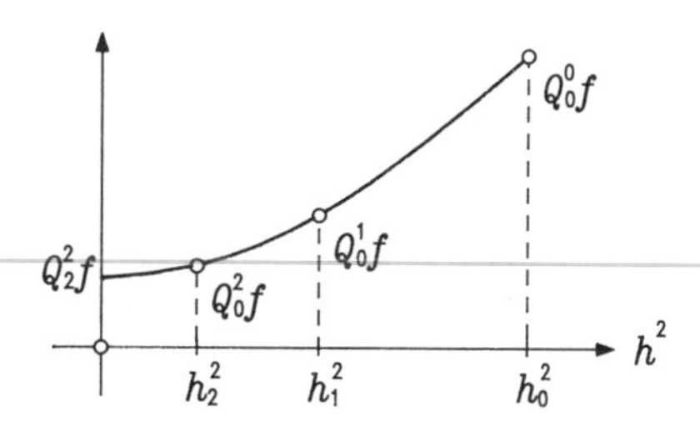

We now show how to combine the values $Q_0^\kappa f$ for $\kappa = 0, \ldots, k$ to extrapolate a new approximate value for $h = 0$. To this end, we interpolate the pairs $(h_0, Q_0^0 f), \ldots, (h^k, Q_0^k f)$ by a polynomial of degree at most k in h^2. This polynomial is uniquely defined, and by 5.2.1, can be written in the Lagrange form

$$\tilde{p}(h^2) = \sum_{\kappa=0}^{k} Q_0^\kappa f \prod_{\substack{\iota=0\\ \iota\neq\kappa}}^{k} \frac{h^2 - h_\iota^2}{h_\kappa^2 - h_\iota^2}.$$

We are now interested in the value $\tilde{p}(0)$ of this interpolation polynomial. To calculate it, we use the algorithm of Aitken-Neville 5.5.2. Beginning with the pairs $(h_0, Q_0^0 f)$ and $(h_1, Q_0^1 f)$, the linear interpolating polynomial $q \in \mathrm{P}_1$ leads to the value

$$q(0) = \frac{1}{h_1^2 - h_0^2} \begin{vmatrix} Q_0^0 f & h_0^2 \\ Q_0^1 f & h_1^2 \end{vmatrix} = \frac{h_1^2 Q_0^0 f - h_0^2 Q_0^1 f}{h_1^2 - h_0^2} =: Q_1^0 f,$$

which for $h_0 := h$, $h_1 := \frac{h_0}{2}$ and $Q_1^0 := T_1^0$ can also be found in the Romberg Table 2.1. Setting

$$Q_{\imath+1}^{\kappa} f := \frac{1}{h_{\kappa+\imath+1}^2 - h_\kappa^2} \begin{vmatrix} Q_\imath^\kappa f & h_\kappa^2 \\ Q_\imath^{\kappa+1} f & h_{\kappa+\imath+1}^2 \end{vmatrix} = \frac{h_\kappa^2 Q_\imath^{\kappa+1} f - h_{\kappa+\imath+1}^2 Q_\imath^\kappa f}{h_\kappa^2 - h_{\kappa+\imath+1}^2},$$

$0 \le \kappa \le k-1$, $0 \le \imath \le k - \kappa - 1$, we see that the values produced by the Aitken-Neville algorithm formally agree with those in the Romberg Table, and we have

$$\tilde{p}(0) = Q_k^0 f.$$

Now in the Romberg method we take $h_\kappa := 2^{-\kappa}$ and form $Q_{\imath+1}^\kappa := T_{\imath+1}^\kappa$ according to the rule

$$T_{\imath+1}^\kappa f = \frac{4^{\imath+1} T_\imath^{\kappa+1} f - T_\imath^\kappa f}{4^{\imath+1} - 1}$$

as in 2.1 ($\kappa =: k$, $\imath + 1 =: j$).

The computation of the values $Q_k^0 f$ using the Romberg method and their interpretation as extrapolated values for $h = 0$ is the desired generalization of Romberg's method. We now show that this extrapolation method produces a sequence of numbers which, under a weak assumption on the sequence (h_k) and assuming the function f is Riemann integrable, converges whenever the sequence $(Q_0^k f)$ converges to $\int_a^b f(x)dx$.

2.4 Convergence. The convergence of the extrapolation method was first established by R. Bulirsch [1964]. We now present a proof of this convergence theorem.

Convergence Theorem. *Let $f : [0,1] \to \mathbb{R}$ be Riemann integrable. Suppose the sequence $(h_k)_{k \in \mathbb{N}}$ of step sizes satisfies $h_0 \le 1$ and $\left(\frac{h_k}{h_{k+1}}\right)^2 \ge c$ with $c > 1$. Let Q_0 be a quadrature operator such that $Q_0^k f := (Q_0 f)(h_k)$ satisfies $\lim_{k\to\infty} Q_0^k f = \int_0^1 f(x)dx$. Then*

$$\lim_{k\to\infty} Q_k^0 f = \int_0^1 f(x)dx$$

for the sequence $(Q_k^0 f)_{k \in \mathbb{N}}$ produced by the extrapolation method.

Proof. We begin by presenting in (a)–(c) below some properties of the Lagrange Polynomials 5.2.1.

(a) Let

$$\ell_{k\kappa} := \prod_{\substack{\iota=0\\ \iota\neq\kappa}}^{k} \frac{h_\iota^2}{h_\iota^2 - h_\kappa^2} \qquad \text{for } 0 \le \kappa \le k.$$

Then by 2.3,

$$Q_k^0 f = \sum_{\kappa=0}^{k} (Q_0^\kappa f)\ell_{k\kappa} \ \text{ with } \sum_{\kappa=0}^{k} \ell_{k\kappa} = 1 \ \text{ by 5.2.1.}$$

(b) We have $\lim_{k\to\infty} \ell_{k\kappa} = 0$ for $\kappa = 0, 1, \ldots$. Indeed, this is a necessary condition for the convergence of the series $\sum_{k=0}^{\infty} \ell_{k\kappa}$ as we can see by applying the quotient test:

$$\lim_{k\to\infty} \left| \frac{\ell_{k+1,\kappa}}{\ell_{k\kappa}} \right| = \lim_{k\to\infty} \left| \frac{h_{k+1}^2}{h_{k+1}^2 - h_\kappa^2} \right| = 0.$$

(c) There exists a constant Λ such that for all $k \in \mathbb{N}$,

$$\sum_{\kappa=0}^{k} |\ell_{k\kappa}| < \Lambda.$$

To see this, we write

$$\sum_{\kappa=0}^{k} |\ell_{k\kappa}| = \prod_{\iota=0}^{k-1} \left| 1 - \frac{h_k^2}{h_\iota^2} \right|^{-1} + \left| 1 - \frac{h_{k-1}^2}{h_k^2} \right|^{-1} \prod_{\iota=0}^{k-2} \left| 1 - \frac{h_{k-1}^2}{h_\iota^2} \right|^{-1} + \\ \cdots + \left| 1 - \frac{h_0^2}{h_1^2} \right|^{-1} \cdots \left| 1 - \frac{h_0^2}{h_k^2} \right|^{-1},$$

and note that by hypothesis,

$$\prod_{\iota=0}^{s-1} \left| 1 - \frac{h_s^2}{h_\iota^2} \right|^{-1} \le \prod_{\iota=0}^{s-1} \left| 1 - \left(\frac{1}{c}\right)^{s-\iota} \right|^{-1} = \prod_{\iota=1}^{s} \left| 1 - \left(\frac{1}{c}\right)^{\iota} \right|^{-1}$$

for $s = 1, 2, \ldots$. Since $c > 1$, it follows that $\sum_{\iota=1}^{\infty} (\frac{1}{c})^\iota$ converges, and therefore, as is well known, also $\prod_{\iota=1}^{\infty} |1 - (\frac{1}{c})^\iota|$. This implies that there exists a bound Λ' such that

$$\prod_{\iota=0}^{s-1} \left| 1 - \frac{h_s^2}{h_\iota^2} \right|^{-1} \le \Lambda'.$$

We can now give the estimate

$$\sum_{\kappa=0}^{k} |\ell_{k\kappa}| \le \Lambda' \left(1 + \frac{1}{c-1} + \frac{1}{c-1}\frac{1}{c^2-1} + \cdots + \frac{1}{c-1} \cdots \frac{1}{c^k-1} \right).$$

By the quotient rule, it is easy to see that

$$\lim_{k\to\infty}\left(1+\frac{1}{c-1}+\cdots+\frac{1}{c-1}\cdots\frac{1}{c^k-1}\right)=:C$$

exists, and we conclude that

$$\sum_{\kappa=0}^{k}|\ell_{k\kappa}|<\Lambda'\cdot C=:\Lambda$$

for all $k\in\mathbb{N}$.

We can now return to the proof of the convergence of the sequence $(Q_k^0 f)_{k\in\mathbb{N}}$. We start with the sequence $(Q_0^k f)$ which we are assuming satisfies $\lim_{k\to\infty} Q_0^k f=\int_0^1 f(x)dx=:Jf$. This means that for any arbitrary $\varepsilon>0$, there exists a number $K\in\mathbb{N}$, such that

$$|Q_0^k f-Jf|<\frac{\varepsilon}{2\Lambda}$$

for all $k>K$. It follows that for $k>K$,

$$|Q_k^0 f-Jf|\overset{(a)}{=}\Big|\,Q_k^0 f-\sum_{\kappa=0}^{k}\ell_{k\kappa}(Jf)\,\Big|\overset{(a)}{=}\Big|\sum_{\kappa=0}^{k}\ell_{k\kappa}(Q_0^\kappa f-Jf\,\Big|<$$

$$<\sum_{\kappa=0}^{K}|\ell_{k\kappa}|\,|Q_0^\kappa f-Jf|+\sum_{\kappa=K+1}^{k}|\ell_{k\kappa}|\frac{\varepsilon}{2\Lambda}\overset{(c)}{<}\sum_{\kappa=0}^{K}|\ell_{k\kappa}|\,|Q_0^\kappa f-Jf|+\frac{\varepsilon}{2}.$$

Now (b) implies $|\ell_{k\kappa}|\ |Q_0^\kappa f-Jf|<\frac{\varepsilon}{2(K+1)}$ for all $0\le\kappa\le K$ provided $k>K'$. We have shown that

$$|Q_k^0 f-Jf|<\frac{\varepsilon}{2(K+1)}(K+1)+\frac{\varepsilon}{2}=\varepsilon$$

for all $k>\max(K,K')$, and so $\lim_{k\to\infty}Q_k^0 f=Jf$. □

Application. The theorem guarantees that for every Riemann integrable function, the sequence of approximations to Jf obtained by extrapolation converges to Jf provided that $\lim_{k\to\infty}Q_0^k f=Jf$. This happens, for example, if we choose $Q_0^\cdot:=T_0^\cdot$, since in this case the values $T_0^k f$ can be estimated from above and below by Riemann upper- and lower-sums with step size h_k, and so the sequence of trapezoidal sums converges to Jf whenever f is Riemann integrable. The same holds, for example, for Simpson's rule, as well as for any other quadrature method which appears in the Romberg Table. We will return to this topic in connection with the convergence of certain quadrature methods in Section 4.

Remark. While the theorem above guarantees convergence under very weak hypotheses, it says nothing about the speed of convergence. This depends primarily on the analytic properties of f. The error analysis in 2.2 shows how the assumptions that f is sufficiently often differentiable and the existence of an Euler-MacLaurin expansion, respectively, affect the order of convergence.

Practical Hint. The natural choice $h_k = 2^{-k}$, $k \in \mathbb{N}$, for the sequence of step sizes in the Romberg method quickly leads to the need to evaluate the function at a very large number of points. This effect can be reduced by using the sequence $h_{2\kappa} := 3^{-\kappa}$, $h_{2\kappa+1} := \frac{1}{2}3^{-\kappa}$ for $\kappa \in \mathbb{N}$ of Rutishauser, or the sequence $h_0 := 1$, $h_{2\kappa-1} := 2^{-\kappa}$, $h_{2\kappa} := \frac{1}{3}2^{-\kappa+1}$ for $\kappa \in \mathbb{Z}_+$ of Bulirsch, which has proved useful in practice.

2.5 Problems. 1) Show that the method of Archimedes for the computation of π mentioned in the introduction to Section 2 converges with order $O(\frac{1}{n^2})$.

2) Suppose we want to calculate the integrals

$$J_1 f := \int_1^2 \frac{dx}{x}, \qquad J_2 f := \int_0^1 \sqrt{x}dx.$$

a) Program the Romberg method, starting with the trapezoidal rule.

b) Carry out the calculation for $k = 0, 1, \ldots, 7$, and determine the error of each approximation.

c) Determine numerically the order of convergence in each of the columns, and explain the results.

3) Find the quadrature formulae which appear in the first three columns of the Romberg table when the method is started with the midpoint rule. Are these positive quadrature formulae?

4) Let $f \in C_4[a,b]$, and suppose $(Qf)(h)$ is the trapezoidal rule with step size $h = \frac{b-a}{6m}$. Using $(Qf)(2h)$ and $(Qf)(3h)$, construct a quadrature formula $\hat{Q}$ such that $(\hat{Q}f)(h) - \int_a^b f(x)dx = O(h^4)$.

5) Find the (two-stage) Romberg method which results from using the step size sequence of Bulirsch.

6) The same situation which led here to extrapolation methods for integration also holds for numerical differentiation. Starting with the formula (*) in 5.3.3, show that the error of the approximation to the 1st derivative can be expanded in powers of h^2, and create a Romberg method for improving these approximate values.

3. Gauss Quadrature

So far in this chapter, we have considered only quadrature formulae with prescribed nodes. In this section we examine formulae of the form

$$\int_a^b f(x)dx = \sum_{\nu=1}^{n} \gamma_{n\nu} f(x_{n\nu}) + R_n f,$$

where both the coefficients $\gamma_{n1}, \ldots, \gamma_{nn}$ and the nodes $x_{n1}, \ldots, x_{nn}$ are free parameters. Our aim is to determine these parameters to obtain a formula with the highest possible accuracy. This is the idea of the

3.1 The Method of Gauss. Since we now have the $2n$ parameters $\gamma_{n\nu}$ and $x_{n\nu}$, $1 \le \nu \le n$, we try to make the quadrature formula exact for all polynomials of degree at most $2n-1$; i.e., we require

$$(*) \qquad \int_a^b p(x)dx = \sum_{\nu=1}^{n} \gamma_{n\nu} p(x_{n\nu})$$

for all $p \in \mathrm{P}_{2n-1}$.

We begin by writing the polynomial $p^* \in \mathrm{P}_{n-1}$ which takes on the values $p(x_{n1}), \ldots, p(x_{nn})$ at n nodes $x_{n1}, \ldots, x_{nn}$ in its Lagrange Form 5.2.1:

$$p^*(x) = \sum_{\nu=1}^{n} \ell_{n-1,\nu}(x) p(x_{n\nu}).$$

Now by Lemma 5.1.2, every polynomial $p \in \mathrm{P}_{2n-1}$ can be written in the form

$$p(x) = p^*(x) + (x - x_{n1}) \cdots (x - x_{nn}) q(x)$$

with $q \in \mathrm{P}_{n-1}$. Then the conditions $(*)$ require that the equations

$$\sum_{\nu=1}^{n} \gamma_{n\nu} p(x_{n\nu}) =$$

$$= \sum_{\nu=1}^{n} \left(\int_{-1}^{+1} \ell_{n-1,\nu}(x)dx \right) p(x_{n\nu}) + \int_{-1}^{+1} (x - x_{n1}) \cdots (x - x_{nn}) q(x) dx$$

must hold for all $q \in \mathrm{P}_{n-1}$ and any choice of the values $p(x_{n\nu})$, $1 \le \nu \le n$.

Now if we take $p(x_{n\nu}) = 0$ for $1 \le \nu \le n$, we see that the

Orthogonality Condition

$$\int_a^b (x - x_{n1}) \cdots (x - x_{nn}) q(x) dx = 0$$

for all $q \in \mathrm{P}_{n-1}$ is a necessary condition for $(*)$.

But, as shown in 4.5.4, this condition is precisely the orthogonality condition which characterizes the Legendre polynomials. Setting $[a, b] := = [-1, +1]$, we see that the Orthogonality Condition is satisfied if we choose the nodes to be the zeros of the Legendre polynomial

$$(x - x_{n1}) \cdots (x - x_{nn}) = \hat{L}_n(x),$$

which by Theorem 4.5.5 are simple, real, and lie in the interval $(-1, +1)$.

Now the choice $p(x_{n\nu}) = \delta_{n\nu}$ shows that the coefficients of the quadrature formula $(*)$ must be given by

$$\gamma_{n\nu} = \int_{-1}^{+1} \ell_{n-1,\nu}(x)dx.$$

These two necessary conditions are in fact also sufficient in order that $(*)$ hold, and we have the

Nodes and Coefficients of the Gauss Quadrature Formula. *If we choose the nodes $x_{n\nu}$, $1 \leq \nu \leq n$, of the quadrature formula*

$$\int_{-1}^{+1} f(x)dx = \sum_{\nu=1}^{n} \gamma_{n\nu} f(x_{n\nu}) + R_n f$$

as the zeros of the Legendre polynomial L_n, and the coefficients as

$$\boxed{\gamma_{n\nu} = \int_{-1}^{+1} \ell_{n-1,\nu}(x)dx,}$$

then

$$\int_{-1}^{+1} f(x)dx = \sum_{\nu=1}^{n} \gamma_{n\nu} f(x_{n\nu}), \text{ i.e. } R_n f = 0,$$

for all polynomials $f := p \in \mathrm{P}_{2n-1}$.

Terminology. Gauss quadrature formulae for approximating integrals of the form $\int_{-1}^{+1} w(x)f(x)dx$ with $w(x) := 1$ for $x \in [-1, +1]$ are called *Gauss-Legendre formulae.* Other weight functions w will be treated in 3.4 and 3.5.

Other Integration Intervals. There was no loss of generality in assuming that $[a, b] := [-1, +1]$ above, since any finite interval $[a, b]$ can be transformed to $[-1, +1]$ by the transformation

$$t = 2\frac{x-a}{b-a} - 1.$$

It is useful to consider the following alternative approach to computing the coefficients $\gamma_{n\nu}$. Consider the Lagrange polynomial $\ell_{n-1,\nu} \in \mathrm{P}_{n-1}$ satisfying $\ell_{n-1,\nu}(x_{n\mu}) = \delta_{\nu\mu}$. Then $\ell^2_{n-1,\nu} \in \mathrm{P}_{2n-2} \subset \mathrm{P}_{2n-1}$, and so

$$\gamma_{n\nu} = \sum_{\mu=1}^{n} \gamma_{n\mu} \ell^2_{n-1,\nu}(x_{n\mu}) = \int_{-1}^{+1} \ell^2_{n-1,\nu}(x)dx,$$

which immediately implies the

Positivity of the Coefficients

$$\boxed{\gamma_{n\nu} = \int_{-1}^{+1} \ell^2_{n-1,\nu}(x)dx > 0}$$

for $1 \leq \nu \leq n$. Gauss quadrature formulae are thus positive. It is easy to see that this assertion also holds for arbitrary weight functions, and that, in general,

$$\gamma_{n\nu} = \int_{-1}^{+1} w(x)\ell_{n-1,\nu}(x)dx = \int_{-1}^{+1} w(x)\ell^2_{n-1,\nu}(x)dx > 0.$$

Tables of nodes and coefficients for $1 \le n \le 5$ can be found in 3.6.

3.2 Gauss Quadrature as Interpolation Quadrature. Every Gauss quadrature formula can also be interpreted as an interpolation quadrature formula. To see this, suppose $f \in C_{2n}[-1,+1]$, and consider the simple Hermite interpolant 5.5.3 with nodes at the zeros $x_{n1}, \ldots, x_{nn}$ of the Legendre polynomial L_n. Then we have

$$f(x) = \sum_{\nu=1}^{n} [\psi_{2n-1,\nu}(x) f(x_{n\nu}) + \chi_{2n-1,\nu}(x) f'(x_{n\nu})] + r(x)$$

with

$$\chi_{2n-1,\nu}(x) = \ell_{n-1,\nu}^2(x)(x - x_\nu),$$
$$\psi_{2n-1,\nu}(x) = \ell_{n-1,\nu}^2(x) \cdot (c_{2n-1,\nu} x + d_{2n-1,\nu})$$

and
$$r(x) = \frac{f^{(2n)}(\xi^*)}{(2n)!}(x - x_{n1})^2 \cdots (x - x_{nn})^2, \quad \xi^* \in (-1,+1).$$

It follows that

$$\int_{-1}^{+1} f(x)dx = \sum_{\nu=1}^{n} \Big(\int_{-1}^{+1} \psi_{2n-1,\nu}(x)dx \Big) f(x_{n\nu}) + \\ + \sum_{\nu=1}^{n} \Big(\int_{-1}^{+1} \chi_{2n-1,\nu}(x)dx \Big) f'(x_{n\nu}) + R_n f.$$

Here

$$\int_{-1}^{+1} \chi_{2n-1,\nu}(x)dx = \Big(\prod_{\substack{\mu=1 \\ \mu\neq\nu}}^{n} (x_{n\nu} - x_{n\mu}) \Big)^{-1} \int_{-1}^{+1} \hat{L}_n(x) \ell_{n-1,\nu}(x)dx = 0,$$

for $1 \le \nu \le n$, since $\ell_{n-1,\nu} \in P_{n-1}$.

Now since the quadrature error $R_n f = 0$ for $f \in P_{2n-1}$, we conclude that this is a Gauss quadrature formula.

3.3 Error Formula. If $f \in C_{2n}[-1,+1]$, then the remainder formula for the interpolating quadrature formula 3.2 is given by

$$R_n f = \int_{-1}^{+1} \frac{f^{(2n)}(\xi^*)}{(2n)!}(x - x_{n1})^2 \cdots (x - x_{nn})^2 dx = \frac{f^{(2n)}(\xi)}{(2n)!} \|\hat{L}_n\|_2^2,$$

$\xi \in (-1,+1)$. Now our discussion of the Legendre polynomials in 4.5.4 implies

$$\|\hat{L}_n\|^2 = \frac{n!}{(2n)!}\frac{1}{c_n} = \frac{n!}{(2n)!} 2^n (n!) \Big(\frac{2}{2n+1} \Big)^{1/2},$$

and thus, for the Gauss-Legendre quadrature formula with n nodes, we have the

Quadrature Error

$$\boxed{R_n f = \frac{2^{2n+1}(n!)^4}{[(2n)!]^3(2n+1)} f^{(2n)}(\xi),}$$

with $\xi \in (-1, +1)$.

In order to calculate or estimate the value of this error expression, we need to find or approximate the $(2n)$-th derivative of f. Of course, it also makes sense to use Gauss quadrature even if f is not $(2n)$-times differentiable. The corresponding error bound in this case can be obtained with the help of the Peano Kernel Formula 5.2.4

$$R_n f = \int_{-1}^{+1} K_m(t) f^{(m+1)}(t)dt$$

which can be applied for all $0 \leq m \leq 2n-1$.

Example. Consider the Gauss-Legendre quadrature formula with the two nodes $x_{21} = -\frac{1}{3}\sqrt{3}$ and $x_{22} = \frac{1}{3}\sqrt{3}$ which (cf. 4.5.4) are the zeros of L_2. The coefficients of this formula are $\gamma_{21} = \gamma_{22} = 1$ as can be found in the table in 3.6 below. Now for $f \in C_2[-1, +1]$, the remainder formula with $m = 1$ becomes

$$R_2 f = \int_{-1}^{+1} f(x)dx - \sum_{\nu=1}^{2} \gamma_{2\nu} f(x_{2\nu}) = \int_{-1}^{+1} K_1(t) f''(t)dt$$

with $K_1(t) = R_2 q_1(\cdot, t)$ and $q_1(x,t) = (x-t)_+ = \begin{cases} (x-t) & \text{for } x \geq t \\ 0 & \text{for } x < t. \end{cases}$

Now

$$R_2 q_1(\cdot, t) = \int_{-1}^{+1} (x-t)_+ dx - \sum_{\nu=1}^{2} \gamma_{2\nu}(x_{n\nu} - t)_+; \quad \text{with}$$

$$\int_{-1}^{+1} (x-t)_+ dx = \int_t^1 (x-t)dx = \frac{(1-t)^2}{2} \quad \text{and so}$$

$$K_1(t) = \begin{cases} \frac{(1+t)^2}{2} & \text{for} \quad -1 \leq t \leq -\frac{1}{3}\sqrt{3} \\ \frac{(1-t)^2}{2} + (t - \frac{1}{3}\sqrt{3}) & \text{for} \quad -\frac{1}{3}\sqrt{3} \leq t \leq \frac{1}{3}\sqrt{3} \\ \frac{(1-t)^2}{2} & \text{for} \quad \frac{1}{3}\sqrt{3} \leq t \leq 1. \end{cases}$$

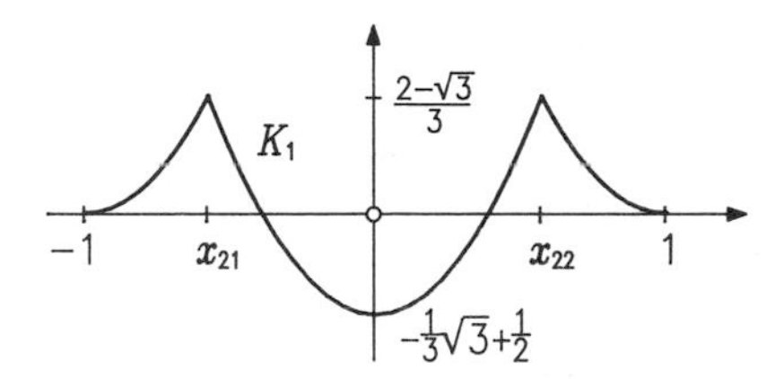

As in this example, it can be shown that for all $0 \leq m \leq 2n-2$, the Peano Kernel K_m corresponding to any Gauss-Legendre quadrature formula with n nodes in the interval $[-1,+1]$ must change sign on that interval. The kernel K_{2n-1} is, however, of one sign, and in this case the Peano Kernel Representation leads to the quadrature error formula given above in which $f^{(2n)}(\xi)$ appears. In all other cases we must be satisfied with estimating $R_n f$:

$$|R_n f| \leq \max_{t\in[-1,+1]} |f^{(m+1)}(t)| \int_{-1}^{+1} |K_m(t)|dt.$$

The values

$$e_{m+1} := \int_{-1}^{+1} |K_m(t)|dt$$

are called *Peano error constants*, and can be computed once and for all. In the example above, $e_2 = 0.081$, and $|R_2 f| \leq 0.081 \max_{x\in[-1,+1]} |f''(x)|$.

For more on Peano kernels for Gauss formulae, see e.g. A. Stroud and D. Secrest [1966]. This book also contains a table of the Peano error constants.

3.4 Modified Gauss Quadrature. For many applications, it is useful to consider a variant of Gauss quadrature in which we fix some of the nodes, and consider the remaining to be free.

The case where one or both of the end points of the integration interval $[-1,+1]$ are fixed is of particular practical importance. In these cases we have the

(i) *Gauss-Radau Formula*: when one end point is a node.

(ii) *Gauss-Lobatto Formula*: when both end points of the interval are nodes.

Case (i): Suppose $x_{n1} := -1$. We now seek to determine the $x_{2n}, \ldots, x_{nn}$ so that all $p \in \mathrm{P}_{2n-2}$ are integrated exactly. The orthogonality condition in 3.1 must now be modified to require that

$$\int_{-1}^{+1} (x+1)(x-x_{n2})\cdots(x-x_{nn})q(x)dx = 0$$

for every polynomial $q \in \mathrm{P}_{n-2}$. Now the desired nodes $x_{n2}, \ldots, x_{nn}$ can be found as the zeros of the polynomial which is orthogonal to all such q with respect to the weight function $w(x) := x+1$. By Theorem 4.5.5, these zeros are simple, real, and lie in $(-1,+1)$.

This leads to the unsymmetric

Gauss-Radau Formulae

$$\int_{-1}^{+1} f(x)dx = \frac{2}{n^2} f(-1) + \sum_{\nu=2}^{n} \gamma_{n\nu} f(x_{n\nu}) + R_n f$$

in the case where $x_{nn} = -1$, and

$$\int_{-1}^{+1} f(x)dx = \frac{2}{n^2} f(+1) + \sum_{\nu=1}^{n-1} \gamma_{n\nu} f(x_{n\nu}) + R_n f$$

in the case where $x_{nn} := +1$. Both formulae satisfy $R_n f = 0$ for all f in P_{2n-2}. The nodes and coefficients of these two formulae are mirror images of each other.

Case (ii): Here we take $x_{n1} := -1$ and $x_{nn} := +1$, and the remaining nodes are the zeros of the polynomial which is orthogonal to all $q \in \mathrm{P}_{n-3}$ with respect to $w(x) := 1 - x^2$. This gives the symmetric

Gauss–Lobatto Formula

$$\int_{-1}^{+1} f(x)dx = \frac{2}{n(n-1)}[f(1) + f(-1)] + \sum_{\nu=2}^{n-1} \gamma_{n\nu} f(x_{n\nu}) + R_n f$$

with $R_n f = 0$ for all $f \in \mathrm{P}_{2n-3}$.

As we already observed in 3.1, these modified Gauss formulae are also positive. The nodes and coefficients for some typical formulae are given in 3.6 below. Additional tables of points and coefficients, as well as error formulae, can be found in the book of A. Stroud and D. Secrest [1966].

The polynomials which are orthogonal on the interval $[-1,+1]$ with respect to the weight function $w(x) := (1-x)^\alpha (1+x)^\beta$, $\alpha, \beta > -1$, are called *Jacobi polynomials*, and are denoted by $P_n^{(\alpha,\beta)}$. The zeros of $P_{n-1}^{(0,1)}$ and $P_{n-1}^{(1,0)}$ are the free nodes of the Radau formulae, while the zeros of $P_{n-2}^{(1,1)}$ are the free nodes of the Lobatto formula.

We now consider Gauss quadrature formulae for

3.5 Improper Integrals. In this section we consider quadrature formulae

$$\int_a^b w(x)f(x)dx = \sum_{\nu=1}^{n} \gamma_{n\nu} f(x_{n\nu}) + R_n f$$

for approximating improper integrals of the form $\int_a^b w(x)f(x)dx$ where either the interval is no longer bounded, or the weight function is not bounded throughout $[a, b]$. We shall consider three specific examples of these kinds of formulae, assuming in each cases that the integral exists.

In these cases, quadrature formulae can be obtained by following the approach used in the previous section. The idea is to choose the nodes so that the Orthogonality Condition 3.1

$$\int_a^b w(x)(x - x_{n1}) \cdots (x - x_{nn}) q(x) dx = 0$$

is satisfied for all polynomials $q \in \mathrm{P}_{n-1}$. Each choice of w leads to a different family of quadrature formulae of Gauss type.

As we have already noted in 3.1, all of these quadrature formulae will be positive; i.e., $\gamma_{n\nu} > 0$ for $1 \leq \nu \leq n$. We give tables of nodes and coefficients in 3.6 below for all three types of formulae treated here. We leave the development of error bounds to the reader. More complete tables as well as a detailed treatment of the error terms can be found in A. Stroud and D. Secrest [1966] or in H. Engels [1980].

Case (i):

$$\int_0^\infty e^{-x} f(x) dx.$$

In this case the polynomials which are orthogonal on $[0, \infty)$ with respect to the weight function $w(x) := e^{-x}$ are the classical *Laguerre Polynomials* given by

$$\Lambda_n(x) = \frac{1}{n!} \hat{\Lambda}_n(x) \text{ with } \hat{\Lambda}_n(x) := (-1)^n e^x \frac{d^n}{dx^n}(x^n e^{-x})$$

and normalized so that

$$\int_0^\infty e^{-x} \Lambda_n^2(x) dx = 1 \text{ and } \hat{\Lambda}_n(x) = x^n + \cdots.$$

If $f \in \mathrm{C}_{2n}[0, \infty)$, then the quadrature error is

$$R_n f = \frac{(n!)^2}{(2n)!} f^{(2n)}(\xi), \quad \xi \in (0, \infty).$$

Case (ii):

$$\int_{-\infty}^\infty e^{-x^2} f(x) dx.$$

Here the polynomials which are orthogonal on $(-\infty, \infty)$ with respect to the weight function $w(x) := e^{-x^2}$ are the classical *Hermite Polynomials* given by

$$H_n(x) = \left(\frac{2^n}{n! \sqrt{\pi}}\right)^{1/2} \hat{H}_n(x) \text{ with } \hat{H}_n(x) := \frac{(-1)^n}{2^n} e^{x^2} \frac{d^n}{dx^n}(e^{-x^2})$$

and normalized so that

$$\int_{-\infty}^{+\infty} e^{-x^2} H_n^2(x)dx = 1 \text{ and } \hat{H}_n(x) = x^n + \cdots.$$

If $f \in C_{2n}(-\infty, +\infty)$, then the quadrature error is

$$R_n f = \frac{n!\sqrt{\pi}}{2^n(2n)!} f^{(2n)}(\xi), \quad \xi \in (-\infty, +\infty).$$

Case (iii):

$$\int_{-1}^{+1} \frac{f(x)}{\sqrt{1-x^2}} dx.$$

For this weight function, the polynomials which are orthogonal on $[-1, 1]$ with respect to the weight function $w(x) = (1-x^2)^{-1/2}$ are the Chebyshev polynomials of the first kind $T_n(x) = \cos(n \arccos(x))$ discussed in 4.4.7. These are, incidentally, Jacobi polynomials with $\alpha = \beta = -\frac{1}{2}$. It follows that the nodes of the Gauss quadrature formula in this case are given by $x_{n\nu} = \cos \frac{2\nu-(2n+1)}{2n}\pi$, $1 \le \nu \le n$. It turns out that the coefficients $\gamma_{n\nu}$ in this case are all equal, and are: $\gamma_{n\nu} = \frac{\pi}{n}$ for $1 \le \nu \le n$. The proof of this fact can be found e.g. in H. Engels ([1980], p. 318).

If $f \in C_{2n}[-1, +1]$, then the quadrature error is given by

$$R_n f = \frac{\pi}{(2n)! 2^{2n-1}} f^{(2n)}(\xi), \quad \xi \in (-1, +1).$$

3.6 Nodes and Coefficients of Gauss Quadrature Formulae.

(i) **Symmetric Formulae:** $x_{n\nu} = -x_{n,n-\nu+1}$

$\gamma_{n\nu} = \gamma_{n,n-\nu+1}$

Gauss-Legendre $[a, b] := [-1, +1]$

$n = 1$	Trapezoidal rule	
$n = 2$	$x_{22} = 0.577\,350$	$\gamma_{22} = 1$
$n = 3$	$x_{32} = 0$	$\gamma_{32} = 0.888\,889$
	$x_{33} = 0.774\,597$	$\gamma_{33} = 0.555\,556$
$n = 4$	$x_{43} = 0.339\,981$	$\gamma_{43} = 0.652\,145$
	$x_{44} = 0.861\,136$	$\gamma_{44} = 0.347\,855$
$n = 5$	$x_{53} = 0$	$\gamma_{53} = 0.568\,889$
	$x_{54} = 0.538\,469$	$\gamma_{54} = 0.478\,629$
	$x_{55} = 0.906\,180$	$\gamma_{55} = 0.236\,927$

Gauss-Hermite $[a,b] := [-\infty,+\infty]$, $w(x) := e^{-x^2}$

$n=1$ $x_{11}=0$ $\gamma_{11}=1.772\,454$

$n=2$ $x_{22}=0.707\,107$ $\gamma_{22}=0.886\,227$

$n=3$ $x_{32}=0$ $\gamma_{32}=1.181\,636$
$x_{33}=1.224\,745$ $\gamma_{33}=0.295\,409$

$n=4$ $x_{43}=0.524\,648$ $\gamma_{43}=0.804\,914$
$x_{44}=1.650\,680$ $\gamma_{44}=0.813\,128\cdot 10^{-1}$

$n=5$ $x_{53}=0$ $\gamma_{53}=0.945\,309$
$x_{54}=0.958\,572$ $\gamma_{54}=0.393\,619$
$x_{55}=2.020\,183$ $\gamma_{55}=0.199\,532\cdot 10^{-1}$

Gauss-Lobatto $[a,b] := [-1,+1]$

$n=3$ Simpson's Rule

$n=4$ $x_{43}=0.447\,214$ $\gamma_{43}=0.833\,333$
$x_{44}=1$ $\gamma_{44}=0.166\,667$

$n=5$ $x_{53}=0$ $\gamma_{53}=0.711\,111$
$x_{54}=0.654\,654$ $\gamma_{54}=0.544\,444$
$x_{55}=1$ $\gamma_{55}=0.100\,000$

(ii) **Unsymmetric Formulae:**

Gauss-Laguerre $[a,b] := [0,\infty]$, $w(x) := e^{-x}$.

$n=1$ $x_{11}=1$ $\gamma_{11}=1$

$n=2$ $x_{21}=0.585\,786$ $\gamma_{21}=0.853\,553$
$x_{22}=3.414\,214$ $\gamma_{22}=0.146\,447$

$n=3$ $x_{31}=0.415\,775$ $\gamma_{31}=0.711\,093$
$x_{32}=2.294\,280$ $\gamma_{32}=0.278\,518$
$x_{33}=6.289\,945$ $\gamma_{33}=0.103\,893\cdot 10^{-1}$

$n=4$ $x_{41}=0.322\,548$ $\gamma_{41}=0.603\,154$
$x_{42}=1.745\,761$ $\gamma_{42}=0.357\,419$
$x_{43}=4.536\,620$ $\gamma_{43}=0.388\,879\cdot 10^{-1}$
$x_{44}=9.395\,071$ $\gamma_{44}=0.539\,295\cdot 10^{-3}$

$n=5$ $x_{51}=0.263\,560$ $\gamma_{51}=0.521\,756$
$x_{52}=1.413\,403$ $\gamma_{52}=0.398\,667$
$x_{53}=3.596\,426$ $\gamma_{53}=0.759\,424\cdot 10^{-1}$
$x_{54}=7.085\,810$ $\gamma_{54}=0.361\,176\cdot 10^{-2}$
$x_{55}=12.640\,801$ $\gamma_{55}=0.233\,700\cdot 10^{-4}$

Gauss-Radau $[a,b] := [-1,+1]$

$n=2$	$x_{21} = -1$	$\gamma_{21} = 0.500\,000$
	$x_{22} = 0.333\,333$	$\gamma_{22} = 1.500\,000$
$n=3$	$x_{31} = -1$	$\gamma_{31} = 0.222\,222$
	$x_{32} = -0.289\,898$	$\gamma_{32} = 1.024\,972$
	$x_{33} = 0.689\,898$	$\gamma_{33} = 0.752\,806$
$n=4$	$x_{41} = -1$	$\gamma_{41} = 0.125\,000$
	$x_{42} = -0.575\,319$	$\gamma_{42} = 0.657\,689$
	$x_{43} = 0.181\,066$	$\gamma_{43} = 0.776\,387$
	$x_{44} = 0.822\,824$	$\gamma_{44} = 0.440\,924$
$n=5$	$x_{51} = -1$	$\gamma_{51} = 0.800\,000 \cdot 10^{-1}$
	$x_{52} = -0.720\,480$	$\gamma_{52} = 0.446\,208$
	$x_{53} = -0.167\,181$	$\gamma_{53} = 0.623\,653$
	$x_{54} = 0.446\,314$	$\gamma_{54} = 0.562\,712$
	$x_{55} = 0.885\,792$	$\gamma_{55} = 0.287\,427$

3.7 Problems. 1) For the Gauss-Legendre quadrature formula with two nodes, find the Peano kernel and Peano error constant for $f \in C_3[-1,+1]$.

2) Show that for all $1 \le m \le 2n-1$, the Peano kernel K_m of a Gauss-Legendre quadrature formula with n nodes in $(-1,+1)$ has a set of exactly $(2n-m-1)$ zeros. *Hint:* Start with the Kernel Formula 7.1.4, and argue that $K'_m(t) = -K_{m-1}(t)$. Then apply Rolle's Theorem.

3)a) Find the nodes and coefficients of the Gauss-Lobatto quadrature formula for $n=4$.

b) Find the nodes and coefficients of the Gauss-Hermite quadrature formula for $n=2$.

4) Establish the error formula 3.5(i) for Gauss-Laguerre quadrature.

5) Compute the integrals

$$J_1 = \int_{\pi/3}^{\pi/3+\pi/2} \sin^2(x)dx \quad \text{and} \quad J_2 = \int_{-1}^{+1} |x - \tfrac{1}{2}|(x - \tfrac{1}{2})dx$$

a) using Simpson's rule with 3, 5, 7, 9 nodes;

b) using Gauss-Legendre with 2, 3, 4, 5 nodes.

c) Compare the amount of computation needed and the convergence for J_1 and J_2.

6) Develop a quadrature formula of the form

$$Qf = \gamma_1 f(x_1) + \gamma_2[f(1) - f(-1)]$$

which exactly integrates polynomials of the highest possible degree. Use this to construct a composite formula which corresponds to integration on each of n equal length subintervals of $[a,b]$. Find the order of convergence with respect to $h = \frac{b-a}{n}$, assuming f is sufficiently differentiable.

4. Special Quadrature Methods

In Section 3.5 above we have shown how Gauss quadrature methods can be applied to approximate improper integrals. In this section we discuss this problem further, as well as the problem of approximating integrals of periodic functions. We begin with

4.1 Integration over an Infinite Interval. As we have seen in 3.5, the Gauss-Laguerre and Gauss-Hermite quadrature formulae can be used to approximate integrals over the intervals $[0, \infty)$ or $(-\infty, +\infty)$. These formulae involve

Integrals with Weight Functions. We can always artificially introduce the weight functions $\exp(-x)$ or $\exp(-x^2)$ by simply writing

$$\int_0^\infty f(x)dx = \int_0^\infty \exp(-x)\varphi(x)dx \quad \text{with} \quad \varphi(x) := f(x)\exp(x),$$

or respectively,

$$\int_{-\infty}^{+\infty} f(x)dx = \int_{-\infty}^{+\infty} \exp(-x^2)\varphi(x)dx \quad \text{with} \quad \varphi(x) := f(x)\exp(x^2),$$

after which the methods of 3.5 can be applied. As an alternative approach, we can consider

Truncation of the Integration Interval. The idea here is to calculate the integral over a finite part of the integration interval in such a way that the integral over the remainder of the interval is small.

To illustrate this idea, suppose we want to compute the integral

$$\int_0^\infty e^{-x^2} dx = \frac{1}{2}\sqrt{\pi} \doteq 0.886227.$$

to a prescribed accuracy. To accomplish this, we split the integral into two parts

$$\int_0^\infty e^{-x^2} dx = \int_0^X e^{-x^2} dx + \int_X^\infty e^{-x^2} dx.$$

Since $x^2 \geq Xx$ for $x \geq X$,

$$\int_X^\infty e^{-x^2} dx < \int_X^\infty e^{-Xx} dx = \frac{1}{X} e^{-X^2}.$$

Choosing $X = 3$, for example, we see that the remainder integral satisfies $\int_3^\infty e^{-x^2} dx < 5 \cdot 10^{-5}$. Thus, if we compute the integral over $[0, 3]$ to an accuracy of $5 \cdot 10^{-5}$, then we have an approximation which is accurate up to one digit in the fourth place. For example, if we use the midpoint rule with 5 nodes and $h = 0.6$, we get the approximate value 0.886216.

Another approach to handling infinite intervals is based on

Transformation to a Finite Integration Interval. The interval $[0, \infty)$ can be transformed to the interval $(0, 1]$ by using either of the transformations $t = \frac{x}{1+x}$ or $t = e^{-x}$. Similarly, the interval $(-\infty, +\infty)$ can be transformed to $(-1, +1)$ by using the transformation $t = \frac{e^x - 1}{e^x + 1}$. If the transformed integrand is continuous, we are led to a standard quadrature problem. If not, we have to deal with a singular integrand.

Thus for example, if we use the transformation $t = e^{-x}$, we get

$$\int_0^\infty f(x)dx = \int_0^1 \frac{f(-\log t)}{t} dt,$$

and the first case arises if the function $g(t) := \frac{f(-\log t)}{t}$ is continuous on $[0, 1]$. This happens, for example, for $f(x) = e^{-x}$ and we get $\int_0^\infty e^{-x} dx = \int_0^1 dt = 1$.

In the other case we are led to

4.2 Singular Integrands. We have already seen an example of a singular integral in 3.5 (iii); namely, $\int_{-1}^{+1} \frac{f(x)}{\sqrt{1-x^2}} dx$. In this section we discuss two ways of handling singular integrands. The first involves

Separation of the Singularity. If we can separate out the singularity and compute its integral exactly, then the problem reduces to computing an ordinary integral. As an example, consider

$$\int_0^1 \frac{e^x}{x^{2/3}} dx = \int_0^1 x^{-2/3} dx + \int_0^1 \frac{e^x - 1}{x^{2/3}} dx = 3 + \int_0^1 (e^x - 1)x^{-2/3} dx.$$

Since the function $g(x) := (e^x - 1)x^{-2/3}$ with $g(0) := 0$ is continuous on $[0, 1]$, the last integral can be computed by standard quadrature methods.

Decomposition as a Product. If the integrand can be written in the form $f(x) = \varphi(x)\psi(x)$, then it may make sense to approximate only one factor. For example, if φ is singular, and we approximate ψ by a polynomial $p \in \mathrm{P}_n$, then we are led to a formula of the form

$$\int_a^b \varphi(x)\psi(x)dx = \sum_{\nu=0}^n \gamma_{n\nu}\psi(x_{n\nu}) + R_n f,$$

where the coefficients depend on φ. To find these coefficients, we need to be able to compute the integrals $\int_a^b \varphi(x)x^\nu dx$ for $0 \le \nu \le n$ exactly.

Example 1. Let $[a, b] := [0, 1]$ and $\varphi(x) := x^{-1/2}$, and consider approximating ψ by a polynomial $p \in \mathrm{P}_2$ interpolating at the nodes $x_{20} = 0$, $x_{21} = \frac{1}{2}$, $x_{22} = 1$.

In this case we need to find the values $\int_0^1 x^{-1/2}x^\nu dx$ for $\nu = 0, 1, 2$ in order to determine the coefficients γ_{20}, γ_{21} and γ_{22}. This leads to the system of equations

$$\gamma_{20} + \gamma_{21} + \gamma_{22} = \int_0^1 x^{-1/2}x^0 dx = 2$$
$$\frac{1}{2}\gamma_{21} + \gamma_{22} = \int_0^1 x^{-1/2}x dx = \frac{2}{3}$$
$$\frac{1}{4}\gamma_{21} + \gamma_{22} = \int_0^1 x^{-1/2}x^2 dx = \frac{2}{5},$$

and we get the quadrature formula

$$(*) \qquad \int_0^1 x^{-1/2}\psi(x)dx = \frac{2}{15}[6\psi(0) + 8\psi(\frac{1}{2}) + \psi(1)] + R_2\psi.$$

To get an error bound, we can start with an estimate of the interpolation error $\|\psi - p\|_\infty$, cf. Problem 4.

Ignoring the Singularity. The singularity of the integrand will not cause any difficulty in the numerical evaluation of a quadrature formula provided that the nodes of the formula are chosen to avoid the singularity, although, of course, the resulting approximation may not be particularly good. This idea can be applied, for example, whenever the singularity occurs at the end of the integration interval, in which case we could use a Gauss-Legendre formulae.

Another possibility for dealing with a singularity of the integrand f at a point x^* is to simply replace the true value of f there by $f(x^*) = 0$. It can be shown that under appropriate assumptions on f, e.g. if it is monotone in a neighborhood of the singularity, then in spite of the false value of f, standard quadrature formulae can lead to a sequence of approximations which converges to the the true integral (Davis and Rabinowitz [1975], 2.12.7).

Example 2. Suppose we approximate

$$\int_0^1 \frac{1}{\sqrt{x}}\exp(\sqrt{x})dx = 2(e-1)$$

using the following quadrature formulae:

(1) Simpson's rule setting $f(0) := 0$;
(2) Gauss-Legendre quadrature;
(3) Product integration of $\int_0^1 \frac{1}{\sqrt{x}}\psi(x)dx$, where we decompose the interval into $[0,1] = [0,2h] \cup [2h,1]$, $h := \frac{1}{n}$. On $[2h,1]$ we use Simpson's rule, and on $[0,2h]$ we use the quadrature rule

$$Qf = \frac{2}{15}\sqrt{2h}[6\psi(0) + 8\psi(h) + \psi(2h)]$$

which is obtained from the method in Example 1 after transformation to the interval $[0, 2h]$.

The following table shows the results of applying these methods using $(n+1)$ nodes for various values of n. These results should be compared with the true value $2(e-1) \doteq 3.43656$.

n	Simpson's rule	Gauss-Legendre	Product Integration
2	2.365 17	3.188 68	3.325 76
4	2.718 78	3.278 48	3.381 06
8	2.948 08	3.344 95	3.408 79
16	3.100 42	3.386 82	3.422 70
32	3.203 41	3.410 57	3.429 66
64	3.273 94	3.423 27	3.433 14

4.3 Periodic Functions. In this section we discuss a special phenomenon which arises in numerically integrating a periodic function over an interval of length equal to the period; namely, that the trapezoidal rule is significantly more accurate than might be expected. This can be explained by using the Euler-MacLaurin Expansion 1.3. Assuming $f \in C_{2m}(-\infty, +\infty)$ is periodic with period $(b-a)$, that is, $f(x) = f(x + (b-a))$ for all $x \in \mathbb{R}$, then $f^{(\mu)}(a) = f^{(\mu)}(b)$ for $\mu = 0, \ldots, 2m$. Then the Euler-MacLaurin expansion reduces to

$$\int_a^b f(x)dx = T_n f - h^{2m} \frac{B_{2m}}{(2m)!} f^{(2m)}(\xi),$$

where $\xi \in (a, b)$. This implies that the quadrature error of the trapezoidal rule is of order $O(h^{2m})$, or equivalently $O(\frac{1}{n^{2m}})$. We can state this formally as the following

Theorem. *If $f \in C_{2m}(-\infty, +\infty)$ is periodic, then the sequence of approximations obtained by applying the trapezoidal rule on an interval of length equal to the period converges to the true value of the integral with the order $O(h^{2m})$.*

A Second Explanation. The enhanced accuracy of the trapezoidal rule when applied to periodic functions over a full period can also be explained by the following simple argument. First, since $f(a) = f(b)$, it is easy to see that the trapezoidal rule reduces to the rectangle rule in this case. Suppose now that we choose equally-spaced nodes $x_\nu = x_0 + \nu \frac{b-a}{n}$, $0 \le \nu \le n$. Then extending the nodes and coefficients periodically by setting $x_{\nu+n} := x_\nu$, $\gamma_{\nu+n} := \gamma_\nu$, we can use any one of these nodes $x_{\nu 0}$ as the first when applying an arbitrary quadrature formula. This leads to the formula

$$Q_n^{\nu_0} f = \sum_{\nu=\nu_0}^{\nu_0+n} \gamma_\nu f(x_\nu).$$

Now let $\sum_0^n \gamma_\nu = b - a$. This guarantees that our quadrature formula exactly integrates the function $f(x) := 1$. Then it is easy to see that the arithmetic mean of the n numbers $Q_n^{\nu_0} f$, $0 \le \nu_0 \le n-1$, originating from the same quadrature formula, is exactly the value $\frac{b-a}{n}\sum_0^{n-1} f(x_\nu)$, obtained by applying the trapezoidal rule.

4.4 Problems. 1) Suppose we want to approximate the improper integral $J := \int_0^\infty \frac{\sin(x)}{1+x^2}dx$ by transformation of the integration interval.

a) Find X such that $|\int_X^\infty \frac{\sin(x)}{1+x^2}dx| < 10^{-4}$. Hint: choose X as a multiple of π and use the alternating property of the sign of the sine function.

b) Compute an approximation to J to an accuracy of $2\cdot 10^{-4}$ by applying the Romberg method on $[0, X]$, starting with the step size $h_0 = \frac{\pi}{2}$.

2) Apply each of the methods in 4.1 to find the improper integral $\int_0^\infty \frac{e^{-x}}{(1+x^2)}dx$, (without doing any numerical computation).

3) Approximate the integral $J := \int_0^{\pi/2} \log(\sin x)dx = -\frac{\pi}{2}\log 2$ by separating out the singularity using the identity $\log(\sin x) = \log(x) + \log\frac{\sin x}{x}$. Use Simpson's rule to approximate the resulting proper integral to 5 digits, and compare with the exact value.

4) Find a bound on the error $R_2\psi$ for the quadrature formula for a product presented in Example 1 in 4.2 by estimating the interpolation error $\|\psi - p\|_\infty$. Improve this bound by a careful treatment of the remainder term.

5) Show: For $n \ge 2$, the trapezoidal rule produces the exact values of the integrals $\int_0^{2\pi}\cos(x)dx$ and $\int_0^{2\pi}\sin(x)dx$.

6) The rate of convergence of the trapezoidal rule is increased whenever $f^{(2\kappa-1)}(a) = f^{(2\kappa-1)}(b)$ for $\kappa = 1, 2, \ldots, k$, without requiring that f be periodic. For $f(x) := \exp(x^2(1-x)^2)$ with $x \in [0,1]$, find the rate of convergence of $T_n f$ to $\int_0^1 f(x)dx$ as a function of h, and check the result numerically.

7) Use the trapezoidal rule on the formula given in 4.4.8 to find the coefficients $c_0, \ldots, c_8$ of the expansion of the exponential function f defined by $f(x) := e^x$ on $[-1, +1]$ in terms of Chebyshev polynomials. Note that the c_k are integrals over a full period of a periodic function, and compute them to an accuracy of $\pm 5\cdot 10^{-9}$.

5. Optimality and Convergence

The concept of "optimality of a quadrature formula" can be given several different meanings. We start with an interpolatory quadrature formula of the form

$$\int_{-1}^{+1} f(x)dx = \sum_{\nu=1}^{n} \gamma_{n\nu} f(x_{n\nu}) + R_n f,$$

with error term

$$R_n f = \frac{1}{n!} \int_{-1}^{+1} f^{(n)}(\xi) \prod_{\nu=1}^{n} (x - x_{n\nu}) dx, \ -1 < \xi < 1, \ \xi = \xi(x),$$

obtained by integrating the remainder term of the interpolant.

We now consider the following problem of

5.1 Norm Minimization. Applying the Hölder Inequality 4.1.4, it follows that

$$|R_n f| \leq \frac{1}{n!} \left[\int_{-1}^{+1} |f^{(n)}(\xi)|^p dx \right]^{\frac{1}{p}} \left[\int_{-1}^{+1} |\prod_{\nu=1}^{n} (x - x_{n\nu})|^q dx \right]^{\frac{1}{q}}$$

for all $p, q \geq 1$ and $\frac{1}{p} + \frac{1}{q} = 1$.

Now as was done in 5.4.1 in the case of interpolation, we can try to make $|R_n f|$ as small as possible, independently from f; i.e., we choose the nodes $x_{n1}, \ldots, x_{nn}$ to minimize the quantity

$$\|\Phi\|_q = \left[\int_{-1}^{+1} | \prod_{\nu=1}^{n} (x - x_{n\nu}) \, |^q \, dx \right]^{\frac{1}{q}}.$$

We now discuss the three most important cases where (i) $p = 1$, $q = \infty$, (ii) $p = \infty$, $q = 1$ and (iii) $p = q = 2$.

Case (i). The problem of minimizing $\|\Phi\|_\infty$ amounts to making the remainder term in the underlying interpolation formula as small as possible. This problem was treated earlier in 4.4.7 and 5.4.1. As we saw there, the minimimum is assumed if we take the nodes to be $x_{n\nu} = -\cos(\frac{2\nu-1}{2n}\pi)$, $1 \leq \nu \leq n$, which are the zeros of the Chebyshev polynomials of the first kind. The resulting quadrature formulae are called *Polya formulae.*

Case (ii). The problem of minimizing $\|\Phi\|_1$ is solved by taking the nodes to be the zeros of the *Chebyshev polynomials of the second kind.* These polynomials form an orthonormal system on $[-1, +1]$ with respect to the weight function $w(x) = (1 - x^2)^{1/2}$ (cf. Prob. 4 in 4.5.9). The n-th polynomial can be written as

$$U_n(x) = \sqrt{\frac{2}{\pi}} \; \frac{\sin((n+1)\arccos x)}{\sin(\arccos x)},$$

and its zeros are $x_{n\nu} = -\cos(\frac{\nu}{n+1}\pi)$, $1 \leq \nu \leq n$. The associated quadrature formula is called a *Filippi formula.*

The quadrature formula which results when the nodes (-1) and $(+1)$ are added to those of a Filippi formula is called a *Clenshaw-Curtis formula.* For a discussion of the advantages and disadvantages of the formulae discussed in cases (i) and (ii), see the monographs of H. Braß [1977] and H. Engels [1980].

Case (iii). As we have seen in 4.5.4, $\|\Phi\|_2$ is minimized by a Legendre polynomial. This leads to Gauss-Legendre quadrature formulae. Hence, these quadrature formulae are optimal in a certain sense and also provide the maximal order of exactness for the integration of polynomials.

5.2 Minimizing Random Errors. In practice, the function values $f(x_{n\nu})$ are often subject to random errors $d_{n\nu}$, in particular if they are experimentally obtained measured values. This suggests that we should try to choose a quadrature formula to minimize the effect of such errors.

In this case, the quadrature error can be written as

$$R_n f = \int_{-1}^{+1} f(x)dx - \left[\sum_{\nu=1}^{n} \gamma_{n\nu} f(x_{n\nu}) + \sum_{\nu=1}^{n} \gamma_{n\nu} d_{n\nu}\right],$$

and so

$$|R_n f| \leq \left| \int_{-1}^{+1} f(x)dx - \sum_{\nu=1}^{n} \gamma_{n\nu} f(x_{n\nu}) \right| + \sum_{\nu=1}^{n} |\gamma_{n\nu}|\, |d_{n\nu}|.$$

Now bounding the last term by

$$\sum_{\nu=1}^{n} |\gamma_{n\nu}|\, |d_{n\gamma}| \leq \left(\sum_{\nu=1}^{n} \gamma_{n\nu}^2\right)^{\frac{1}{2}} \left(\sum_{\nu=1}^{n} d_{n\nu}^2\right)^{\frac{1}{2}},$$

we have succeeded in separating the random errors from the coefficients, and it makes sense to try to minimize the sum of the squares of the coefficients subject to the side condition that $\sum_{\nu=1}^{n} \gamma_{n\nu} = 2$. This problem can be solved by using Lagrange multipliers, and it can be shown (see Problem 2) that the minimum is assumed when all of the coefficients are equal; i.e., $\gamma_{n\nu} = \frac{2}{n}$ for $\nu = 1, \ldots, n$.

This leads to the problem of choosing the nodes of a quadrature formula, given prescribed coefficients. This is the natural companion to the problem of choosing the coefficients, given prescribed nodes. Quadrature formulae with equal coefficients were first studied in 1874 by P. L. Chebyshev. Such formulae are called *Chebyshev quadrature formulae.* It turns out that the problem of making these formulae exact for polynomials of highest possible degree can no longer be solved by taking the nodes to be the zeros of some polynomial which belongs to an orthogonal system on $[-1, +1]$ with respect to a weight function. As a consequence, it follows that Chebyshev quadrature formulae with simple real nodes lying in $[-1, +1]$ exist only for $n = 1, \ldots, 7$ and $n = 9$. If sufficient accuracy cannot be obtained with one of the Chebyshev quadrature formulae directly, then we can divide the integration interval $[a, b]$ into pieces, and treat each piece separately.

For $n = 2$ and $n = 3$, the corresponding Chebyshev quadrature formulae are obtained by requiring that they exactly integrate polynomials of degree

2 and 3, respectively. This leads to the formulae

$$n = 2: \quad Qf = f(-\frac{\sqrt{3}}{3}) + f(\frac{\sqrt{3}}{3})$$

$$n = 3: \quad Qf = \frac{2}{3}[f(-\frac{\sqrt{2}}{2}) + f(0) + f(\frac{\sqrt{2}}{2})].$$

5.3 Optimal Quadrature Formulae. By an optimal quadrature formula we mean one which gives the best error bound among all formulae of a given type and over some prescribed class of functions. As a starting point, we represent the quadrature error

$$R_n f = \int_a^b f(x)dx - \sum_{\nu=0}^{n} \gamma_{n\nu} f(x_{n\nu})$$

using a Peano kernel. Assuming $f \in \mathrm{C}_{m+1}[a,b]$ and $R_n f = 0$ for $f \in \mathrm{P}_m$, we can write

$$R_n f = \int_a^b K_m(t) f^{(m+1)}(t)dt.$$

Applying Hölder's inequality, it follows that

$$|R_n f| \le \left[\int_a^b |f^{(m+1)}(t)|^p dt\right]^{\frac{1}{p}} \left[\int_a^b |K_m(t)|^q dt\right]^{\frac{1}{q}}$$

for all $1 \le p,\ q \le \infty$ and $\frac{1}{p} + \frac{1}{q} = 1$. Now the term involving the Peano kernel is independent of f, and we can try to minimize it over the choice of coefficients and nodes.

The two most interesting cases are

(i) $p = \infty$, $q = 1$, where

$$|R_n f| \le \|f^{(m+1)}\|_\infty \int_a^b |K_m(t)|dt$$

and

(ii) $p = q = 2$, where

$$|R_n f| \le \|f^{(m+1)}\|_2 \left[\int_a^b |K_m(t)|^2 dt\right]^{\frac{1}{2}}.$$

Before proceeding further, we observe that there is a

Connection with Splines. Indeed, the Peano kernel K_m can be written in the form

$$K_m(t) = \frac{(b-t)^{m+1}}{(m+1)!} - s(t),$$

where $s(t) := \frac{1}{m!}\sum_{\nu=0}^{n}\gamma_{n\nu}(x_{n\nu}-t)_+^m$ is a spline in $S_m(\{x_{n\nu}\}_{\nu=0,\ldots,n})$. Thus, the problem of minimizing

$$\left[\int_a^b |K_m(t)|^q dt\right]^{\frac{1}{q}} = \|K_m\|_q$$

is equivalent to finding the best approximation of the polynomial $\frac{(b-t)^{m+1}}{(m+1)!}$ in t by a spline with respect to the norm $\|\cdot\|_q$.

The problem of finding optimal quadrature formulae has not been solved in general, even in the two special cases (i) and (ii). To illustrate what is known, we now discuss formulae corresponding to

m=0 and equally-spaced nodes. Let $x_{n\nu} = a + \nu h$, $h = \frac{b-a}{n}$, $0 \le \nu \le n$. As we shall see, in this case we can deal with all $1 \le q \le \infty$ at once.

The problem is the following: Find the coefficients $\gamma_{n\nu}$ of a quadrature formula defined for $f \in C_1[a,b]$ such that the formula is exact for constants and $\|K_0\|_q$ is minimal. For the exactness condition $R_n f = 0$ for $f \in P_0$ to be satisfied, we need $\sum_{\nu=0}^{n}\gamma_{n\nu} = b-a$.

The Peano kernel is

$$K_0(t) = (b-t) - \sum_{\nu=0}^{n}\gamma_{n\nu}(x_{n\nu}-t)_+^0 = (b-t) - \sum_{\substack{\nu \text{ with} \\ x_{n\nu}\ge t}}\gamma_{n\nu},$$

or in other words,

$$\begin{aligned}
K_0(t) &= (b-t) - \textstyle\sum_{\nu=0}^{n}\gamma_{n\nu} = a-t && \text{for} \quad t = a;\\
K_0(t) &= (b-t) - \textstyle\sum_{\nu=1}^{n}\gamma_{n\nu} = a-t+\gamma_{n0} && \text{for} \quad a < t \le x_{n1};\\
K_0(t) &= (b-t) - \textstyle\sum_{\nu=2}^{n}\gamma_{n\nu} = a-t+(\gamma_{n0}+\gamma_{n1}) && \text{for} \quad x_{n1} < t \le x_{n2};\\
&\vdots && \vdots\\
K_0(t) &= (b-t) - \gamma_{nn} && \text{for} \quad x_{n,n-1} < t \le x_{nn}.
\end{aligned}$$

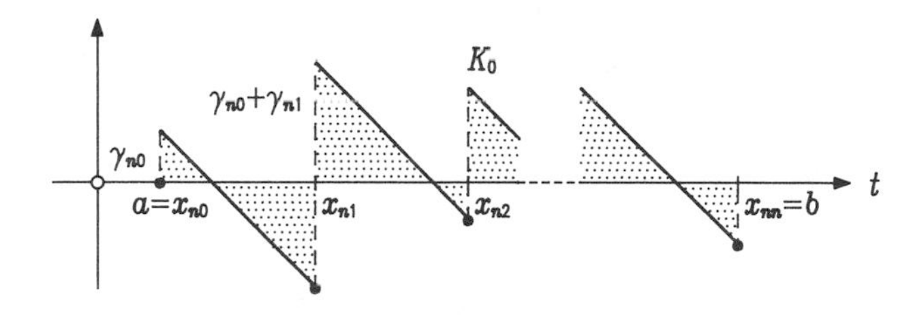

For $q = 1$ the problem amounts to minimizing the sum of the areas shown in the figure above, while for $1 < q < \infty$ we need to minimize the integral of the corresponding function consisting piecewise of monomials of degree q. A simple argument, which we leave to the reader, shows that the minimum is attained when the marked area is made up of $2n$ triangles of equal area. In the case $q = \infty$, we can arrive at the same result in another way: We have to minimize $\max_{t\in[a,b]} |K_0(t)|$ which gives $|K_0(x_{n\nu})| = \frac{h}{2}$ for $1 \le \nu \le n$.

Here the optimal formula is given by the trapezoidal rule with coefficients $\gamma_{n0} = \gamma_{nn} = \frac{h}{2}$, $\gamma_{n\nu} = h$ for $1 \le \nu \le n-1$.

Sard's Formulae. L. F. Meyers and A. Sard [1950] found the optimal quadrature formulae in case (ii), where $\|K_m\|_2$ is to be minimized, assuming equally-spaced nodes. The problem is to minimize the function

$$F(\gamma_{n0}, \dots, \gamma_{nn}) := \int_a^b |K_m(t)|^2 dt$$

over the coefficients $\gamma_{n0}, \dots, \gamma_{nn}$, subject to the side conditions

$$\sum_{\nu=0}^{n} \gamma_{n\nu}\, x_{n\nu}^{\mu} = \frac{b^{\mu+1} - a^{\mu+1}}{\mu+1}, \qquad 0 \le \mu \le m.$$

Meyers and Sard give a large number of such formulae. For example, for $m = 1$ and $n = 1$ the optimal formula $Q_n^m f$ is the trapezoidal rule, while for $m = 1$, $n = 2$ and $[a, b] := [-1, +1]$, it is

$$Q_2^1 f = \frac{1}{8}[3f(-1) + 10f(0) + 3f(1)].$$

Other optimal formulae include Simpson's formula for $m = 2$ and $n = 2$, Newton's $\frac{3}{8}$-rule for $m = 2$ and $n = 3$, and the formula

$$Q_4^2 f = \frac{1}{120}[21f(-1) + 76f(-\frac{1}{2}) + 46f(0) + 76f(\frac{1}{2}) + 21f(1)]$$

for $m = 2$ and $n = 4$.

We emphasize that these *Sard formulae* are not designed to exactly integrate polynomials of the highest possible degree, and in fact do not do so in general. For example, $Q_2^1 f$ is exact for all polynomials in P_1, but not for all in P_2. I. J. Schoenberg [1964] showed, however, that $Q_n^m f$ is exact for all natural splines in $S_{2m+1}(\{x_{n\nu}\})$.

The study of optimal quadrature formulae leads to many other interesting problems. Although their study provides useful insight into the theory of quadrature, optimal quadrature formulae are of relatively minor practical importance.

5.4 Convergence of Quadrature Formulae. Except for our treatment of extrapolation in Section 2, which was valid for general Riemann integrable functions, so far we have discussed the convergence of a sequence of quadrature values $(Q_n f)$ to the true integral as $n \to \infty$ or $h \to 0$ only for integrands with a certain amount of differentiability.

The question of convergence when the integrand is only continuous is certainly of interest, since in practice we often have to numerically integrate a function whose smoothness properties are unknown.

We now want to discuss this question in sufficient generality to include the cases where we have a sequence of Gauss quadrature formulae with an increasing number of nodes, or a sequence of Newton-Cotes formulae with increasing degrees. Thus, we consider quadrature formulae of the form

$$\int_a^b w(x)f(x)dx = \sum_{\nu=0}^{n} \gamma_{n\nu} f(x_{n\nu}) + R_n f = Q_n f + R_n f,$$

and make the following

Definition. A sequence $(Q_\nu f)_{\nu \in \mathbb{N}}$ of quadrature formulae is called *convergent for continuous functions* if $\lim_{\nu\to\infty} R_\nu f = 0$ for every $f \in \mathrm{C}[a, b]$.

The key result in this connection is the

Convergence Theorem. *A sequence $(Q_n f)_{n\in\mathbb{N}}$ of quadrature formulae converges for all continuous functions provided that*

(i) it converges for every polynomial,

and

(ii) there is a constant Γ such that

$$\sum_{\nu=0}^{n} |\gamma_{n\nu}| < \Gamma \quad \textit{for all } n \in \mathbb{N}.$$

Proof. The proof is based on the Weierstrass Approximation Theorem 4.2.2. It asserts that for every $\varepsilon > 0$, there is a $k \in \mathbb{N}$ and a polynomial $p \in \mathrm{P}_k$ such that $\|f - p\|_\infty < \varepsilon$. Now since $(Q_n p)_{n\in\mathbb{N}}$ converges for every polynomial p, it follows that $|R_n p| < \varepsilon$ whenever k is sufficiently large. But then

$$|R_n f| \leq \left| \int_a^b w(x)[f(x) - p(x)]dx \right| + |R_n p| + \left| \sum_{\nu=0}^{n} \gamma_{n\nu}[f(x_{n\nu}) - p(x_{n\nu})] \right|,$$

and setting $\int_a^b w(x)dx =: W$, we get

$$|R_n f| < \varepsilon W + \varepsilon + \varepsilon\Gamma = \varepsilon(W + 1 + \Gamma)$$

for all sufficiently large $k \in \mathbb{N}$. □

More on the Theorem. In the above convergence theorem we have only asserted that the stated conditions are sufficient for convergence of a sequence of quadrature formulae. This form of the theorem goes back to a paper of V. Steklov in 1916. In fact, the conditions (i) and (ii) are both necessary and sufficient, as shown by G. Polya [1933]. Polya's proof of the necessity involves the construction of a continuous function φ for which the sequence $(Q_n\varphi)$ diverges if condition (ii) is not satisfied. The necessity of (i) is obvious. A more modern and elegant proof can be based on the Banach-Steinhaus Theorem; see e.g. C. Cryer [1982].

Corollary. The condition (ii) of the convergence theorem is satisfied whenever the quadrature operators Q_n have all positive coefficients and exactly integrate constants. Indeed, in this case

$$\sum_{\nu=0}^{n} |\gamma_{n\nu}| = \sum_{\nu=0}^{n} \gamma_{n\nu} = W.$$

Application. The corollary implies the convergence of all Gauss quadrature formulae for finite integration intervals since, as shown in 3.1, such formulae are always positive. This includes all Gauss-Legendre, Gauss-Lobatto, Gauss-Radau, and Gauss-Chebyshev formulae.

As we already remarked in 2.2, not all Newton-Cotes formulae are positive, and so the convergence theorem proved above is not applicable. In fact, any sequence of Newton-Cotes formulae diverges. Indeed, we have already noted that condition (ii) is necessary, and it can be shown that for Newton-Cotes formulae, $(\sum_{\nu=0}^{n} |\gamma_{n\nu}|)_{n\in\mathbb{N}}$ diverges, and so the condition is not satisfied.

In 2.4 we have already shown that the extrapolation methods discussed there converge for all Riemann-integrable functions. For positive quadrature formulae, it is not hard to extend the above convergence theorem to this class of functions and we have the

Generalization. *A positive quadrature method which converges for all polynomials, also converges for all Riemann integrable functions.*

Proof. See H. Braß ([1977], p. 35, Theorem 10). □

Example. Consider the Runge function $f(x) := (1+x^2)^{-1}$ on $[-5,+5]$ which we discussed in 5.4.2 in connection with the question of the convergence of a sequence of interpolating polynomials. The integral of this function over $[-5,5]$ can be rewritten as $\int_{-5}^{+5} \frac{dx}{1+x^2} = 5 \cdot Jf$, where

$$Jf := \int_{-1}^{+1} \frac{dt}{1+25t^2} = \frac{2}{5}\arctan(5) \doteq 0.54936.$$

The following table shows the values of Jf as computed by various quadrature rules using $n+1$ nodes.

n	Newton-Cotes	Gauss-Legendre	Extrapolation	Simpson
2	1.359	0.95833	1.03846	1.35897
4	0.475	0.70694	0.53006	0.53006
6	0.774	0.61612	0.51208	0.64403
8	0.300	0.57870	0.53713	0.52348
10	0.935	0.56245	0.55003	0.56983
12	-0.063	0.55524	0.55151	0.54036
14	1.580	0.55201	0.55003	0.55470
16	-1.248	0.55055	0.54925	0.54666
18	3.775	0.54990	0.54921	0.55084
20	-5.370	0.54960	0.54933	0.54858

Supplement. Finally, we should mention that the quadrature rules of Polya, Filippi and Clenshaw-Curtis given in 5.1 are positive (R. Askey and J. Fitch [1968]), and so are also convergent. For these formulae, this result can also be deduced from the fact that they are based on orthogonal polynomials with respect to the weight functions $(1-x^2)^{-1/2}$ and $(1-x^2)^{1/2}$ (H. Braß [1977], Theorem 62).

5.5 Quadrature Operators. Our approach to deriving quadrature formulae of the form $Q_n f = \sum_{\nu=0}^{n} \gamma_{n\nu} f(x_{n\nu})$ has been to approximate f by a function $\tilde{f}$, and then estimate $Jf := \int_a^b f(x)dx$ by applying the integration operator J to $\tilde{f}$. This led to the error expression

$$R_n f = Jf - J\tilde{f} = J(f - \tilde{f}),$$

and to the error bound

$$|R_n f| \leq \|J\| \, \|f - \tilde{f}\|.$$

Another possible approach is to think of the quadrature formula as the result of applying the quadrature operator Q_n to f, and to bound the error

$$R_n f = Jf - Q_n f = (J - Q_n)f$$

by

$$|R_n f| \leq \|J - Q_n\| \, \|f\|.$$

Convergence would then follow whenever $\lim_{n\to\infty} \|J - Q_n\| = 0$.

The following simple counterexample shows, however, that this kind of assertion cannot hold in the space $(\mathrm{C}[a,b], \|\cdot\|_\infty)$. Consider the rectangle rule $Q_n f = h \sum_{\nu=0}^{n-1} f(a + \nu h)$ with $h = \frac{b-a}{n}$. We already know that this formula converges for every continuous function as $h \to 0$; we now consider $\lim_{n\to\infty} \|J - Q_n\|_\infty$.

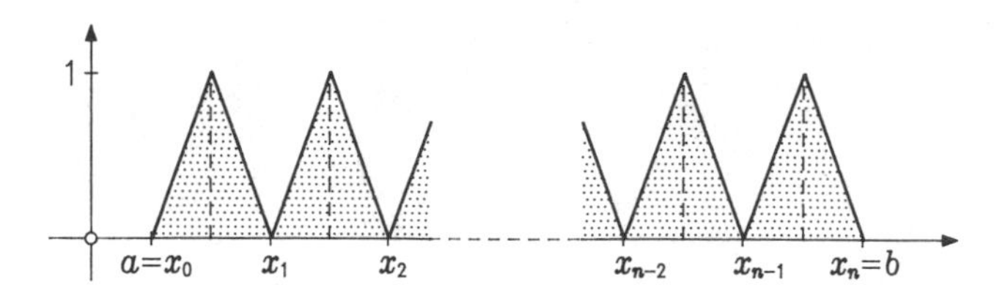

In order to examine $\|J - Q_n\|_\infty$ for a given n, consider the continuous piecewise linear saw tooth function $\varphi_n : [a,b] \to [0,1]$ which takes on the values

$$\begin{aligned} \varphi_n(a + \nu h) &:= 0 && \text{for } \nu = 0,\ldots,n \text{ and} \\ \varphi_n(a + (\nu + \tfrac{1}{2})h) &= 1 && \text{for } \nu = 0,\ldots,n-1. \end{aligned}$$

Then $J\varphi_n = \frac{b-a}{2}$ and $Q_n\varphi_n = 0$, and thus

$$\begin{aligned} \|J - Q_n\|_\infty &= \sup_{\|f\|_\infty = 1} \|(J - Q_n)f\|_\infty \geq \|(J - Q_n)\varphi_n\|_\infty = \\ &= \|J\varphi_n\|_\infty = \frac{b-a}{2} \end{aligned}$$

for all $n = 1, 2, \ldots$. This implies that $lim_{n\to\infty}\|J - Q_n\|_\infty \neq 0$, and so convergence cannot be established in this way.

5.6 Problems. 1) Using the method of 5.1, find nodes and coefficients for the following quadrature formulae:

a) Polya formula for $n = 3$; b) Filippi formula for $n = 3$;

c) Clenshaw-Curtis formula for $n = 4$.

2) Using the method of 5.2 and Lagrange multipliers, find the coefficients $\gamma_{n1}, \ldots, \gamma_{nn}$ of the Chebyshev quadrature formula which minimizes $\sum_{\nu=1}^n \gamma_{n\nu}^2$ subject to the side condition $\sum_{\nu=1}^n \gamma_{n\nu} = 2$.

3) Derive the optimal Sard formula $Q_2^1 f$ given in 5.3. Hint: Use Lagrange multipliers.

4) Show: A necessary condition for a sequence $(Q_n f)$ of quadrature formulae on an interval $[a,b]$ to converge for every function $f \in \mathrm{C}[a,b]$ is that there exists a dense subset M of $[a,b]$ such that every neighborhood of a point $x \in \mathrm{M}$ contains some node $x_{n\nu}$ for sufficiently large n.

5) It might be conjectured that, in general, the convergence of a sequence of quadrature formulae for functions in $\mathrm{C}[a,b]$ implies its convergence for Riemann integrable functions. Use the sequence $(Q_n f)_{n\in\mathbb{Z}_+}$ with

$$Q_n f := f(0) - f(\frac{1}{n}) + \frac{1}{n-1}\sum_{\nu=2}^{n} f(\frac{\nu}{n})$$

on the interval $[0,1]$ to construct a counterexample.

6) Derive a quadrature rule for functions $f \in C[0,1]$ by integrating the Bernstein polynomial $B_n f$ of 4.2.2, and show that the resulting sequence of quadrature formulae converges.

6. Multidimensional Integration

Moving from the problem of computing one dimensional integrals to the multidimensional case leads to a series of new problems. While in one dimension we encountered three possible types of integration intervals – finite, semi-infinite, and infinite – now we have to deal with a wide variety of domains. In addition, as is already evident in two dimensions, the functions being integrated can have singularities not only at a point, but even on an entire manifold. These complications make the multidimensional case considerably more difficult than the univariate one, and accounts for the fact that the theory of multidimensional quadrature is by no means so complete as in the one dimensional case. Indeed, there remain numerous open questions.

In this section we present several typical multidimensional quadrature methods. We concentrate mostly on two dimensions, particularly where the generalization to higher dimensions is clear. Because of space limitations, we can only discuss a small part of the theory; for a more complete, but still introductory treatment, see the book of P. J. Davis and P. Rabinowitz ([1975], Chap. 5); for a comprehensive study, see the books of A. H. Stroud [1971] and H. Engels [1980].

6.1 Tensor Products. We consider first integration over a rectangular domain $\overline{G} := \{(x,y) \in \mathbb{R}^2 \mid a \le x \le b, c \le y \le d\}$ in the plane, and seek to numerically approximate the integral $\int_c^d \int_a^b f(x,y)dxdy$.

In analogy with the development of Newton-Cotes formulae in one dimension, it is natural to start with an interpolant of the integrand on $\overline{G}$. If we replace f by the interpolating polynomial p of 5.6.2, then we get the formula

$$\int_c^d \int_a^b f(x,y)dxdy = \int_c^d \int_a^b \sum_{\substack{0 \le \nu \le n \\ 0 \le \kappa \le k}} f(x_\nu, y_\kappa)\ell_{n\nu}(x)\ell_{k\kappa}(y)dxdy + Rf,$$

where the nodes (x_ν, y_κ) are located at the corners of a grid defined by $a = x_0 < x_1 < \cdots < x_n = b$ and $c = y_0 < y_1 < \cdots < y_k = d$.

The Cubature Formula

$$Qf := \sum_{\substack{0 \le \nu \le n \\ 0 \le \kappa \le k}} f(x_\nu, y_\kappa) \cdot \int_a^b \ell_{n\nu}(x)dx \cdot \int_c^d \ell_{k\kappa}(y)dy =$$

$$= \sum_{\kappa=0}^{k} \left[\sum_{\nu=0}^{n} f(x_\nu, y_\kappa) \int_a^b \ell_{n\nu}(x)dx \right] \int_c^d \ell_{k\kappa}(y)dy$$

is a product rule based on the $(n+1)(k+1)$ nodes, and can be thought of as arising by using a Newton-Cotes quadrature rule to do the integrations in the x and y directions, respectively.

As an example, suppose $n = 2$ and $k = 2$, and suppose the nodes in the x and y directions are equally spaced with $h_x := \frac{b-a}{2}$ and $h_y := \frac{d-c}{2}$. Now using Simpson's Rule 7.1.4 in both directions, we get the explicit formula

$$Qf := \frac{h_x h_y}{9} \left[f(x_0,y_0) + f(x_2,y_0) + f(x_0,y_2) + f(x_2,y_2) + \right. \\ \left. + 4[f(x_0,y_1) + f(x_1,y_0) + f(x_2,y_1) + f(x_1,y_2)] + 16 f(x_1,y_1) \right].$$

In case $n = 2m$, $m > 1$, and $k = 2\ell$, $\ell > 1$, the coefficients of the corresponding cubature formula, up to the factor $\frac{h_x h_y}{9}$, can be found in the following table:

$$\begin{array}{ccccccccc} 1 & 4 & 2 & 4 & \cdots & 2 & 4 & 1 \\ 4 & 16 & 8 & 16 & \cdots & 8 & 16 & 4 \\ \vdots & & & \vdots & & \vdots & \vdots & \vdots \\ 4 & 16 & 8 & 16 & \cdots & 8 & 16 & 4 \\ 2 & 8 & 4 & 8 & & 4 & 8 & 2 \\ 4 & 16 & 8 & 16 & & 8 & 16 & 4 \\ 1 & 4 & 2 & 4 & \cdots & 2 & 4 & 1 \end{array}$$

Similar product formulae can be constructed from any univariate quadrature formulae.

Example 1. Suppose we want to find a numerical approximation to the integral $Jf := \int_0^1 \int_0^1 \frac{dx\,dy}{1-xy} = \frac{\pi^2}{6} \doteq 1.644934$, whose integrand has a point singularity at $x = y = 1$. The following table shows the results of applying Gauss-Legendre quadrature operators with the same number of equally-spaced nodes in both directions; cf. Example 2 in 4.2.

number of nodes	approximate value Qf	$Qf - Jf$
1	1.333	-0.312
2	1.523	-0.122
3	1.581	-0.064
4	1.606	-0.039
5	1.619	-0.026
6	1.626	-0.019
7	1.631	-0.014
8	1.634	-0.011

Error Bound. A simple error bound for a product rule can be easily obtained from the error bounds for the quadrature formulae used to construct it. Suppose we choose

$$(Q_x f)(y) = \sum_{\nu=0}^{n} \gamma_{\nu x} f(x_\nu, y) \text{ in the } x \text{ direction and}$$

$$(Q_y f)(x) = \sum_{\kappa=0}^{n} \gamma_{\kappa y} f(x, y_\kappa) \text{ in the } y \text{ direction.}$$

Then the cubature operator $Q := Q_y Q_x$ satisfies

$$(*) \qquad \int_c^d \int_a^b f(x,y)dxdy = Q_y[(Q_x f)(y) + (R_x f)(y)] + R_y F$$
$$= Qf + Q_y R_x f + R_y F,$$

where $F(y) := \int_a^b f(x,y)dx$, and R_x and R_y are the corresponding remainder terms for the quadrature formulae.

Now if we have bounds

$$\left| \int_a^b f(x,y)dx - (Q_x f)(y) \right| = |(R_x f)(y)| \leq E_x$$

for all $c \leq y \leq d$ and

$$\left| \int_c^d F(y)dy - Q_y F \right| = |R_y F| \leq E_y,$$

and if moreover $\sum_{\nu=0}^{n} |\gamma_{\nu x}| \leq \Gamma_1$ and $\sum_{\kappa=0}^{k} |\gamma_{\kappa y}| \leq \Gamma_2$, then $(*)$ implies the error bound

$$\left| \int_c^d \int_a^b f(x,y)dxdy - Qf \right| \leq E_y + \Gamma_2 E_x.$$

In particular, if the coefficients of the quadrature formulae used are positive, and if constants are exactly integrated, then $\sum_{\kappa=0}^{k} |\gamma_{\kappa y}| = d - c$ and we get the error bound $E_y + (d-c)E_x$.

Reversing the order of integration, the same analysis can be applied, and starting with the bounds $|(R_y f)(x)| \leq \hat{E}_y$ for all $a \leq x \leq b$ and $|R_x \hat{F}| \leq \hat{E}_x$ where $\hat{F}(x) := \int_c^d f(x,y)dy$, we get the bound $\hat{E}_x + (b-a)\hat{E}_y$.

Bounds for the Product Trapezoidal Rule. The following error bounds hold for the trapezoidal rule:

$$|(R_x f)(y)| \leq \frac{h_x^2}{12} \max_{(x,y)\in\overline{G}} \left| \frac{\partial^2 f}{\partial x^2} \right| (b-a) =: E_x$$

and

$$|R_y F| \leq \frac{h_y^2}{12} \max_{c \leq y \leq d} |F''(y)|(d-c) \leq \frac{h_y^2}{12} \max_{(x,y)\in\overline{G}} \left| \frac{\partial^2 f}{\partial y^2} \right| (b-a)(d-c) =: E_y.$$

Now using the trapezoidal rule in both directions with step sizes h_x and h_y, we get a product rule Q satisfying

$$|\int_c^d \int_a^b f\,dxdy - Qf\,| \leq$$
$$\leq \frac{(b-a)(d-c)}{12}\left[h_x^2 \max_{(x,y)\in \overline{G}} \left|\frac{\partial^2 f}{\partial x^2}\right| + h_y^2 \max_{(x,y)\in \overline{G}} \left|\frac{\partial^2 f}{\partial y^2}\right|\right].$$

Remark. In constructing product integration formulae, it is not necessary to use the same quadrature formulae in both directions. Thus, for example, if the integrand is periodic in one of the directions, then it would be advantageous to use the trapezoidal rule 4.3 in that direction, regardless of which rule is used in the other direction.

Example 2. To approximate the value $Jf := \int_0^1 (\int_0^\pi \frac{e^y}{1+\cos 2x+\cos y} dx) dy \doteq$ $\doteq 3.6598795$, we compare using the trapezoidal rule in the x direction and Simpson's rule in the y direction, and using Simpson's rule in both directions:

n, k	Qf Trapezoidal-Simpson	$Qf - Jf$	Qf Simpson-Simpson	$Qf - Jf$
2	4.5137043	0.8538248	5.3733365	1.7134570
4	3.7359409	0.0760614	3.4959030	-0.1639765
8	3.6609053	0.0010258	3.6370275	-0.0228520
16	3.6598934	0.0000139	3.6596225	-0.0002570

6.2 Integration over Standard Domains. In many cases it is possible to transform a given integration domain into a rectangle, after which one of the product rules in 6.1 can be applied. For example, to compute an integral over the unit circle, we can use polar coordinates to write

$$\int_{-1}^{+1}\left(\int_{-\sqrt{1-x^2}}^{\sqrt{1+x^2}} f(x,y)dy\right)dx = \int_0^{2\pi}\int_0^1 f(r\cos\varphi, r\sin\varphi) r\,dr\,d\varphi,$$

and the integral reduces to one over the rectangle $0 \leq r \leq 1$, $0 \leq \varphi \leq 2\pi$.

In general, however, it is useful to develop special cubature formulae for some of the standard domains which are likely to arise such as a triangle in the plane or its generalization in more than two dimensions, a simplex.

In the case of rectangular domains in any number of dimensions, there is a very natural way to construct product formulae. The situation for triangles and simplices is different. For example in the two dimensional case, we showed in 5.6.2 that there is a unique interpolating polynomial of degree at most n in x and of degree at most k in y based on grid points in a rectangle, and this leads to product formulae as in 6.1. For other integration domains, however, it is more useful to work with the monomials 1, x, y, x^2,

xy, y^2, etc., and to look for a cubature formula which exactly integrates all monomials of the form $x^\nu y^\kappa$, $0 \le \nu$, $0 \le \kappa$ and $\nu + \kappa \le \ell$. We say that these kinds of cubature formulae have the *degree of accuracy* ℓ. The generalization to arbitrary dimensions d is straightforward, as we now show.

In d variables, there are $\binom{\ell+d}{d}$ monomials of the form $x_1^{\nu_1} \cdots x_d^{\nu_d}$ of degree $\nu_1 + \cdots + \nu_d \le \ell$. The question now arises of whether there exist integration formulae using at most $\binom{\ell+d}{d}$ nodes in the domain, with the property that all such monomials are exactly integrated. The *Theorem of Chakalov* [1957] not only answers this question in the affirmative, but also guarantees simultaneously that for an arbitrary integration domain, there exist integration formulae all of whose coefficients are positive.

In addition to this general result, we are interested in integration formulae using a minimal number of nodes. In this connection, there are a considerable number of isolated results for particular integration domains, dimensions and degrees of accuracy, but by no means is there a complete theory. The book of A. H. Stroud [1971] contains many such integration formulae. As examples, we give formulae for the triangle and for the simplex in $\mathbb{R}^3$ with degree of accuracy $\ell = 2$. The formulae use only $(d+1)$ nodes (see p. 307, loc. cit.).

Integration over a Triangle. For the standard triangle with the vertices $(0,0)$, $(0,1)$ and $(1,0)$, we have the cubature formulae:

(i) $Qf = \frac{1}{6}[f(\frac{1}{2},0) + f(0,\frac{1}{2}) + f(\frac{1}{2},\frac{1}{2})]$;

(ii) $Qf = \frac{1}{6}[f(\frac{1}{6},\frac{1}{6}) + f(\frac{2}{3},\frac{1}{6}) + f(\frac{1}{6},\frac{2}{3})]$.

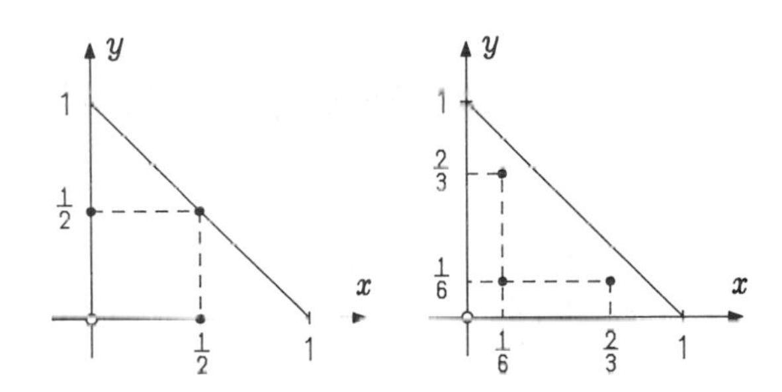

The reader can check for himself that the monomials $1, x, y, x^2, xy, y^2$ are exactly integrated by both cubature formulae.

Integration over a Tetrahedron. We consider the standard tetrahedron $(0,0,0)$, $(1,0,0)$, $(0,1,0)$, $(0,0,1)$. Here we may use the formulae:

$$Qf = \frac{1}{24}[f(r,r,r) + f(s,r,r) + f(r,s,r) + f(r,r,s)];$$

$$\text{(i) } r = \tfrac{5-\sqrt{5}}{20}, \quad s = \tfrac{5+3\sqrt{5}}{20} \quad \text{and} \quad \text{(ii) } r = \tfrac{5+\sqrt{5}}{20}, \quad s = \tfrac{5-3\sqrt{5}}{20}.$$

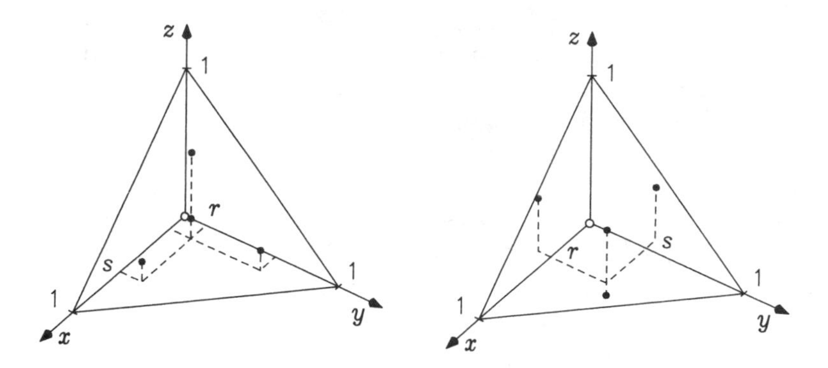

Both integration formulae exactly integrate all of the monomials $1, x, y, z, x^2, xy, xz, y^2, yz, z^2$ of highest degree 2. We note that in case (ii), all of the nodes lie outside of the tetrahedron.

6.3 The Monte-Carlo Method. A completely different approach to numerical integration can be based on the methods of statistics. The resulting formulae, called *Monte-Carlo Methods*, play an especially important role for integrals of very high dimension. These methods are easy to describe in the one dimensional case, where we follow the development in the book of P. J. Davis and P. Rabinowitz ([1975], p. 288 – 314). This book is particularly useful for its extensive list of references.

The number $Jf := \int_0^1 f(x)dx$ can be considered to be the average value of the numbers $f(x)$ as x runs over the interval $[0,1]$. Now if $x_1, \ldots, x_n$ are randomly chosen points in $[0,1]$, then the average

$$\overline{f}_n := \frac{1}{n}\sum_{\nu=1}^{n} f(x_\nu)$$

is an approximation to Jf.

Assuming that the number of randomly chosen nodes can be made arbitrarily large, then the *strong law of large numbers* can be used to describe the behavior of the sequence $(\frac{1}{n}\sum_{\nu=1}^{n} f(x_\nu))_{n\in\mathbb{Z}_+}$, giving the following

Limit Result. *Let μ be a probability density; i.e., $\int_{-\infty}^{+\infty}\mu(x)dx = 1$. Then the integral $If := \int_{-\infty}^{+\infty} f(x)\mu(x)dx$ satisfies*

$$\text{prob}(\lim_{n\to\infty}\frac{1}{n}\sum_{\nu=1}^{n} f(x_\nu) = If) = 1.$$

If $If := Jf = \int_0^1 f(x)dx$, then we may choose

$$\mu(x) := \begin{cases} 1 & \text{for } 0 \le x \le 1 \\ 0 & \text{otherwise.} \end{cases}$$

Using the *central limit theorem* of statistics, we can estimate the probability that a Monte-Carlo approximation has a given accuracy. In particular, we have the

Error probability. Let

$$\sigma^2 := \int_{-\infty}^{+\infty} [f(x) - Jf]^2 \mu(x) dx = \int_{-\infty}^{+\infty} f^2(x)\mu(x)dx - (Jf)^2$$

be the variance of the numbers $f(x)$. Then by the central limit theorem, we have

$$\text{prob}\left(\mid \frac{1}{n}\sum_{\nu=1}^{n} f(x_\nu) - Jf \mid \leq \frac{\lambda\sigma}{\sqrt{n}} \right) = \frac{1}{\sqrt{2\pi}} \int_{-\lambda}^{+\lambda} \exp\left(-\frac{x^2}{2}\right) dx + O\left(\frac{1}{\sqrt{n}}\right).$$

A similar formula holds for multiple integrals. For fixed λ, the bound $\frac{\lambda\sigma}{\sqrt{n}}$ behaves like $\frac{1}{\sqrt{n}}$. This slow convergence of Monte-Carlo methods restricts their usefulness. In practice, they are only used when other methods cannot be applied, for example when the dimension is high (say greater than 10). For integrals in very high dimensions, however, the Monte-Carlo method is the only available general method.

Practical Application. The main difficulty in applying the Monte-Carlo method is the generation of random numbers. In order to avoid the cumbersome use of tables, in practice one usually uses sequences of *pseudo-random numbers*. By this we mean mathematically well-defined sequences of numbers constructed from some formula which generates sequences of random numbers. Such sequences have the additional advantage that they are reproducable.

As an example of a sequence of pseudo-random numbers, consider the sequence

$$x_{n+1} = ax_n + c \pmod{m},$$

where a, c and m are prescribed integers, and x_0 is the starting value. The elements of the sequence are the remainders which arise when dividing the numbers $ax_n + c$ by m. The sequence (x_n) is periodic, and the period is at most m; consequently, m has to be chosen very large in comparison with the number of random numbers needed.

Example. Consider computing an approximation to

$$Jf := \int_0^1 \int_0^1 \int_0^1 \int_0^1 e^{xy} \cos(\frac{\pi}{2}uv) dx dy du dv \doteq 1.150073.$$

using the Monte-Carlo method. For comparison purposes, we computed the value given above for this integral by splitting the integral into two 2-dimensional integrals, and applying the Romberg method.

We denote the sequence of pseudo-random numbers $x_1, y_1, u_1, v_1, x_2, y_2, \ldots$ by $z_1, z_2, \ldots$. These were computed by working on the interval $[0, m]$, and using the recurrence formula

$$z_{n+1} = az_n \pmod{m}, \quad n \geq 0,$$

starting with $z_0 := 1$. Here we have $a := 8[\sqrt{m}/8] + 3$ and $m := 2^{\mu}$, where the integer μ should be chosen to match the properties of the computer being used; it should not be smaller than 16; we used $\mu := 16$. (Here $[\sqrt{m}/8] :=$ denotes the largest integer $\leq \sqrt{m}/8$.)

points	approximate value	points	approximate value
2	0.805 882	2^7	1.152 769
2^2	0.964 270	2^8	1.147 233
2^3	1.027 190	2^9	1.120 108
2^4	0.968 520	2^{10}	1.131 058
2^5	1.101 655	2^{11}	1.142 133
2^6	1.149 216	2^{12}	1.149 970

Other numerical examples can be found in P. J. Davis – P. Rabinowitz ([1975], p. 297); see also Problem 5.

6.4 Problems. 1) Apply the extrapolation method to two-dimensional integration. In particular, consider the product trapezoidal rule $T_0^k f$, where the cubature error can be expanded in powers of h^2 whenever f has sufficiently many continuous partial derivatives.

a) By halving the step size at each step, in both directions, find an explicit formula for the rule T_2^0. What special feature arises in comparison with the one-dimensional case?

b) Test the Romberg method on Example 2 in 6.1.

2) a) Verify the degree of accuracy of integration formulae presented in 6.2 for integration over a triangle and over a tetrahedron.

b) What is the degree of accuracy of the approximation formula

$$Qf = \frac{4}{3}[f(1,0,0)+f(-1,0,0)+f(0,1,0)+f(0,-1,0)+f(0,0,1)+f(0,0,-1)]$$

for integration over the cube with boundary faces $x = \pm 1$, $y = \pm 1$, $z = \pm 1$?

3) Derive a formula

$$\int \cdots \int_{[-1,+1]^d} f(x) dx_1 \cdots dx_d = \gamma[f(\pm u, 0, \ldots, 0) + f(0, \pm u, 0, \ldots, 0) + \cdots + f(0, \ldots, 0, \pm u)] + Rf$$

for integration over the d-dimensional cube with degree of accuracy 3. Here $f(\pm u, 0, \ldots, 0) := f(u, 0, \ldots, 0) + f(-u, 0, \ldots, 0)$, etc.

4) In order to derive a cubature formula on the unit circle of the form

$$Qf = \gamma_1 f(0,\rho) + \gamma_2 f(-\frac{\sqrt{3}}{2}\rho, -\frac{\rho}{2}) + \gamma_3 f(\frac{\sqrt{3}}{2}\rho, -\frac{\rho}{2}),$$

$0 < \rho < 1$, with degree of accuracy 2, first find coefficients so that the degree of accuracy 1 is obtained. As a second step, find ρ. How can the location of the nodes be altered, without reducing the degree of accuracy of the formula?

5) Compute an approximate value for $\int_0^1 x^2 dx$ using the Monte-Carlo method:

a) Use 2^j nodes for $j = 1, \ldots, 16$, obtained from the pseudo-random number sequence given in Example 6.3 with a suitable μ.

b) How is the length of the period of the sequence of pseudo-random numbers – in this case $2^{\mu-2}$ – reflected in the results?

c) How large must the number of nodes be chosen so that the error of the approximation is at most $1 \cdot 10^{-2}$, respectively $1 \cdot 10^{-3}$, with a probability of 95 % (i.e. for $\lambda = 1.960$)?

8
Iteration

The problem of solving an equation or a system of equations is one of the basic problems in mathematics and its applications. This problem can be formulated as follows: Find a solution x of the operator equation $Fx = 0$ in a given normed linear space $(\mathrm{X}, \|\cdot\|)$. Here the operator F is a mapping $F : \mathrm{D} \to \mathrm{X}$, $\mathrm{D} \subset \mathrm{X}$. An element $\xi \in \mathrm{D}$ for which $F\xi = 0$ holds is called a *zero* of F.

Example 1. To determine the orbit of a planet, one needs to solve the "Kepler Equation". This involves finding the "eccentric anomaly" E as a solution of the equation

$$E = e \cdot \sin(E) + \frac{2\pi}{U} t.$$

Here U is the time to transverse the orbit, t is the time in days since passing the perihedron, and e is the numerical eccentricity of the orbital ellipse.

This is a zero-finding problem of the type mentioned above with $\mathrm{X} = \mathbb{R}$ and $Fx := x - e\sin(x) - \frac{2\pi}{U}t$.

Example 2. In the numerical solution of boundary-value problems for differential equations, discretization always leads to a system of equations of the form:

$$f(x) = \begin{pmatrix} f_1(x_1, \ldots, x_m) \\ \vdots \\ f_m(x_1, \ldots, x_m) \end{pmatrix} = \begin{pmatrix} y_1 \\ \vdots \\ y_m \end{pmatrix} =: y.$$

This is again a zero-finding problem where we are trying to find a solution $\xi \in \mathbb{R}^m$ corresponding to a given y.

A solution of the equation $Fx = 0$ can be determined in a finite number of steps only in very rare cases. In general, we have to employ *iterative methods*.

In this chapter we consider operator equations of the form $x = \Phi\, x$, and study iterative ways to solve them. Throughout the chapter we shall restrict our attention to equations in spaces of finite dimension.

1. The General Iteration Method

Let $x = (x_1, \ldots, x_m)^T \in \mathbb{K}^m$ with $\mathbb{K} := \mathbb{R}$ or $\mathbb{K} := \mathbb{C}$. Given a mapping $\Phi : \mathrm{D} \to \mathbb{K}^m$, $D \subset \mathbb{K}^m$, we consider the equation

$$x = \Phi\, x,$$

whose solution is to be found using the

Iteration Method

$$x^{(\kappa+1)} = \Phi\, x^{(\kappa)}, \quad \kappa \in \mathbb{N},$$

with a given starting point $x^{(0)}$.

We now present two examples of typical iterations, assuming the existence of a solution ξ. In later sections we will deal with existence and convergence simultaneously.

1.1 Examples of Convergent Iterations. Suppose $m := 1$ and $\mathbb{K} := \mathbb{R}$, and suppose $\varphi \in \mathrm{C}[a, b]$. For functions φ of one or more variables, we write $x = \varphi(x)$ for the equation to be solved.

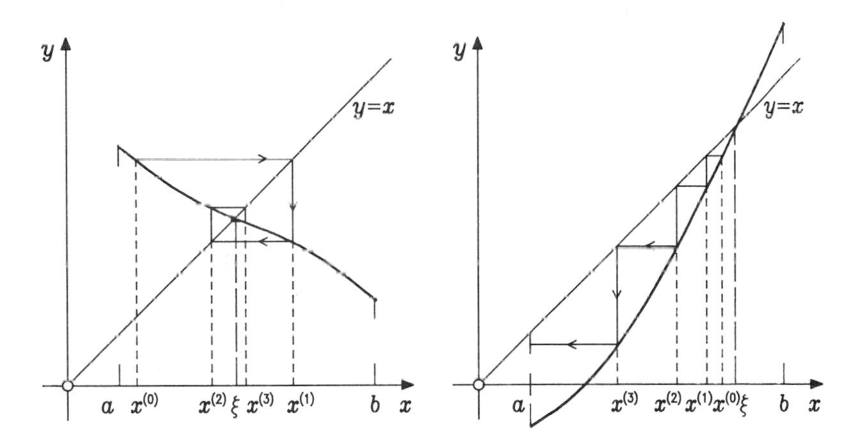

The figure on the left shows a convergent sequence of values $(x^{(\kappa)})_{\kappa \in \mathbb{N}}$ obtained from the iteration equation. The figure on the right shows a different function φ for which the sequence of values $x^{(\kappa)}$ monotonely diverges from the solution ξ, even though the initial point $x^{(0)}$ is very near the solution. We leave it to the reader to construct functions φ leading to monotone convergence and alternating divergence.

1.2 Convergence of Iterative Methods. We say that the sequence produced by an iterative method converges to a solution ξ, provided that

$$\lim_{\kappa \to \infty} x^{(\kappa)} = \xi.$$

This is synonymous with convergence in the sense

$$\lim_{\kappa\to\infty} \|x^{(\kappa)} - \xi\| = 0$$

or

$$\lim_{\kappa\to\infty} x_\mu^{(\kappa)} = \xi_\mu \quad \text{for} \quad 1 \le \mu \le m.$$

In order to get a sufficient condition for convergence, we now suppose that $(\mathrm{X}, \|\cdot\|)$ is a Banach space, and that $\Phi : \mathrm{X} \to \mathrm{X}$ is a mapping of X into itself. In addition, suppose the operator Φ is *contractive*; i.e., for some $\alpha < 1$,

$$\|\Phi\, x - \Phi\, z\| \le \alpha \|x - z\|$$

for all elements $x,\ z \in \mathrm{X}$.

Contraction Theorem (Banach Fixed-Point Theorem). *Suppose $\Phi : \mathrm{X} \to \mathrm{X}$ is a contractive mapping. Then it has exactly one fixed point $\xi = \Phi\, \xi$, and moreover, for any choice $x^{(0)}$ of a starting point, the iteration $x^{(\kappa+1)} = \Phi\, x^{(\kappa)}$ converges to the fixed point.*

Proof. This is a well-known result in analysis, but we present a full proof since we need one of the intermediate estimates later in developing error bounds.

1) The sequence $(x^{(\kappa)})_{\kappa\in\mathbb{N}}$ is a Cauchy sequence since

$$\begin{aligned} \|x^{(\kappa+1)} - x^{(\kappa)}\| &= \|\Phi\, x^{(\kappa)} - \Phi\, x^{(\kappa-1)}\| \le \\ &\le \alpha \|x^{(\kappa)} - x^{(\kappa-1)}\| \le \dots \le \alpha^\kappa \|x^{(1)} - x^{(0)}\| \end{aligned}$$

implies that for $\lambda > \kappa$,

$$\begin{aligned} \|x^{(\lambda)} - x^{(\kappa)}\| &\le \|x^{(\lambda)} - x^{(\lambda-1)}\| + \|x^{(\lambda-1)} - x^{(\lambda-2)}\| + \\ &+ \dots + \|x^{(\kappa+1)} - x^{(\kappa)}\| \le \\ &\le (\alpha^{\lambda-1} + \alpha^{\lambda-2} + \dots + \alpha^\kappa)\|x^{(1)} - x^{(0)}\| = \\ &= \alpha^\kappa \frac{1-\alpha^{\lambda-\kappa}}{1-\alpha}\|x^{(1)} - x^{(0)}\| \le \frac{\alpha^\kappa}{1-\alpha}\|x^{(1)} - x^{(0)}\|. \end{aligned}$$

It follows that $\|x^{(\lambda)} - x^{(\kappa)}\| < \varepsilon$ whenever κ is sufficiently large. Hence, the sequence $(x^{(\kappa)})_{\kappa\in\mathbb{N}}$ is a Cauchy sequence, and so the limit

$$\lim_{\kappa\to\infty} x^{(\kappa)} =: \xi$$

exists.

2) ξ is a fixed point since

$$\|\xi - \Phi\, \xi\| \le \|\xi - x^{(\kappa)}\| + \|x^{(\kappa)} - \Phi\, \xi\| \le \|\xi - x^{(\kappa)}\| + \alpha \|x^{(\kappa-1)} - \xi\|,$$

and thus $\|\xi - \Phi\,\xi\| < \varepsilon$ whenever κ is sufficiently large, and so $\xi = \Phi\,\xi$.

3) ξ is the only fixed point, since the assumption $\eta = \Phi\,\eta$ implies

$$\|\xi - \eta\| = \|\Phi\,\xi - \Phi\,\eta\| \leq \alpha\|\xi - \eta\|,$$

and since $\alpha < 1$, this means that $\xi = \eta$. □

Remark. If equality $x^{(\kappa+1)} = x^{(\kappa)}$ holds at the $(\kappa + 1)$-th iteration step, then we stop the iteration, and $x^{(\kappa)} = x^{(\kappa+1)} = \Phi\,x^{(\kappa)}$ is the solution.

Example 1. If we use the iteration method to solve the linear system of equations $(I - A)x = b$ with a square matrix A, then each step of the iteration has the form

$$x^{(\kappa+1)} = Ax^{(\kappa)} + b.$$

In this case it follows from the Contraction Theorem that convergence to the unique solution ξ will take place for any arbitrary choice of a starting vector $x^{(0)}$ whenever $\|Ax - Az\| \leq \|A\|\,\|x - z\|$ with a contraction factor $\alpha := \|A\|$ satisfying $\alpha < 1$.

Extension. In many practical cases, the operator Φ is only defined on a closed subset $\mathrm{D} \subset \mathrm{X}$. If $\Phi : \mathrm{D} \to \mathrm{D}$ and is contractive on D, then the proof of the Contraction Theorem can be carried over verbatim. Now if we choose $x^{(0)} \in \mathrm{D}$, then $x^{(1)} = \Phi\,x^{(0)} \in \mathrm{D}$, and in general $x^{(\kappa)} \in \mathrm{D}$ for all $\kappa \geq 2$, and so the iterates remain in D. We conclude that there exists a unique fixed point $\xi = \Phi\,\xi$ with $\lim_{\kappa\to\infty} x^{(\kappa)} = \xi$.

Example 2. Let $\mathrm{X} := \mathbb{R}$, $\mathrm{D} := [1, 2]$, and suppose $\varphi : \mathrm{D} \to \mathrm{D}$ is defined by

$$\varphi(x) := \frac{1}{2}x + \frac{1}{x}.$$

Then since

$$|\varphi(x) - \varphi(z)| = |\frac{1}{2} - \frac{1}{xz}|\,|x - z| \leq \frac{1}{2}|x - z|,$$

φ is contractive with $\alpha = \frac{1}{2}$. Thus, the iteration

$$x^{(\kappa+1)} = \frac{1}{2}x^{(\kappa)} + \frac{1}{x^{(\kappa)}}$$

converges for any $x^{(0)} \in [1, 2]$ to the solution $\xi = \sqrt{2}$ of the nonlinear equation $x = \varphi(x)$.

Note. If, as in this example, φ is a real-valued continuous function of a single variable of the form $\varphi : [a, b] \to [a, b]$, $-\infty < a < b < +\infty$, then the existence of a solution of the equation $x = \varphi(x)$ follows immediately from the Intermediate-value Theorem.

Local and Global Convergence. If the sequence of iterates only converges for starting points $x^{(0)}$ in some neighborhood $\mathrm{U} \subset \mathrm{D}$ of the fixed point ξ, then we say that the iteration is *locally convergent*. This is the case if the mapping Φ is only contractive on U. When $x^{(0)}$ can be chosen arbitrarily in the domain D, then we say that the method is *globally convergent.*

1.3 Lipschitz Constants. If the mapping Φ is Lipschitz bounded, i.e. $\|\Phi x - \Phi z\| \leq K\|x-z\|$ for all $x, z \in$ D, and if the Lipschitz constant $K < 1$, then the mapping is contractive.

The proof of the Lipschitz boundedness of a mapping can be difficult. If we are dealing with a mapping defined by a real-valued function $\varphi = (\varphi_1, \ldots, \varphi_m)$ of real variables $x = (x_1, \ldots, x_m)$ which is continuously differentiable on a bounded closed convex set D, then the mapping is always Lipschitz bounded on D. Indeed, by the mean-value theorem,

$$\|\varphi(x) - \varphi(z)\| \leq \sup_{0<\theta<1} \|J_\varphi(z + \theta(x-z))\| \, \|x - z\|,$$

where J_φ is the Jacobian matrix

$$J_\varphi(\zeta) = \left(\frac{\partial \varphi_\mu(\zeta)}{\partial x_\nu}\right)_{\mu,\nu=1}^{m},$$

and so we can take $K := \|J_\varphi\|$.

Thus, if we use the vector norms $\|\cdot\|_1$ or $\|\cdot\|_\infty$, then by 2.4.3, we can use

$$\|J_\varphi\|_1 = \max_{x \in \mathrm{D}}\left(\max_{\nu} \sum_{\mu=1}^{m} \left|\frac{\partial \varphi_\mu(x)}{\partial x_\nu}\right|\right)$$

or

$$\|J_\varphi\|_\infty = \max_{x \in \mathrm{D}}\left(\max_{\mu} \sum_{\nu=1}^{m} \left|\frac{\partial \varphi_\mu(x)}{\partial x_\nu}\right|\right)$$

as Lipschitz constants. For the vector norm $\|\cdot\|_2$, instead of the spectral norm of J_φ, we can use the matrix norm

$$\|J_\varphi\|_F = \max_{x \in \mathrm{D}}\left(\sum_{\mu,\nu=1}^{m} \left|\frac{\partial \varphi_\mu}{\partial x_\nu}\right|^2\right)^{\frac{1}{2}}$$

introduced in Example 2.4.3 as a Lipschitz constant in the inequality

$$\|\varphi(x) - \varphi(z)\|_2 \leq \|J_\varphi\|_F \, \|x - z\|_2.$$

If one of the norms $\|J_\varphi\|_1$, $\|J_\varphi\|_F$ or $\|J_\varphi\|_\infty$ is smaller than one, then we are assured of the convergence of the sequence of iterates with respect to the associated vector norm, and in view of the equivalence of all norms on a finite dimensional vector space, we also have convergence with respect to all other vector norms.

1.4 Error Bounds. It follows from the proof of the Contraction Theorem that the difference of two iterates with $\lambda > \kappa$ satisfies

$$\|x^{(\lambda)} - x^{(\kappa)}\| \leq \frac{\alpha^\kappa}{1-\alpha}\|x^{(1)} - x^{(0)}\|.$$

Passing to the limit as $\lambda \to \infty$ and setting $\lim_{\lambda\to\infty} x^{(\lambda)} = \xi$, we get the

a-priori Bound

$$\|\xi - x^{(\kappa)}\| \le \frac{\alpha^\kappa}{1-\alpha}\|x^{(1)} - x^{(0)}\|.$$

Using this a-priori bound, after only one step of the iteration we already have an upper bound for the number of iterations needed to achieve a prescribed accuracy.

Analogously, we can modify the first step in the proof of the Contraction Theorem to show that for $\rho < \kappa$,

$$\|x^{(\kappa+1)} - x^{(\kappa)}\| \le \alpha^{\kappa-\rho}\|x^{(\rho+1)} - x^{(\rho)}\|,$$

which leads to

$$\|x^{(\lambda)} - x^{(\kappa)}\| \le (\alpha^{\lambda-1} + \cdots + \alpha^{\kappa-\rho})\|x^{(\rho+1)} - x^{(\rho)}\| =$$

$$= \alpha^{\kappa-\rho}\frac{1-\alpha^{\lambda-(\kappa-\rho)}}{1-\alpha}\|x^{(\rho+1)} - x^{(\rho)}\| \le \frac{\alpha^{\kappa-\rho}}{1-\alpha}\|x^{(\rho+1)} - x^{(\rho)}\|.$$

Taking $\rho = \kappa - 1$, it follows that

$$\|x^{(\lambda)} - x^{(\kappa)}\| \le \frac{\alpha}{1-\alpha}\|x^{(\kappa)} - x^{(\kappa-1)})\|,$$

and arguing as before we get the

a-posteriori Bound

$$\|\xi - x^{(\kappa)}\| \le \frac{\alpha}{1-\alpha}\|x^{(\kappa)} - x^{(\kappa-1)}\|.$$

This estimate can be used to check the accuracy of the approximation $x^{(\kappa)}$ *after* carrying out the iteration.

Example. Consider solving the transcendental equation $x = \exp(-\frac{1}{2}x)$. Then starting with $x^{(0)} = 0.8$, we get $x^{(1)} \doteq 0.670320$. Since $\varphi'(x) = -\frac{1}{2}\exp(-\frac{1}{2}x)$ is negative, we can expect that the solution will always lie between any two successive iterates. The number $|\varphi'(x^{(1)})| \doteq 0.357$ can be taken as an upper bound for the contraction constant α. Now the a-priori bound shows that for $\kappa = 10$, the accuracy of the κ-th iterate is bounded by

$$|\xi - x^{(\kappa)}| \le \frac{\alpha^\kappa}{1-\alpha}|x^{(1)} - x^{(0)}| \lesssim 6.78 \cdot 10^{-6}.$$

The iteration leads to the following values:

κ	$x^{(\kappa)}$	κ	$x^{(\kappa)}$
0	0.8	6	0.703 646 98
1	0.670 320	7	0.703 404 27
2	0.715 224	8	0.703 489 64
3	0.699 344	9	0.703 459 61
4	0.715 224	10	0.703 470 17
5	0.702 957		

where the correct answer is $\xi \doteq 0.70346742$. Thus, the true error for $\kappa = 10$ is $|\xi - x^{(10)}| = 2.75 \cdot 10^{-6}$.

To get the corresponding a-posteriori bound for $\kappa = 10$, we observe that $\alpha \leq |\varphi'(x^{(10)})| \doteq 0.352$, and we obtain

$$|\xi - x^{(10)}| \leq \frac{0.352}{0.648}|x^{(10)} - x^{(9)}| \doteq 5.74 \cdot 10^{-6}.$$

Both the a-priori and a-posteriori bounds are good approximations of the true error.

1.5 Convergence. Suppose now that $\varphi \in C_1[a,b]$, $\varphi : [a,b] \to [a,b]$ and that $K = \max|\varphi'(x)| < 1$. It follows from the mean-value theorem that $(x^{(\kappa)})$ is a *monotonely* convergent sequence when $\varphi'(x) > 0$ throughout the entire interval, and that it is an *alternating* convergent sequence when $\varphi'(x) < 0$; see Example 1.4 for the second case.

Rate of Convergence. In order to estimate the rate of convergence, we now study the convergence behavior of the sequence $(\delta^{(\kappa)})_{\kappa \in \mathbb{N}}$ of deviations $\delta^{(\kappa)} := x^{(\kappa)} - \xi$.

By the mean-value theorem,

$$\delta^{(\kappa+1)} = \varphi(x^{(\kappa)}) - \xi = \varphi'(\xi + \theta\delta^{(\kappa)})\delta^{(\kappa)}, \quad 0 < \theta < 1.$$

If the iteration does not stop, then $\delta^{(\kappa)} \neq 0$, and so

$$\lim_{\kappa \to \infty} \frac{\delta^{(\kappa+1)}}{\delta^{(\kappa)}} = \varphi'(\xi).$$

If $\varphi'(\xi) \neq 0$, we call this *linear convergence*, which means that asymptotically, in *each* step of the iteration, the value of the deviation is reduced by a constant factor, in this case $|\varphi'(\xi)| < 1$.

If, however, $\varphi'(\xi) = 0$, then we call the convergence *superlinear*. Then, for example, if $\varphi \in C_2[a,b]$, we have

$$\lim_{\kappa \to \infty} \frac{\delta^{(\kappa+1)}}{(\delta^{(\kappa)})^2} = \frac{1}{2}\varphi''(\xi),$$

which means that we have at least *quadratic convergence*.

Analogously, for the general equation

$$x = \Phi x,$$

we say that we have (at least) linear convergence if

$$\delta^{(\kappa+1)} = O(\|\delta^{(\kappa)}\|),$$

and (at least) quadratic convergence if

$$\delta^{(\kappa+1)} = O(\|\delta^{(\kappa)}\|^2).$$

It is not hard to see that for a differentiable function φ of several real variables, linear convergence holds when the Jacobian J_φ is not equal to the zero matrix, and that we have at least quadratic convergence whenever φ is twice continuously differentiable and J_φ vanishes.

1.6 Problems. 1) Use iteration to solve the Kepler equation given in Example 1 in the introduction to this chapter for the typical values $e = 0.1$ and $\frac{2\pi}{U}t = 0.85$.

2) Suppose y is defined by the formula

$$y := \sqrt{z + \sqrt{z + \sqrt{z + \cdots}}} \quad \text{for } z \in \mathbb{R}_+.$$

a) Find an iterative method for computing the number y, and determine for which choices of the starting value the iteration will converge.

b) Compute y.

3) Show that the sequence of iterates

$$x^{(\kappa+1)} = \frac{x^{(\kappa)}}{1 + (x^{(\kappa)})^2}$$

converges for an arbitrary choice of $x^{(0)} \in \mathbb{R}$. Carry out 10 iteration steps starting with $x^{(0)} = 1$, and discuss the order of convergence.

4) Show:

a) Assuming $\varphi \in C_1[a,b]$ and that $M := \min_{x\in[a,b]} |\varphi'(x)| > 1$, then every iteration starting with $x^{(0)} \neq \xi$ diverges.

b) The iteration $x^{(\kappa+1)} = \varphi^{-1}(x^{(\kappa)})$, $\kappa \in \mathbb{N}$, defined in terms of the inverse function φ^{-1} converges for an arbitrary starting point $x^{(0)} \in [a,b]$ to the solution ξ of $x = \varphi(x)$.

c) Construct a convergent iterative method for solving $x = \varphi(x)$ with $\varphi(x) := \tan(x)$, $[a,b] := [\frac{5}{4}\pi, \frac{3}{2}\pi]$, and carry out several steps.

5) Suppose we want to use iteration to find the solution of the transcendental system of equations

$$x_1 = \frac{1}{10}x_1^2 + \sin(x_2)$$

$$x_2 = \cos(x_1) + \frac{1}{10}x_2^2$$

lying in the interval $0.7 \leq x_1 \leq 0.9$, $0.7 \leq x_2 \leq 0.9$. What can you say about the convergence in terms of the Lipschitz constants $\|J_\varphi\|_p$ for $p = 1, \infty, F$? Compute the solution vector to an accuracy of $\pm 1 \cdot 10^{-5}$.

2. Newton's Method

In view of our discussion of the rate of convergence of general iterative methods, the question naturally arises as to whether or not it is possible to achieve superlinear convergence by an appropriate choice of the iteration formula. In this section we show how this can be done for the problem of solving the equation $f(x) = 0$, where $f = (f_1, \ldots, f_m)$ is a vector-valued differentiable function of the real variables $x = (x_1, \ldots, x_m)$. The method is due to Newton, and also plays an important role in the solution of general operator equations. Here we consider only the simpler case described above, since a study of the general case requires introducing generalized derivatives.

In the first four subsections of this chapter we discuss Newton's method for solving the equation $f(x) = 0$ where f is a function of one real variable. The case $m > 1$ will be dealt with in Section 2.5.

2.1 Accelerating the Convergence of an Iterative Method. Suppose we want to find a solution of the equation

$$f(x) = 0 \quad \text{for } f \in C_1[a, b].$$

As above, we assume that there is at least one solution $\xi \in [a, b]$.

Our aim is to make use of the fact, shown in 1.5, that if $\varphi'(\xi) = 0$, then the general iterative method will converge superlinearly in a neighborhood of ξ. To this end, suppose that $g \in C_1[a, b]$, and consider the equation

$$g(x)f(x) = 0.$$

Assuming $g(x) \neq 0$ for $x \in [a, b]$, it is clear that this equation has exactly the same solutions as the equation $f(x) = 0$. We now consider the corresponding fixed point equation

$$x = x + g(x)f(x) =: \varphi(x),$$

and attempt to determine g so that $\varphi'(\xi) = 0$. Since

$$\varphi'(x) = 1 + g'(x)f(x) + g(x)f'(x),$$

assuming $f'(\xi) \neq 0$ and $f(\xi) = 0$, we are led to

$$g(\xi) = -(f'(\xi))^{-1}.$$

We can achieve this condition by choosing

$$g(x) := -(f'(x))^{-1}$$

in a neighborhood of the zero ξ. This gives

Newton's Iterative Method

$$x^{(\kappa+1)} = x^{(\kappa)} - (f'(x^{(\kappa)}))^{-1} f(x^{(\kappa)}),$$

which by the above discussion, converges superlinearly in a neighborhood of ξ for any given $f \in C_1[a, b]$.

Newton's method leads to a convergent sequence of iterates provided that $x^{(0)}$ is chosen to be a sufficiently good inital approximation. Indeed, since

$$\varphi(x) = x - \frac{f(x)}{f'(x)}, \quad \text{implies} \quad \varphi'(x) = \frac{f(x)f''(x)}{[f'(x)]^2},$$

then since $f(\xi) = 0$, we must have $|\varphi'(x)| < 1$ for all values x in a neighborhood $U(\xi)$ of the desired zero ξ. This means that φ is a contractive mapping of this neighborhood into itself.

If $f \in C_2[a, b]$, then in fact we even have quadratic convergence locally. Indeed, since

$$\xi - x^{(\kappa+1)} = \xi - x^{(\kappa)} + \frac{f(x^{(\kappa)})}{f'(x^{(\kappa)})}$$

and

$$f(\xi) = f(x^{(\kappa)}) + f'(x^{(\kappa)})(\xi - x^{(\kappa)}) + \frac{1}{2} f''(x^{(\kappa)} + \theta(\xi - x^{(\kappa)}))(\xi - x^{(\kappa)})^2$$

with $0 < \theta < 1$, using $f(\xi) = 0$ we see that

$$\xi - x^{(\kappa+1)} = -\frac{1}{2} \frac{1}{f'(x^{(\kappa)})} f''(x^{(\kappa)} + \theta(\xi - x^{(\kappa)}))(\xi - x^{(\kappa)})^2,$$

and hence

$$\lim_{\kappa \to \infty} \frac{|\xi - x^{(\kappa+1)}|}{|\xi - x^{(\kappa)}|^2} = \frac{1}{2} \left| \frac{f''(\xi)}{f'(\xi)} \right|.$$

This implies $\delta^{(\kappa+1)} = O(|\delta^{(\kappa)}|^2)$.

2.2 Geometric Interpretation. In this section we give a geometric interpretation of Newton's method. Suppose $f \in C_1[a, b]$, and let $x^{(\kappa)}$ be an approximate value for the solution ξ of the equation $f(x) = 0$. Then

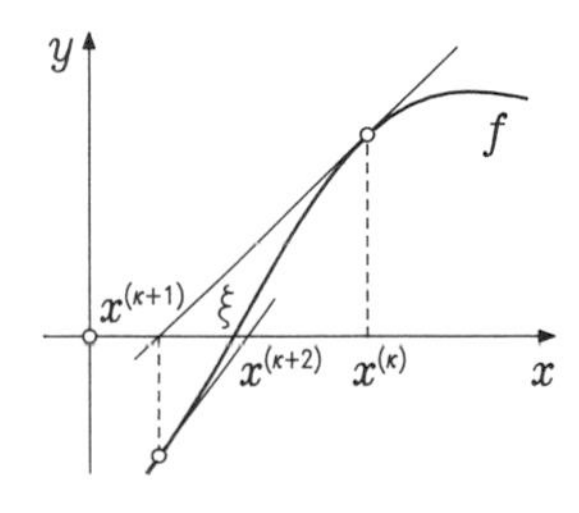

$y = f(x^{(\kappa)}) + f'(x^{(\kappa)})(x - x^{(\kappa)})$ is the linear part of the expansion of f about $x^{(\kappa)}$; i.e., it is the tangent to f at the point $(x^{(\kappa)}, f(x^{(\kappa)}))$. Setting $y = 0$, we find that this line intersects the x-axis at the point

$$x^{(\kappa+1)} := x^{(\kappa)} - \frac{f(x^{(\kappa)})}{f'(x^{(\kappa)})}.$$

In general, this is a better approximation to ξ than $x^{(\kappa)}$.

This process can be regarded as a linearization of the problem: the solution of the nonlinear equation is replaced by a sequence of problems involving linear equations.

In 1669, NEWTON developed a method for computing a root of a cubic equation, based on an iterative linearization process. He published his method as a means to solve the Kepler equation mentioned in the introduction to this chapter. In about 1690, JOSEPH RAPHSON formulated Newton's ideas for the case of a polynomial in a form which is closer to the formula given above. Thus, the method is often referred to as the *Newton-Raphson method.*

Example. Applying Newton's method to solve the equation $x - \exp(-\frac{1}{2}x) = 0$ (cf. Example 1.4 where the same equation was solved with the general iterative method), leads to the following values:

κ	$x^{(\kappa)}$	$\lvert\xi - x^{(\kappa)}\rvert / \lvert\xi - x^{(\kappa-1)}\rvert^2$
0	0.8	--
1	0.703	$6.378 \cdot 10^{-2}$
2	0.703 467 4	$6.506\,02 \cdot 10^{-2}$
3	0.703 467 422 498 391 6	$6.505\,234\,2 \cdot 10^{-2}$
4	0.703 467 422 498 391 652 049 818 601 8	--

The numbers in the $x^{(\kappa)}$ column in the table were all computed to a machine accuracy of 28 decimal digits. The last column clearly shows the quadratic rate of convergence, and indeed,

$$\lim_{\kappa\to\infty} \frac{|\xi - x^{(\kappa)}|}{|\xi - x^{(\kappa-1)}|^2} = \frac{1}{2} \left| \frac{f''(\xi)}{f'(\xi)} \right| \doteq 6.505\,233\,0 \cdot 10^{-2}.$$

Here Newton's method has produced a result which is correct up to machine accuracy after only four steps.

2.3 Multiple Zeros. We now want to free ourselves from the assumption $f'(\xi) \neq 0$ made above. Suppose that $f \in C_\ell[a,b]$, $\ell > 1$, and that it has an ℓ-fold isolated zero at $\xi \in [a,b]$; i.e.,

$$f(\xi) = f'(\xi) = \cdots = f^{(\ell-1)}(\xi) = 0 \text{ and } f^{(\ell)}(\xi) \neq 0.$$

We now show that Newton's method

$$x^{(\kappa+1)} = \varphi(x^{(\kappa)}) \text{ with } \varphi(x) = x - \frac{f(x)}{f'(x)}$$

still converges in a neighborhood of ξ. It is easy to see that setting $\varphi(\xi) := \xi$, φ is continuous and even differentiable in a neighborhood of ξ, and that $\varphi'(\xi) = 1 - \frac{1}{\ell}$. Since $\ell > 1$, we have $0 < \varphi'(\xi) < 1$, and thus the iteration converges locally at least linearly.

In fact, it is also possible to recover the superlinear convergence. To this end, we define an iterative method corresponding to the function

$$\varphi(x) := \begin{cases} x - \ell\frac{f(x)}{f'(x)} & \text{for } x \neq \xi \\ \xi & \text{for } x = \xi, \end{cases}$$

and note that again $\varphi'(\xi) = 0$.

Now if the function f has an ℓ-fold zero at ξ, then the iteration

$$x^{(\kappa+1)} = x^{(\kappa)} - \ell\frac{f(x^{(\kappa)})}{f'(x^{(\kappa)})}$$

provides local superlinear convergence.

Critical Note. In practice, we will usually not know the multiplicity ℓ of a zero. Usually, however, we can assume that $\ell > 1$ whenever Newton's method gives only linear convergence.

2.4 The Secant Method. If we replace the quantity $f'(x^{(\kappa)})$ appearing in Newton's method by the divided difference $\frac{f(x^{(\kappa)})-f(x^{(\kappa-1)})}{x^{(\kappa)}-x^{(\kappa-1)}}$, then we are led to the iterative method

$$x^{(\kappa+1)} = \frac{x^{(\kappa-1)}f(x^{(\kappa)}) - x^{(\kappa)}f(x^{(\kappa-1)})}{f(x^{(\kappa)}) - f(x^{(\kappa-1)})}.$$

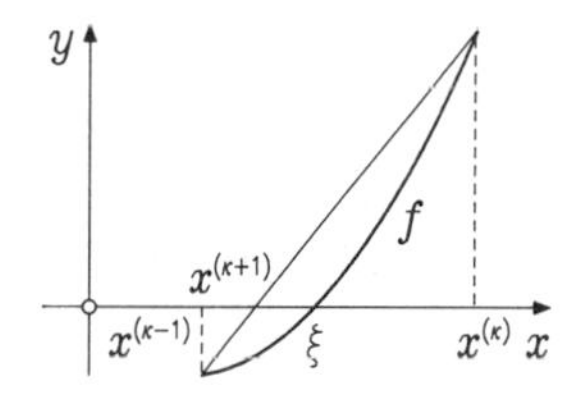

This method is called the *secant method*, and geometrically amounts to determining a new approximation $x^{(\kappa+1)}$ to ξ as the intersection with the x-axis of the straight line joining the points $(x^{(\kappa-1)}, f(x^{(\kappa-1)}))$ and $(x^{(\kappa)}, f(x^{(\kappa)}))$.

The rule for carrying out one step of the secant method is called *regula falsi*, and the secant method simply amounts to carrying out this step iteratively.

Rate of Convergence. The secant method cannot be written in the form $x^{(\kappa+1)} = \varphi(x^{(\kappa)})$, and so the previous convergence analysis cannot be applied. We now show, however, that under the same conditions as for Newton's method, we get local convergence at a rate lying between linear and quadratic. Consider

$$x^{(\kappa+1)} - \xi = x^{(\kappa)} - \xi - f(x^{(\kappa)})\frac{x^{(\kappa)} - x^{(\kappa-1)}}{f(x^{(\kappa)}) - f(x^{(\kappa-1)})}.$$

Now since $f(\xi) = 0$, using the divided difference notation in 5.2.3, we have

$$\begin{aligned} x^{(\kappa+1)} - \xi &= (x^{(\kappa)} - \xi)\left(1 - \frac{[x^{(\kappa)}\xi]f}{[x^{(\kappa)}x^{(\kappa-1)}]f}\right) = \\ &= (x^{(\kappa)} - \xi)(x^{(\kappa-1)} - \xi)\frac{[x^{(\kappa)}x^{(\kappa-1)}\xi]f}{[x^{(\kappa)}x^{(\kappa-1)}]f}. \end{aligned}$$

If $f \in C_2[a, b]$, then by 5.2.6 there exist two points $\eta, \hat{\eta} \in [a, b]$ such that $[x^{(\kappa)}x^{(\kappa-1)}\xi]f = \frac{1}{2}f''(\eta)$ and $[x^{(\kappa)}x^{(\kappa-1)}]f = f'(\hat{\eta})$. In a neighborhood $\hat{U}(\xi)$ of a simple zero of f with $f'(\xi) \neq 0$, we have the uniform bound

$$\left| \frac{[x^{(\kappa)}x^{(\kappa-1)}\xi]f}{[x^{(\kappa)}x^{(\kappa-1)}]f} \right| = \left| \frac{1}{2}\frac{f''(\eta)}{f'(\hat{\eta})} \right| \leq c_1,$$

and hence the deviations $\delta^{(\kappa)} = x^{(\kappa)} - \xi$ satisfy the inequality

$$|\delta^{(\kappa+1)}| \leq c_1|\delta^{(\kappa)}|\,|\delta^{(\kappa-1)}|.$$

Setting $d^{(\kappa)} := c_1|\delta^{(\kappa)}|$, we see that

$$d^{(\kappa+1)} \leq d^{(\kappa)}d^{(\kappa-1)}.$$

Now suppose $d^{(0)} \leq d$ and $d^{(1)} \leq d$ with $d < 1$. Then

$$d^{(2)} \leq d^2,\ d^{(3)} \leq d^3,\ d^{(4)} \leq d^5,\ d^{(5)} \leq d^8,$$

and in general, $d^{(\kappa)} \leq d^{a_\kappa}$, where the sequence (a_κ) is defined by

$$a_0 = a_1 = 1 \quad \text{and} \quad a_{\kappa+1} = a_\kappa + a_{\kappa-1} \quad \text{for} \quad \kappa \geq 1.$$

These are the famous *Fibonacci numbers*, named after LEONARDO OF PISA (1175–1230), who was also called Fibonacci.

Once we have chosen a_0 and a_1, all remaining terms in the sequence are uniquely defined. The recursion formula defining the Fibonacci numbers is a difference equation which can be explicitly solved. It is easy to check that the sequence defined by

$$a_\kappa = \frac{1}{\sqrt{5}}[b_1^{\kappa+1} - b_2^{\kappa+1}],$$

where

$$b_1 = \frac{1+\sqrt{5}}{2} \quad \text{and} \quad b_2 = \frac{1-\sqrt{5}}{2}$$

are the two roots of the equation $b^2 = b + 1$, satisfies the recursion relation defining the Fibonacci numbers, with starting values $a_0 = a_1 = 1$.

It now follows that $d^{(\kappa)} \leq d^{a_\kappa} = d^{\frac{1}{\sqrt{5}} b_1^{\kappa+1}} \cdot d^{-\frac{1}{\sqrt{5}} b_2^{\kappa+1}}$, and using $|b_2| < 1$ and $d^{-\frac{1}{\sqrt{5}} b_2^{\kappa+1}} \leq c_2$, we get the estimate

$$d^{(\kappa)} \leq c_2 \left(d^{\frac{1}{\sqrt{5}} b_1}\right)^{b_1^\kappa}.$$

We have shown that the deviation $d^{(\kappa)}$ converges with an order of at least $b_1 \doteq 1.618$.

Example. We now use the secant method to solve the equation $x - \exp(-\frac{1}{2}x) = 0$ considered in Examples 1.4 and 2.2.

κ	$x^{(\kappa)}$	$\lvert\xi - x^{(\kappa)}\rvert / \lvert\xi - x^{(\kappa-1)}\rvert^{1.618}$
0	0.8	--
1	0.7	--
2	0.703 5	0.206 150
3	0.703 467 4	0.172 692
4	0.703 467 422 498 4	0.182 599
5	0.703 467 422 498 391 652 049 8	0.180 042
6	0.703 467 422 498 391 652 049 818 601 8	--

The numbers in the $x^{(\kappa)}$ column were all computed to a machine accuracy of 28 digits. The last column shows that the convergence is of order $O(|\xi - x^{(\kappa)}|^{1.618})$. The example discussed in Problem 4 provides further insight into the behavior of the secant method.

We remark in passing that the number b_1 introduced above is the proportion of the golden section.

2.5 Newton's Method for $m > 1$. We start again with the linear part

$$y = f(x^{(\kappa)}) + J_f(x^{(\kappa)})(x - x^{(\kappa)}),$$

$$J_f(x^{(\kappa)}) = \left(\frac{\partial f_\mu(x^{(\kappa)})}{\partial x_\nu}\right)_{\mu,\nu=1}^{m},$$

of the expansion of f about $x^{(\kappa)}$. Now setting $y = 0$ and assuming that $\det(J_f(x^{(\kappa)})) \neq 0$, we get the new approximate value

$$x^{(\kappa+1)} := x^{(\kappa)} - J_f^{-1}(x^{(\kappa)})(f(x^{(\kappa)})).$$

This is the iteration formula for Newton's method. In using this formula in practice, we write it in the form

$$J_f(x^{(\kappa)})(x^{(\kappa+1)} - x^{(\kappa)}) = -f(x^{(\kappa)}),$$

and thus each step of the method involves solving a linear system of equations.

Variants. There are numerous other iterative methods for both the cases $m = 1$ and $m > 1$ which can be regarded as variants of Newton's method. For example, we can get a sharper form of Newton's method if we take higher order terms in the Taylor expansion of f in a neighborhood of one of its zero ξ (cf. also Problem 3). Similarly, interpolation of higher order can be used to improve the secant method; the resulting methods can also be generalized to higher dimensions. A detailed discussion of various iterative methods for solving equations, including the classical case of an equation $f(x) = 0$ where f is a function of several real variables, can be found in the book of A. M. Ostrowski [1973].

2.6 Roots of Polynomials. If $f \in P_n$, then the equation $f(x) = 0$ corresponds to the classical problem of finding the roots of an algebraic equation. In this case, Newton's method requires the computation of the values of the polynomial f and its derivative f' at the points $x^{(\kappa)}$. These can be computed using Horner's Algorithm 5.5.1.

If several or all of the zeros $\xi_1, \ldots, \xi_n$ of a polynomial $p \in P_n$ are to be computed, then we can make use of the fact that the polynomial can always be written in the form

$$\begin{aligned} p_n(x) &= a_0 + a_1 x + \cdots + a_n x^n = \\ &= a_n(x - \xi_1) \cdots (x - \xi_n), \end{aligned}$$

and thus once we have computed one of its zeros ξ_ν, we can remove the factor $(x - \xi_\nu)$ to get a polynomial of one degree lower. In performing Newton's method to compute ξ_ν, the polynomial $p_{n-1}(x) := p_n(x)/(x - \xi_\nu)$ will be automatically produced by the Horner algorithm. Indeed, as we observed in 5.5.1, Horner's scheme for computing $p_n(\xi)$ produces the expansion

$$p_n(x) = a_0' + (x - \xi)(a_1' + a_2' x + \cdots + a_n' x^{n-1}),$$

with $a_n' = a_n$, and thus if $\xi = \xi_\nu$ is a zero of p_n with $p_n(\xi) = a_0' = 0$, then p_n can be written in the form

$$p_n(x) = (x - \xi_\nu) p_{n-1}(x).$$

Except for ξ_ν, p_{n-1} has the same zeros as p_n.

To compute another zero $\xi_{\nu+1}$ of p_n, we can now use Newton's method on the polynomial p_{n-1}. In general, we will be working with an approximate value for ξ_ν, and thus the coefficients of p_{n-1} will be subject to some error which will be propagated in the computation of the zero $\xi_{\nu+1}$. This propagated error can be reduced by performing several steps of Newton's method using the original polynomial p_n, and starting with the approximate value for $\xi_{\nu+1}$ obtained from p_{n-1}. A careful analysis of the error behavior of Newton's method can be found in the book of J. Wilkinson [1963].

Other Methods for Computing Roots. The problem of finding the roots of an algebraic equation provided the motivation for Newton's creation of a method for finding zeros. Even before Newton, and on into the 20-th century, this classical problem was considered to be the most important of the zero problems. Consequently, numerous special methods for the computation of all of the zeros of a polynomial have been developed, along with a series of useful criteria for determining the number of real zeros, and localizing the position of both real and complex roots. The most accessible of these methods can be found in the book of J. Stoer and Bulirsch [1983].

We do not have space in this book to present additional special methods for computing the roots of polynomials. These days, it is simpler to use a plotter to get an idea of the number and location of the real zeros of a polynomial, rather than one of these special methods. To get a rough estimate of where the complex roots lie, we can use the Gerschgorin disks described in 3.2.2. Newton's method can be used on polynomials defined over the field of complex numbers in exactly the same way as for polynomials with real coefficients. In looking for complex roots of a polynomial with real coefficients, we have to start the iteration with a complex initial point $x^{(0)}$, since otherwise only real values are produced. Finally, we should also remark that while eigenvalues can be found by finding the roots of a polynomial, in general it is better to use one of the methods discussed in Chapter 3 to find eigenvalues.

2.7 Problems. 1) a) Given a positive real number α, we can compute $\sqrt[k]{\alpha}$ by solving the equation $x^k - \alpha = 0$ using Newton's method. Describe the method, and discuss for which initial values $x^{(0)}$ it converges. Compute $\sqrt[3]{17}$ to an accuracy of $\pm 1 \cdot 10^{-7}$.

b) In order to compute the number $\frac{1}{\alpha}$ without using division, we can solve the equation $\frac{1}{x} - \alpha = 0$ using Newton's method. Find an interval of convergence, and compute π^{-1} to an accuracy of $\pm 1 \cdot 10^{-7}$. Note that $\pi \doteq 3.1415926535$.

2) a) If f has an ℓ-fold zero at ξ, then ξ is a simple zero of $f^{\frac{1}{\ell}}$. What does Newton's method look like applied to $f^{\frac{1}{\ell}}$?

b) Give a sufficient conditions for the monotone (resp. alternating) convergence of Newton's method, and illustrate it graphically.

3) a) If we determine an improved approximation $x^{(\kappa+1)}$ of a zero from the approximation $x^{(\kappa)}$ by using not only the linear part, but also the terms of second order in the Taylor expansion of f about $x^{(\kappa)}$, the resulting method is called a Newton method of order 2. Describe the method, and interpret it geometrically.

b) The secant method can be extended in the same way by using an interpolation polynomial of degree 2. Develop this iterative method.

4) If we solve the equation $x - \cos(x) = 0$ using the secant method and starting with the initial values $x^{(0)} = 0.8$ and $x^{(1)} = 0.7$, then after seven iterations we get a value ξ which is accurate to 28 digits. But the sequence of numbers $(|\xi - x^{(\kappa)}|/|\xi - x^{(\kappa-1)}|^{1.618})$ exhibits a much weaker convergence. Follow through numerically the series of estimates which led to the Fibonacci sequence to check that these inequalities are sharp.

5) Consider the two-step Newton method

$$y^{(\kappa)} = x^{(\kappa)} - \frac{f(x^{(\kappa)})}{f'(x^{(\kappa)})}, \quad x^{(\kappa+1)} = y^{(\kappa)} - \frac{f(y^{(\kappa)})}{f'(y^{(\kappa)})},$$

and show:

a) If the method converges, then

$$\lim_{\kappa\to\infty} \frac{x^{(\kappa-1)} - \xi}{(y^{(\kappa)} - \xi)(x^{(\kappa)} - \xi)} = \frac{f''(\xi)}{f'(\xi)}.$$

b) The convergence is cubic:

$$\lim_{\kappa\to\infty} \frac{x^{(\kappa+1)} - \xi}{(x^{(\kappa)} - \xi)^3} = \frac{1}{2}\left[\frac{f''(\xi)}{f'(\xi)}\right]^2.$$

6) Solve the transcendental system of equations in Problem 4, Section 1.6 using Newton's method.

7) Show that if we have computed m zeros $\xi_1, \ldots, \xi_m$ of a polynomial

$$p_n(x) = a_n \prod_{\nu=1}^{n} (x - \xi_\nu), \quad n > m,$$

then the formula

$$x^{(\kappa+1)} = x^{(\kappa)} - \frac{1}{\frac{p_n'(x^{(\kappa)})}{p_n(x^{(\kappa)})} - \sum_{\nu=1}^{m} \frac{1}{x^{(\kappa)} - \xi_\nu}}$$

describes a Newton method for finding the remaining zeros $\xi_{m+1}, \ldots, \xi_n$.

3. Iterative Solution of Linear Systems of Equations

The general iteration method $x^{(\kappa+1)} := \Phi\, x^{(\kappa)}$, $\kappa \in \mathbb{N}$, for determing a fixed point of Φ can also be applied to solve linear systems of equations of the form $Ax = b$. In particular, iterative methods are especially useful when A is large and sparse. These kinds of matrices arise frequently in the discretization of problems involving differential equations. In order to write the linear system of equations $Ax = b$ in the form of a fixed point equation, we consider the equivalent reformulation $x = (I - A)x + b$ and set $C := I - A$. Then the iteration function φ is defined by $\varphi(x) := Cx + b$. If ξ is a solution of $x = \varphi(x)$, then the identity

$$x^{(\kappa+1)} - \xi = \varphi(x^{(\kappa)}) - \varphi(\xi) = C(x^{(\kappa)} - \xi) = C^{\kappa}(x^{(1)} - \xi)$$

implies that the sequence of iterates $(x^{(\kappa)})_{\kappa \in \mathbb{N}}$ with $x^{(\kappa+1)} := \varphi(x^{(\kappa)})$ and $x^{(1)} \neq \xi$ converges to ξ precisely when $\lim_{\kappa \to \infty} C^{(\kappa)} = 0$ componentwise. Hence, we first investigate under what conditions such sequences of matrices form a null sequence.

3.1 Sequences of Iteration Matrices. Let C be an arbitrary $m \times m$ matrix, and let $\rho(C)$ be its spectral radius. We now establish the following

Convergence Criterion. *The sequence $(C^{\kappa})_{\kappa \in \mathbb{N}}$ is a null sequence if and only if $\rho(C) < 1$.*

Proof. Suppose first that $\rho(C) \geq 1$. Then there exists an eigenvalue λ with $|\lambda| \geq 1$ and a vector $x \neq 0$ with $Cx = \lambda x$. Since $C^{\kappa}x = \lambda^{\kappa}x$ and $\lim_{\kappa \to \infty} \lambda^{\kappa} \neq 0$, it follows that (C^{κ}) cannot be a null sequence, i.e., the condition $\rho(C) < 1$ is necessary .

Now let $\rho(C) < 1$. Since $(TCT^{-1})^{\kappa} = TC^{\kappa}T^{-1}$ for every similarity transformation T, it suffices to show that $\lim_{\kappa \to \infty}(TCT^{-1})^{\kappa} = 0$. The matrix C can be transformed to Jordan normal form J using a similarity transformation. We now show that $\lim_{\kappa \to \infty} J^{\kappa} = 0$ if all of the eigenvalues $\lambda_1, \lambda_2, \ldots, \lambda_m$ have absolute value smaller than one. To this end, let

$$J_{\mu} = \begin{pmatrix} \lambda_{\mu} & 1 & & 0 \\ & \ddots & \ddots & \\ & & \ddots & 1 \\ 0 & & & \lambda_{\mu} \end{pmatrix}$$

be a Jordan block corresponding to the eigenvalue λ_{μ} in the Jordan normal form J of C. Since obviously

$$J^{\kappa} = \begin{pmatrix} J_1^{\kappa} & & & 0 \\ & J_2^{\kappa} & & \\ & & \ddots & \\ 0 & & & J_k^{\kappa} \end{pmatrix}$$

with $1 \le k \le m$, it suffices to investigate the convergence of each Jordan block J_μ.

We write J_μ in the form $J_\mu = \lambda_\mu I + S$ with

$$S := \begin{pmatrix} 0 & 1 & & & 0 \\ & \ddots & \ddots & & \\ & & \ddots & \ddots & \\ 0 & & & \ddots & 1 \\ & & & & 0 \end{pmatrix},$$

and form $J_\mu^\kappa = (\lambda_\mu I + S)^\kappa$. Applying the binomial formula and noting that $S^m = 0$, we get the equation

$$J_\mu^\kappa = \sum_{\nu=0}^{m-1} \binom{\kappa}{\nu} \lambda_\mu^{\kappa-\nu} S^\nu.$$

For every fixed ν, we have the estimate

$$\left| \binom{\kappa}{\nu} \lambda_\mu^{\kappa-\nu} \right| \le |\lambda_\mu^{\kappa-\nu} \kappa^\nu|,$$

and using $|\lambda_\mu| < 1$ we get the convergence $\lim_{\kappa\to\infty} |\binom{\kappa}{\nu} \lambda_\mu^{\kappa-\nu}| = 0$. □

Since in general it is not easy to compute the spectral radius of a matrix, we now show that any natural matrix norm leads to an upper bound for the spectral norm (cf. also 2.4.4, Problem 4).

Lemma. *Let $C \in \mathbb{C}^{(m,m)}$. Then $\rho(C) \le \|C\|$ for every natural matrix norm.*

Proof. If λ is an eigenvalue of C with associated eigenvector x, then

$$\frac{\|Cx\|}{\|x\|} = |\lambda|,$$

and thus $\|C\| \ge |\lambda|$. □

Proposition. Let $C \in \mathbb{C}^{(m,m)}$ and $c \in \mathbb{C}^m$. An iterative method of the form $x^{(\kappa+1)} = \varphi(x^{(\kappa)})$ with $\varphi(x) := Cx + c$, $x \in \mathbb{C}^m$, converges for every starting point $x^{(0)}$ if and only if $\rho(C) < 1$. A sufficient condition for this is that there exist a natural matrix norm with $\|C\| < 1$.

Extension. We have seen that the convergence $\lim_{\kappa\to\infty} C^\kappa = 0$ follows from $\|C\| < 1$. Conversely, it can be shown that there always exists a natural matrix norm with $\|C\| < 1$ if $\lim_{\kappa\to\infty} C^\kappa = 0$.

Proof. Let $\|\cdot\|$ be a vector norm on $\mathbb{C}^m$, and suppose $T \in \mathbb{C}^{(m,m)}$ is a nonsingular matrix. Then $\|x\|_T := \|Tx\|$ defines a vector norm. The corresponding induced matrix norm $\|C\|_T$ is then given by

$$\|C\|_T := \sup_{\|x\|_T=1} \|Cx\|_T = \sup_{\|Tx\|=1} \|TCx\| = \sup_{\|x\|=1} \|(TCT^{-1})x\| = \|TCT^{-1}\|.$$

In view of this, it suffices to establish the assertion for any matrix similar to C. We first transform C to Jordan normal form

$$J = \begin{pmatrix} J_1 & & & 0 \\ & J_2 & & \\ & & \ddots & \\ 0 & & & J_k \end{pmatrix} = SCS^{-1}, \quad 1 \le k \le m,$$

and then apply a further similarity transformation using a diagonal matrix $D = \operatorname{diag}(1, \varepsilon^{-1}, \ldots, \varepsilon^{1-m})$ with $\varepsilon > 0$. This leads to the similar matrix

$$\hat{J} = DJD^{-1} = \begin{pmatrix} \hat{J}_1 & & & 0 \\ & \hat{J}_2 & & \\ & & \ddots & \\ 0 & & & \hat{J}_k \end{pmatrix}, \quad \hat{J}_\mu = \begin{pmatrix} \lambda_\mu & \varepsilon & & 0 \\ & \ddots & \ddots & \\ 0 & & \ddots & \varepsilon \\ & & & \lambda_\mu \end{pmatrix}$$

for $\mu = 1, 2, \ldots, k$. We now consider the maximum norm $\|\cdot\|_\infty$ on $\mathbb{C}^m$, and the corresponding natural matrix norm given by the maximal row sums. Then with $T := DS$, we have

$$\|C\|_T = \|DSC(DS)^{-1}\| = \|\hat{J}\| \le \rho(C) + \varepsilon.$$

Since we have assumed that $\lim_{\kappa\to\infty} C^\kappa = 0$, by the Convergence Criterion, $\rho(C) < 1$. This means that $\varepsilon > 0$ can be chosen so that $\|C\|_T < 1$. □

In the following sections we will investigate the various ways of choosing C and c in order to get convergent iterative methods for solving linear systems of equations.

3.2 The Jacobi Method. Suppose $A \in \mathbb{R}^{(n,n)}$ is a nonsingular matrix, and let $b \in \mathbb{R}^n$. In order to solve the linear system of equations $Ax = b$ iteratively, we split $A = (a_{\mu\nu})$ into

$$A = -L + D - R$$

with

$$L := -\begin{pmatrix} 0 & & & \\ a_{21} & \ddots & 0 & \\ \vdots & \ddots & \ddots & \\ a_{n1} & \cdots & a_{nn-1} & 0 \end{pmatrix}, \quad D := \begin{pmatrix} a_{11} & & & \\ & \ddots & & 0 \\ & & \ddots & \\ 0 & & & \ddots \\ & & & & a_{nn} \end{pmatrix},$$

$$R := -\begin{pmatrix} 0 & a_{12} & \cdots & a_{1n} \\ & \ddots & \ddots & \vdots \\ 0 & & \ddots & a_{n-1n} \\ & & & 0 \end{pmatrix}.$$

Since A is nonsingular, by possibly permuting rows and columns, we can always assure that $a_{\mu\mu} \neq 0$ for all $1 \leq \mu \leq n$. Thus, we can assume that D is nonsingular. Now the matrix C and the vector c defined by

$$C := D^{-1}(L+R), \quad c := D^{-1}b,$$

describe an iterative method which we call the Jacobi method. Writing the iteration formula $x^{(\kappa+1)} = \varphi(x^{(\kappa)})$, $\varphi(x) := Cx + c$, componentwise as

$$x_\mu^{(\kappa+1)} = \frac{1}{a_{\mu\mu}}\Bigl(b_\mu - \sum_{\substack{\nu=1\\ \nu\neq\mu}}^{n} a_{\mu\nu}x_\nu^{(\kappa)}\Bigr), \quad 1 \leq \mu \leq n,$$

we see that to compute the component $x_\mu^{(\kappa+1)}$ of the vector $x^{(\kappa+1)}$, we need *all* of the components of the previous iterate $x^{(\kappa)} = (x_1^{(\kappa)}, \ldots, x_n^{(\kappa)})^T$.

The general considerations in the previous section immediately lead to a sufficient condition for the convergence of the Jacobi method.

Corollary. The Jacobi method converges for every starting vector $x^{(0)} \in \mathbb{R}^n$ provided that the *strong row-sum property*

$$\sum_{\substack{\nu=1\\ \nu\neq\mu}}^{n} |a_{\mu\nu}| < |a_{\mu\mu}|, \quad 1 \leq \mu \leq n$$

holds.

Extension. In Proposition 3.1 we have seen that the iterative method converges if $\|C\| < 1$ for some natural matrix norm. If instead of the maximal absolute row-sum norm we take the maximal absolute column-sum norm, then a sufficient condition for convergence is provided by the *strong column-sum criterion*

$$\sum_{\substack{\mu=1\\ \mu\neq\nu}}^{n} |a_{\mu\nu}| < |a_{\nu\nu}|, \quad 1 \leq \nu \leq n.$$

In practical applications, it frequently happens that neither the strong row-sum or strong column-sum criterion is satisfied.

Example. Suppose we are looking for a function $y \in C_2[0,1]$ on the interval $[0,1]$ satisfying the differential equation

$$\frac{d^2}{dx^2}y(x) + y(x) = g(x)$$

along with the boundary condition $y(0) = y(1) = 0$. To solve this problem numerically, we may discretize the interval $[0,1]$ as $I_h := \{x_\nu \in [0,1] \mid x_\nu = \nu h,\ h := \frac{1}{n},\ 0 \leq \nu \leq n\}$, and approximate (cf. 5.3.3) the derivatives $\frac{d^2}{dx^2}y(x)$ at the data points x_ν by the difference quotients $\frac{1}{h^2}(y(x_{\nu+1})-2y(x_\nu)+y(x_{\nu-1}))$. This leads to the linear system of equations

$$\begin{pmatrix} (-2+h^2) & 1 & & & \\ 1 & (-2+h^2) & 1 & & 0 \\ & \ddots & \ddots & \ddots & \\ 0 & & \ddots & \ddots & 1 \\ & & & 1 & (-2+h^2) \end{pmatrix} \begin{pmatrix} y_1 \\ \vdots \\ \vdots \\ y_{n-1} \end{pmatrix} = h^2 \begin{pmatrix} g(x_1) \\ \vdots \\ \vdots \\ g(x_{n-1}) \end{pmatrix}$$

for the approximations y_ν to the function values $y(x_\nu)$, $1 \le \nu \le n-1$.

Clearly, in this simple problem, neither the strong row-sum or strong column-sum criterion is satisfied.

Even for the still simpler differential equation

$$\frac{d^2}{dx^2}y(x) = g(x),$$

we may not be able to apply the above criteria for convergence. In this case, however, we at least know that the coefficient matrix $A = (a_{\mu\nu})$ satisfies the weak forms of our criteria:

$$\sum_{\substack{\nu=1 \\ \nu\neq\mu}}^{n-1} |a_{\mu\nu}| \le |a_{\mu\mu}|, \quad 1 \le \mu \le n-1,$$

and

$$\sum_{\substack{\mu=1 \\ \mu\neq\nu}}^{n-1} |a_{\mu\nu}| \le |a_{\nu\nu}|, \quad 1 \le \nu \le n-1.$$

We now investigate the question of whether these weaker conditions suffice for the convergence of our iterative method.

Definition. A matrix $A \in \mathbb{C}^{(n,n)}$, $A = (a_{\mu\nu})$, is called *decomposable* if there exist nonempty subsets N_1 and N_2 of the index set $\mathrm{N} := \{1, 2, \ldots, n\}$ with the properties

(a) $\mathrm{N}_1 \cap \mathrm{N}_2 = \emptyset$, (b) $\mathrm{N}_1 \cup \mathrm{N}_2 = N$,
(c) $a_{\mu\nu} = 0$ for all $\mu \in \mathrm{N}_1$ and $\nu \in \mathrm{N}_2$.

A matrix which is not decomposable is called *indecomposable*.

Example. (a) Consider a matrix of the form

$$A = \begin{pmatrix} a_{11} & a_{12} & \cdots & a_{1k} & 0 & \cdots & 0 \\ a_{21} & a_{22} & \cdots & a_{2k} & 0 & \cdots & 0 \\ \vdots & \vdots & & \vdots & \vdots & & \vdots \\ a_{k1} & a_{k2} & \cdots & a_{kk} & 0 & \cdots & 0 \\ a_{k+1\,1} & a_{k+1\,2} & \cdots & a_{k+1\,k} & a_{k+1\,k+1} & \cdots & a_{k+1\,n} \\ \vdots & \vdots & & \vdots & a_{k+2\,k+1} & \cdots & a_{k+2\,n} \\ \vdots & \vdots & & \vdots & \vdots & & \\ a_{n1} & a_{n2} & \cdots & a_{nk} & a_{n\,k+1} & \cdots & a_{nn} \end{pmatrix}.$$

The subsets $N_1 := \{1, 2, \ldots, k\}$, $N_2 := \{k+1, \ldots, n\}$ have the required properties given in the definition, and thus A is decomposable. We leave it to the reader to show that a system of equations $Ax = b$ with decomposable coefficient matrix A can always be divided into smaller subproblems.

(b) A matrix A with nonzero sub- and super-diagonal elements $a_{\mu+1\mu}$ and $a_{\mu\mu+1}$, $1 \le \mu \le n-1$, is indecomposable.

Indeed, suppose N_1 and N_2 are nonempty subsets of N with $N_1 \cap N_2 = \emptyset$ and $N_1 \cup N_2 = N$, and suppose that $k = \max N_1$. Then $k+1 \in N_2$, and $a_{kk+1} \ne 0$. An analogous argument can be carried out with $k = \max N_2$, and it follows that A is indecomposable.

We now show that the hypotheses of the Corollary can be somewhat weakened to obtain the following

Convergence Theorem. *Suppose $A \in \mathbb{R}^{(n,n)}$ is an indecomposable matrix satisfying the weak row-sum criterion*

$$\sum_{\substack{\nu=1\\ \nu\neq\mu}}^{n} |a_{\mu\nu}| \le |a_{\mu\mu}|, \quad 1 \le \mu \le n,$$

and

$$\sum_{\substack{\nu=1\\ \nu\neq\mu_0}}^{n} |a_{\mu_0\nu}| < |a_{\mu_0\mu_0}|$$

for some index μ_0, $1 \le \mu_0 \le n$. Then for every starting vector, the Jacobi method converges to the solution of the linear system of equations $Ax = b$ with coefficient matrix A.

Proof. By Lemma 3.1, all eigenvalues of the matrix $C := D^{-1}(L+R)$ have absolute value less than or equal to one. It remains to be shown that no eigenvalue has value one (cf. the Convergence Criterion 3.1).

We assume the contrapositive, i.e., that the matrix C has an eigenvalue $\lambda \in \mathbb{C}$ with $|\lambda| = 1$. Suppose the associated eigenvector x is normalized so that $\|x\|_\infty = 1$.

Since $\sum_{\substack{\nu=1\\ \nu\neq\mu}}^{n} c_{\mu\nu}x_\nu = (\lambda - c_{\mu\mu})x_\mu = \lambda x_\mu$, it follows that the inequality

$$(*) \qquad |x_\mu| = |\lambda|\,|x_\mu| \le \sum_{\substack{\nu=1\\ \nu\neq\mu}}^{n} \frac{|a_{\mu\nu}|}{|a_{\mu\mu}|}|x_\nu| \le \sum_{\substack{\nu=1\\ \nu\neq\mu}}^{n} \frac{|a_{\mu\nu}|}{|a_{\mu\mu}|} \le 1$$

holds for all $1 \le \mu \le n$. We now define the subsets $N_1 := \{\mu \in \mathbb{N} \mid |x_\mu| = 1\}$ and $N_2 := N \setminus N_1$. The set N_1 is trivially nonempty. By the weak row-sum criterion and $(*)$, it follows that $\mu_0 \in N_2$, and thus, $N_2 \ne \emptyset$. Since we have assumed that A is indecomposable, there exist indices $\tilde{\mu} \in N_1$ and $\tilde{\nu} \in N_2$ with

$$\frac{|a_{\tilde{\mu}\tilde{\nu}}|}{|a_{\tilde{\mu}\tilde{\mu}}|} \ne 0.$$

Now since $|x_{\tilde{\nu}}| < 1$, this chain of inequalities implies the strict inequality

$$|x_{\tilde{\mu}}| = |\lambda|\,|x_{\tilde{\mu}}| \le \sum_{\substack{\nu=1\\ \nu\ne\tilde{\mu}}}^{n} \frac{|a_{\tilde{\mu}\nu}|}{|a_{\tilde{\mu}\tilde{\mu}}|}|x_\nu| < \sum_{\substack{\nu=1\\ \nu\ne\tilde{\mu}}}^{n} \frac{|a_{\tilde{\mu}\nu}|}{|a_{\tilde{\mu}\tilde{\mu}}|} \le 1.$$

But this contradicts $|x_{\tilde{\mu}}| = 1$, and thus $\tilde{\mu}$ cannot be in N_1. This contradiction establishes the theorem. □

A *weak column-sum criterion* can be defined analogously to the weak row-sum criterion, and the Convergence Theorem also holds under the correspondingly modified hypotheses. We leave the details to the reader.

3.3 The Gauss-Seidel Method. Examining the componentwise formulae for the Jacobi Method 3.2, we see that in iteratively computing the component $x_\mu^{(\kappa+1)}$ of the vector $x^{(\kappa+1)}$, we can insert the components $x_1^{(\kappa+1)}$, $x_2^{(\kappa+1)}, \ldots, x_{\mu-1}^{(\kappa+1)}$, which have already been computed into the right-hand side of the equations. This leads to the iteration formula

$$(*)\qquad x_\mu^{(\kappa+1)} = \frac{1}{a_{\mu\mu}}\left(b_\mu - \sum_{\nu=1}^{\mu-1} a_{\mu\nu}x_\nu^{(\kappa+1)} - \sum_{\nu=\mu+1}^{n} a_{\mu\nu}x_\nu^{(\kappa)}\right), \quad 1\le\mu\le n,$$

where we define $\sum_{\nu=1}^{0} a_{\mu\nu}x_\nu^{(\kappa+1)} := 0$. Decomposing the matrix A into $A = -L + D - R$ as in 3.2, we can now describe this iterative method formally as

$$x^{(\kappa+1)} := Cx^{(\kappa)} + c$$

with $C := (D-L)^{-1}R$ and $c := (D-L)^{-1}b$. Assuming as in 3.2 that A is nonsingular, we can assume that $a_{\mu\mu} \ne 0$, $1 \le \mu \le n$, and it follows that the matrix $(D-L)$ is nonsingular. The iterative method defined in $(*)$ is called the *Gauss-Seidel method* . The question of convergence of the Gauss-Seidel method is in general not simple, and the natural conjecture that the Gauss-Seidel method converges at least as fast as the Jacobi method can only be established under additional hypotheses on the matrix A. We have the following

Convergence Theorem. *Suppose $A \in \mathbb{R}^{(n,n)}$ is a nonsingular matrix satisfying either the strong row-sum criterion or the strong column-sum criterion. Then the Gauss-Seidel method converges to a solution of the linear system of equations $Ax = b$, and the convergence is at least as fast as for the Jacobi method.*

Proof. We prove the assertion under the assumption that the strong row-sum criterion is satisfied. Suppose the iteration matrices of the Jacobi method and Gauss-Seidel methods are denoted by $C_J := D^{-1}(L+R)$ and $C_G := = (D-L)^{-1}R$, respectively. If the strong row-sum criterion is satisfied, then

$$\|C_J\|_\infty = \max_{1\le\mu\le n} \sum_{\substack{\nu=1\\ \nu\ne\mu}}^{n} \frac{|a_{\mu\nu}|}{|a_{\mu\mu}|} < 1.$$

Now suppose $y \in \mathbb{R}^n$ is arbitrary, and that $z := (C_G)y$. We now show by complete induction that all components z_μ of the vector z satisfy

$$|z_\mu| \le \sum_{\substack{\nu=1\\ \nu\neq\mu}}^{n} \frac{|a_{\mu\nu}|}{|a_{\mu\mu}|}\, \|y\|_\infty.$$

To this end, we rewrite the equation $z = (C_G)y$ as $(D - L)z = Ry$. Now

$$|z_1| \le \sum_{\nu=2}^{n} \frac{|a_{1\nu}|}{|a_{11}|}|y_\nu| \le \sum_{\nu=2}^{n} \frac{|a_{1\nu}|}{|a_{11}|}\|y\|_\infty,$$

and using $(*)$ and the induction hypothesis, we get

$$\begin{aligned}
|z_\mu| &\le \frac{1}{|a_{\mu\mu}|}\left(\sum_{\nu=1}^{\mu-1} |a_{\mu\nu}|\,|z_\nu| + \sum_{\nu=\mu+1}^{n} |a_{\mu\nu}|\,|y_\nu|\right) \le \\
&\le \frac{1}{|a_{\mu\mu}|}\left(\sum_{\nu=1}^{\mu-1} |a_{\mu\nu}|\,\|C_J\|_\infty + \sum_{\nu=\mu+1}^{n} |a_{\mu\nu}|\right)\|y\|_\infty \le \\
&\le \sum_{\substack{\nu=1\\ \nu\neq\mu}}^{n} \frac{|a_{\mu\nu}|}{|a_{\mu\mu}|}\,\|y\|_\infty.
\end{aligned}$$

Thus, $\|(C_G)y\|_\infty = \|z\|_\infty \le \|C_J\|_\infty\|y\|_\infty$, and $\|C_G\|_\infty \le \|C_J\|_\infty < 1$. Then Lemma 3.1 and Proposition 3.1 imply the assertion of the theorem. □

We now give an example to show that without appropriate assumptions on the matrix A, either of the two iterative methods can converge while the other does not.

Example. (a) Let

$$A = \begin{pmatrix} 1 & -2 & 2\\ -1 & 1 & -1\\ -2 & -2 & 1 \end{pmatrix}.$$

In this case the iteration matrices C_J and C_G are given by

$$C_J = \begin{pmatrix} 0 & 2 & -2\\ 1 & 0 & 1\\ 2 & 2 & 0 \end{pmatrix} \quad\text{and}\quad C_G = \begin{pmatrix} 0 & 2 & -2\\ 0 & 2 & -1\\ 0 & 8 & -6 \end{pmatrix}.$$

The spectral radii of these matrices are $\rho(C_J) = 0$ and $\rho(C_G) = 2(1+\sqrt{2})$. By the Convergence Criterion 3.1, the Jacobi method converges to a solution of a linear system of equations with coefficient matrix A, but the Gauss-Seidel method does not.

(b) For the reverse situation where the Gauss-Seidel method converges but the Jacobi method does not, consider the matrix A given by

$$A = \frac{1}{2}\begin{pmatrix} 2 & 1 & 1\\ -2 & 2 & -2\\ -1 & 1 & 2 \end{pmatrix}.$$

Then the associated iteration matrices

$$C_J = \frac{1}{2}\begin{pmatrix} 0 & -1 & -1 \\ 2 & 0 & 2 \\ 1 & -1 & 0 \end{pmatrix} \quad \text{and} \quad C_G = \frac{1}{2}\begin{pmatrix} 0 & -1 & -1 \\ 0 & -1 & 1 \\ 0 & 0 & -1 \end{pmatrix}$$

have spectral radii $\rho(C_J) = \frac{1}{2}\sqrt{5}$ and $\rho(C_G) = \frac{1}{2}$.

In the following section we compare the Jacobi and Gauss-Seidel methods in more detail for certain special matrices.

3.4 The Theorem of Stein and Rosenberg. In this section we analyze the spectral radius of a large class of matrices which includes many matrices arising in the discretization of differential equations (cf. 3.2). Our aim is to establish the Theorem of Stein and Rosenberg, which will settle the question of when the Gauss-Seidel method should be used instead of the Jacobi method. We introduce the class of matrices of interest in this section in the following

Definition. A matrix $B \in \mathbb{R}^{(m,n)}$, $B = (b_{\mu\nu})$, is called *nonnegative* if all $b_{\mu\nu}$, $1 \leq \mu \leq m$ and $1 \leq \nu \leq n$, are nonnegative. We write $B \geq 0$ in this case.

The key tool in this section is the

Theorem of Perron and Frobenius. *Suppose $B \in \mathbb{R}^{(n,n)}$, $B = (b_{\mu\nu})$, $n > 1$, is indecomposable, and that $B \geq 0$. Then B has the properties:*
(i) B possesses an eigenvalue $\lambda_B > 0$ with $\rho(B) = \lambda_B$.
(ii) There exists an eigenvector $y > 0$ corresponding to $\lambda_B = \rho(B)$.
(iii) The eigenvalue $\lambda_B = \rho(B)$ is simple.
(iv) For every indecomposable matrix $F \in \mathbb{R}^{(n,n)}$ satisfying $B \geq F \geq 0$, $\rho(B) \geq \rho(F)$ holds.

Here $B \geq F$ means that $B - F \geq 0$.

Proof. This result was published independently by O. Perron [1907] and G. Frobenius [1912]. We prove it only for *symmetric matrices*. The general case is treated in R. S. Varga ([1962], p. 30 ff.).

Suppose now that B is symmetric.

(i). Since B has real eigenvalues $\lambda_1 \leq \lambda_2 \leq \cdots \leq \lambda_n$, it follows that $\rho(B) = \max\{|\lambda_1|, \lambda_n\}$. Thus either $\rho(B) = -\lambda_1$ or $\rho(B) = \lambda_n$. If $x^1 \in \mathbb{R}^n$ is an eigenvector of B corresponding to λ_1, then the extremal property of the Rayleigh quotient

$$\frac{\overline{x}^T Bx}{\overline{x}^T x} = \frac{\overline{\langle x, Bx\rangle}}{\langle x, x\rangle} = \frac{\langle x, Bx\rangle}{\langle x, x\rangle}$$

(cf. 3.3.3 and 4.1.3) and the fact that $B \geq 0$ imply

$$|\lambda_1| = \frac{|\langle x^1, Bx^1\rangle|}{\langle x^1, x^1\rangle} \leq \frac{\langle |x^1|, B|x^1|\rangle}{\langle |x^1|, |x^1|\rangle} \leq \lambda_n.$$

Here $|x^1|$ denotes the vector $|x^1| := (|x_1^1|, \ldots, |x_n^1|)^T$. This inequality already implies $\rho(B) = \lambda_n$. Now if $\lambda_n = 0$, then it follows that all of the eigenvalues of B are zero. Since B is symmetric, it follows that it is the zero matrix, and hence is trivially decomposable. This contradiction shows that $\lambda_n > 0$.

(ii) Let $x^n \in \mathbb{R}^n$ be the eigenvector corresponding to the eigenvalue λ_n. By the extremal property of the Rayleigh quotient, we again get that

$$\lambda_n \geq \frac{\langle |x^n|, B|x^n| \rangle}{\langle |x^n|, |x^n| \rangle},$$

and in addition

$$\lambda_n = |\lambda_n| = \frac{|\langle x^n, Bx^n \rangle|}{\langle x^n, x^n \rangle} \leq \frac{\langle |x^n|, B|x^n| \rangle}{\langle |x^n|, |x^n| \rangle}.$$

This implies that there exists a vector $y \in \mathbb{R}^n$, $y := |x^n| \geq 0$, with

$$(*) \qquad \lambda_n = \frac{|\langle y, By \rangle|}{\langle y, y \rangle}.$$

We now show that y is an eigenvector associated with $\lambda_n = \rho(B)$. Since B is symmetric, there is a complete system of orthonormal eigenvectors $x^1, x^2, \ldots, x^n \in \mathbb{R}^n$. We write

$$(**) \qquad y = \sum_{\mu=1}^{n} \alpha_\mu x^\mu$$

with certain fixed coefficients $\alpha_\mu \in \mathbb{R}$. Inserting this expansion in $(*)$, we get

$$\lambda_n = \sum_{\mu=1}^{n} \left(\alpha_\mu^2 \Big/ \sum_{\nu=1}^{n} \alpha_\nu^2 \right) \lambda_\mu.$$

This immediately implies that for all eigenvalues λ_μ, $1 \leq \mu \leq n-1$, with $\lambda_\mu < \lambda_n$, the corresponding coefficient α_μ in the expansion $(**)$ must vanish, since otherwise equality could not hold.

We conclude that since y is a linear combination of eigenvectors associated with the eigenvalue λ_n, it is itself an eigenvector associated with this eigenvalue. If any components of y vanish, we can assume that they are

$$y_{k+1} = y_{k+2} = \cdots = y_n = 0$$

for some index $1 \leq k \leq n-1$. Then the last $(n-k)$ equations of the eigenvalue equation $By = \lambda_n y$ become

$$\sum_{\nu=1}^{k} b_{\mu\nu} y_\nu = 0, \quad k+1 \leq \mu \leq n.$$

Since $y_\nu > 0$ for $1 \le \nu \le k$ and $B \ge 0$, it follows that

$$b_{\mu\nu} = 0$$

for $k+1 \le \mu \le n$ and $1 \le \nu \le k$. We now define $N_1 := \{1, 2, \ldots, k\}$ and $N_2 := \{k+1, k+2, \ldots, n\}$. These sets satisfy all of the conditions in Definition 3.2, and so B is decomposable, contradicting our assumption. We have shown that $y > 0$.

(iii) We assume the opposite; namely that $\lambda_B = \rho(B)$ is a multiple eigenvalue. Then by the symmetry of B, there are at least two eigenvectors x^1 and x^2 corresponding to the eigenvalue λ_B, and they are orthogonal to each other. As in the proof of (ii), we conclude that $x^1_\mu \ne 0$ and $x^2_\mu \ne 0$ for all $1 \le \mu \le n$, since otherwise B would be decomposable. We now normalize x^1 and x^2 so that $x^1_1 = x^2_1 = 1$. Let $N^\kappa_1 := \{\mu | x^\kappa_\mu > 0\}$ and $N^\kappa_2 := \{\nu | x^\kappa_\nu < 0\}$. By (*), we see that

$$\frac{1}{\langle x^\kappa, x^\kappa\rangle} \mid \sum_{\mu,\nu=1}^{n} b_{\mu\nu} x^\kappa_\mu x^\kappa_\nu \mid = \frac{1}{\langle x^\kappa, x^\kappa\rangle} \sum_{\mu,\nu=1}^{n} b_{\mu\nu} |x^\kappa_\mu \cdot x^\kappa_\nu|$$

for $\kappa = 1, 2$. This implies that for all $\mu \in N^\kappa_1$ and $\nu \in N^\kappa_2$, the matrix elements $b_{\mu\nu}$ must vanish. But since B is not decomposable and $N_1 \ne \emptyset$, it follows that $N_2 = \emptyset$. We conclude that the vectors x^1 and x^2 can have only positive components, which implies $\langle x^1, x^2\rangle > 0$, contradicting our assumption that the two vectors are orthogonal.

(iv) Suppose λ_F is the eigenvalue of F with $\lambda_F = \rho(F)$, and x^F is an associated positive eigenvector. By the extremal property of the Rayleigh quotient,

$$\rho(F) = \lambda_F = \frac{\langle x^F, Fx^F\rangle}{\langle x^F, x^F\rangle} \le \frac{\langle x^F, Bx^F\rangle}{\langle x^F, x^F\rangle} \le \rho(B). \qquad \square$$

We can now use the Theorem of Perron and Frobenius to establish the

Theorem of Stein and Rosenberg. *Suppose the iteration matrix $C_J \in \mathbb{R}^{(n,n)}$ in the Jacobi method is nonnegative. Then precisely one of the following assertions on the spectral radii $\rho(C_J)$ and $\rho(C_G)$ of the Jacobi and Gauss-Seidel methods holds:*

(i) $\rho(C_J) = \rho(C_G) = 0$, *(ii)* $0 < \rho(C_G) < \rho(C_J) < 1$,

(iii) $1 = \rho(C_G) = \rho(C_J)$, *(iv)* $1 < \rho(C_J) < \rho(C_G)$.

Before proving this theorem, we need some additional facts.
Given arbitrary lower and upper $n \times n$ triangular matrices L and R with zero entries on the main diagonals, consider the real functions

$$q(\sigma) := \rho(\sigma L + R) \qquad \text{and} \qquad s(\sigma) := \rho(L + \sigma R)$$

defined for all $\sigma \ge 0$. These functions possess the following

Properties.

(*i*) $q(0) = s(0) = 0, \quad q(1) = s(1) = \rho(L+R), \quad s(\sigma) = \sigma\, q(\frac{1}{\sigma})$ for $\sigma > 0$.

(ii) If $L+R \geq 0$, then $\rho(L+R) > 0$ implies the functions q and s are strictly monotone increasing, and $\rho(L+R) = 0$ implies $q = s = 0$.

Proof. The properties (i) are obvious. We prove (ii) under the additional assumption that the matrix $(L + R)$ is ***indecomposable.*** The proof for the general case is the object of Problem 7.

By the Theorem of Perron and Frobenius, $\rho(L+R) > 0$. Thus neither L nor R is the zero matrix. In addition, it is easy to see that the assumption that $L+R$ is indecomposable implies $Q(\sigma) := \sigma L + R$ also has this property for every $\sigma > 0$. Now from part (iv) of the proof of the Theorem of Perron and Frobenius, we deduce that $q(\sigma) = \rho(\sigma L + R)$ is a strictly monotone increasing function. The proof for the function s is similar. □

We are now ready for the

Proof of the Theorem of Stein and Rosenberg. We again restrict ourselves to the case where $C_J = D^{-1}(L+R)$ is ***indecomposable.*** Since $D^{-1}L$ and $D^{-1}R$ are nonnegative matrices with zero entries on the main diagonals, it follows from the identities

$$(I - D^{-1}L)^{-1} = I + D^{-1}L + (D^{-1}L)^2 + \cdots + (D^{-1}L)^{n-1}$$

and $C_G = (D-L)^{-1}R = (I - D^{-1}L)^{-1}D^{-1}R$, that C_G is also nonnegative. With some additional work along the lines of the proof of the Theorem of Perron and Frobenius (cf. R. S. Varga ([1962], p. 46 ff.)), it can be shown that even without the hypothesis that C_G is indecomposable, there exists an eigenvalue λ_G with $\rho(C_G) = \lambda_G$ and an associated eigenvector x^G with $x^G \geq 0$. We leave it to the reader to show that $\lambda_G > 0$ and $x^G > 0$ follow from the indecomposability of C_J (Problem 8).

We now write the eigenvalue equation for C_G in the form

$$D^{-1}(\lambda_G L + R)x^G = \lambda_G x^G \quad \text{or} \quad D^{-1}(L + \frac{1}{\lambda_G}R)x^G = x^G.$$

The matrices $Q(\lambda_G) := D^{-1}(\lambda_G L + R)$ and $S(\frac{1}{\lambda_G}) = D^{-1}(L + \frac{1}{\lambda_G}R)$ are nonnegative and indecomposable. The corresponding functions q and s satisfy

$$q(\lambda_G) = \lambda_G, \quad s(\frac{1}{\lambda_G}) = 1.$$

We now treat the four cases $\rho(C_J) = 0$, $0 < \rho(C_J) < 1$, $\rho(C_J) = 1$ and $\rho(C_J) > 1$ separately.

Case (i) $\rho(C_J) = 0$: The monotonicity of q along with $q(1) = \rho(C_J) = 0$ and $q(\lambda_G) = \lambda_G \geq 0$ implies that $\rho(C_G) = 0$.

Case (ii) $0 < \rho(C_J) < 1$: By $s(1) = \rho(C_J)$, $s(\frac{1}{\lambda_G}) = 1$, and the monotonicity of s, we have $0 < \lambda_G < 1$. On the other hand, q is strictly monotone increasing and $q(1) = \rho(C_J)$. Thus $q(\lambda_G) = \lambda_G$ implies that $0 < \rho(C_G) < \rho(C_J)$.

Case (iii) $\rho(C_J) = 1$: By the strict monotonicity of s and the facts that $s(1) = 1 = \rho(C_J)$ and $s(\frac{1}{\lambda_G}) = 1$, we get $\lambda_G = 1$.

Case (iv) $1 < \rho(C_J)$: The proof in this case is similar to Case (ii). □

Under the assumption that the iteration matrix C_J of the Jacobi method is nonnegative, the Theorem of Stein and Rosenberg asserts that the Gauss-Seidel method converges precisely when the Jacobi method converges. The convergence of the Gauss-Seidel method in this case is asymptotically faster than that of the Jacobi method. This completely explains the convergence behavior for systems of equations $Ax = b$ with $A \geq 0$ and positive diagonal elements.

3.5 Problems. 1) Sufficient conditions for the convergence of the Jacobi method for solving a linear system of equations $Ax = b$, $A \in \mathbb{R}^{(n,n)}$ and $b \in \mathbb{R}^n$, are provided by the strong row-sum criterion, the strong column-sum criterion, and also by the ***strong square-sum criterion***

$$\sum_{\mu=1}^{n} \sum_{\substack{\nu=1 \\ \nu \neq \mu}}^{n} \left| \frac{a_{\mu\nu}}{a_{\mu\mu}} \right|^2 < 1.$$

Show by appropriate simple examples that these three criteria are not equivalent.

2) Solve the system of equations

$$\begin{array}{rcrcrcrcrcrcl}
10x_1 & & x_2 & & & & & & & & & = & 10 \\
-x_1 & + & 10x_2 & - & x_3 & & & & & & & = & 10 \\
 & - & x_2 & + & 10x_3 & - & x_4 & & & & & = & 0 \\
 & & & - & x_3 & + & 10x_4 & - & x_5 & & & = & 10 \\
 & & & & & & & - & x_5 & + & 10\,x_6 & = & 10 \\
 & & & & & & x_4 & + & 10x_5 & - & x_6 & = & 0
\end{array}$$

a) using the Jacobi method, b) using the Gauss-Seidel method.

3) Show that a matrix A is decomposable if and only if there exists a permutation matrix P with

$$P^{-1}AP = \begin{pmatrix} A_1 & 0 \\ A_2 & A_3 \end{pmatrix},$$

where A_1 and A_3 are square matrices.

4) Let $A \in \mathbb{R}^{(n,n)}$ be symmetric and indecomposable. In addition, suppose A has positive diagonal elements, and that it satisfies the weak row-sum criterion. Show that all eigenvalues of A are positive.

5) Construct matrices A for which the Jacobi method converges, but the Gauss-Seidel method does not, and conversely.

6) Let $A, B \in \mathbb{R}^n$ be nonnegative, indecomposable, symmetric matrices with nonzero entries on their diagonals. Show that if $A \geq B$ and $a_{\mu\nu} \neq b_{\mu\nu}$ for at least one μ and ν, then $\rho(A) > \rho(B)$.

7) Prove the property 3.5 (ii) without the additional hypothesis that $L + U$ is indecomposable. (Hint: Use Problem 3).)

8) Let $C_J := D^{-1}(L + R)$ be nonnegative and indecomposable. Show that for the corresponding Gauss-Seidel method, $\rho(C_G) > 0$ and that the eigenvector of C_G associated with $\rho(C_G) =: \lambda_G$ is positive.

4. More on Convergence

As in the previous sections, we again consider the system of equations $Ax = b$ with $A \in \mathbb{R}^{(n,n)}$ and $b \in \mathbb{R}^n$, where we suppose that the matrix A has nonvanishing diagonal elements $a_{\mu\mu}$. Our aim in this section is to show how to modify the iterative methods discussed above for solving such systems so as to improve their speed of convergence. This turns out to be of considerable practical importance since, even for very simple model problems, both the Jacobi and Gauss-Seidel methods may converge very slowly.

We begin by studying the Jacobi Method 3.2, whose iteration equation can be written as

$$x^{(\kappa+1)} = D^{-1}(L + R)x^{(\kappa)} + D^{-1}b,$$

or equivalently as

$$\begin{aligned} x^{(\kappa+1)} &= x^{(\kappa)} + D^{-1}(L - D + R)x^{(\kappa)} + D^{-1}b = \\ &= x^{(\kappa)} - D^{-1}(Ax^{(\kappa)} - b). \end{aligned}$$

We shall denote the defect vector at the κ-th step by $d^{(\kappa)} := Ax^{(\kappa)} - b$. Now the Jacobi method has the following interpretation: to compute $x^{(\kappa+1)}$ from the previous iterate $x^{(\kappa)}$, we subtract the vector $D^{-1}d^{(\kappa)}$.

4.1 Relaxation for the Jacobi Method. In view of the above interpretation of the Jacobi method as a process of correcting the iterates by subtracting a quantity depending on the defect, it is natural to introduce the following modified version involving a *relaxation parameter* $\omega \in \mathbb{R}$:

$$x^{(\kappa+1)} = x^{(\kappa)} - \omega D^{-1}(Ax^{(\kappa)} - b).$$

The resulting method

$$x^{(\kappa+1)} = C_J(\omega)x^{(\kappa)} + c(\omega)$$

is described by the iteration matrix $C_J(\omega) := (1-\omega)I + \omega D^{-1}(L+R)$ and the vector $c(\omega) := \omega D^{-1}b$, and is called the *simultaneous relaxation method.* We shall refer to it as the SR method. Our goal now is to determine the parameter ω so that the spectral norm $\rho(C_J(\omega))$ is minimal. To this end, it is useful to express the eigenvalues of $C_J(\omega)$ in terms of those of $C_J(1) = C_J$.

Remark. Suppose the matrix $C_J = D^{-1}(L+R)$ has the eigenvalues $\lambda_1, \lambda_2, \ldots, \lambda_n$ with associated eigenvectors $x^1, x^2, \ldots, x^n$. Then the matrix $C_J(\omega)$ has the eigenvalues $\lambda_\mu(\omega) := 1 - \omega + \omega\lambda_\mu$, $1 \le \mu \le n$, with the same eigenvectors. This follows immediately from

$$C_J(\omega)x^\mu = (1-\omega)x^\mu + \omega D^{-1}(L+R)x^\mu = ((1-\omega) + \omega\lambda_\mu)x^\mu.$$

Without any further assumptions on $L+R$, we now obtain the following sufficient

Convergence Condition for the SR Method. Suppose the Jacobi method converges. Then so does the simultaneous relaxation method for all $0 < \omega \le 1$.

Proof. Let $\lambda_\mu = r_\mu e^{i\theta_\mu}$ be the possibly complex eigenvalues of C_J. From $\rho(C_J) < 1$ it follows that $r_\mu < 1$ for $1 \le \mu \le n$. Now the eigenvalues $\lambda_\mu(\omega)$ of $C_J(\omega)$ satisfy

$$\begin{aligned}|\lambda_\mu(\omega)|^2 &= |1-\omega+\omega r_\mu e^{i\theta_\mu}|^2 = (1-\omega)^2 + 2\omega r_\mu(1-\omega)\cos\theta_\mu + \omega^2 r_\mu^2 \le \\ &\le (1-\omega+\omega r_\mu)^2 < 1\end{aligned}$$

for $0 < \omega < 1$. □

If all of the eigenvalues of the matrix C_J are real, as is the case for example if this matrix is symmetric, then we can derive an explicit formula for the optimal relaxation parameter.

Theorem. *Suppose the iteration matrix C_J has only real eigenvalues $\lambda_1 \le \lambda_2 \le \cdots \le \lambda_n$ lying in the interval $(-1, +1)$. Then the spectral radius $\rho(C_J(\omega))$ of the matrix $C_J(\omega)$ is minimal for*

$$\omega_{\min} = \frac{2}{2 - \lambda_1 - \lambda_n}.$$

Proof. We consider the function $f_\omega(\lambda) := 1 - \omega + \omega\lambda$, and try to find the parameter ω in order to minimize $\max_{1\le i\le n} |f_\omega(\lambda_i)|$. This can be regarded as a discrete approximation problem with respect to the Chebyshev norm, and the theory of 4.4 can be applied with minor modifications. First, it is easy to see that the points λ_1 and λ_n must form an alternant. Then by the Alternation Theorem, the optimal parameter $\omega_{\min}$ is characterized by the property that

$$f_{\omega_{\min}}(\lambda_1) = -f_{\omega_{\min}}(\lambda_n).$$

This leads directly to

$$\omega_{\min} = \frac{2}{2 - \lambda_1 - \lambda_n}. \qquad \square$$

Remark. The Alternation Theorem 4.4.3 also simultaneously gives the following formula for the spectral radius

$$\rho(C_J(\omega_{\min})) = |f_{\omega_{\min}}(\lambda_1)| = |f_{\omega_{\min}}(\lambda_n)|,$$

and in fact

$$\rho(C_J(\omega_{\min})) = \frac{\lambda_n - \lambda_1}{2 - \lambda_n - \lambda_1}.$$

Corollary. If $\lambda_1 \neq -\lambda_n$, then the spectral radius $\rho(C_J(\omega_{\min}))$ corresponding to the optimal relaxation parameter $\omega_{\min}$ satisfies

$$\rho(C_J(\omega_{\min})) < \rho(C_J).$$

Proof. From $\lambda_1 \neq -\lambda_n$ it follows that $\omega_{\min} = \frac{2}{2-\lambda_1-\lambda_n} \neq 1$. Now from $C_J(1) = C_J$ and the uniqueness of the minimizing value $\omega_{\min}$, we conclude that the assertion holds. $\square$

The optimal relaxation parameter $\omega_{\min}$ lies in the open interval $(0.5, \infty)$. When $0.5 < \omega_{\min} < 1$, we speak of *simultaneous under-relaxation*, and when $\omega_{\min} > 1$, of *simultaneous over-relaxation.* The eigenvalues λ_1 and λ_n of C_J can be computed only in rare cases, but a nearly optimal relaxation parameter can be found as soon as we have reasonably sharp estimates for λ_1 and λ_n.

4.2 Relaxation for the Gauss-Seidel Method. The simultaneous relaxation method discussed in 4.1 can also be applied to the Gauss-Seidel method. The only difference here is that on the right-hand side of the iteration equation, we should insert the components of $x^{(\kappa+1)}$ which have already been computed. This leads to the equation

$$x^{(\kappa+1)} = C_G(\omega)x^{(\kappa)} + c(\omega),$$

where

$$C_G(\omega) := (I - \omega D^{-1}L)^{-1}((1-\omega)I + \omega D^{-1}R) \quad \text{and}$$
$$c(\omega) := \omega(I - \omega D^{-1}L)^{-1}D^{-1}b.$$

As a first step in finding the optimal parameter ω, we establish the

Theorem of W. Kahan [1958]. *The spectral radius of the matrix $C_G(\omega)$ satisfies the inequality*

$$\rho(C_G(\omega)) \geq |\omega - 1|$$

for all $\omega \in \mathbb{R}$, and equality holds precisely when all of the eigenvalues of $C_G(\omega)$ have value $|\omega - 1|$.

Proof. By a well-known theorem of linear algebra, the eigenvalues $\lambda_\mu(\omega)$ of the matrix $C_G(\omega)$ satisfy the identity

$$\prod_{\mu=1}^{n} \lambda_\mu(\omega) = \det C_G(\omega).$$

The special form of the matrix $C_G(\omega)$ leads immediately to

$$\det C_G(\omega) = \det((1-\omega)I - \omega D^{-1}R) = (1-\omega)^n.$$

This in turn implies the estimate

$$\rho(C_G(\omega)) = \max_{1\le\mu\le n} |\lambda_\mu(\omega)| \ge |1-\omega|,$$

where equality holds in the inequality precisely when all eigenvalues $\lambda_\mu(\omega)$ have the value $|1-\omega|$. □

By the Theorem of Kahan, the condition $0 < \omega < 2$ is necessary for the convergence of the relaxation method

$$x^{(\kappa+1)} = C_G(\omega)x^{(\kappa)} + c(\omega).$$

We refer to this as *under-relaxation* when $0 < \omega < 1$, and as *over-relaxation* when $1 < \omega < 2$. In the literature both cases are often referred to as the *SOR method.*

The bound in the Theorem of Kahan holds for arbitrary iteration matrices $C_G(\omega)$. This bound can be sharpened in special cases.

Theorem. *Suppose the matrix $A \in \mathbb{R}^{(n,n)}$ is symmetric, and that its diagonal elements are positive. Then the SOR method converges if and only if A is positive definite and $0 < \omega < 2$.*

Proof. Suppose we begin the iteration $x^{(\kappa+1)} = C_G(\omega)x^{(\kappa)} + c(\omega)$ with a start vector $x^{(0)} \neq 0$. If ξ is a solution of the system of equations $Ax = b$, then the error $d^{(\kappa)} := x^{(\kappa)} - \xi$ satisfies the iterative equation

$$(*) \qquad (D - \omega L)d^{(\kappa+1)} = ((1-\omega)D + \omega R)d^{(\kappa)}, \quad \kappa \in \mathbb{N}.$$

Now let $z^{(\kappa)} := d^{(\kappa)} - d^{(\kappa+1)}$ and $A = D - R - L$. Then for $\kappa \in \mathbb{N}$ we have

$$(D - \omega L)z^{(\kappa)} = \omega A d^{(\kappa)}$$

and

$$\omega A d^{(\kappa+1)} = (1-\omega)Dz^{(\kappa)} + \omega R z^{(\kappa)}.$$

Multiplying on the left in the first equation by $(d^{(\kappa)})^T$, in the second equation by $(d^{(\kappa+1)})^T$, and so on, we arrive at the identity

$$\langle d^{(\kappa)}, Dz^{(\kappa)}\rangle - (1-\omega)\langle d^{(\kappa+1)}, Dz^{(\kappa)}\rangle - \omega\langle d^{(\kappa)}, Lz^{(\kappa)}\rangle - \omega\langle d^{(\kappa+1)}, Rz^{(\kappa)}\rangle =$$
$$= \omega(\langle d^{(\kappa)}, Ad^{(\kappa)}\rangle - \langle d^{(\kappa+1)}, Ad^{(\kappa+1)}\rangle).$$

Observing that $R^T = L$, using $(*)$, and rearranging, we get

$$(**) \qquad (2-\omega)\langle z^{(\kappa)}, Dz^{(\kappa)}\rangle = \omega(\langle d^{(\kappa)}, Ad^{(\kappa)}\rangle - \langle d^{(\kappa+1)}, Ad^{(\kappa+1)}\rangle).$$

Now suppose A is positive definite and that $0 < \omega < 2$. For $d^{(0)}$ we choose an eigenvector of the matrix $C_G(\omega)$ corresponding to the eigenvalue λ. Then $d^{(1)} = C_G(\omega)d^{(0)} = \lambda d^{(0)}$, and by $(**)$, we have

$$\frac{2-\omega}{\omega}|1-\lambda|^2\langle d^{(0)}, Dd^{(0)}\rangle = (1-|\lambda|^2)\langle d^{(0)}, Ad^{(0)}\rangle.$$

The factor $(2-\omega)/\omega$ is positive. Moreover, the expressions $\langle d^{(0)}, Dd^{(0)}\rangle$ and $\langle d^{(0)}, Ad^{(0)}\rangle$ are also positive. This implies that $|\lambda| \leq 1$. Now if $|\lambda| = 1$, then it would follow that $d^{(0)} = d^{(1)}$, and thus that $z^{(0)} = 0$. Then from the relationship derived above between $z^{(\kappa)}$ and $d^{(\kappa)}$, we deduce that $Ad^{(0)} = 0$. This is a contradiction in view of our hypotheses that $d^{(0)} \neq 0$ and that A is positive definite. Thus we must have $|\lambda| < 1$, and the iterative method converges.

For the converse, suppose now that the iterative method converges. By the Theorem of Kahan, $0 < \omega < 2$. Moreover, $\lim_{\kappa\to\infty} d^{(\kappa)} = 0$ for every starting vector $d^{(0)}$. Then $(**)$ implies the inequality

$$\begin{aligned}\langle d^{(\kappa)}, Ad^{(\kappa)}\rangle &= \frac{2-\omega}{\omega}\langle z^{(\kappa)}, Dz^{(\kappa)}\rangle + \langle d^{(\kappa+1)}, Ad^{(\kappa+1)}\rangle \geq \\ &\geq \langle d^{(\kappa+1)}, Ad^{(\kappa+1)}\rangle.\end{aligned}$$

Suppose now that A is not positive definite. Then there exists a nonzero vector $d^{(0)}$ with $\langle d^{(0)}, Ad^{(0)}\rangle \leq 0$. Thus $z^{(0)} = d^{(0)} - d^{(1)} = d^{(0)} - C_G(\omega)d^{(0)} = (I - C_G(\omega))d^{(0)}$, and all eigenvalues of $C_G(\omega)$ have absolute value smaller than one since the iteration converges. This implies $z^{(0)} \neq 0$, and we have the strict inequality

$$\langle d^{(1)}, Ad^{(1)}\rangle < \langle d^{(0)}, Ad^{(0)}\rangle \leq 0.$$

But $\langle d^{(\kappa+1)}, Ad^{(\kappa+1)}\rangle \leq \langle d^{(\kappa)}, Ad^{(\kappa)}\rangle \leq \langle d^{(1)}, Ad^{(1)}\rangle < 0$ and moreover $\lim_{\kappa\to\infty} d^{(\kappa)} = 0$. This is impossible, and we conclude that A must be positive definite. □

The calculation of the optimal relaxation parameter for the SOR method requires a considerable amount of work. We consider just one special case in the following section. For more general results, see J. Stoer and R. Bulirsch [1983].

4.3 Optimal Relaxation Parameters. Discretizing differential equations often leads to a linear system of equations whose coefficient matrix has a special structure.

Example. Suppose G is an L-shaped domain in the plane, and consider the Dirichlet problem

$$\frac{\partial^2 u}{\partial x^2}(x,y) + \frac{\partial^2 u}{\partial y^2}(x,y) = f(x,y), \quad (x,y) \in \mathrm{G},$$

$$\text{with boundary condition} \quad u(x,y) = 0, \quad \text{for } (x,y) \in \Gamma,$$

where Γ denotes the boundary of G.

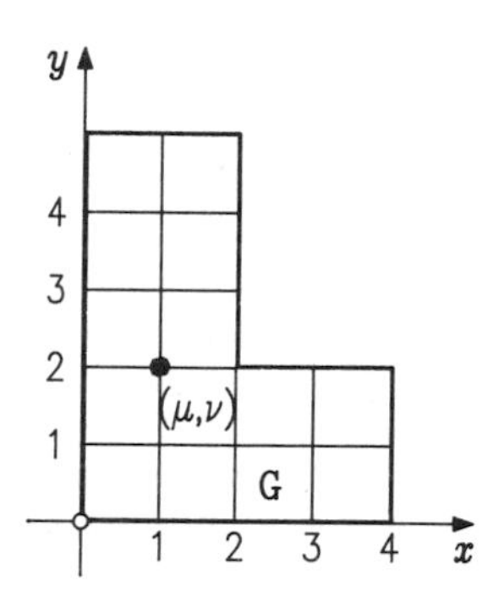

Now replacing the derivatives by difference quotients at the grid points (μ,ν) and taking account of the boundary condition, we get the following equations for the approximate values $\tilde{u}(\mu,\nu)$ of the function u at the interior grid points of the domain G:

$$\begin{aligned}
-4\tilde{u}(1,1) + \tilde{u}(2,1) + \tilde{u}(1,2) &= f(1,1) \\
\tilde{u}(1,1) - 4\tilde{u}(2,1) + \tilde{u}(3,1) &= f(2,1) \\
\tilde{u}(2,1) - 4\tilde{u}(3,1) &= f(3,1) \\
-4\tilde{u}(1,2) + \tilde{u}(1,1) + \tilde{u}(1,3) &= f(1,2) \\
-4\tilde{u}(1,3) + \tilde{u}(1,2) + \tilde{u}(1,4) &= f(1,3) \\
-4\tilde{u}(1,4) + \tilde{u}(1,3) &= f(1,4).
\end{aligned}$$

Abbreviating $\tilde{u}_{\mu\nu} := \tilde{u}(\mu,\nu)$ and $f_{\mu\nu} = f(\mu,\nu)$ and rearranging, this leads to the linear system of equations

$$\begin{pmatrix} -4 & 1 & 0 & 0 & 1 & 0 \\ 1 & -4 & 1 & 0 & 0 & 0 \\ 0 & 1 & -4 & 1 & 0 & 0 \\ 0 & 0 & 1 & -4 & 0 & 0 \\ 1 & 0 & 0 & 0 & -4 & 1 \\ 0 & 0 & 0 & 0 & 1 & -4 \end{pmatrix} \begin{pmatrix} \tilde{u}_{11} \\ \tilde{u}_{12} \\ \tilde{u}_{13} \\ \tilde{u}_{14} \\ \tilde{u}_{21} \\ \tilde{u}_{31} \end{pmatrix} = \begin{pmatrix} f_{11} \\ f_{12} \\ f_{13} \\ f_{14} \\ f_{21} \\ f_{31} \end{pmatrix}.$$

Now by appropriate simultaneous row and column interchanges, we can transform the coefficient matrix A to the special form

$$\begin{pmatrix} D_1 & A_{12} \\ A_{21} & D_2, \end{pmatrix}$$

where D_1 and D_2 are diagonal matrices. Indeed, if we apply the permutation matrix

$$P = \begin{pmatrix} 1 & 0 & 0 & 0 & 0 & 0 \\ 0 & 0 & 1 & 0 & 0 & 0 \\ 0 & 0 & 0 & 0 & 0 & 1 \\ 0 & 0 & 0 & 1 & 0 & 0 \\ 0 & 1 & 0 & 0 & 0 & 0 \\ 0 & 0 & 0 & 0 & 1 & 0 \end{pmatrix},$$

we obtain

$$PAP^T = \begin{pmatrix} \mathbf{-4} & \mathbf{0} & \mathbf{0} & 0 & 1 & 1 \\ \mathbf{0} & \mathbf{-4} & \mathbf{0} & 1 & 1 & 0 \\ \mathbf{0} & \mathbf{0} & \mathbf{-4} & 0 & 0 & 1 \\ 0 & 1 & 0 & \mathbf{-4} & \mathbf{0} & \mathbf{0} \\ 1 & 1 & 0 & \mathbf{0} & \mathbf{-4} & \mathbf{0} \\ 1 & 0 & 1 & \mathbf{0} & \mathbf{0} & \mathbf{-4} \end{pmatrix}.$$

It will be useful to work with matrices of this form, so we make the following

Definition. A matrix $A \in \mathbb{R}^{(n,n)}$ has *property* A if there exists a permutation matrix P such that

$$PAP^T = \begin{pmatrix} D_1 & A_{12} \\ A_{21} & D_2 \end{pmatrix}$$

with diagonal matrices D_1 and D_2.

The class of matrices with property A was introduced by D. M. Young [1971] in his study of iterative methods. For such matrices we can establish the following

Lemma. *Suppose $A = (a_{\mu\nu})$ is a $n \times n$ matrix with $a_{\mu\mu} \neq 0$, $1 \leq \mu \leq n$, satisfying property A. For each $\tau \in \mathbb{C}$ with $\tau \neq 0$, consider the matrix*

$$M(\tau) := \hat{D}^{-1}(\tau\hat{L} + \frac{1}{\tau}\hat{R}),$$

where $\hat{A} = PAP^T$ has the decomposition $\hat{A} = -\hat{L} + \hat{D} - \hat{R}$. Then the eigenvalues of M are independent of the value of τ.

Proof. By property A, there exists a permutation matrix P such that

$$PAP^T = \begin{pmatrix} D_1 & A_{12} \\ A_{21} & D_2 \end{pmatrix} = -\hat{L} + \hat{D} - \hat{R}$$

Now let $\tau \in \mathbb{C}$, $\tau \neq 0$. Then $M(\tau)$ has the form

$$M(\tau) = \begin{pmatrix} D_1^{-1} & 0 \\ 0 & D_2^{-1} \end{pmatrix} \begin{pmatrix} 0 & -\tau^{-1}A_{12} \\ -\tau A_{21} & 0 \end{pmatrix} = \begin{pmatrix} 0 & -\tau^{-1}D_1^{-1}A_{12} \\ -\tau D_2^{-1}A_{21} & 0 \end{pmatrix}$$
$$= \begin{pmatrix} I_1 & 0 \\ 0 & \tau I_2 \end{pmatrix} M(1) \begin{pmatrix} I_1 & 0 \\ 0 & \tau I_2 \end{pmatrix}^{-1},$$

and hence $M(\tau)$ and $M(1)$ are similar and so have the same eigenvalues. $\square$

Our goal now is to show that for matrices with property A, the spectral radius of the iteration matrix of the SOR method can be expressed in terms of the spectral radius of the matrix corresponding to the Jacobi method.

Theorem. *Suppose $A = (a_{\mu\nu})$ is a $n \times n$ matrix with $a_{\mu\mu} \neq 0$, $1 \leq \mu \leq n$, and that it has the following additional properties:*

(i) A has property A,

(ii) $C_J := D^{-1}(L + R)$ has only real eigenvalues,

(iii) $\rho(C_J) < 1$.

Then for all $0 < \omega < 2$, the spectral radius $\rho(C_G(\omega))$ of the iteration matrix $C_G(\omega)$ of the SOR method can be written as

$$(*) \qquad \rho(C_G(\omega)) = \begin{cases} \frac{1}{4}[\omega\rho(C_J) + (4(1-\omega) + \omega^2\rho^2(C_J))^{1/2}]^2 \\ \qquad\qquad \text{for } 0 < \omega < 2[1 + (1 - \rho^2(C_J))^{1/2}]^{-1}, \\ \omega - 1 \qquad \text{for } 2[1 + (1 - \rho^2(C_J))^{1/2}]^{-1} \leq \omega < 2. \end{cases}$$

Proof. By hypothesis (i), there exists a permutation matrix P such that

$$\hat{A} = PAP^T = \begin{pmatrix} D_1 & A_{12} \\ A_{21} & D_2 \end{pmatrix}$$

with diagonal matrices D_1 and D_2. Since the matrices A and $\hat{A}$ have the same eigenvalues, we may as well assume that A already has the form of $\hat{A}$.

The eigenvalues of $C_G(\omega)$ are the zeros of the characteristic polynomial

$$p_\omega(\lambda) := \det((1 - \omega - \lambda)I + \omega D^{-1}(R + \lambda L)).$$

We now distinguish two possibilities.

Case 1: $\rho(C_G(\omega)) = 0$. In this case $0 = p_\omega(0) = (1-\omega)^n$, and so $\omega = 1$. This implies $\rho(C_J) = 0$. Indeed, if τ is a nonzero eigenvalue of the matrix $C_J = D^{-1}(L + R)$, then applying the Lemma to the identity

$$\begin{aligned} p_1(\tau^2) &= \det(-\tau^2 I + D^{-1}(R + \tau^2 L)) = \tau^n \det(D^{-1}(\frac{1}{\tau}R + \tau L) - \tau I) = \\ &= \tau^n \det(M(\tau) - \tau I) = \tau^n \cdot K \cdot \det(M(1) - \tau I) = \\ &= \tau^n \cdot K \cdot \det(D^{-1}(L + R) - \tau I) = \tau^n \cdot K \cdot \det(C_J - \tau I) = 0 \end{aligned}$$

where K is a nonzero constant, we see that $\tau^2 \neq 0$ is an eigenvalue of $C_G(1)$. This contradicts $\rho(C_G(\omega)) = 0$ and $\omega = 1$, and we conclude that $(*)$ holds in this case.

Case 2: $\rho(C_G(\omega)) \neq 0$. Let λ be a nonzero eigenvalue of the matrix $C_G(\omega)$. Then λ is a zero of the characteristic polynomial p_ω, and again using the Lemma, the chain of equalities

$$\begin{aligned}
0 = p_\omega(\lambda) &= \det((1-\omega-\lambda)I + \omega D^{-1}(R+\lambda L)) = \\
&= \omega^n \cdot \lambda^{\frac{n}{2}} \cdot \det(D^{-1}(\sqrt{\lambda}L + \frac{1}{\sqrt{\lambda}}R) - \frac{(\lambda+\omega-1)}{\omega\sqrt{\lambda}}I) = \\
&= \omega^n \cdot \lambda^{\frac{n}{2}} \cdot K \cdot \det(M(1) - \frac{\lambda+\omega-1}{\omega\sqrt{\lambda}}I) = \\
&= \omega^n \cdot \lambda^{\frac{n}{2}} \cdot K \cdot \det(C_J - \frac{\lambda+\omega-1}{\omega\sqrt{\lambda}}I)
\end{aligned}$$

implies that

$$(**) \qquad \tau := \frac{\lambda+\omega-1}{\omega\sqrt{\lambda}}$$

is an eigenvalue of the matrix C_J. Moreover, the Lemma also implies

$$\begin{aligned}
0 = \det(C_J - \tau I) &= \det(M(1) - \tau I) = K_1 \cdot \det(D^{-1}(-\tau L - \frac{1}{\tau}R) - \tau I) = \\
&= (-1)^n \cdot K_1 \cdot \det(M(\tau) - (-\tau)I) = (-1)^n K_1 \cdot K_2 \cdot \det(C_J - (-\tau)I)
\end{aligned}$$

with nonzero constants K_i, $i = 1, 2$. Thus both τ and $(-\tau)$ are eigenvalues of C_J, and we can assume that $\tau \geq 0$. Conversely, if $\tau > 0$ is an eigenvalue of C_J, then there always exists an eigenvalue λ of the matrix $C_G(\omega)$ which is related to τ by the formula (**). The values of λ can be computed from the quadratic equation

$$\lambda^2 - 2\lambda(1 - \omega + \frac{1}{2}\omega^2\tau^2) + (1-\omega)^2 = 0.$$

For $0 < \omega < 2$, the two solutions

$$\lambda_{1/2} = (1 - \omega + \frac{1}{2}\omega^2\tau^2) \pm \omega\tau\sqrt{\frac{1}{4}\omega^2\tau^2 + (1-\omega)}$$

are real or complex depending on whether

$$0 < \omega \leq 2[1 + (1-\tau^2)^{1/2}]^{-1} \quad \text{or} \quad 2[1 + (1-\tau^2)^{1/2}]^{-1} < \omega < 2.$$

In the real case,

$$(***) \qquad \lambda = \frac{1}{4}[\omega\tau + (4(1-\omega) + \omega^2\tau^2)^{1/2}]^2$$

is the root of largest absolute value, while in the complex case,

$$|\lambda| = \omega - 1,$$

independently of the value of τ. Thus in the real case, we get the spectral radius of $C_G(\omega)$ by putting $\tau = \rho(C_J)$ in the formula $(***)$. This completes the proof of the theorem. □

To illustrate this result, we return to the above

Example. The eigenvalues of the scaled iteration matrix

$$C_J = \frac{1}{2}\begin{pmatrix} 0 & 1 & 0 & 0 & 1 & 0 \\ 1 & 0 & 1 & 0 & 0 & 0 \\ 0 & 1 & 0 & 1 & 0 & 0 \\ 0 & 0 & 1 & 0 & 0 & 0 \\ 1 & 0 & 0 & 0 & 0 & 1 \\ 0 & 0 & 0 & 0 & 1 & 0 \end{pmatrix}$$

are $\tau_1 = 0.5\sqrt{3} \doteq 0.86603$, $\tau_2 = 0.5$, $\tau_3 = \tau_4 = 0$, $\tau_5 = -\tau_2$, $\tau_6 = -\tau_1$. This implies that the spectral radius is $\rho(C_J) \doteq 0.86603$. The graph of the function $f(\omega) := \rho(C_G(\omega))$, $0 < \omega < 2$, has the following typical shape:

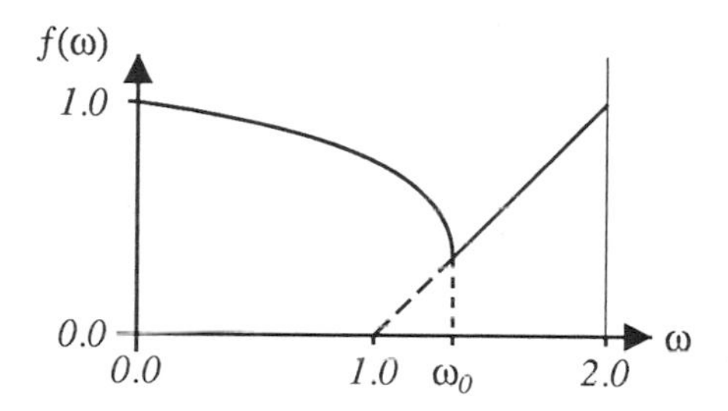

Here $\omega_0 := 2[1 + (1 - \rho^2(C_J))^{1/2}]^{-1}$.

The above theorem permits a more precise comparison of the Jacobi and Gauss-Seidel methods. In particular, if we take $\omega = 1$, then the formula $(*)$ leads immediately to the following

Corollary. *Suppose the hypotheses of the previous theorem are satisfied. Then*

$$\rho(C_G) = \rho^2(C_J).$$

This assertion can be interpreted as follows: for the same accuracy, the Jacobi method requires twice as many iterations as the Gauss-Seidel method.

Extension. In the interval $0 < \omega \leq \omega_0 := 2[1 + (1 - \rho^2(C_J))^{1/2}]^{-1}$, the function $f(\omega) := \frac{1}{4}[\omega\rho(C_J) + (4(1-\omega) + \omega^2\rho^2(C_J))^{1/2}]^2$ is strictly monotone decreasing. Hence, it takes its minimum at ω_0, and the optimal relaxation parameter ω_0 and the associated spectral radius $\rho(C_G(\omega_0))$ are thus

$$\omega_0 = 2[1 + (1 - \rho^2(C_J))^{1/2}]^{-1},$$
$$\rho(C_G(\omega_0)) = [\rho(C_J)(1 + (1 - \rho^2(C_J))^{1/2})^{-1}]^2 = \omega_0 - 1.$$

In using the SOR method in practice, it usually requires too much work to compute the optimal relaxation parameter, so instead we usually try to find good estimates for it based on bounds for the spectral radius of C_J. We do not have space here for the details; see the book of D. M. Young [1971].

4.4 Problems. 1) Show that the conditions A is positive definite and $0 < \omega < 2$ used in Theorem 4.2 are sufficient for the convergence of the SOR method, and thus the Gauss-Seidel method converges for positive definite matrices.

2) Show that $\lambda = 0$ is an eigenvalue of the matrix $C_G = C_G(1)$ in the Gauss-Seidel method.

3) Carry over the proof of Theorem 4.2 to the case of a matrix $A \in \mathbb{C}^{(n,n)}$ which is Hermitian with positive diagonal elements.

4) Use the SOR method to solve the system of equations $Ax = b$ for N=4, 8, 16, 32 and the relaxation parameters $\omega = 1, 1.2, 1.4, 1.6, 1.8, 1.9$, where $h := \mathrm{N}^{-1}$, $k := \mathrm{N} - 1$, $A \in \mathbb{R}^{(k^2,k^2)}$, $B \in R^{(k,k)}$, and $b \in \mathbb{R}^{k^2}$ with

$$A = \begin{pmatrix} B_k & -I_k & & 0 \\ -I_k & B_k & -I_k & \\ & \ddots & \ddots & \ddots \\ & & \ddots & \ddots & -I_k \\ 0 & & & -I_k & B_k \end{pmatrix}, \quad B_k = \begin{pmatrix} 4 & -1 & & 0 \\ -1 & 4 & \ddots & \\ & \ddots & \ddots & -1 \\ 0 & & -1 & 4 \end{pmatrix}.$$

Here I_k is the identity matrix in $\mathbb{R}^{(k,k)}$, and the vector b is given by

$$b = h^2 \begin{pmatrix} f^1 \\ f^2 \\ \vdots \\ f^k \end{pmatrix}, \quad f^\mu = \begin{pmatrix} f_1^\mu \\ f_2^\mu \\ \vdots \\ f_k^\mu \end{pmatrix} \in \mathbb{R}^k \quad \text{and} \quad f_\nu^\mu := 5\pi^2 \sin(2\pi\nu h)\sin(\pi\mu h),$$

$1 \le \mu, \nu \le k$. Starting with the vector $x^{(0)} = 0$, how many iterations $\kappa = \kappa(\mathrm{N}, \omega)$ are needed to get $\|Ax^{(\kappa)} - b\|_\infty \le 10^{-3}$?

5) Solve the system of equations in Example 4.3 with the right-hand side $f_{\mu\nu} = 1$ for all μ and ν using the Jacobi method and the SOR method with optimal relaxation parameter. When is the SOR method significantly faster with respect to the refinement of the grid?

9
Linear Optimization

"Since everything in the entire world is best possible, and since it is all the work of the wisest of creators, there is nothing in this world which is not blessed with either a maximum or minimum property. Thus, there can be no doubt that all of our worldly processes can as easily be derived by the method of maxima and minima as from their basic properties themselves." This observation of Leonhard Euler – freely translated from an article in Commentationes Mechanicae – makes crystal clear the key role that the problem of maximizing or minimizing a function plays in mathematics and its applications. In this chapter we will restrict our attention primarily to the special case of linear functions with linear side conditions. Nevertheless, the theory presented here has many applications, since in fact there are a huge number of natural problems which are linear, and, moreover, nonlinear problems can often be linearized. A central role in our considerations will be played by the simplex method, one of the most used methods in all of numerical analysis.

1. Introductory Examples and the General Problem

Linear optimization uses the geometry of m-dimensional Euclidean space. But with today's modern computers, we can solve linear optimization problems with very large values of m, and thus it is also possible to deal with infinite dimensional problems by approximating them by problems in $\mathbb{R}^m$. Hence, in this section we will give examples of optimization problems in function spaces, as well as in Euclidean m-space.

1.1 Optimal Production Planning. Suppose a manufacturer produces m products $A_1, A_2, \dots, A_m$ using n different raw materials $B_1, B_2, \dots, B_n$. In addition, suppose that the product A_μ contains $a_{\nu\mu}$ units of the raw material B_ν, and that the net income from selling one unit of A_μ is c_μ monetary units. Suppose the amount of raw material B_ν available is given by b_ν units. For simplicity, we assume that the market can absorb an infinite amount of each product A_μ, and that the amount offered for sale has no effect on the price. Now the problem is as follows: Find the amount x_μ of the product A_μ which should be produced in order to maximize profit. This problem can be formulated mathematically as one of determining a maximum of the *objective function*

$$f(x_1, x_2, \ldots, x_m) := \sum_{\mu=1}^{m} c_\mu x_\mu,$$

subject to the *side conditions*

$$\sum_{\mu=1}^{m} a_{\nu\mu} x_\mu \le b_\nu, \ 1 \le \nu \le n,$$

and taking account of the *sign conditions*

$$x_\mu \ge 0, \quad 1 \le \mu \le m.$$

In terms of the matrices and vectors $A := (a_{\mu\nu})$, $c^T := (c_1, \ldots, c_m)$, $b^T = (b_1, \ldots, b_n)$ and $x^T := (x_1, \ldots, x_m)$, we have the

Optimization Problem: Maximize the objective function $f(x) := c^T x$, subject to the side conditions

$$(*) \qquad \begin{aligned} Ax &\le b, \\ x &\ge 0. \end{aligned}$$

Here the symbols "$\le$" and "$\ge$" operating on vectors are to be interpreted componentwise. In general, the number of inequality side conditions is much larger than the number of variables $x_1, x_2, \ldots, x_m$, and so here we assume $m < n$.

The optimization problem $(*)$ can be given a simple geometric interpretation. Consider the following

Example. Let $m = 2$ and $n = 6$. In addition, let A, b, and c be given by

$$A := \begin{pmatrix} -6 & 5 \\ -7 & 12 \\ 0 & 1 \\ 19 & 14 \\ 1 & 0 \\ 4 & -7 \end{pmatrix}, \quad b := \begin{pmatrix} 30 \\ 84 \\ 9 \\ 266 \\ 10 \\ 28 \end{pmatrix}, \quad c := \begin{pmatrix} 3 \\ 1 \end{pmatrix}.$$

Then the inequalities $Ax \le b$ and $x \ge 0$ describe a *polyhedron* in $\mathbb{R}^2$. We are interested in the objective function

$$s := f(x_1, x_2) = 3x_1 + x_2.$$

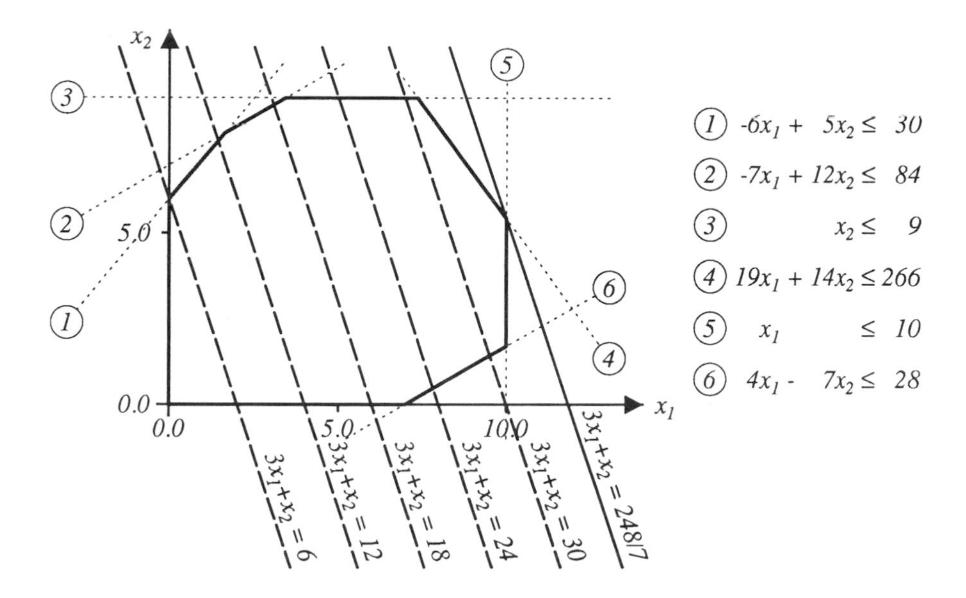

As we vary the parameter s, this defines a family of straight lines in $\mathbb{R}^2$. The straight line with the largest value of s which intersects the polyhedron corresponds to the solution of the optimization problem. In the example at hand, the solution (x_1, x_2) is uniquely defined. It lies at the *vertex* (10, 5.4) of the polyhedron. The corresponding value of the objective function is $s = f(10, 5.4) = 35.4$. The fact that the maximal point lies at a vertex of the polyhedron is of essential importance, and we will discuss it further later.

1.2 A Semi-Infinite Optimization Problem. The problem of approximating a function with respect to a given norm (cf. Chap. 4, Sect. 4) is a special case of an optimization problem. For example, suppose $d \in C[a, b]$, and that P_{m-1} is the space of polynomials of maximal degree $m-1$. Then the problem of finding the best approximation of d with respect to the Chebyshev norm $\|\cdot\|_\infty$ by polynomials in P_{m-1} can be reformulated as follows:

Find a vector $(a_0, a_1, \ldots, a_{m-1}, a_m)^T \in \mathbb{R}^{m+1}$ such that the component a_m is minimal, subject to the side conditions

$$\sum_{\mu=0}^{m-1} a_\mu t^\mu - a_m \le d(t),$$

$$-\sum_{\mu=0}^{m-1} a_\mu t^\mu - a_m \le -d(t), \quad t \in [a, b].$$

Now setting $c := (0, \ldots, 0, 1)^T \in \mathbb{R}^{m+1}$, $d_1(t) := d(t)$, $d_2(t) = -d(t)$, and $x_\mu := a_\mu$, $0 \le \mu \le m$, then we obtain the following analog of the linear optimization problem $(*)$ in 1.1:

Minimize $f(x) = c^T x$, $x = (x_0, x_1, \ldots, x_m)^T \in \mathbb{R}^{m+1}$, under the side conditions

$$\begin{aligned} x_0 + x_1 t + \cdots + x_{m-1} t^{m-1} - x_m &\leq d_1(t), \\ -x_0 - x_1 t - \cdots - x_{m-1} t^{m-1} - x_m &\leq d_2(t) \end{aligned}$$

for all $t \in [a, b]$.

The only essential difference between this problem and (*) in 1.1 is the fact that here the number of side conditions is no longer finite since t runs over the interval $[a, b]$. Here we have a linear optimization problem in Euclidean space $\mathbb{R}^{m+1}$, but with an infinite number of side conditions. This case is called a *semi-infinite optimization problem.* Many problems in approximation theory can be expressed as semi-infinite optimization problems. For more details, see the book of R. Hettich and P. Zencke [1982]. In order to solve a semi-infinite optimization problem approximately, we may replace the interval $[a, b]$ by a discrete grid

$$a \leq t_1 < t_2 < \cdots < t_{n+1} \leq b,$$

and then minimize the objective function $f(x) = c^T x$ subject to the finitely many side conditions

$$\begin{aligned} x_0 + x_1 t_\nu + \cdots + x_{m-1} t_\nu^{m-1} - x_m &\leq d_1(t_\nu), \\ -x_0 - x_1 t_\nu - \cdots - x_{m-1} t_\nu^{m-1} - x_m &\leq d_2(t_\nu), \end{aligned}$$

for $1 \leq \nu \leq n + 1$. Except for the missing conditions on the signs of the x_μ, this is a linear optimization problem of the type (*) discussed in 1.1. The exchange method of Remez which was discussed in 4.4.6 can be used to compute especially convenient grid points t_ν. We refer to the literature for details.

1.3 A Linear Control Problem. Suppose a metal block in the form of a cube is to be heated in an oven so that in a prescribed amount of time, a given temperature profile is to be optimally approximated. This is a typical problem in control theory. We now show how it can be easily formulated as an optimization problem in a function space.

Suppose $T > 0$ is the final time, and that the desired temperature distribution is given by a continuous function z on the domain $\Omega \subset \mathbb{R}^3$, whose boundary $\partial\Omega$ is assumed to be piecewise smooth. We are looking for a measurable and almost everywhere bounded control function u defined on $[0, T]$ so that the associated solution y of the heat equation

$$\frac{\partial}{\partial t} y(x_1, x_2, x_3, t) - \sum_{\nu=1}^{3} \frac{\partial^2}{\partial x_\nu} y(x_1, x_2, x_3, t) = 0$$

for $(x_1, x_2, x_3) \in \Omega$ and $0 < t < T$, subject to the boundary condition

$$\alpha \frac{\partial}{\partial \vec{n}} y(x_1, x_2, x_3, t) + y(x_1, x_2, x_3, t) = u(t), \; (x_1, x_2, x_3) \in \partial\Omega, \; 0 < t < T,$$

and the initial condition

$$y(x_1, x_2, x_3, 0) = 0, \qquad (x_1, x_2, x_3) \in \Omega$$

provides a minimum of the quantity

$$a := \max_{(x_1, x_2, x_3) \in \overline{\Omega}} |z(x_1, x_2, x_3) - y(x_1, x_2, x_3, T)|.$$

Here $\alpha > 0$ is a constant, and $\vec{n}$ is the outwards-pointing normal vector for the domain Ω. For technical reasons, we assume that $0 \le u(t) \le 1$ for almost all $t \in [0, T]$.

It can be shown that this problem has a solution, and that in the case $a > 0$, the optimal control u is characterized by the property that for every $\varepsilon > 0$ in the time interval $[0, T - \varepsilon]$, $|u(t)| = 1$ with at most a finite number of jumps. Such a control is called a *bang-bang control.* This property of the solution of the control problem suggests that to construct an approximate solution, we can divide the interval $[0, T]$ into subintervals of equal length, and then look for an optimal control u which is constant on each subinterval, and is less than or equal to one in absolute value. This is a linear approximation problem of the type discussed in Chap. 4, Sect. 4, and can be rewritten as a semi-infinite optimization problem in the same way as was done in 1.2. After an appropriate discretization of the heat equation, we are then led to a linear optimization problem in a finite dimensional Euclidean space.

1.4 The General Problem. Having given several introductory examples, we now rewrite the general problem of linear optimization introduced in 1.1 in

Standard Form. Minimize

$$f(x) = c^T x,$$

subject to the equality side conditions

$$Ax = b, \quad x \ge 0.$$

Here $c \in \mathbb{R}^m$, $A \in \mathbb{R}^{(n,m)}$, and $b \in \mathbb{R}^n$ are given, and we can assume now that $m > n$. If $m \le n$, then, in general, the system of equations $Ax = b$ has either exactly one solution, or none at all. Neither case is of interest for our optimization problem. A wide variety of optimization problems, formulated in various ways, can be rewritten in standard form. In the following remarks we show how this can be done for some typical problems.

Remarks. (i) Side conditions of the form

$$\sum_{\mu=1}^{m} a_{\nu\mu} x_\mu \leq b_\nu, \qquad 1 \leq \nu \leq n,$$

can be transformed to conditions of the type

$$\sum_{\mu=1}^{m} a_{\nu\mu} x_\mu + x_{m+\nu} = b_\nu, \qquad x_\mu \geq 0,\ 1 \leq \nu \leq n,$$

by introducing auxiliary variables $x_{m+1}, x_{m+2}, \ldots, x_{m+n}$, also called *slack variables*. The inequality $Ax \leq b$ must then be replaced by the equation $\tilde{A}\tilde{x} = b$ with

$$\tilde{A} := \begin{pmatrix} a_{11} & \cdots & a_{1m} & 1 & 0 & \cdots & 0 \\ a_{21} & \cdots & a_{2m} & 0 & 1 & & \vdots \\ \vdots & & \vdots & \vdots & & \ddots & 0 \\ a_{n1} & \cdots & a_{nm} & 0 & 0 & \cdots & 1 \end{pmatrix}.$$

(ii) Side conditions in the form $\sum_{\mu=1}^{m} a_{\nu\mu} x_\mu \geq b_\nu$ can be written in the standard form $\sum_{\mu=1}^{m} (-a_{\nu\mu}) x_\mu \leq (-b_\nu)$.

(iii) If some variable x_μ in the problem is not subject to a sign condition, then we can replace it by $x_{1\mu} - x_{2\mu}$, and require that $x_{1\mu} \geq 0$ and $x_{2\mu} \geq 0$.

(iv) A maximization problem with the objective function g is equivalent to a minimization problem with the objective function $f := -g$.

In the sequel, we denote the set of all vectors which satisfy the side conditions in the standard optimization problem by

$$\mathrm{M} := \{x \in \mathbb{R}^m \mid Ax = b,\ x \geq 0\}.$$

We refer to such vectors as *admissible vectors* for the standard optimization problem. This set defines a *polyhedron* in $\mathbb{R}^m$. We establish some properties of this set in the following section.

1.5 Problems. 1) Suppose a market offers only two types of vegetables P_1 and P_2. What purchases should a mathematically competent customer make in order to prepare a meal with at least 50 kilo-calories and at least 1200 units of vitamins at the least possible cost? The following table gives the calories, vitamin content, and price (per lb) of each of the vegetables:

Type	P_1	P_2
Kilo-calories	200	100
Vitamins	2000	3000
Price in $	.50	.60

Solve this problem by constructing a mathematical model, and determine the solution geometrically.

2) Consider the following optimization problem: Minimize the objective function $f(x_1, x_2, x_3) = x_1 + 4x_2 + x_3$ subject to the side conditions that $2x_1 - 2x_2 + x_3 = 4$, $x_1 - x_3 = 1$ and the sign conditions $x_2 \geq 0$, $x_3 \geq 0$.

a) Write this problem in standard form.

b) Does the problem have a solution?

c) Find the solution, if it exists.

3) Solve the following semi-infinite optimization problem: Minimize the objective function $f(x_1, x_2) = x_1 + 2x_2$ subject to the side conditions that $x_1 t + x_2 t^2 \geq -1 + 2t$, $0 \leq t \leq 1$, $x_1 \geq 0$, $x_2 \geq 0$.

4) Graphically solve the following optimization problem: Maximize the sum $(x_1 + x_2)$ subject to the side conditions $x_1 + x_2 \leq p$, $x_1 + 3x_2 \leq 4$, $p \in [0, \infty)$, $x_1 \geq 0$, $x_2 \geq 0$.

2. Polyhedra

In Example 1.1 we have seen that the solution of the linear optimization problem discussed there corresponds to a vertex of the polyhedron describing the constraints. In this section we show that this property holds in general. The results obtained here will be used in the following section to develop the simplex method.

We begin by showing that if M is bounded and nonempty, then the linear optimization problem always has at least one solution. These assumptions on M along with the fact that it is closed imply that it is compact. It follows that the continuous function f must take its extreme values on M, and so there exists a solution. In fact, existence can also be shown under somewhat weaker hypotheses (cf. Problem 1).

We now develop some properties of polyhedra.

2.1 Characterization of Vertices. If the set M of admissible vectors corresponding to a standard optimization problem contains two given vectors x and y, then it also contains all vectors of the form $\lambda x + (1-\lambda)y$, $0 \leq \lambda \leq 1$. Thus M is convex (cf. 4.3.3).

Definition. An element x of the convex set M is called an *extreme point of* M, if the condition $x = \lambda y + (1 - \lambda)z$ for $y, z \in \mathrm{M}$ and $0 < \lambda < 1$ implies $x = y = z$. We denote the set of extreme points of M by $\mathrm{E_M}$.

Examples . (i) Suppose $\mathrm{M} := \{x \in \mathbb{R}^n \mid \|x\|_1 \leq 1\}$. Then the set of extreme points is $\mathrm{E_M} := \{(1, 0, \ldots, 0)^T, (0, 1, 0, \ldots, 0)^T, \ldots, (0, \ldots, 0, 1)^T\}$.

(ii) In Example 1.1 the set of extreme points of M consists of the vectors

$$\begin{pmatrix} 0 \\ 0 \end{pmatrix}, \begin{pmatrix} \frac{60}{37} \\ \frac{294}{37} \end{pmatrix}, \begin{pmatrix} \frac{24}{7} \\ 9 \end{pmatrix}, \begin{pmatrix} \frac{140}{19} \\ 9 \end{pmatrix}, \begin{pmatrix} 10 \\ \frac{38}{7} \end{pmatrix}, \begin{pmatrix} 10 \\ \frac{12}{7} \end{pmatrix}, \begin{pmatrix} 7 \\ 0 \end{pmatrix}, \begin{pmatrix} 0 \\ 6 \end{pmatrix}.$$

These vectors represent points in the polyhedron M which are actually vertices. Thus when dealing with polyhedra associated with a standard optimization problem, we will also call such points *vertices of the polyhedron* M.

Let $\mathrm{M} = \{x \in \mathbb{R}^m \mid Ax = b,\ x_\mu \geq 0,\ 1 \leq \mu \leq m\}$. In order to characterize the vertices of M, define $\mathrm{I}(x) := \{\mu \in \{1,2,\ldots,m\} \mid x_\mu > 0\}$.

Characterization Theorem. *Suppose $A \in \mathbb{R}^{(n,m)}$, $A = (a^1,\ldots,a^m)$ with $a^\mu \in \mathbb{R}^n$ for $1 \leq \mu \leq m$. Then the following two assertions are equivalent:*

(i) x is a vertex of M;

(ii) The vectors a^μ, $\mu \in \mathrm{I}(x)$, are linearly independent.

Proof. Suppose $x \in \mathrm{M}$ is a vertex, and that the components of x are numbered so that $\mathrm{I}(x) = \{1,2,\ldots,r\}$. We may assume that $r \geq 1$, since otherwise the assertion (ii) is trivial. Since $x \in \mathrm{M}$, we have $\sum_{\mu=1}^r x_\mu a^\mu = b$. Now if the vectors $a^1, a^2, \ldots, a^r$ are linearly dependent, then there exists a nontrivial combination with $\sum_{\mu=1}^r \alpha_\mu a^\mu = 0$, $(\alpha_1,\alpha_2,\ldots,\alpha_r) \neq 0$. Since the components of x_μ are positive for $i \in \mathrm{I}(x)$, we can find a sufficiently small number $\varepsilon > 0$ such that $x_\mu \pm \varepsilon\alpha_\mu > 0$. Now set

$$y_+ := (x_1 + \varepsilon\alpha_1, \ldots, x_r + \varepsilon\alpha_r, 0, \ldots, 0)^T \in \mathbb{R}^m,$$
$$y_- := (x_1 - \varepsilon\alpha_1, \ldots, x_r - \varepsilon\alpha_r, 0, \ldots, 0)^T \in \mathbb{R}^m.$$

Then $y_+ \geq 0$, $y_- \geq 0$, and moreover,

$$\sum_{\mu=1}^m (y_\pm)_\mu a^\mu = \sum_{\mu=1}^r (y_\pm)_\mu a^\mu = \sum_{\mu=1}^r x_\mu a^\mu \pm \varepsilon \sum_{\mu=1}^r \alpha_\mu a^\mu = \sum_{\mu=1}^r x_\mu a^\mu = b.$$

It follows that y_+ and y_- are elements of M, and the formula

$$\frac{1}{2}y_+ + \frac{1}{2}y_- = (x_1, x_2, \ldots, x_r, 0, \ldots, 0)^T = x$$

expresses x as a nontrival combination of two other elements y_+ and y_- in M. This means that x cannot be a vertex of M.

For the converse, suppose the vectors a^μ, $\mu \in \mathrm{I}(x) = \{1,2,\ldots,r\}$ are linearly independent and that $x \in \mathrm{M}$. We now consider writing x in the form $x = \lambda y + (1-\lambda)z$, $0 < \lambda < 1$ for some $y, z \in \mathrm{M}$. Clearly, $\mathrm{I}(x) = \mathrm{I}(y) \cup \mathrm{I}(z)$, and since $Ay = Az = b$, it follows that $0 = \sum_{\mu=1}^m (y_\mu - z_\mu)a^\mu = \sum_{\mu=1}^r (y_\mu - z_\mu)a^\mu$. Because of the linear independence of $a^1, a^2, \ldots, a^r$, we get $y_\mu = z_\mu$ for $1 \leq \mu \leq m$, and thus x is a vertex of M. □

Corollary. Since rank $(A) \leq n$ for every vertex $x \in \mathrm{M}$, we have $|\mathrm{I}(x)| \leq n$. Here $|\mathrm{I}(x)|$ denotes the number of elements in the set $\mathrm{I}(x)$. Moreover, since there are only $\binom{m}{n}$ possible ways to choose n indices from a set of m indices, it follows that M has at most a finite number of vertices.

2.2 Existence of Vertices. With the help of the Characterization Theorem in 2.1, we now show that a polyhedron M must have vertices.

Existence Theorem. *Every nonempty polyhedron* $\mathrm{M} \subset \mathbb{R}^m$ *possesses vertices.*

Proof. Since the set $\mathrm{I} := \{|\mathrm{I}(z)| \mid z \in \mathrm{M}\} \subset \{1, 2, \ldots, m\}$ is discrete and finite, there exists a number $\underline{\gamma} \geq 0$ with $\underline{\gamma} = \min\{\gamma \mid \gamma \in \mathrm{I}\}$ and hence also an element $x \in \mathrm{M}$ with $|\mathrm{I}(x)| = \underline{\gamma}$. We will show that x is a vertex of M.

If $\underline{\gamma} = 0$, then x is obviously a vertex, since then the set of column vectors of the matrix A corresponding to positive components of x is empty. By definition, an empty set of vectors is automatically linearly independent. It remains to consider the case $\underline{\gamma} > 0$. We can restrict our attention to sets $\mathrm{I}(x)$ of the form $\mathrm{I}(x) = \{1, 2, \ldots, \underline{\gamma}\}$. The proof now proceeds by contradiction. Suppose that the vectors $a^1, a^2, \ldots, a^{\underline{\gamma}} \in \mathbb{R}^n$ are linearly dependent. Then there exist numbers $\alpha_\mu \in \mathbb{R}$, $1 \leq \mu \leq \underline{\gamma}$, with $(\alpha_1, \ldots, \alpha_{\underline{\gamma}}) \neq 0$ such that $\sum_{\mu=1}^{\underline{\gamma}} \alpha_\mu a^\mu = 0$. Set

$$\lambda := \min\{\frac{x_\mu}{|\alpha_\mu|} \mid \alpha_\mu \neq 0,\ 1 \leq \mu \leq \underline{\gamma}\},$$

and consider the index $\tilde{\mu}$ for which the minimum occurs; i.e., $\lambda = x_{\tilde{\mu}}/|\alpha_{\tilde{\mu}}|$. Then the vector

$$\tilde{x} := (x_1 - \lambda\alpha_1, x_2 - \lambda\alpha_2, \ldots, x_{\underline{\gamma}} - \lambda\alpha_{\underline{\gamma}}, 0, \ldots, 0)^T \in \mathbb{R}^m$$

lies in M since

$$A\tilde{x} = Ax - \lambda \sum_{\mu=1}^{\underline{\gamma}} \alpha_\mu a^\mu = Ax = b.$$

Moreover, by our construction of λ, $\tilde{x} \geq 0$, and

$$|\mathrm{I}(\tilde{x})| \leq |\mathrm{I}(x) \setminus \{\tilde{\mu}\}| = \underline{\gamma} - 1.$$

This contradicts the minimal property of $\underline{\gamma}$. □

The vertices are the most important points in a polyhedron. If we know them, we can completely describe the polyhedron. In this connection we have the

Representation Theorem. *Every point* x *in a nonempty bounded polyhedron* $\mathrm{M} \subset \mathbb{R}^m$ *can be written as a convex combination of the vertices of* M*; i.e., for every point* $x \in \mathrm{M}$, *there exist vertices* $z^1, z^2, \ldots, z^\ell \in \mathrm{E_M}$ *and real numbers* $0 \leq \lambda_\mu \leq 1$, $1 \leq \mu \leq \ell$, *with* $\sum_{\mu=1}^{\ell} \lambda_\mu = 1$ *such that*

$$x = \sum_{\mu=1}^{\ell} \lambda_\mu z^\mu.$$

Proof. Let $x \in \mathrm{M}$ and $r := |\mathrm{I}(x)|$. By the definition of $\mathrm{I}(x)$, it follows that $\sum_{\mu \in I(x)} x_\mu a^\mu = b$. We now proceed by induction on r. For $r = 0$, x is obviously a vertex of M, and the assertion holds trivially. Now suppose $r > 0$. If the vectors a^μ, $\mu \in \mathrm{I}(x)$ are linearly independent, then x is again a vertex. We may thus assume that there exists a nontrivial linear combination of the form

$$\sum_{\mu \in \mathrm{I}(x)} \alpha_\mu a^\mu = 0, \qquad \sum_{\mu \in \mathrm{I}(x)} \alpha_\mu^2 > 0.$$

For each ε, let $x(\varepsilon)$ be the vector in $\mathbb{R}^m$ with components

$$x_\mu(\varepsilon) := \begin{cases} x_\mu + \varepsilon \alpha_\mu & \text{for } \mu \in \mathrm{I}(x) \\ 0 & \text{for } \mu \notin \mathrm{I}(x). \end{cases}$$

Since M is closed, bounded and convex, there exist two numbers $\varepsilon_1 < 0$ and $\varepsilon_2 > 0$ such that for all $\varepsilon_1 \leq \varepsilon \leq \varepsilon_2$, the vectors $x(\varepsilon)$ lie in M, while $x(\varepsilon) \notin \mathrm{M}$ for $\varepsilon < \varepsilon_1$ and for $\varepsilon > \varepsilon_2$. In addition, for all $\mu \notin \mathrm{I}(x)$ we have $x_\mu(\varepsilon_1) = x_\mu(\varepsilon_2) = 0$. Now corresponding to ε_1 and ε_2, there must be indices $\tilde{\mu}, \tilde{\tilde{\mu}} \in \mathrm{I}(x)$ with $x_{\tilde{\mu}}(\varepsilon_1) = x_{\tilde{\tilde{\mu}}}(\varepsilon_2) = 0$. But then $|\mathrm{I}(x(\varepsilon_1))| < r$ and $\mathrm{I}(x(\varepsilon_2))| < r$. By the induction hypothesis, we can represent $x(\varepsilon_1)$ and $x(\varepsilon_2)$ as convex combinations of the vertices, and every component x_μ of the vector x must satisfy

$$x_\mu = \frac{\varepsilon_2}{\varepsilon_2 - \varepsilon_1}(x_\mu + \varepsilon_1 \alpha_\mu) + \left(1 - \frac{\varepsilon_2}{\varepsilon_2 - \varepsilon_1}\right)(x_\mu + \varepsilon_2 \alpha_\mu).$$

Thus, x can also be represented as a convex combination of the vertices in $\mathrm{E_M}$. □

We are now ready for the most important result of this section.

2.3 The Main Result. In Example 1.1 we saw that for the linear optimization problem (*) in 1.1, the maximal value of the objective function was assumed at a vertex of the polyhedron of admissible vectors. We now prove that this holds in general.

Theorem. *Suppose the set* $\mathrm{M} := \{x \in \mathbb{R}^m \mid Ax = b,\ x \geq 0\}$ *of admissible vectors corresponding to the general optimization problem 1.4 in standard form is nonempty and bounded. Then the objective function* $f(x) = c^T x$ *takes its minimum at a vertex of* M.

Proof. Since M is closed and bounded, and hence compact, f must take a minimum for some point $\tilde{x} \in \mathrm{M}$. By the Representation Theorem 2.2, $\tilde{x}$ can be written as a convex combination of the vertices; i.e., $\tilde{x} = \sum_{\mu=1}^{\ell} \lambda_\mu \tilde{x}^\mu$ with $\tilde{x}^\mu \in \mathrm{E_M}$ and $\sum_{\mu=1}^{\ell} \lambda_\mu = 1$, $\lambda_\mu \in [0,1]$, for $1 \leq \mu \leq \ell$. In addition, since

$$\min\{c^T x \mid x \in \mathrm{M}\} = c^T \tilde{x} = \sum_{\mu=1}^{\ell} \lambda_\mu c^T \tilde{x}^\mu \geq \min\{c^T \tilde{x}^\mu \mid 1 \leq \mu \leq \ell\},$$

there must be a vertex $\tilde{x}^{\tilde{\mu}}$ with $\tilde{\mu} \in \{1, 2, \ldots, \ell\}$ such that $c^T \tilde{x} = c^T \tilde{x}^{\tilde{\mu}}$. □

In principle, this theorem already shows how to solve the Optimization Problem 1.4. We need to look for the minimum of the objective function f among the set of vertices E_M. In addition, the Characterization Theorem 2.1 says that all vertices of the polyhedron M can be found by identifying all sets of $k \le n$ linearly independent vectors a^{μ_1}, $a^{\mu_2}, \ldots, a^{\mu_k}$ in the set of column vectors of the matrix A. This approach is not practical in general, however, since the number of vertices can be larger than $\binom{m}{n}$ (cf. Corollary 2.1), and it is well known that these binominal numbers grow very quickly with increasing n. For example, $\binom{30}{10} = 30,045,015$. We will see later how the search of the vertices of a polyhedron for a minimum of the objective function f can be accomplished more quickly. For this, we need an algebraic characterization of the vertices of M.

2.4 An Algebraic Characterization of Vertices. The Characterization Theorem 2.1 implies that if rank $A = n$, then every vertex of M has at most n positive components. It can also have fewer than n, of course. This observation leads us to the following

Definition. Suppose $A \in \mathbb{R}^{(n,m)}$ with rank $A = n$, and suppose that $B = (a^{\mu_1}, a^{\mu_2}, \ldots, a^{\mu_n})$ is a submatrix of A with rank $B = n$. We call a vector $x \in \mathbb{R}^m$ a *basis point of B* provided $x_\mu = 0$ for $\mu \notin \{\mu_1, \mu_2, \ldots, \mu_n\}$ and $\sum_{\nu=1}^{n} x_\nu a^{\mu_\nu} = b$. If $x \in \mathbb{R}^m$ is a basis point of B, then we call its components x_{μ_1}, $x_{\mu_2}, \ldots, x_{\mu_n}$ *basis variables*. In general, we call $x \in \mathbb{R}^m$ a *basis point* whenever there exists a submatrix B of A such that x is a basis point of B.

Obviously, since rank $(A) = n$, there always exist basis points. The polyhedron M can be completely characterized in terms of basis points, as we show in the following

Equivalence Theorem. *Suppose* $M = \{x \in \mathbb{R}^m \mid Ax = b, \, x \ge 0\}$ *is a polyhedron, where* $A \in \mathbb{R}^{(n,m)}$ *with rank* $(A) = n$. *Then the vector* $x \in \mathbb{R}^m$ *is a vertex of* M *if and only if it is a basis point.*

Proof. Suppose x is a vertex of M with positive components $x_{\mu_1} > 0$, $x_{\mu_2} > 0, \ldots, x_{\mu_p} > 0$ and $p \le m$. Then $\sum_{\nu=1}^{p} x_{\mu_\nu} a^{\mu_\nu} = b$. We now show that $p \le n$. Since rank $(A) = n$, it suffices to verify that the vectors a^{μ_1}, $a^{\mu_2}, \ldots, a^{\mu_p}$ are linearly independent. Suppose they are not. Then there would exist a linear combination of the form $\sum_{\nu=1}^{p} \alpha_{\mu_\nu} a^{\mu_\nu} = 0$ with $\sum_{\nu=1}^{p} \alpha_{\mu_\nu}^2 \ne 0$. Then for appropriate $\varepsilon > 0$, the vectors $y_\pm \in \mathbb{R}^m$ with components

$$(y_\pm)_\kappa := \begin{cases} x_{\mu_\nu} \pm \varepsilon \alpha_{\mu_\nu} & \text{for } \kappa = \mu_\nu, \ 1 \le \nu \le p, \\ 0 & \text{for } \kappa \ne \mu_\nu, \ 1 \le \nu \le p \end{cases}$$

would be elements of M, so that $x = \frac{1}{2} y_+ + \frac{1}{2} y_-$, contradicting the assumption that x is a vertex. If $p < n$, then we can extend $a^{\mu_1}, \ldots, a^{\mu_p}$ by $(n - p)$

additional column vectors $a^{\mu_{p+1}}, \dots, a^{\mu_n}$ of A to a system of μ_n linearly independent vectors and take $B = (a^{\mu_1}, \dots, a^{\mu_n})$.

Conversely, if $x \in \mathbb{R}^m$ is a basis point, then its components must be

$$x_\kappa = \begin{cases} x_{\mu_\nu} & \text{for } \kappa = \mu_\nu,\ 1 \le \nu \le n \\ 0 & \text{for } \kappa \ne \mu_\nu,\ 1 \le \nu \le n, \end{cases}$$

and we can write $\sum_{\nu=1}^n x_{\mu_\nu} a^{\mu_\nu} = b$ in terms of the linearly independent vectors $a^{\mu_1}, a^{\mu_2}, \dots, a^{\mu_n}$. Then x can be represented as $x = \lambda y + (1-\lambda) z$ with $y, z \in \mathrm{M}$ and $0 < \lambda < 1$. Since $x \ge 0$, $y \ge 0$ and $z \ge 0$, we have $y_\kappa = z_\kappa = 0$ for $\kappa \ne \mu_\nu$, $1 \le \mu \le n$. It follows that $\sum_{\nu=1}^n y_{\mu_\nu} a^{\mu_\nu} = b$ and $\sum_{\nu=1}^n z_{\mu_\nu} a^{\mu_\nu} = b$, and further that $\sum_{\nu=1}^n (y_{\mu_\nu} - z_{\mu_\nu}) a^{\mu_\nu} = 0$. The linear independence of the vectors $a^{\mu_1}, a^{\mu_2}, \dots, a^{\mu_n}$ implies $y = z$. Thus x is a vertex of M. □

We have already shown that a vertex $x \in \mathrm{E_M}$ can have less than n positive basis variables. Such a vertex is called *degenerate*. In looking for a minimum of the objective function f, we have to handle degenerate vertices in a special way. We return to this point later. We now have all of the tools needed to present a method for solving linear optimization problems in $\mathbb{R}^m$.

2.5 Problems. 1) Show that assuming $\mathrm{M} \ne \emptyset$ and $\inf \{c^T x \mid x \in \mathrm{M}\} > -\infty$, the standard optimization problem always has a solution.

2) Suppose $A \in \mathbb{R}^{(n,m)}$ and $b \in \mathbb{R}^n$. Show that the two sets

$$\tilde{\mathrm{M}} := \{x \in \mathbb{R}^m \mid Ax \le b,\ x \ge 0\},$$

$$\mathrm{M} := \{\begin{pmatrix} x \\ y \end{pmatrix} \in \mathbb{R}^{m+n} \mid Ax + y = b,\ x \ge 0,\ y \ge 0\}$$

have the following properties:

a) If $\binom{x}{y}$ is an extreme point of M, then x is an extreme point of $\tilde{\mathrm{M}}$.

b) If x is an extreme point of $\tilde{\mathrm{M}}$, then $\binom{x}{y}$ with $y := b - Ax$ is an extreme point of M.

3) Show that the set of all solutions of a linear optimization problem is convex.

4) Consider the following optimization problem: Minimize $c^T x$ subject to the side conditions $Ax = b$, $0 \le x \le h$.

a) Write this problem in standard form.

b) Characterize the vertices of the set

$$\mathrm{M} := \{x \in \mathbb{R}^m \mid Ax = b,\ 0 \le x \le h\}.$$

c) Show that if $\mathrm{M} \ne \emptyset$, then the problem has a solution.

5) Consider the following optimization problem: Minimize $x_1 + x_2$ subject to the side conditions $x_1 + x_2 + x_3 = 1$, $2x_1 + 3x_2 = 1$, $x_1 \ge 0$, $x_2 \ge 0$, $x_3 \ge 0$.

a) Give an upper bound on the number of vertices of the following polyhedron: $M := \{(x_1, x_2, x_3)^T \in \mathbb{R}^3 \mid x_1 + x_2 + x_3 = 1,\ 2x_1 + 3x_2 = 1,\ x \geq 0\}$.

b) Find the vertices of M.

c) Find a solution of the optimization problem.

3. The Simplex Method

The most frequently used method for solving linear optimization problems is the simplex method. It was introduced in 1947/48 by George B. Danzig (cf. G. B. Dantzig [1963]). In his development, he divided the process into two steps:

Phase I: Determine a vertex of M,

Phase II: Move from one vertex to a neighboring one in such a way as to reduce the value of the objective function, and decide whether an additional vertex exchange would further reduce the value of the objective function, or whether the optimal solution of the optimization problem has already been found (stopping criterion).

We now discuss both steps in detail.

3.1 Introduction. Suppose the vertex $x \in M$ has basis variables $x_{\mu_1}, x_{\mu_2}, \ldots, x_{\mu_n}$. Then the vectors $a^{\mu_1}, a^{\mu_2}, \ldots, a^{\mu_n}$ form a basis for $\mathbb{R}^n$. Suppose that relative to this basis, the vectors $a^1, a^2, \ldots, a^n$ can be represented as

$$a^\kappa = \sum_{\nu=1}^{n} \alpha_{\nu\kappa} a^{\mu_\nu}, \quad 1 \leq \kappa \leq m,$$

$$b = \sum_{\nu=1}^{n} \alpha_{\nu 0} a^{\mu_\nu}.$$

Trivially, we have $\alpha_{\iota\mu_\nu} = \delta_{\iota\nu}$ for $1 \leq \iota \leq n$, $1 \leq \nu \leq n$. We now insert the coefficients of these expansions in a tableau as follows:

(*)

μ_1	$\alpha_{11} \cdots \alpha_{1m}$	α_{10}
μ_2	$\alpha_{21} \cdots \alpha_{2m}$	α_{20}
$\vdots$	$\vdots \qquad \vdots$	$\vdots$
μ_n	$\alpha_{n1} \cdots \alpha_{nm}$	α_{n0}

The columns of the tableau (*) are $d^\kappa := (\alpha_{1\kappa}, \ldots, \alpha_{n\kappa})^T$, $0 \le \kappa \le m$. Any arbitrary admissible vector $\tilde{x} \in \mathrm{M}$ can be written as

$$b = \sum_{\kappa=1}^{m} \tilde{x}_\kappa a^\kappa = \sum_{\kappa=1}^{m} \tilde{x}_\kappa \sum_{\nu=1}^{n} \alpha_{\nu\kappa} a^{\mu_\nu} = \sum_{\nu=1}^{n} \Big(\sum_{\kappa=1}^{m} \alpha_{\nu\kappa} \tilde{x}_\kappa\Big) a^{\mu_\nu}.$$

This can be rewritten in the equivalent form $\sum_{\kappa=1}^{m} \tilde{x}_\kappa d^\kappa = d^0$.

Furthermore, the vertex x satisfies the equation

$$b = \sum_{\nu=1}^{n} x_{\mu_\nu} a^{\mu_\nu}.$$

Since the vectors $a^{\mu_1}, \ldots, a^{\mu_n}$ are linearly independent, by comparing coefficients for $\nu = 1, 2, \ldots, n$, we obtain the following connection between the basis variables for the vertex x and the components of an arbitrary admissible vector $\tilde{x}$:

$$\tilde{x}_{\mu_\nu} = x_{\mu_\nu} - \sum_{\substack{\kappa=1 \\ \kappa \ne \mu_1, \ldots, \kappa \ne \mu_n}}^{m} \alpha_{\nu\kappa} \tilde{x}_\kappa.$$

Using this expansion, calculating the value of the objective function f at $\tilde{x} \in \mathrm{M}$ as a function of its value at the given vertex x corresponding to the tableau (*), we get

$$\begin{aligned} f(\tilde{x}) &= \sum_{\nu=1}^{n} c_{\mu_\nu} \tilde{x}_{\mu_\nu} + \sum_{\substack{\kappa=1 \\ \kappa \ne \mu_1, \ldots, \kappa \ne \mu_n}}^{m} c_\kappa \tilde{x}_\kappa = \\ &= \sum_{\nu=1}^{n} c_{\mu_\nu} x_{\mu_\nu} + \sum_{\substack{\kappa=1 \\ \kappa \ne \mu_1, \ldots, \kappa \ne \mu_n}}^{m} \Big(c_\kappa - \sum_{\nu=1}^{n} c_{\mu_\nu} \alpha_{\nu\kappa}\Big) \tilde{x}_\kappa = \\ &= f(x) + \sum_{\substack{\kappa=1 \\ \kappa \ne \mu_1, \ldots, \kappa \ne \mu_n}}^{m} (c_\kappa - z_\kappa) \tilde{x}_\kappa, \end{aligned}$$

where we have used the abbreviation $z_\kappa := \sum_{\nu=1}^{n} c_{\mu_\nu} \alpha_{\nu\kappa}$.

Two cases arise:

(i) $c_\kappa - z_\kappa \ge 0$ for all $\kappa \notin \mathrm{I}(x)$,
(ii) $c_\kappa - z_\kappa < 0$ for a $\kappa \notin \mathrm{I}(x)$.

In case (i), $x \in \mathrm{M}$ is a solution of the optimization problem since in view of $\tilde{x} \ge 0$, the value of the objective function cannot be further reduced.

On the other hand, if $c_{\kappa_0} - z_{\kappa_0} = \min_{\kappa \notin \mathrm{I}(x)} (c_\kappa - z_\kappa) < 0$, so that we have case (ii), then the variable with index κ_0 is a candidate for an exchange.

Since the quantities $c_\kappa - z_\kappa$ play a role in later vertex exchanges, we expand the tableau (*) by adding an $(n+1)$-st row containing the values $\alpha_{n+1\nu} := c_\nu - z_\nu$, $0 \le \nu \le m$. Here we are supposing that $c_0 := 0$. The tableau now has the following shape:

μ_1	α_{11}	$\cdots$	α_{1n}	α_{10}
μ_2	α_{21}	$\cdots$	α_{2n}	α_{20}
$\vdots$	$\vdots$		$\vdots$	$\vdots$
μ_m	α_{m1}	$\cdots$	α_{mn}	α_{m0}
	$\alpha_{m+1\,1}$	$\cdots$	$\alpha_{m+1\,n}$	$\alpha_{m+1\,0}$

Example. Consider Example 1.1, and introduce slack variables x_μ, $3 \leq \mu \leq 8$. Clearly, the rank of the matrix of the associated standard problem is 6. The point $x = (0, 0, 30, 84, 9, 266, 10, 28)$ is a vertex of M. Our aim is to minimize the objective function $f(x) := c^T x$ corresponding to $c := (-3, -1, 0, 0, 0, 0, 0, 0)^T$. This leads to the following tableau:

3	-6	5	1	0	0	0	0	0	30
4	-7	12	0	1	0	0	0	0	84
5	0	1	0	0	1	0	0	0	9
6	19	14	0	0	0	1	0	0	266
7	1	0	0	0	0	0	1	0	10
8	4	-7	0	0	0	0	0	1	28
	-3	-1	0	0	0	0	0	0	0

Here we have case (ii), and we may exchange the variable x_1.

3.2 The Vertex Exchange Without Degeneracy. In the discussion above, the point $\tilde{x} \in$ M was arbitrary. We now describe the passage from a vertex $x \in$ M with basis variables $x_{\mu_1}, x_{\mu_2}, \ldots, x_{\mu_n}$ to a point $\tilde{x} \in$ M such that one of the basis variables $x_{\mu_{\tilde{\nu}}}$ of x is replaced by a variable $x_{\tilde{\kappa}}$ with $\tilde{\kappa} \notin \mathrm{I}(x)$ so that $\tilde{x}$ is a vertex of M.

Suppose $\tilde{\kappa} \notin \mathrm{I}(x)$, and that $a^{\tilde{\kappa}} = \sum_{\nu=1}^{n} \alpha_{\nu\tilde{\kappa}} a^{\mu_\nu}$. In addition, we have $b = \sum_{\nu=1}^{n} x_{\mu_\nu} a^{\mu_\nu}$. These assumptions imply that for every $\varepsilon \geq 0$,

$$(*) \qquad \sum_{\nu=1}^{n} (x_{\mu_\nu} - \varepsilon \alpha_{\nu\tilde{\kappa}}) a^{\mu_\nu} + \varepsilon a^{\tilde{\kappa}} = b.$$

Using this, define a vector $z(\varepsilon) \in \mathbb{R}^m$ by

$$(**) \qquad z_\kappa(\varepsilon) := \begin{cases} x_{\mu_\nu} - \varepsilon \alpha_{\nu\tilde{\kappa}} & \text{for } \kappa = \mu_\nu,\ 1 \leq \nu \leq n, \\ \varepsilon & \text{for } \kappa = \tilde{\kappa}, \\ 0 & \text{otherwise}. \end{cases}$$

We now choose the number $\varepsilon > 0$ so that $z(\varepsilon)$ is a vertex of M. To achieve this, one component z_κ with $\kappa = \mu_\nu$, $1 \le \nu \le n$, must be zeroed out, while all other components remain larger or equal to zero.

We now assume that the vertex $x \in \mathrm{E_M}$ is nondegenerate. Then $x_{\mu_\nu} > 0$ for all $\nu = 1, 2, \ldots, n$, and thus $\mathrm{I}(x) = \{\mu_1, \mu_2, \ldots, \mu_n\}$. We have to check that the set

$$\{\nu \in \{1, 2, \ldots, n\} \mid \alpha_{\nu\tilde{\kappa}} > 0\}$$

is nonempty. If it were empty, then (*) and (**) would imply that the set M is unbounded, since $z(\varepsilon) \in \mathrm{M}$ for every $\varepsilon \ge 0$.

We now make the additional hypothesis that M is bounded. This allows us to choose

$$\tilde{\varepsilon} := \min_{1 \le \nu \le n} \Big\{ \frac{x_{\mu_\nu}}{\alpha_{\nu\tilde{\kappa}}} \;\Big|\; \alpha_{\nu\tilde{\kappa}} > 0 \Big\}.$$

Next we exchange the index $\mu_{\tilde{\nu}}$ such that $\tilde{\varepsilon} = x_{\mu_{\tilde{\nu}}} / \alpha_{\tilde{\nu}\tilde{\kappa}}$ with the index $\tilde{\kappa}$, and consider

$$\mathrm{I}(z(\tilde{\varepsilon})) = (\mathrm{I}(x) \setminus \{\mu_{\tilde{\nu}}\}) \cup \{\tilde{\kappa}\}.$$

The element $z(\tilde{\varepsilon})$ lies in M. We now show that it is actually a vertex of M.

Suppose the vectors a^μ, $\mu \in \mathrm{I}(z(\tilde{\varepsilon}))$ are linearly dependent. Then some linear combination satisfies

$$\sum_{\mu \in \mathrm{I}(x) \setminus \{\mu_{\tilde{\nu}}\}} \lambda_\mu a^\mu + \lambda_{\tilde{\kappa}} a^{\tilde{\kappa}} = 0,$$

where we can assume that $\lambda_{\tilde{\kappa}} \neq 0$. By the normalization $\lambda_{\tilde{\kappa}} = 1$, we get

$$\begin{aligned} 0 = a^{\tilde{\kappa}} + \sum_{\mu \in \mathrm{I}(x) \setminus \{\mu_{\tilde{\nu}}\}} \lambda_\mu a^\mu &= \sum_{\nu=1}^{n} \alpha_{\nu\tilde{\kappa}} a^{\mu_\nu} + \sum_{\mu \in \mathrm{I}(x) \setminus \{\mu_{\tilde{\nu}}\}} \lambda_\mu a^\mu = \\ &= \alpha_{\tilde{\nu}\tilde{\kappa}} a^{\mu_{\tilde{\nu}}} + \sum_{\mu \in \mathrm{I}(x) \setminus \{\mu_{\tilde{\nu}}\}} (\lambda_\mu + a_{\mu\tilde{\kappa}}) a^\mu. \end{aligned}$$

Since the vectors a^μ, $\mu \in \mathrm{I}(x)$, are linearly independent, we have $a_{\tilde{\nu}\tilde{\kappa}} = 0$. This is a contradiction to the construction of $\tilde{\varepsilon}$, and we conclude that the elements a^μ, $\mu \in \mathrm{I}(z(\tilde{\varepsilon}))$, are linearly independent. By the Characterization Theorem 2.1, this implies that $z(\tilde{\varepsilon})$ is a vertex of M.

This completes the description of how to pass from one vertex x to another vertex $\tilde{x} := z(\tilde{\varepsilon})$ in the nondegenerate case by exchanging basis variables. The new vertex $\tilde{x}$ is degenerate if and only if there is more than one choice for the index $\tilde{\nu}$.

We now expand the tableau with an additional column containing the elements $x_{\mu_\nu} \cdot \alpha_{\nu\tilde{\kappa}}^{-1} = \alpha_{\nu 0} \cdot \alpha_{\nu\tilde{\kappa}}^{-1}$, where we write "$\infty$" in the corresponding entry whenever $\alpha_{\nu\tilde{\kappa}} = 0$.

μ_1	α_{11}	$\cdots$	α_{1n}	α_{10}	$\alpha_{10} \cdot \alpha_{1\tilde{\kappa}}^{-1}$
μ_2	α_{21}	$\cdots$	α_{1n}	α_{10}	$\alpha_{20} \cdot \alpha_{2\tilde{\kappa}}^{-1}$
$\vdots$	$\vdots$		$\vdots$	$\vdots$	$\vdots$
μ_m	α_{m1}	$\cdots$	α_{1n}	α_{10}	$\alpha_{m0} \cdot \alpha_{m\tilde{\kappa}}^{-1}$
	$\alpha_{m+1\,1}$	$\cdots$	$\alpha_{m+1\,n}$	$\alpha_{m+1\,0}$	

In Example 3.1, the additional column in the tableau ($\tilde{\kappa} = 1$) is

$$\alpha_{10} \cdot \alpha_{11}^{-1} = -5, \ \alpha_{20} \cdot \alpha_{21}^{-1} = -12, \ \alpha_{30} \cdot \alpha_{31}^{-1} = \infty, \ \alpha_{40} \cdot \alpha_{41}^{-1} = 14,$$
$$\alpha_{50} \cdot \alpha_{51}^{-1} = 10, \ \alpha_{60} \cdot \alpha_{61}^{-1} = 7.$$

This tells us that the basis variable x_8 ($\tilde{\nu} = 6$) should be exchanged with the variable x_1. The new vertex $\tilde{x} = z(\tilde{\varepsilon}) = (7,0,72,133,9,133,3,0)^T$ is nondegenerate. The new value for the objective function is

$$f(\tilde{x}) = f(x) + \sum_{\kappa \notin I(x)} (c_\kappa - z_\kappa)\tilde{x}_\kappa = 0 + 7(-3-0) = -21.$$

We now describe in general the computations which are required for a vertex exchange. It will suffice to explain how to go from a tableau with the basis variables $x_{\mu_1}, x_{\mu_2}, \ldots, x_{\mu_n}$ to another tableau with the basis variables $\tilde{x}_{\kappa_1}, \tilde{x}_{\kappa_2}, \ldots, \tilde{x}_{\kappa_n}$.

Suppose we want to exchange the basis variable $x_{\mu_{\tilde{\nu}}}$ with the variable $\tilde{x}_{\tilde{\kappa}}$. We assume that $\alpha_{\tilde{\nu}\tilde{\kappa}} > 0$. Now $a^{\tilde{\kappa}} = \sum_{\substack{\nu=1 \\ \nu \neq \tilde{\nu}}}^n a_{\nu\tilde{\kappa}} a^{\mu_\nu} + \alpha_{\tilde{\nu}\tilde{\kappa}} a^{\mu_{\tilde{\nu}}}$ implies

$$a^{\mu_{\tilde{\nu}}} = \alpha_{\tilde{\nu}\tilde{\kappa}}^{-1} a^{\tilde{\kappa}} - \sum_{\substack{\nu=1 \\ \nu \neq \tilde{\nu}}}^n a_{\nu\tilde{\kappa}} \cdot \alpha_{\tilde{\nu}\tilde{\kappa}}^{-1} a^{\mu_\nu}.$$

Inserting this in

$$a^\mu = \sum_{\nu=1}^n \alpha_{\nu\mu} a^{\mu_\nu}, \quad b = \sum_{\nu=1}^n \alpha_{\nu 0} a^{\mu_\nu},$$

we get

$$a^\mu = \sum_{\substack{\nu=1 \\ \nu \neq \tilde{\nu}}}^n (\alpha_{\nu\mu} - \alpha_{\nu\tilde{\kappa}} \cdot \alpha_{\tilde{\nu}\tilde{\kappa}}^{-1} \cdot \alpha_{\tilde{\nu}\mu}) a^{\mu_\nu} + \alpha_{\tilde{\nu}\mu} \cdot \alpha_{\tilde{\nu}\tilde{\kappa}}^{-1} a^{\tilde{\kappa}}$$

and

$$b = \sum_{\substack{\nu=1\\ \nu\neq\tilde{\nu}}}^{n} (\alpha_{\nu 0} - \alpha_{\nu\tilde{\kappa}} \cdot \alpha_{\tilde{\nu}\tilde{\kappa}}^{-1} \cdot \alpha_{\tilde{\nu}0}) a^{\mu_\nu} + \alpha_{\tilde{\nu}0} \cdot \alpha_{\tilde{\nu}\tilde{\kappa}}^{-1} a^{\tilde{\kappa}}.$$

We can now read off the new values of the tableau:

$$\boxed{\tilde{\alpha}_{\nu\mu} := \begin{cases} \alpha_{\nu\mu} - \alpha_{\nu\tilde{\kappa}} \cdot \alpha_{\tilde{\nu}\tilde{\kappa}}^{-1} \cdot \alpha_{\tilde{\nu}\mu} & \text{for } \nu \neq \tilde{\nu}, \\ \alpha_{\tilde{\nu}\mu} \cdot \alpha_{\tilde{\nu}\tilde{\kappa}}^{-1} & \text{for } \nu = \tilde{\nu}, \end{cases}}$$

for $1 \le \nu \le n$ and $0 \le \mu \le m$. The new last row of the tableau is calculated from

$$\begin{aligned}
\tilde{\alpha}_{m+1\,\mu} &= c_\mu - \tilde{z}_\mu = c_\mu - \sum_{\substack{\nu=1\\ \nu\neq\tilde{\nu}}}^{n} c_{\mu_\nu} \tilde{\alpha}_{\nu\mu} - c_{\tilde{\kappa}} \tilde{\alpha}_{\tilde{\nu}\mu} = \\
&= c_\mu - \sum_{\substack{\nu=1\\ \nu\neq\tilde{\nu}}}^{n} c_{\mu_\nu} \alpha_{\nu\mu} + \sum_{\substack{\nu=1\\ \nu\neq\tilde{\nu}}}^{n} c_{\mu_\nu} \cdot \alpha_{\nu\tilde{\kappa}} \cdot \alpha_{\tilde{\nu}\tilde{\kappa}}^{-1} \cdot \alpha_{\tilde{\nu}\mu} - c_{\tilde{\kappa}} \cdot \alpha_{\tilde{\nu}\mu} \cdot \alpha_{\tilde{\nu}\tilde{\kappa}}^{-1} = \\
&= c_\mu - z_\mu + c_{\mu_{\tilde{\nu}}} \alpha_{\tilde{\nu}\mu} - \alpha_{\tilde{\nu}\mu} \cdot \alpha_{\tilde{\nu}\tilde{\kappa}}^{-1} \big(c_{\tilde{\kappa}} - \sum_{\substack{\nu=1\\ \nu\neq\tilde{\nu}}}^{n} c_{\mu_\nu} \alpha_{\nu\tilde{\kappa}}\big) = \\
&= \alpha_{n+1\mu} - \alpha_{\tilde{\nu}\mu} \cdot \alpha_{\tilde{\nu}\tilde{\kappa}}^{-1} (c_{\tilde{\kappa}} - z_{\tilde{\kappa}}) = \alpha_{n+1\,\mu} - \alpha_{\tilde{\nu}\mu} \cdot \alpha_{\tilde{\nu}\tilde{\kappa}}^{-1} \alpha_{n+1\tilde{\kappa}},
\end{aligned}$$

and is thus

$$\boxed{\tilde{\alpha}_{n+1\mu} := \alpha_{n+1\,\mu} - \alpha_{\tilde{\nu}\mu} \cdot \alpha_{\tilde{\nu}\tilde{\kappa}}^{-1} \alpha_{n+1\tilde{\kappa}}}$$

for $\mu = 0, 1, \ldots, m$.

Returning to Example 3.1, the following is the complete tableau after the first exchange step ($\tilde{\kappa} = 1$, $\tilde{\nu} = 6$):

0	-11/2	1	0	0	0	0	3/2	72	-144/11
0	-1/4	0	1	0	0	0	7/4	133	-532
0	1	0	0	1	0	0	0	9	9
0	189/4	0	0	0	1	0	-19/4	133	532/189
0	-7/4	0	0	0	0	1	-1/4	3	-12/7
1	-7/4	0	0	0	0	0	1/4	7	-4
0	-25/4	0	0	0	0	0	3/4	21	

Clearly, in the second step we should choose $\tilde{\kappa} = 2$ and $\tilde{\nu} = 4$. We leave the computation of further exchange steps to the reader.

We now summarize the strategy for exchanging vertices in the

Theorem. *Suppose $x \in \mathrm{E_M}$ is a nondegenerate vertex with $f(x) =: z_0$. Then*

(i) *If $d_\kappa := c_\kappa - z_\kappa \geq 0$ for all $\kappa \notin \mathrm{I}(x)$, then the minimal value of the objective function f is assumed at x.*

(ii) *If, on the other hand, $d_{\tilde{\kappa}} := c_{\tilde{\kappa}} - z_{\tilde{\kappa}} < 0$ for some index $\tilde{\kappa} \notin \mathrm{I}(x)$, then there are two cases:*

  (ii_1) *If no $\nu \in \{1,2,\ldots,m\}$ with $\alpha_{\nu\tilde{\kappa}} > 0$ exists, then the set M of admissible vectors is unbounded.*

  (ii_2) *If there is an index $\tilde{\nu} \in \{1,2,\ldots,m\}$ with $\alpha_{\tilde{\nu}\tilde{\kappa}} > 0$ and such that $\frac{\alpha_{\tilde{\nu}0}}{\alpha_{\tilde{\nu}\tilde{\kappa}}} = \min\{\frac{\alpha_{\nu 0}}{\alpha_{\nu\tilde{\kappa}}} \mid \alpha_{\nu\tilde{\kappa}} > 0\}$, then replacing the basis variable $x_{\mu_{\tilde{\nu}}}$ by $x_{\tilde{\kappa}}$, we get a vertex for which the objective function takes on a smaller value. This vertex is degenerate if and only if the index $\tilde{\nu}$ is not uniquely defined.*

This leads to the following *algorithm*:

(i) Start with the simplex tableau for a nondengerate vertex;

(ii) Solution test: If $d_\kappa \geq 0$ for all $\kappa \notin \mathrm{I}(x)$, stop the algorithm and return the solution x;

(iii) Choice of the exchange column: Choose an index $\tilde{\kappa} \notin \mathrm{I}(x)$ such that $d_{\tilde{\kappa}} = \min\{d_\kappa \mid \kappa \notin \mathrm{I}(x)\}$;

(iv) M is unbounded if $\alpha_{\nu\tilde{\kappa}} \leq 0$ for all $\nu = 1,2,\ldots,n$;

(v) Choice of the basis variable $x_{\mu_{\tilde{\nu}}}$ to be exchanged: Choose an index $\tilde{\nu} \in \{1,2,\ldots,n\}$ with

$$\frac{\alpha_{\tilde{\nu}0}}{\alpha_{\tilde{\nu}\tilde{\kappa}}} = \min\{\frac{\alpha_{\nu 0}}{\alpha_{\nu\tilde{\kappa}}} \mid \alpha_{\nu\tilde{\kappa}} > 0\}.$$

(vi) Computation of a new simplex tableau: Set $\mathrm{I}(x) := (\mathrm{I}(x) \setminus \{\mu_{\tilde{\nu}}\}) \cup \{\tilde{\kappa}\}$;

$$\alpha_{\nu\mu} := \begin{cases} \alpha_{\nu\mu} - \alpha_{\nu\tilde{\kappa}}\alpha_{\tilde{\nu}\tilde{\kappa}}^{-1}\alpha_{\tilde{\nu}\mu} & \text{for } \nu \neq \tilde{\nu} \\ \alpha_{\tilde{\nu}\mu}\alpha_{\tilde{\nu}\tilde{\kappa}}^{-1} & \text{for } \nu = \tilde{\nu} \end{cases},$$

$$d_\mu := d_\mu - \alpha_{\tilde{\nu}\mu}\alpha_{\tilde{\nu}\tilde{\kappa}}^{-1} d_{\tilde{\kappa}}.$$

(vii) Computation of the last column.

The number $\alpha_{\tilde{\nu}\tilde{\kappa}}$ is called the *pivot element* of the exchange step.

So far we have assumed that the process is started with a nondegenerate vertex of the polyhedron M. We discuss how to find such a starting vertex in the next subsection.

3.3 Finding a Starting Vertex. We begin with the optimization problem with side conditions in inequality form:

Minimize $f(x) = c^T x$ subject to the side conditions $Ax \leq b$, $x \geq 0$.

By introducing slack variables as in 1.4, this problem can be reduced to standard form. Clearly, the vector $(0,\ldots,0,b_1,\ldots,b_n)^T$ is a vertex of this problem, and it is nondegenerate if $b_\mu > 0$, $1 \leq \mu \leq n$.

The problem of finding a starting vertex is more difficult if the optimization problem is in the general standard form without any special structure on the matrix A. In this case we consider the following auxillary problem:

Minimize the objective function $f^*(x, y_1, \ldots, y_n) := \sum_{\nu=1}^n y_\nu$ subject to the side conditions $Ax + y = b$, $x \geq 0$ and $y \geq 0$.

This problem always has a solution since the objective function f^* is bounded below on the set $\mathrm{M}^* := \{\binom{x}{y} \in \mathbb{R}^{m+n} \mid Ax + y = b,\ x \geq 0,\ y \geq 0\}$ (cf. 2.5, Problem 1).

Obviously, $\binom{x}{y} := \binom{0}{b}$ is a vertex of the polyhedron M^*, and it is nondegenerate if $b > 0$. Exchanging vertices as in Theorem 3.2, we can find a vertex $\binom{x^*}{y^*} \in \mathrm{E}_{\mathrm{M}^*}$ where f^* assumes its minimum. Now there are two cases:

(i) $y^* \neq 0$.

In this case the set of admissible vectors $\mathrm{M} := \{x \in \mathbb{R}^m \mid Ax = b, x \geq 0\}$ for the original problem is empty. Indeed, if $\mathrm{M} \neq \emptyset$ and $x \in \mathrm{M}$, then $\binom{x}{0}$ is a solution of our auxiliary problem with $f^*(x, 0) = 0$. But then the vertex $\binom{x^*}{y^*}$ is also a solution with $f^*(x^*, y^*) > 0$, which gives a contradiction.

(ii) $y^* = 0$.

In this case x^* is a vertex in M, and it is nondegenerate if no component of y^* is a basis variable of the vertex $\binom{x^*}{y^*}$ of M^*.

It remains only to consider the case where some components of y^* are basis variables for the vertex $\binom{x^*}{y^*} \in \mathrm{E}_{\mathrm{M}^*}$. Suppose the indices of these components of y^* are $\mathrm{I}(y^*) = \{\mu_1^*, \ldots, \mu_k^*\}$. We now attempt to reach a nondegenerate vertex $x \in \mathrm{E}_\mathrm{M}$ by further exchange steps. Suppose $\mu_{\tilde{\nu}}$ lies in $\mathrm{I}(y^*)$, and suppose that in the corresponding row of the simplex tableau, $\alpha_{\tilde{\nu}\tilde{\kappa}} \neq 0$ for some $\tilde{\kappa}$ belonging to a component of x^*. Then we take

$$\tilde{\alpha}_{\nu\mu} := \begin{cases} \alpha_{\nu\mu} - \alpha_{\nu\tilde{\kappa}} \cdot \alpha_{\tilde{\nu}\tilde{\kappa}}^{-1} \cdot \alpha_{\tilde{\nu}\mu} & \text{for } \nu \neq \tilde{\nu} \\ \alpha_{\tilde{\nu}\mu} \cdot \alpha_{\tilde{\nu}\tilde{\kappa}}^{-1} & \text{for } \nu = \tilde{\nu} \end{cases}.$$

This exchange step replaces the basis variable $y_{\mu_{\tilde{\nu}}}^*$ by $x_{\tilde{\kappa}}^*$. Since $\alpha_{\tilde{\nu}0} = 0$ or $d_{\tilde{\nu}} = 0$, respectively, it follows that the value of the basis variable and the value of d_μ remains unchanged, respectively.

By repeated application of this step, we try to replace all of the basis variables $y_{\mu_{\tilde{\nu}}}^*$ with $\mu_{\tilde{\nu}} \in \mathrm{I}(y^*)$ by some $x_{\tilde{\kappa}}^*$. If this is possible, then we arrive at a nondegenerate vertex of M. If not, then there exists an index $\tilde{\nu} \in \mathrm{I}(y^*)$ such that $\alpha_{\tilde{\nu}\kappa} = 0$ for all indices κ belonging to x^*. In this case, however, the rank of A must be smaller than n. Now in view of $Ax^* = b$, this means that the rows of A are linear dependent, and so some of the equations are *redundant*.

We note that for optimization problems in standard form, we may assume that $b \geq 0$. This can always be achieved by multiplying the equations with (-1) where necessary.

In this section we have shown how to find a starting vertex. It remains to discuss how to handle degenerate vertices.

3.4 Degenerate Vertices. So far we have been assuming that in carrying out the vertex exchange process, no degeneracies arise. Suppose now that x is an arbitrary vertex of M with basis variables $x_{\mu_1}, x_{\mu_2}, \dots, x_{\mu_n}$, and suppose the new basis variable is to be $x_{\tilde{\kappa}}$, $\tilde{\kappa} \notin \mathrm{I}(x)$. As before, let

$$\tilde{\varepsilon} := \min_{1 \le \nu \le n} \Big\{ \frac{x_{\mu_\nu}}{\alpha_{\nu\tilde{\kappa}}} \;\Big|\; \alpha_{\nu\tilde{\kappa}} > 0 \Big\}.$$

Since x is possibly degenerate, it can happen that either $\tilde{\varepsilon} > 0$ or $\tilde{\varepsilon} = 0$.

If $\tilde{\varepsilon} > 0$, then we carry out the exchange step as described in 3.2, and obtain a vertex $\tilde{x}$, $\tilde{x} \ne x$, with $f(\tilde{x}) < f(x)$.

The case $\tilde{\varepsilon} = 0$ requires special treatment. In this case we choose an index $\tilde{\nu}$ for which $\alpha_{\tilde{\nu}\tilde{\kappa}} > 0$ and $x_{\mu_{\tilde{\nu}}} \cdot \alpha_{\tilde{\nu}\tilde{\kappa}} = 0$. Then making the exchange using the *pivot element* $\alpha_{\tilde{\nu}\tilde{\kappa}}$ leads to a new basis for the vertex x, and the value of the objective function will not be changed. If $\tilde{\varepsilon} = 0$ occurs several times in a row, then the method may stay at the vertex x for several steps. In time its basis variable will be exchanged, and it can happen that after a certain number of steps, we get a basis for x which we have already used before. In this case, the method is stuck in a *cycle*. We call this a *cyclical exchange*. This situation is of little significance in practice, since for large optimization problems we will be using a digital computer, and roundoff errors will generally prevent cycles from ocurring. It is also possible to adjust the rules of the exchange process so that cycles cannot occur; see e.g. the book of L. Collatz and W. Wetterling ([1966], p. 19 ff.). An example where cycles occur can be found in the monograph of S. I. Gass ([1964], p. 119 ff.).

3.5 The Two-Phase Method. We continue to discuss the optimization problem

$$\text{Minimize } f(x) = c^T x \quad \text{subject to the}$$
$$\text{side conditions } Ax = b,\ x \ge 0$$

where $A \in \mathbb{R}^{(n,m)}$, $b \in \mathbb{R}^n$ and $c \in \mathbb{R}^m$.

As discussed in 3.2 and 3.3, the simplex method for solving this problem proceeds in two phases as follows:

Phase I. Compute a solution of the auxiliary problem 3.3. This leads to a vertex for the standard optimization problem.

Phase II. Using vertex exchanges as in 3.2 and 3.4, compute a vertex of M which solves the problem.

Both phases use the *simplex algorithm*, which we now summarize.

It can happen that at the start we have a vertex x with basis variables x_{μ_ν}, $1 \le \nu \le n$ such that $\{a^{\mu_\nu} \mid \nu \in \mathrm{I}(x)\} = \{e^{\mu_\nu} \in \mathbb{R}^n \mid 1 \le \nu \le n\}$. This can occur in two ways: either for the standard problem itself (in which case Phase I is superfluous), or for the auxiliary problem (in which case Phase I is done in advance).

Each step of the algorithm proceeds as follows:

(i) Set $\alpha_{\nu\kappa} := a_\nu^\kappa$, $\alpha_{\nu 0} := b_\nu$, $\alpha_{n+1\kappa} := c_\kappa$, $\alpha_{n+10} := c^T x$ for $1 \le \nu \le n$ and $1 \le \kappa \le m$,

(ii) For all $\mu_{\tilde\nu} \in \mathrm{I}(x)$ with $c_{\mu_{\tilde\nu}} = 0$ set

$$\alpha_{\nu\mu} := \begin{cases} \alpha_{\nu\mu} - \alpha_{\tilde\nu\mu}\alpha_{\nu\mu_{\tilde\nu}} & \text{for } \nu \ne \tilde\nu,\ \mu \notin \mathrm{I}(x) \\ \alpha_{\tilde\nu\mu} & \text{for } \nu = \tilde\nu,\ \mu \notin \mathrm{I}(x), \end{cases}$$

$$\alpha_{n+1\mu} := \alpha_{n+1\mu} - \alpha_{\tilde\nu\mu}\alpha_{n+1\mu_{\tilde\nu}},$$

(iii) If $\alpha_{n+1\mu} \ge 0$ for all $\mu \notin \mathrm{I}(x)$, then the vector x with components $\alpha_{\mu_{\tilde\nu}0}$ for $\mu_{\tilde\nu} \in \mathrm{I}(x)$, and zero otherwise, is a solution of the optimization problem. The value of the objective function is $(-\alpha_{n+1\,0})$; stop.

(iv) Choose the exchange column $\tilde\kappa$ so that

$$\alpha_{n+1\tilde\kappa} = \min\{\alpha_{n+1\kappa} \mid \kappa \notin \mathrm{I}(x)\}.$$

(v) If $\alpha_{\nu\tilde\kappa} \le 0$ for $1 \le \nu \le n$, then the objective function is not bounded on M; stop.

(vi) Choose the exchange row $\tilde\nu$ so that

$$\frac{\alpha_{\tilde\nu 0}}{\alpha_{\tilde\nu\tilde\kappa}} = \min\{\alpha_{\nu 0} \cdot \alpha_{\nu\tilde\kappa}^{-1} \mid \alpha_{\nu\tilde\kappa} > 0\}.$$

(vii) Set

$$\alpha_{\nu\mu} := \begin{cases} \alpha_{\nu\mu} - \alpha_{\tilde\nu\mu} \cdot \alpha_{\tilde\nu\tilde\kappa}^{-1} \cdot \alpha_{\nu\tilde\mu} & \text{for } \nu \ne \tilde\nu \\ \alpha_{\tilde\nu\mu} \cdot \alpha_{\tilde\nu\tilde\kappa}^{-1} & \text{for } \nu = \tilde\nu, \end{cases}$$

$$\alpha_{n+1\mu} := \alpha_{n+1\mu} - \alpha_{\tilde\nu\mu} \cdot \alpha_{\tilde\nu\tilde\kappa}^{-1} \cdot \alpha_{n+1\tilde\kappa},$$

$$\mu_{\tilde\nu+\mu} := \mu_{\tilde\nu+\mu+1} \quad \text{for } 0 \le \kappa \le n - \tilde\nu - 1,$$

$$\mu_n := \mu_{\tilde\kappa},$$

$$\mathrm{I}(x) := \{\mu_1, \mu_2, \ldots, \mu_n\}.$$

Go to step (iii).

Remarks. i) The loop in step (ii) will be executed at most m times.

ii) In step (iv) the exchange column is uniquely defined.

iii) If in step (vi) the exchange row is not uniquely defined, then the new vertex will be degenerate.

The computational effort required for the simplex method can be substantially reduced if at each step we alter only those quantities in the tableau which are relevant. We discuss the corresponding modified algorithm in the following subsection.

3.6 The Modified Simplex Method. Let x be an admissible vector in M, and let $\{\mu_1, \mu_2, \dots, \mu_n\} \subset \{1, 2, \dots, m\}$. We now partition the matrix $A = (a^\mu)_{\mu=1,2,\dots,m}$ into submatrices $B := (a^{\mu_\nu})_{\nu=1,2,\dots,n}$, $B \in \mathbb{R}^{(n,n)}$, and $D := (a^\mu)_{\substack{\mu=1,2,\dots,m \\ \mu\neq\mu_\nu}}$, $D \in \mathbb{R}^{(m-n,m)}$. Similarly, we partition the vectors $x, c \in \mathbb{R}^m$ into $x_B := (x_{\mu_\nu}) \in \mathbb{R}^n$, $x_D := (x_\mu)_{\mu\neq\mu_\nu} \in \mathbb{R}^{m-n}$ and $c_B \in \mathbb{R}^n$, $c_D \in \mathbb{R}^{n-m}$.

Using this notation, we can rewrite the standard optimization problem as

$$\text{Minimize} \quad c_B^T \cdot x_B + c_D^T \cdot x_D \ \text{ subject to the}$$
$$\text{side conditions} \quad Bx_B + Dx_D = b, \quad x_B \geq 0, \quad x_D \geq 0.$$

If the matrix B is invertible, then the objective function and the side conditions can be rewritten:

$$(*) \qquad \begin{aligned} &\text{Minimize} \quad (c_D^T - c_B^T B^{-1} D) x_D + c_B^T B^{-1} b \ \text{ subject to the} \\ &\text{side conditions} \quad x_B + B^{-1} D x_D = B^{-1} b \ \text{ and the} \\ &\text{sign conditions} \ x_B \geq 0, \ x_D \geq 0. \end{aligned}$$

Now if x is a vertex of M with basis variables $x_{\mu_1}, \dots, x_{\mu_n}$, then $x_D = 0$ and $x_B = B^{-1}b$. We refer to the vector $r := (c_D^T - c_B^T B^{-1} D) x_D$ for $x \in$ M as the *cost vector* for x. Now corresponding to a vertex $x \in \mathrm{E_M}$ with basis variables $\mu_1, \mu_2, \dots, \mu_n$, the simplex tableau for problem $(*)$ has the following form:

I	$B^{-1}D$	$B^{-1}b$
0	$c_D^T - c_B^T B^{-1} D$	$c_B B^{-1} b$

We note that in every step of the method, we need only invert a $n \times n$ matrix B and perform a multiplication with a $n \times (m-n)$ matrix D. Moreover, in successive steps, the matrices B to be inverted differ only in the column which was just exchanged. By comparision, for the simplex method discussed in 3.5, the product $B^{-1}A$ had to be formed in each step.

We now describe the *modified simplex algorithm*.

Suppose that $x \in \mathrm{E_M}$ is a vertex with basis variables $x_{\mu_1}, \dots, x_{\mu_n}$. Set $B := (a^{\mu_1}, \dots, a^{\mu_n})$, and carry out the following steps:

(i) Compute $B^{-1} =: (w^1, w^2, \dots, w^n)$ and $w^0 := B^{-1}b$,

(ii) Compute $\lambda := c_B^T B^{-1}$ and set $r := c_D^T - \lambda D$,

(iii) Find the exchange column with the index $\tilde{\kappa}$ such that $r_{\tilde{\kappa}} := \min\{r_\kappa \mid \kappa \notin \{\mu_1, \dots, \mu_n\}\}$ and compute $\alpha^{\tilde{\kappa}} := (\alpha_{1\tilde{\kappa}}, \alpha_{2\tilde{\kappa}}, \dots, \alpha_{n\tilde{\kappa}})^T := B^{-1} a^{\tilde{\kappa}}$,

(iv) Boundedness test: If $\alpha^{\tilde{\kappa}} \leq 0$, then the objective function is not bounded on M; stop.

(v) Choose the exchange row by finding an index $\tilde{\nu}$ with

$$\frac{w^0_{\tilde{\nu}}}{\alpha_{\tilde{\nu}\tilde{\kappa}}} = \min\{w^0_\nu \cdot \alpha^{-1}_{\nu\tilde{\kappa}} \mid \alpha^{-1}_{\nu\tilde{\kappa}} > 0\}.$$

(vi) Set

$$w^\kappa_\nu := \begin{cases} w^\kappa_\nu - w^{\tilde{\kappa}}_\nu \cdot \alpha^{-1}_{\tilde{\nu}\tilde{\kappa}} \cdot w^\kappa_{\tilde{\nu}} & \text{for } \nu \neq \tilde{\nu}, \\ w^\kappa_{\tilde{\nu}} \cdot \alpha^{-1}_{\tilde{\nu}\tilde{\kappa}} & \text{for } \nu = \tilde{\nu} \end{cases}$$

and $1 \leq \nu \leq m$, $0 \leq \kappa \leq m$;

$$B^{-1} = (w^1, \ldots, w^n) \quad \text{and } \mu_{\tilde{\nu}} := \tilde{\kappa};$$

Go to step (ii).

In programming the modified simplex method, it turns out to be useful to find the LR decomposition of the matrix B. The theorem on triangular decomposition of a nonsingular matrix in 2.1.3 assures us that there exist nonsingular lower and upper triangular matrices L and R, respectively, and a permutation matrix P such that $P \cdot B = L \cdot R$. Moreover, if we change just one column of B, then the new triangular decomposition can be computed efficiently (cf. Remark 2.1.4).

This suggests the following changes in the modified simplex algorithm:

(i) Find the decomposition $L'B = R$, $L' := L^{-1} \cdot P$, set $\overline{b} := L'b$, and solve the linear system of equations $Rw^0 = \overline{b}$,

(ii) Solve $R^T w = c_B$ and set $\lambda := L'^T w$, $r := c^T_D - \lambda^T D$,

(iii) Set $\overline{w} := L' a^{\tilde{\kappa}}$ and solve $R\alpha_{\tilde{\kappa}} = \overline{w}$.

We never have to explicitly compute the matrix B^{-1}. The systems of equations are easy to solve since the matrices have triangular form. An Algol program for the modified simplex method with LR decomposition can be found in R. H. Bartels, J. Stoer and C. Zenger ([1971], p. 152 ff.).

3.7 Problems. 1) Construct examples in $\mathbb{R}^2$ and in $\mathbb{R}^3$ to illustrate geometrically a degenerate vertex of a polyhedron.

2) Consider the optimization problem: Maximize $c^T x$ under the side conditions $Ax = b$. Suppose $\mathrm{M} = \{x \mid Ax = b\} \neq \emptyset$. Show the equivalence of the following assertions:

(i) The maximal value of the objective function is finite;

(ii) all admissible vectors are solutions of the optimization problem;

(iii) the vector c depends linearly on the row vectors of the matrix A.

3) Consider the optimization problem of minimizing $(x_2 - 3x_3 + 2x_5)$ under the side conditions $x_1 + 3x_2 - x_3 + 2x_5 = 7$, $-2x_2 + 4x_3 + x_4 = 12$, $-4x_2 + 3x_3 + 8x_5 + x_6 = 10$, $x_\mu \geq 0$ for $1 \leq \mu \leq 3$. Show that the set of admissible vectors for this problem contains the vertex $x = (10, 0, 3, 0, 0, 1)$, and write down the associated simplex tableau.

4) Using the exchange method, solve the optimization problem of minimizing the sum $(x_4 + x_5)$ under the side conditions $2x_1 + x_2 + 2x_3 + x_4 = 4$, $3x_1 + 3x_2 + x_3 + x_5 = 3$, $x_\mu \geq 0$ for $1 \leq \mu \leq 5$.

5) Consider the optimization problem of minimizing the objective function $(-0.75x_1 + 150x_2 - 0.02x_3 + 6x_4)$ subject to the side conditions $0.25x_1 - 60x_2 - 0.04x_3 + 9x_4 + x_5 = 0$, $0.5x_1 - 90x_2 - 0.02x_3 + 3x_4 + x_6 = 0$, $x_3 + x_7 = 1$, $x_\mu \geq 0$ for $1 \leq \mu \leq 7$. Find a degenerate vertex, and write down the simplex tableau corresponding to the basis variables x_5, x_6, x_7. Carry out one step of the exchange process with $\tilde{\kappa} = 1$, $\tilde{\nu} = 1$.

6) Solve the following optimization problem using the two-phase method: Minimize x_2 under the side conditions $-x_1 - 2x_3 = 5$, $2x_1 - 3x_2 + x_3 = 3$, $2x_1 - 5x_2 + 6x_3 = 5$.

7) Solve the following optimization problem: minimize $(2x_1 + x_2 + 4x_3)$ under the side conditions $x_1 + x_2 + 2x_3 = 3$, $2x_1 + x_2 + 3x_3 = 5$, $x_1 \geq 0$, $x_2 \geq 0$:

a) using the modified simplex algorithm;

b) using the simplex algorithm with LR decomposition.

8) Write a computer program for the modified simplex method with LR decomposition, and use it to solve the following optimization problems:

a) (Test problem): Minimize the expression $-3x_1 - x_2 - 3x_3$ under the side conditions $2x_1 + x_2 + x_3 \leq 2$, $x_1 + 2x_2 + 3x_3 \leq 5$, $2x_1 + 2x_2 + x_3 \leq 6$, $x_\mu \geq 0$ for $1 \leq \mu \leq 3$.

b) (Example with cycles from the book of S. I. Gass ([1964], p. 106)): The optimization Problem 4.

4. Complexity Analysis

Each step of Phase II of the simplex algorithm corresponds to moving along an edge of a polyhedron from one vertex to a neighboring one. In doing so, the tableau corresponding to the original vertex is modified to get the tableau corresponding to the new vertex. This requires $2m(n+1)+1$ multiplications (divisions) and $m(n+1)$ additions (subtractions). In order to estimate the total computational complexity of the simplex method, we need to have some idea of how many vertices will have to be examined in order to find the optimal vertex minimizing the objective function. In 2.3 we showed that $\binom{m}{n}$ is an upper bound on the number of steps required, where m is the number of inequality side conditions, and n is the number of variables in the objective function. It has been observed in practice, however, that a far smaller number of steps, on the order of $\frac{3}{2}(m - n)$, are required. This observation prompted W. M. Hirsch in 1957 to conjecture that for every linear optimization problem, there is a *variant of the simplex algorithm* which solves the problem in at most $(m - n)$ pivot steps. The *Hirsch conjecture* has so far only been settled in the cases $m - n \leq 5$ and $n = 3$. For the simplex method discussed in 3.5, V. Klee [1965] showed that there always exists a linear optimization problem which cannot be solved in less than $(m-n)(n-1)+1$ exchange steps. Since then, the complexity of methods for solving linear optimization problems has been the subject of intense research. We now describe some of the results.

4.1 The Examples of Klee and Minty. For most variants of the simplex method, examples of optimization problems are known whose solution cannot be carried out with a number of pivot steps which is bounded by a polynomial in m and n. V. Klee and G. Minty [1972] constructed a series of such examples for the simplex algorithm described in 3.5, where in each example all of the vertices of the polyhedron have to be examined in order to find the solution. These examples involve n variables and $m = 2n$ inequality constraints, chosen in such a way that in executing the simplex algorithm, all 2^n vertices of the polyhedron have to be examined, which means that $(2^n - 1)$ pivot steps are needed. We now give a typical optimization problem of this type.

Maximize the objective function $f(x) = (e^n)^T x$ subject to the side conditions

$$\begin{array}{ll} x_1 \geq 0, & x_1 \leq 1\,; \\ x_2 \geq \varepsilon x_1, & x_2 \leq 1 - \varepsilon x_1\,; \\ x_3 \geq \varepsilon x_2, & x_3 \leq 1 - \varepsilon x_2\,; \\ \vdots & \vdots \\ x_n \geq \varepsilon x_{n-1}, & x_n \leq 1 - \varepsilon x_{n-1}\,. \end{array}$$

Here $\varepsilon \in (0, 0.5)$ is an arbitrary number. The construction principal for the polyhedrons can already be observed in the special case $n = 3$, $m = 6$ shown in the figure on the next page. The simplex algorithm must run through all vertices in order to reach the vertex $x_7 = (0,0,1)$ where the maximum of the objective function is attained.

We note that, in general, for problems of this type ($2n$ inequalities and n variables), the number of pivot steps grows exponentially with n. This is certainly quite different from the observed performance of the simplex method on a large number of practical examples, where the algorithm is far more efficient. This leads us to the question of the average running time of the method.

4.2 On the Average Behavior of the Algorithm. The discussion in 4.1 suggests that, using an appropriate statistical model, we should investigate the expected number of pivot steps needed in an algorithm for solving a linear optimization problem. Results of this type were first obtained by K. H. Borgwardt in 1977, S. Smale [1982], [1983], and M. Haimovich [1983]. Here we discuss a result of K. H. Borgwardt [1981], [1982], which shows that a variant of the simplex algorithm – the *shadow vertex algorithm* – has

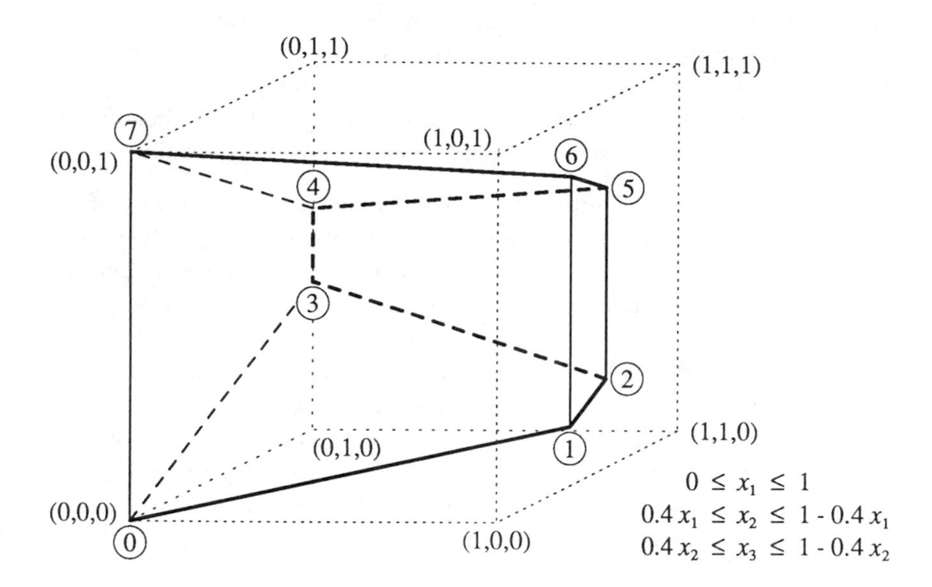

a mean running time in Phase II which is bounded by a polynomial in m and n.

The underlying statistical model assumes the data $A = (a^1, a^2, \ldots, a^n)$ in $\mathbb{R}^{(m,n)}$ and $c \in \mathbb{R}^n$ in the linear optimization problem come from a distribution on $\mathbb{R}^n \setminus \{0\}$ with the following properties:

(i) independence;
(ii) uniformity;
(iii) symmetry under rotations.

We consider solving the optimization problem ($m \geq n$):

$$\text{maximize} \quad f(x) := c^T x \quad \text{subject to the}$$

$$\text{side conditions} \quad \sum_{\nu=1}^{n} a_{\mu\nu} x_\nu \leq 1, \quad 1 \leq \mu \leq m,$$

using the shadow vertex algorithm. Borgwardt [1982] established the following

Theorem. *For all distributions satisfying conditions (i)–(iii), the expected value $T_S^M(m,n)$ of the number of pivot steps needed in Phase II by the shadow vertex algorithm is bounded by*

$$T_S^M(m,n) \leq e\pi\left(\frac{\pi}{2} + \frac{1}{e}\right) n^3 m^{\frac{1}{n-1}}.$$

This theorem leads to the result that the corresponding variant of the simplex algorithm requires a mean number of pivot steps which can be bounded by a polynomial in m and n. For the proof of this assertion, as well as other results on the complexity of algorithms for solving the linear optimization problem, see the book of K. H. Borgwardt [1987].

We now take a look at some questions relating to the runtime in the *worst case.* To this end, we need a more precise definition of runtime.

4.3 Runtime of Algorithms. In 1.4.3 we discussed the complexity of an algorithm in terms of the number of elementary arithmetic steps which have to be carried out. As noted already there, the coding length of the numbers in the calculation also plays an important role, since arithmetic operations with numbers with fewer digits is less work than with numbers with more digits. Since only rational numbers can be dealt with in a digital computer (cf. 1.1.3), we now restrict our attention to the field $\mathbb{Q}$ of rational numbers.

As is well known, in most computers, the integers are coded in binary. The binary representation of a number $n \in \mathbb{Z}$ requires $\lceil log_2(|n| + 1)\rceil$ bits and an additional bit for the sign. Here $\lceil r\rceil$ denotes the smallest integer which is larger or equal to $r \in \mathbb{R}$. The *coding length* of an integer $n \in \mathbb{Z}$ will be denoted by

$$\langle n\rangle := \lceil log_2(|n| + 1)\rceil + 1.$$

Examples. 1) Suppose $r \in \mathbb{Q}$ is given by $r = \frac{p}{q}$, where $p, q \in \mathbb{Z}$ have no common divisors and $q > 0$. Then the coding length of r is given by

$$\langle r\rangle = \langle p\rangle + \langle q\rangle.$$

2) A matrix $A \in \mathbb{Q}^{(m,n)}$, $A = (a_{\mu\nu})$, has coding length

$$\langle A\rangle = \sum_{\mu=1}^{m}\sum_{\nu=1}^{n}\langle a_{\mu\nu}\rangle.$$

3) A linear optimization problem P with data in $\mathbb{Q}$ of the form maximize $c^T x$, subject to the side conditions $Ax \leq b$, has coding length

$$\langle P\rangle := \langle A\rangle + \langle b\rangle + \langle c\rangle.$$

Using this concept, we make the following

Definition. Let A be an algorithm for solving a problem P.

(i) The *runtime* $t_A^S(P)$ *of an algorithm* A for solving problem P is the number of elementary arithmetic steps, multiplied by the coding length of the largest number which appears in executing the algorithm.

(ii) Given an algorithm A to solve problems P in the problem class $\mathcal{P}$, we call the function

$$T_A^S : \mathbb{N} \to \mathbb{N},$$
$$T_A^S(n) := \max\{t_A^S(P) \mid P \in \mathcal{P},\ \langle P \rangle \leq n\}$$

the *runtime function* of A. Here $\langle P \rangle$ denotes the coding length of the data describing the problem P.

(iii) The algorithm A has *polynomial runtime* if there exists a polynomial $p : \mathbb{N} \to \mathbb{N}$ such that for all $n \in \mathbb{N}$, we have the bound

$$T_A^S(n) \leq p(n).$$

In this case we say that A is a *polynomial algorithm.*

The example of Klee and Minty in 1.4 shows that the simplex method does not have polynomial runtime. But from the work of Borgwardt and others, we know that there are algorithms for solving linear optimization problems which have polynomial runtime in mean. Recently a great deal of effort has been expended on finding an algorithm which has polynomial runtime even in the worst case.

4.4 Polynomial Algorithms. It was something of a scientific sensation, and was even reported in daily newspapers, when L. G. Khachiyan [1979] presented his *ellipsoid method* for solving linear optimization problems in polynomial runtime. Although it turned out later that the method was of less practical use than expected, its study has nevertheless led to a deeper understanding of linear optimization theory and related areas. A detailed discussion, with particular emphasis on questions of combinatoric optimization, can be found in the book of M. Grötschel, L. Lovász and A. Schrijver [1988].

An algorithm for solving linear optimization problems in polynomial time which promises to be of practical importance was recently given by N. Karmarkar [1984]. The basic version of the *Karmarkar algorithm* deals with the following optimization problem:

$$\text{Minimize} \quad f(x) := c^T x \quad \text{subject to the}$$
$$\text{side conditions} \quad Ax = 0,\ x \geq 0,\ \sum_{\nu=1}^{n} x_\nu = 1.$$

Here $A \in \mathbb{Q}^{(m,n)}$ and $c \in \mathbb{Q}^n$. For convenience, we introduce the vector $\mathbf{1} := (1, 1, \ldots, 1)^T \in \mathbb{R}^n$, and write the side conditions $\sum_{\nu=1}^{n} x_\nu = 1$ as $\mathbf{1}^T \cdot x = 1$.

This minimization problem is of a very special form. Karmarkar has shown, however, how more general problems can be reduced to it.

The Karmarkar method is based on a well-known approach in *nonlinear optimization*. Starting with an admissible vector $x^{(\mu)} \in \mathbb{R}^n$, we look for a direction $d^{(\mu)} \in \mathbb{R}^n$ such that the value of the objective function is reduced if we make a step in this direction. Then the idea is to choose a step size $\rho^{(\mu)}$, and set $x^{(\mu+1)} := x^{(\mu)} + \rho^{(\mu)} d^{(\mu)}$. The vector $d^{(\mu)} \in \mathbb{R}^n$ and the number $\rho^{(\mu)}$ should be chosen so that a significant reduction in the objective function occurs, and so that we do not leave the set of admissible vectors. A "good" direction (for example in the direction of steepest descent of the objective function) may not produce much improvement if only a small step in this direction can be made. In contrast, a "worse" direction can lead to a larger reduction in the objective function because we may be able to take a larger step without leaving the set of admissible vectors.

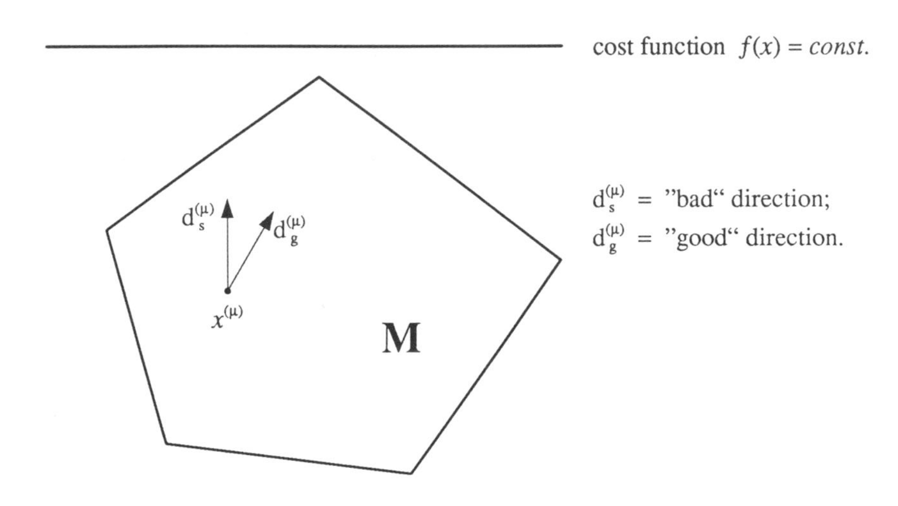

To explain the operation of the Karmarkar algorithm, we set

$$\begin{aligned} \mathrm{U} &:= \{x \in \mathbb{R}^n \mid Ax = 0\} ; \\ \mathrm{H} &:= \{x \in \mathbb{R}^n \mid \mathbf{1}^T \cdot x = 1\} ; \\ \mathrm{S} &:= \mathrm{H} \cap \{x \in \mathbb{R}^n \mid x \geq 0\} ; \\ \mathrm{M} &:= \mathrm{S} \cap \mathrm{U}. \end{aligned}$$

Karmarkar's idea is to perform a projective transformation in each step of the algorithm which maps the simplex S onto itself, which maps the affine subspace $\mathrm{U}^{(\mu-1)}$ into an affine subspace $\mathrm{U}^{(\mu)}$, and which maps the interior point $x^{(\mu-1)}$ to the center $\frac{1}{n}\mathbf{1}$ of S. Then the polyhedron $\mathrm{M}^{(\mu-1)} := = \mathrm{S} \cap \mathrm{U}^{(\mu-1)}$, $\mu > 1$, will be mapped into a new polyhedron $\mathrm{M}^{(\mu)}$. Now from the center $\frac{1}{n}\mathbf{1}$, we can take a relatively large step in any direction without leaving the admissible set $M^{(\mu)}$. This leads to a new point $y^{(\mu+1)} \in M^{(\mu)}$ which we then transform back to get the next iterate $x^{(\mu+1)} \in M^{(\mu+1)}$. To determine the direction and the step size, given $x^{(\mu)}$, it is natural to compute the optimal solution of the transformed problem. This leads to problems, however, since the transformed objective function is no longer linear. Karmarkar linearizes this objective function in an appropriate way, and solves the auxiliary problem:

$$\text{Minimize} \quad f^{(\mu)}(x) := (c^{(\mu)})^T \cdot x \quad \text{subject to the side conditions} \quad x \in \mathrm{U}^{(\mu)} \cap \mathrm{S}.$$

We now show how to choose the vector $c^{(\mu)} \in \mathbb{R}^n$ in each step.

In place of solving this auxiliary problem, we can solve the related auxiliary problem where the simplex S is replaced by the largest sphere K with center $\frac{1}{n}\mathbf{1}$ which is contained in the simplex S. The optimization problem

$$\text{Minimize} \quad f^{(\mu)}(x) = (c^{(\mu)})^T \cdot x \quad \text{under the side conditions} \quad x \in \mathrm{U}^{(\mu)} \cap \mathrm{K}$$

still produces reasonable descent directions which do not change the convergence behavior of the Karmarkar algorithm. Moreover, the solution $y^{(\mu+1)}$ of this second auxiliary problem can be given explicitly. It can happen, however, that due to rounding errors, transforming $y^{(\mu+1)}$ back may lead to a point which no longer lies in the relative interior of the polyhedron. Thus in practice, instead of the sphere K, it is better to take a smaller sphere with the same center.

We now assume that the matrix A has full row rank, and that $x^{(\mu)}$ is a point in the interior of $M^{(\mu)}$. Let

$$D^{(\mu)} := \begin{pmatrix} x_1^{(\mu)} & & 0 \\ & \ddots & \\ 0 & & x_n^{(\mu)} \end{pmatrix} \in \mathbb{R}^{(n,n)},$$

and define a projective transformation $T^{(\mu)} : \mathbb{R}^n \setminus \mathrm{L} \to \mathbb{R}^n$ on the set $\mathrm{L} := \{y \in \mathbb{R}^n \mid \mathbf{1}^T (D^{(\mu)})^{-1} y = 0\}$ by

$$T^{(\mu)}(x) := \frac{1}{\mathbf{1}^T (D^{(\mu)})^{-1} x} (D^{(\mu)})^{-1} x.$$

The mapping $T^{(\mu)}$ has the following

Properties. (i) For all $x \geq 0$, $T^{(\mu)}(x) \geq 0$;
(ii) $x \in \mathrm{H}$ implies $T^{(\mu)}(x) \in \mathrm{H}$;
(iii) $(T^{(\mu)})^{-1}(y) = \frac{1}{\mathbf{1}^T D^{(\mu)} y} D^{(\mu)} y$ for all $y \in \mathrm{S}$;
(iv) $T^{(\mu)}(\mathrm{S}) = \mathrm{S}$;
(v) $\mathrm{M}^{(\mu)} := T^{(\mu)}(\mathrm{M}) = T^{(\mu)}(\mathrm{S} \cap \mathrm{U}) = \mathrm{S} \cap \mathrm{U}^{(\mu)}$,
with $\mathrm{U}^{(\mu)} = \{y \in \mathbb{R}^n \mid AD^{(\mu)} y = 0\}$;
(vi) $T^{(\mu)}(x^{(\mu)}) = \frac{1}{\mathbf{1}^T . \mathbf{1}} (D^{(\mu)})^{-1} x^{(\mu)} = \frac{1}{n}\mathbf{1} \in \mathrm{M}^{(\mu)}$.

We leave the simple proofs of these properties to the reader.

Note that $T^{(\mu)}$ is a bijective mapping of $\mathrm{M}^{(\mu-1)}$ into $\mathrm{M}^{(\mu)}$ which takes the simplex S into itself. The point $x^{(\mu)}$ is mapped to the center $\frac{1}{n}\mathbf{1}$ of S. For the original optimization problem, we have the following chain of equalities:

$$\begin{aligned}
&\min\{c^T \cdot x \mid Ax = 0, \mathbf{1}^T \cdot x = 1, x \geq 0\} = \min\{c^T x \mid x \in \mathrm{M}\} = \\
&= \min\{c^T (T^{(\mu)})^{-1} y \mid y \in \mathrm{M}^{(\mu)} = T^{(\mu)}(\mathrm{M})\} = \\
&= \min\{\frac{1}{\mathbf{1}^T \cdot D^{(\mu)} y} c^T D^{(\mu)} y \mid AD^{(\mu)} y = 0, \mathbf{1}^T \cdot y = 1, y \geq 0\}.
\end{aligned}$$

Now we replace the nonlinear objective function in the last minimization problem by $(c^{(\mu)})^T y$, where

$$c^{(\mu)} := D^{(\mu)} c.$$

In this way we get the following linear auxiliary problem mentioned above:

$$\begin{array}{ll} \text{Minimize} & (c^{(\mu)})^T y \quad \text{under the} \\ \text{side conditions} & AD^{(\mu)} y = 0, \ \mathbf{1}^T \cdot y = 1, \ y \geq 0. \end{array}$$

The largest sphere with center $\frac{1}{n}\mathbf{1}$ which lies in the simplex S has radius $\frac{1}{\sqrt{n(n-1)}}$. In forming the approximate problem, it is convenient to take a sphere with half this radius, and we get the approximate auxiliary problem:

$$\begin{array}{ll} \text{Minimize} & (c^{(\mu)})^T y \quad \text{under the} \\ \text{side conditions} & AD^{(\mu)} y = 0, \ \mathbf{1}^T \cdot y = 1, \ \|y - \dfrac{1}{n}\mathbf{1}\|_2 \leq \dfrac{1}{2} \dfrac{1}{\sqrt{n(n-1)}}. \end{array}$$

The solution of this optimization problem can now be given explicitly.

Lemma. *The auxiliary problem has the solution*

$$y^{(\mu+1)} = \frac{1}{n}\mathbf{1} - \frac{1}{2} \frac{1}{\sqrt{n(n-1)}} \frac{c_\perp^{(\mu)}}{\|c_\perp^{(\mu)}\|_2}$$

with $c_\perp^{(\mu)} := (I - D^{(\mu)} A^T (A (D^{(\mu)})^2 A^T)^{-1} A D^{(\mu)} - \frac{1}{n}\mathbf{1} \cdot \mathbf{1}^T) D^{(\mu)} c.$

Proof. The vector $c_\perp^{(\mu)} \in \mathbb{R}^n$ is the orthogonal projection of $c^{(\mu)}$ onto the subspace $\mathrm{R} := \{x \in \mathbb{R}^n \mid AD^{(\mu)}x = 0, \mathbf{1}^T \cdot x = 0\}$. Indeed, if $x \in \mathrm{R}$, then because of the symmetry of the matrix

$$F^{(\mu)} := I - D^{(\mu)}A^T(A(D^{(\mu)})^2A^T)^{-1}AD^{(\mu)} - \frac{1}{n}\mathbf{1}\cdot\mathbf{1}^T)$$

and the fact that $c^{(\mu)} = D^{(\mu)}c$, we get

$$\langle c^{(\mu)} - c_\perp^{(\mu)}, x\rangle = \langle c^{(\mu)}, x\rangle - \langle D^{(\mu)}c, F^{(\mu)}x\rangle = \langle c^{(\mu)}, x\rangle - \langle D^{(\mu)}c, x\rangle = 0.$$

Now it is easy to see that $\langle AD^{(\mu)}, c_\perp^{(\mu)}\rangle = \langle \mathbf{1}, c_\perp^{(\mu)}\rangle = 0$.
This immediately implies that $(c^{(\mu)})^T \cdot y = (c_\perp^{(\mu)})^T \cdot y$ for all $y \in \mathrm{R}$. Thus, the auxiliary problem is equivalent to the optimization problem:

$$\text{Minimize} \quad (c_\perp^{(\mu)})^T \cdot y \quad \text{subject to the one}$$
$$\text{side condition} \quad \|y - \frac{1}{n}\mathbf{1}\|_2 \le \frac{1}{2}\frac{1}{\sqrt{n(n-1)}}.$$

The minimum on the sphere of this linear function (whose gradient is $c_\perp^{(\mu)}$) can be obtained by taking a step of length equal to the radius of the sphere in the direction of $-c_\perp^{(\mu)}$, and so

$$y^{(\mu+1)} = \frac{1}{n}\mathbf{1} - \frac{1}{2}\frac{1}{\sqrt{n(n-1)}}\frac{1}{\|c_\perp^{(\mu)}\|_2}c_\perp^{(\mu)}. \qquad \square$$

After transforming the vector $y^{(\mu+1)}$ back, we get

$$x^{(\mu+1)} := (T^{(\mu)})(y^{(\mu+1)}) = \frac{1}{\mathbf{1}^T \; D^{(\mu)}y^{(\mu+1)}}D^{(\mu)}y^{(\mu+1)},$$

which is the starting vector for the next step of the iteration.

The *Karmarkar algorithm* is designed to solve the problem:

$$(*) \qquad \text{Find } x \in \mathrm{M} \text{ with } c^T \cdot x \le 0.$$

It is clear that this problem has no solution if and only if the problem of minimizing the objective function $c^T \cdot x$ for $x \in \mathrm{M}$ has a solution with $c^T \cdot x > 0$.
We now formulate an algorithm for solving (∗).

Input: $A \in \mathbb{Q}$, $c \in \mathbb{Q}$.
Assume that $\frac{1}{n}A\mathbf{1} = 0$ and $c^T \cdot \mathbf{1} > 0$;

Output: A vector $x \in \mathrm{M}$ with $c^T \cdot x \le 0$, or the message that (∗) has no solution;

Initialization: $x^{(0)} := \frac{1}{n}\mathbf{1}$, $\mu := 0$, $N := 3n(\langle A\rangle + 2\langle c\rangle - n)$;

(**) *Stopping criterion:* (i) If $\mu = N$, then (*) has no solution.
(ii) If $c^T x^{(\mu)} \leq 2^{-\langle A \rangle - \langle c \rangle}$, then there exists a solution of (*). We get it immediately if $c^T \cdot x^{(\mu)} \leq 0$. Otherwise it can be computed from $x^{(\mu)}$ as a solution of (*) using linear algebra.

Update: $D := \operatorname{diag}(x^{(\mu)})$;
$c_\perp := (I - DA^T(AD^2A^T)^{-1}AD - \frac{1}{n}\mathbf{1} \cdot \mathbf{1}^T)Dc$;
$y^{(\mu+1)} := \frac{1}{n}\mathbf{1} - \frac{1}{2}\frac{1}{\sqrt{n(n-1)}}\frac{c_\perp}{\|c_\perp\|_2}$;
$x^{(\mu+1)} := \frac{1}{\mathbf{1}^T Dy^{(\mu+1)}} Dy^{(\mu+1)}$;
$\mu := \mu + 1$;
go to step (**).

We have not shown that this algorithm can always be carried out. Moreover, we have not justified the stopping criterion. For a discussion of these points, see the original papers. For completeness, however, we now make the following

Remarks. (i) For all $\mu \in \{0, 1, \ldots, N\}$, we have $Ax^{(\mu)} = 0$, $\mathbf{1}^T \cdot x^{(\mu)} = 1$ and $x^{(\mu)} > 0$.

(ii) There is a solution of problem (*) if and only if there exists an index $\mu \in \{0, 1, \ldots, N\}$ with

$$c^T \cdot x^{(\mu)} < 2^{-\langle A \rangle - \langle c \rangle}.$$

The algorithm is polynomial with an appropriate implementation of the updating step.

Theorem of Karmarkar. *The runtime of the Karmarkar algorithm for solving problems of the type (*) is* $O(n^{3.5}(\langle A \rangle - \langle c \rangle)^2)$.

The proof can be found in the original papers. It remains to be seen if this method provides a viable alternative to the simplex algorithm.

4.5 Problems. 1) Find the simplex tableaus for the Klee-Minty Example 4.1 ($n = 3$, $\varepsilon := 0.4$). How many pivots are needed to reach the solution?

2) Show that the number of bits needed for the binary coding of a number $n \in \mathbb{Z}$ is $\lceil log_2(|n| + 1) \rceil + 1$.

3) Using the notation of 4.4, show that $Ax^{(\mu)} = 0$, $\mathbf{1}^T \cdot x^{(\mu)} = 1$ and $x^{(\mu)} > 0$ for all $\mu \in \{1, 2, \ldots, N\}$.

4) Pose a suitably sized problem of type (*) in 4.4, and carry out several steps of the Karmarkar algorithm.

References

This section contains

I) A selection of textbooks and monographs, most of which are explicitly referred to in our text. These include

  Ia) several books containing material from analysis and linear algebra. We assume the reader is basically familiar with this material. In addition, our list includes

  Ib) books dealing with numerical analysis, especially when they supplement the material presented here, or present it in another way.

II) A number of original papers which are cited in the book, either for mathematical or for historical purposes.

Ia) Analysis, Linear Algebra and Biographies.

H.-D. Ebbinghaus [1990]: Numbers. GTM Readings in Mathematics 123, Springer-Verlag, New York.

H. H. Goldstine [1977]: A History of Numerical Analysis, Springer-Verlag, Berlin.

W. H. Greub [1981]: Linear Algebra. Springer-Verlag, New York.

M. Koecher [1985]: Lineare Algebra und analytische Geometrie. Grundwissen Mathematik 2, Springer-Verlag, Berlin.

K. Reich [1985]: Carl Friedrich Gauß 1777–1855. Verlag Moos u. Partner, Gräfelfing.

C. Reid [1973]: Hilbert. Springer-Verlag, New York.

R. Remmert [1990]: Theory of Complex Functions. GTM Readings in Mathematics 122, Springer-Verlag, New York.

I. Runge [1949]: Carl Runge und sein wissenschaftliches Werk. Verlag Vandenhoeck u. Ruprecht, Göttingen.

G. Strang [1988]: Linear Algebra and Its Applications. Academic Press, New York.

W. Walter [1989]: Analysis I. Grundwissen Mathematik 3, Springer-Verlag, Berlin.

W. Walter [1990]: Analysis II. Grundwissen Mathematik 4, Springer-Verlag, Berlin.

Ib) Textbooks and Monographs on Numerical Analysis.

G. A. Baker, Jr. and P. Graves-Morris [1981]: Padé Approximants, Part I: Basic Theory. Encyclopedia of Mathematics and its Applications, Addison-Wesley Publ. Comp., Reading Mass.

A. Ben-Israel and T. N. E. Greville [1974]: Generalized Inverses. John Wiley and Sons, Inc., New York.

C. de Boor [1978]: A Practical Guide to Splines. Springer-Verlag, Berlin.

H. Braß [1977]: Quadraturverfahren. Verlag Vandenhoeck u. Ruprecht, Göttingen.

L. Collatz [1966]: Functional Analysis and Numerical Mathematics. Academic Press, New York.

L. Collatz – W. Wetterling [1975]: Optimization Problems. Springer-Verlag, New York.

C. W. Cryer [1982]: Numerical Functional Analysis. Oxford University Press.

G. B. Dantzig [1963]: Linear Programming and Extensions. Princeton University Press, Princeton.

P. J. Davis [1963]: Interpolation und Approximation. Blaisdell Publ. Comp., New York

P. J. Davis – P. Rabinowitz [1975]: Methods of Numerical Integration. Academic Press, New York.

H. Engels [1980]: Numerical Quadrature und Cubature. Academic Press, New York.

S. I. Gass [1964]: Linear Programming. McGraw-Hill Book Comp., New York.

M. Grötschel, L. Lovásc and A. Schrijver [1988]: Geometric Algorithms and Combinatorial Optimization. Algorithms and Combinatorics 2, Springer-Verlag, Berlin.

G. Hämmerlin [1978]: Numerische Mathematik I, 2. Auflage. Bibliographisches Institut Mannheim.

R. W. Hamming [1962]: Numerical Methods for Scientists and Engineers. McGraw-Hill Book Comp., Inc., New York.

P. Henrici [1964]: Elements of Numerical Analysis. John Wiley and Sons, Inc., New York.

R. Hettich – P. Zencke [1982]: Numerische Methoden der Approximation und semi-infiniten Optimierung. Verlag B. G. Teubner, Stuttgart.

B. Hofmann [1986]: Regularisation of Applied Inverse and Ill-Posed Problems: A Numerical Approach. Teubner, Leipzig.

A. S. Householder [1964]: The Theory of Matrices in Numerical Analysis. Dover Publications, Inc., New York.

V. I. Krylov [1962]: Approximate calculation of integrals. The MacMillan Company, New York

G. G. Lorentz – K. Jetter – S. D. Riemenschneider [1983]: Birkhoff-Interpolation. Addison-Wesley Publ. Comp., Reading Mass.

D. G. Luenberger [1969]: Optimization by Vector Space Methods. John Wiley and Sons, Inc., New York.

G. Meinardus [1967]: Approximation of Functions: Theory and Numerical Methods. Springer-Verlag, Berlin.

R. E. Moore [1966]: Interval Analysis. Prentice-Hall, Inc., Englewood Cliffs N.J.

I. P. Natanson [1964], [1965]: Constructive Function Theory I, II, III. Frederick Ungar Publ. Comp., New York.

A. M. Ostrowski [1973]: Solution of Equations in Euclidean and Banach Spaces. Academic Press, New York.

M. J. D. Powell [1981]: Approximation Theory and Methods. Cambridge University Press.

M. H. Schultz [1973]: Spline Analysis. Prentice Hall, Inc., Englewood Cliffs N. J.

L. L. Schumaker [1981]: Spline Functions: Basic Theory. John Wiley and Sons, Inc., New York.

H. R. Schwarz [1989]: Numerical Analysis: A Comprehensive Introduction. John Wiley and Sons, Inc., New York.

J. Stoer – R. Bulirsch [1983]: Introduction to Numerical Analysis. Springer-Verlag, Berlin

A. H. Stroud [1971]: Approximate Calculation of Multiple Integrals. Prentice-Hall, Inc., Englewood Cliffs N.J.

A. H. Stroud [1974]: Numerical Quadrature and Solution of Ordinary Differential Equations. Springer-Verlag, Berlin.

A. H. Stroud – D. Secrest [1966]: Gaussian Quadrature Formulas. Prentice-Hall, Inc., Englewood Cliffs N.J.

R. S. Varga [1962]: Matrix Iterative Analysis. Prentice-Hall, Inc., Englewood Cliffs, N. J.

G. A. Watson [1980]: Approximation Theory and Numerical Methods. John Wiley and Sons, Inc., New York

J. H. Wilkinson [1965]: The Algebraic Eigenvalue Problem. Clarendon Press, Oxford.

J. H. Wilkinson [1963]: Rounding Errors in Algebraic Processes. Her Majesty's Stationery Office, London.

D. M. Young [1971]: Iterative Solution of Large Linear Systems. Comp. Sci. and Appl. Math., Academic Press, New York.

II) Original Papers.

R. Askey and J. Fitch [1968]: Positivity of the cotes numbers for some ultraspecial abscissas. SIAM J. Numer. Anal. 5, 199–201.

R. H. Bartels, J. Stoer, C. Zenger [1971]: A Realization of the Simplex Method based on Triangular Decompositions. In: Handbook for Automatic Computation, Linear Algebra, J. H. Wilkinson and C. Reinsch, Springer-Verlag, Berlin, 152–190.

S. N. Bernstein [1912]: Sur l'ordre de la meilleure approximation des fonctions continues par les polynômes de degré donné. Mém. Acad. Roy. Belg. 4, 1–104.

K. H. Borgwardt [1981], [1982]: The Expected Number of Pivot Steps Required by a Certain Variant of the Simplex Method is Polynomial. Meth. of Operations Research 43, 35–41 (1981).

K. H. Borgwardt [1987]: The Simplex Method. A Probabilistic Analysis. Algorithms and Combinatorics 1, Springer-Verlag, Berlin

R. Bulirsch [1964]: Bemerkungen zur Romberg-Integration. Num. Math. 6, 6–16.

V. Chakalov [1957]: Formules de cubature mécaniques à coefficients non negatifs. Bull. Sci. Math. [2] 81, 123–134.

D. Coppersmith – S. Winograd [1986]: Matrix Multiplication via Behrend's Theorem. Preprint IBM Yorktown Heights, RC 12104 (# 54531), 8/29/86.

R. Courant [1943]: Variational methods for the solution of problems of equilibrium and vibrations. Bull. Amer. Math. Soc. 49, 1–23.

G. Faber [1914]: Über die interpolatorische Darstellung stetiger Funktionen. Jahresber. d. DMV 23, 192–210.

J. Favard [1940]: Sur l'interpolation. J. Math. Pures Appl. (a) 19, 281–306.

J. G. F. Francis [1961]: The QR-transformation. A unitary analogue to the LR-transformation. Comp. J. 4, 265–271.

G. Frobenius [1912]: Über Matrizen aus nichtnegativen Elementen. Sitzg. ber. Kgl. Preuß. Akad. d. Wiss., Berlin, 456–477.

W. J. Gordon [1969]: Spline-Blended Surface Interpolation through Curve Networks. J. of Math. and Mech. 18, 931–952.

M. Grötschel, L. Lovásc and A. Schrijver [1982]: The Average Number of Pivot Steps Required by the Simplex-Method is Polynomial. Z. Oper. Res. 26, 157–177.

H. Haimovich [1983]: The Simplex Algorithm is very Good! – On the Expected Number of Pivot Steps and Related Properties of Random Linear Programs. Columbia University, New York.

C. A. Hall [1968]: On error bounds for spline interpolation. J. Approx. Theory 1, 209–218.

K. Hessenberg [1941]: Auflösung linearer Eigenwertaufgaben mit Hilfe der Hamilton-Cayleyschen Gleichung. Diss. T. H. Darmstadt.

W. Kahan [1958]: Gauß-Seidel Method of Solving Large Systems of Linear Equations. Dissertation, University of Toronto.

N. Karmarkar [1984]. A New Polynomial-Time Algorithm for Linear Programming. AT & T Bell Laboratories, Murray Hill.

L. G. Khachyan [1979]: A Polynomial Algorithm in Linear Programming. Doklady Akad. Nauk SSSR 244, 1093–1096 (Russian) (English translation: Soviet Mathematics Doklady 20 (1979), 191–194).

V. Klee [1965]: A Class of Linear Programming Problems Requiring a Large Number of Iterations. Numer. Math. 7, 313–321.

V. Klee and G. Minty [1972]: How Good is the Simplex-Algorithm? In: Inequalities III, ed. O. Shisha, Academic Press, New York, 159–175.

L. F. Meyers – A. Sard [1950]: Best approximate integration formulas. J. Math. Phys. 29, 118–123.

R. v. Mises und H. Pollaczek-Geiringer [1929]: Praktische Verfahren der Gleichungsauflösung. Z. angew. Math. Mech. 9, 58–77 und 152–164.

O. Perron [1907]: Zur Theorie der Matrizen. Math. Ann. 64, 248–263.

G. Polya [1933]: Über die Konvergenz von Quadraturverfahren. Math. Z. 37, 264–286.

W. Quade und L. Collatz [1938]: Zur Interpolationstheorie der reellen periodischen Funktionen. Sitzg. ber. d. Preuß. Akad. d. Wiss. Berlin, Phys. Math. Kl. XXX, 383–429.

L. F. Richardson – J. A. Gaunt [1927]: The deferred approach to the limit. Phil. Trans. Royal Soc. London Ser. A 226, 299–349.

C. Runge [1901]: Über empirische Funktionen und die Interpolation zwischen äquidistanten Ordinaten. Z. f. Math. u. Phys. 46, 224–243.

H. Rutishauser [1958]: Solution of eigenvalue problems with the LR-transformation. Appl. Math. Ser. Nat. Bur. Stand. 49, 47–81.

E. Schäfer [1989]: Korovkin's theorems: A unifying version. Functiones et Approximatio XVIII, 43–49.

I. J. Schoenberg [1946a]: Contributions to the problem of approximation of equidistant data by analytic functions, Part A: On the problem of smoothing of graduation, a first class of analytic approximation formulae. Quart. Appl. Math. 4, 45–99.

I. J. Schoenberg [1946b]: Contributions to the problem of approximation of equidistant data by analytic functions, Part B: On the problem of osculatory interpolation, a second class of analytic approximation formulae. Quart. Appl. Math. 4, 112–141.

I. J. Schoenberg [1964]: Spline interpolation and best quadrature. Bull. Amer. Math. Soc. 70, 143–148.

I. J. Schoenberg – A. Whitney [1953]: On Polya Frequency Functions. Trans. Amer. Math. Soc. 74, 246–259.

C. E. Shannon [1938]: A symbolic analysis of switching and relay circuits. Trans. of the Amer. Inst. of Electronic Engineers, Volume 57.

S. Smale [1982], [1983]: The Problem of the Average Speed of the Simplex Method. In: Mathematical Programming; The State of the Art, Bonn 1982, 530–539.
On the Average Speed of the Simplex Method. Math. Progr. 27 (1983), 241–262.

V. Strassen [1969]: Gaussian Elimination is not optimal. Numer. Math. 13, 354–356.

A. N. Tichonov [1963]: On the solution of ill-posed problems using the method of regularisation. Doklady Akad. Nauk SSSR 151, 3 (Russian).

K. Weierstrass [1885]: Über die analytische Darstellbarkeit sogenannter willkürlicher Funktionen reeller Argumente. Sitzg. ber. Kgl. Preuß. Akad. d. Wiss. Berlin, 663–689 and 789–805.

Symbols

$\mathbb{C}$	Field of complex numbers
$\mathbb{N}$	Set of natural numbers $\{0, 1, \cdots\}$
$\mathbb{R}$	Field of real numbers
$\mathbb{R}_+$	Set of positive real numbers
$\mathbb{Z}$	Set of integers $\{\cdots, -1, 0, 1, \cdots\}$
$\mathbb{Z}_+$	Set of positive integers $\{1, 2, \cdots\}$
$[a, b]$	Closed interval of real numbers
(a, b)	Open interval of real numbers
$\square$	End of a proof

Symbols introduced in the text:

$O(\cdot)$ and $o(\cdot)$	Landau symbols	1.4.3	(p. 37)
e^i	i-th unit vector in $\mathbb{R}^n$	2.1.1	(p. 48)
$\|\cdot\|$	Norm of an element	2.4.1	(p. 68)
	resp. of an operator	4.1.5	(p. 124)
$(X, \|\cdot\|)$	Normed vector space	2.4.1	(p. 68)
cond (A)	Condition of a matrix A	2.5.1	(p. 73)
A^+	Pseudoinverse of a $n \times m$ matrix A	2.6.3	(p. 84)
$\langle \cdot, \cdot \rangle$	Inner product	4.1.3	(p. 120)
$\omega_f(\delta)$	Modulus of continuity	4.2.5	(p. 134)
$E_{\mathrm{T}}(\cdot)$	Minimal deviation	4.3.1	(p. 137)
ONS	Orthonormal system	4.5.3	(p. 162)

Index

NOTE: Page numbers in italics following a proper name refer to historical remarks.

Undergraduate Texts in Mathematics

Apostol: Introduction to Analytic Number Theory. Second edition.
Armstrong: Groups and Symmetry.
Armstrong: Basic Topology.
Bak/Newman: Complex Analysis.
Banchoff/Wermer: Linear Algebra Through Geometry.
Brémaud: An Introduction to Probabilistic Modeling.
Bressoud: Factorization and Primality Testing.
Brickman: Mathematical Introduction to Linear Programming and Game Theory.
Cederberg: A Course in Modern Geometries.
Childs: A Concrete Introduction to Higher Algebra.
Chung: Elementary Probability Theory with Stochastic Processes. Third edition.
Curtis: Linear Algebra: An Introductory Approach. Fourth edition.
Dixmier: General Topology.
Driver: Why Math?
Ebbinghaus/Flum/Thomas: Mathematical Logic.
Edgar: Measure, Topology, and Fractal Geometry.
Fischer: Intermediate Real Analysis.
Flanigan/Kazdan: Calculus Two: Linear and Nonlinear Functions.
Fleming: Functions of Several Variables. Second edition.
Foulds: Optimization Techniques: An Introduction.
Foulds: Combinatorial Optimization for Undergraduates.
Franklin: Methods of Mathematical Economics.
Hämmerlin/Hoffmann: Numerical Mathematics.
Readings in Mathematics.
Halmos: Finite-Dimensional Vector Spaces. Second edition.
Halmos: Naive Set Theory.
Iooss/Joseph: Elementary Stability and Bifurcation Theory. Second edition.
James: Topological and Uniform Spaces.
Jänich: Topology.
Kemeny/Snell: Finite Markov Chains.
Klambauer: Aspects of Calculus.
Lang: A First Course in Calculus. Fifth edition.
Lang: Calculus of Several Variables. Third edition.
Lang: Introduction to Linear Algebra. Second edition.
Lang: Linear Algebra. Third edition.
Lang: Undergraduate Algebra. Second edition.
Lang: Undergraduate Analysis.
Lax/Burstein/Lax: Calculus with Applications and Computing. Volume 1.
LeCuyer: College Mathematics with APL.
Lidl/Pilz: Applied Abstract Algebra.
Macki/Strauss: Introduction to Optimal Control Theory.
Malitz: Introduction to Mathematical Logic.
Marsden/Weinstein: Calculus I, II, III. Second edition.
Martin: The Foundations of Geometry and the Non-Euclidean Plane.
Martin: Transformation Geometry: An Introduction to Symmetry.
Millman/Parker: Geometry: A Metric Approach with Models.
Owen: A First Course in the Mathematical Foundations of Thermodynamics.
Peressini/Sullivan/Uhl: The Mathematics of Nonlinear Programming.
Prenowitz/Jantosciak: Join Geometries.
Priestly: Calculus: An Historical Approach.
Protter/Morrey: A First Course in Real Analysis.

Undergraduate Texts in Mathematics

Protter/Morrey: Intermediate Calculus. Second edition.
Ross: Elementary Analysis: The Theory of Calculus.
Samuel: Projective Geometry.
Readings in Mathematics.
Scharlau/Opolka: From Fermat to Minkowski.
Sigler: Algebra.
Simmonds: A Brief on Tensor Analysis.
Singer/Thorpe: Lecture Notes on Elementary Topology and Geometry.
Smith: Linear Algebra. Second edition.
Smith: Primer of Modern Analysis. Second edition.
Stanton/White: Constructive Combinatorics.
Stillwell: Mathematics and Its History.
Strayer: Linear Programming and Its Applications.
Thorpe: Elementary Topics in Differential Geometry.
Troutman: Variational Calculus with Elementary Convexity.
Wilson: Much Ado About Calculus.